**Springer Proceedings in Physics**

Volume 303

Indexed by Scopus

The series Springer Proceedings in Physics, founded in 1984, is devoted to timely reports of state-of-the-art developments in physics and related sciences. Typically based on material presented at conferences, workshops and similar scientific meetings, volumes published in this series will constitute a comprehensive up to date source of reference on a field or subfield of relevance in contemporary physics. Proposals must include the following:

- Name, place and date of the scientific meeting
- A link to the committees (local organization, international advisors etc.)
- Scientific description of the meeting
- List of invited/plenary speakers
- An estimate of the planned proceedings book parameters (number of pages/articles, requested number of bulk copies, submission deadline).

***Please contact:***

For Americas and Europe: Dr. Zachary Evenson; zachary.evenson@springer.com
For Asia, Australia and New Zealand: Dr. Loyola DSilva; loyola.dsilva@springer.com

Mohammad Tariqul Islam · Norbahiah Misran ·
Mandeep Jit Singh
Editors

# Proceedings of the 8th International Conference on Space Science and Communication

IconSpace 2023, 3–4 October, Penang, Malaysia

Springer

*Editors*
Mohammad Tariqul Islam
Institute of Climate Change
Universiti Kebangsaan Malaysia
UKM Bangi, Selangor, Malaysia

Department of Electrical, Electronic and Systems Engineering
Faculty of Engineering and Built Environment
Universiti Kebangsaan Malaysia
UKM Bangi, Selangor, Malaysia

Norbahiah Misran
Institute of Climate Change
Universiti Kebangsaan Malaysia
UKM Bangi, Selangor, Malaysia

Department of Electrical, Electronic and Systems Engineering
Faculty of Engineering and Built Environment
Universiti Kebangsaan Malaysia
UKM Bangi, Selangor, Malaysia

Mandeep Jit Singh
Institute of Climate Change
Universiti Kebangsaan Malaysia
UKM Bangi, Selangor, Malaysia

Department of Electrical, Electronic and Systems Engineering
Faculty of Engineering and Built Environment
Universiti Kebangsaan Malaysia
UKM Bangi, Selangor, Malaysia

ISSN 0930-8989 ISSN 1867-4941 (electronic)
Springer Proceedings in Physics
ISBN 978-981-97-0141-4 ISBN 978-981-97-0142-1 (eBook)
https://doi.org/10.1007/978-981-97-0142-1



This Springer imprint is published by the registered company Springer Nature Singapore Pte Ltd.
The registered company address is: 152 Beach Road, #21-01/04 Gateway East, Singapore 189721, Singapore

# Preface

This book addresses complications of ground-breaking initiatives and solutions for space science and communications research, telecommunications and meteorology, in alignment with the United Nation's Sustainable Development Goals (SDGs). With the theme "Advanced Space Technology: Accelerating Global Agenda", this book provides valid information to understand trends, evaluate needs, and create global development policies and programs in the best interest of all the people. This book was prepared in conjunction with the eighth meeting of the 8th International Conference on Space Science and Communication (IconSpace2023), held in Penang, Malaysia, on 3–4 October 2023. It aims to introduce graduate students, researchers, lecturers, engineers, geospatialist, meteorologists, climatologists, astronomers, and practitioners to the latest applications of space science, telecommunications, meteorology, remote sensing, and related fields. The technical program of this conference includes four keynote speakers, four invited talks, two tutorial sessions, and 42 full papers, which address the most significant topics within space science, communications, remote sensing, and satellite technology. The tracks of the sessions include astronomy, atmospheric and magnetospheric sciences, remote sensing and applications, and satellite and communication technology. The papers were evaluated following the blind peer-review process by renowned experts in their respective fields.

UNIVERSITI
KEBANGSAAN
MALAYSIA
*The National University of Malaysia*

Bangi, Malaysia

Mohammad Tariqul Islam
Norbahiah Misran
Mandeep Jit Singh

# Contents

# Contributors

**Nurul Shazana Abdul Hamid** Department of Applied Physics, Faculty of Science and Technology, Universiti Kebangsaan Malaysia, Bangi, Selangor, Malaysia

**Mohd Adha Abdul Majid** Department of Geoinformation, Faculty of Built Environment and Surveying, Universiti Teknologi Malaysia, Johor Bahru, Malaysia

**Mardina Abdullah** Space Science Centre (ANGKASA), Institute of Climate Change (IPI), Universiti Kebangsaan Malaysia (UKM), Bangi, Selangor, Malaysia; Department of Electrical, Electronic and Systems Engineering, Faculty of Engineering and Built Environment, Universiti Kebangsaan Malaysia (UKM), Bangi, Selangor, Malaysia

**Sabirin Abdullah** Space Science Centre (ANGKASA), Institute of Climate Change (IPI), Universiti Kebangsaan Malaysia (UKM), Bangi, Selangor D.E., Malaysia

**Md. Salah Uddin Afsar** Space Science Centre (ANGKASA), Institute of Climate Change (IPI), Universiti Kebangsaan Malaysia, Bangi, Selangor D. E., Malaysia

**Mohd Riduan Ahmad** Atmospheric and Lightning Research Laboratory, Fakulti Kejuruteraan Elektronik dan Kejuruteraan Komputer, Centre for Telecommunication Research and Innovation (CeTRI), Universiti Teknikal Malaysia Melaka, Durian Tunggal, Melaka, Malaysia

**Tanveer Ahsan** Department of Electrical, Electronic and Systems Engineering, Faculty of Engineering and Built Environment, Universiti Kebangsaan Malaysia, UKM Bangi, Selangor, Malaysia

**Neyamul Alam** Department of Electronic and Telecommunication Engineering, International Islamic University Chittagong (IIUC), Chattogram, Bangladesh

**Touhidul Alam** Pusat Sains Angkasa (ANGKASA), Institut Perubahan Iklim (IPI), Universiti Kebangsaan Malaysia (UKM), Bangi, Selangor, Malaysia

**Osama Ali** Small Satellite Technology and Research Lab, National Center of GIS and Space Application Institute of Space Technology, Islamabad, Pakistan

**Samir Salem Al-Bawri** Space Science Centre, Climate Change Institute, Universiti Kebangsaan Malaysia (UKM), Bangi, Malaysia

**Adlyka Annuar** Department of Applied Physics, Faculty of Science and Technology, Universiti Kebangsaan Malaysia, Bangi, Malaysia

**Farizah Ansarudin** Microwave Technology Research Group, Department of Electrical, Electronic and Systems Engineering, Faculty of Engineering and Built Environment, Universiti Kebangsaan Malaysia, Bangi, Malaysia

**Radial Anwar** School of Applied Science, Telkom University, Bandung, Indonesia

**Christian David C. Aquino** Department of Physics, Mapúa University, Manila, Philippines

**Peb Ruswono Aryan** Vienna University of Technology, Vienna, Austria

**Amira Farhana Azhar** Faculty of Engineering, Jalan Gombak, International Islamic University Malaysia(IIUM), Kuala Lumpur, Selangor, Malaysia

**Rezaul Azim** Research Laboratory for Computational Electromagnetics, Department of Physics, University of Chittagong, Chattogram, Bangladesh

**Shahrul Azmin** Department of Medicine, Faculty of Medicine, Neurology Unit, Universiti Kebangsaan Malaysia Medical Centre, Kuala Lumpur, Malaysia

**Khalida Azudin** Department of Electrical, Electronic and Systems Engineering, Faculty of Engineering and Built Environment, Universiti Kebangsaan Malaysia, Bangi, Malaysia

**Khairayu Badron** Faculty of Engineering, Jalan Gombak, International Islamic University Malaysia(IIUM), Kuala Lumpur, Selangor, Malaysia

**Siti Aminah Bahari** Space Science Centre (ANGKASA), Institute of Climate Change (IPI), Universiti Kebangsaan Malaysia (UKM), Bangi, Selangor, Malaysia

**Mohd Hafiz Baharuddin** Department of Electrical, Electronic and Systems Engineering, Faculty of Engineering and Built Environment, Universiti Kebangsaan Malaysia (UKM), Bangi, Selangor, Malaysia

**Badariah Bais** Department of Electrical, Electronic and Systems Engineering, Faculty of Engineering and Built Environment, Universiti Kebangsaan Malaysia (UKM), Bangi, Selangor, Malaysia

**Mohtady Ehab Hasan Aly Barakat** Faculty of Engineering, Multimedia University, Cyberjaya, Selangor, Malaysia

**Andrei V. Bibaoco** Department of Earth and Space Sciences, Rizal Technological University, Mandaluyong City, Philippines

**Mark Francis James Bona** Senior High School Department, Mapúa University, Intramuros, Manila, Philippines

**Uman Buñi** Senior High School Department, Mapúa University, Intramuros, Manila, Philippines

**Julia Marie Cabral** Senior High School Department, Mapua University Muralla St. Intramuros, Manila, Philippines

**Kah Yoong Chan** Faculty of Engineering, Multimedia University, Cyberjaya, Selangor, Malaysia

**Ungku Azlin Ungku Chulan** Department of Electrical, Electronic and Systems Engineering, Faculty of Engineering and Built Environment, Universiti Kebangsaan Malaysia UKM, Bangi, Selangor, Malaysia

**Ungku Azmi Ungku Chulan** Department of Electrical, Electronic and Systems Engineering, Faculty of Engineering and Built Environment, Universiti Kebangsaan Malaysia UKM, Bangi, Selangor, Malaysia

**Gwo Chin Chung** Faculty of Engineering, Multimedia University, Cyberjaya, Selangor, Malaysia

**Vincent Cordova** Senior High School Department, Mapúa University, Intramuros, Manila, Philippines

**Nishan Das** Research Laboratory for Computational Electromagnetics, Department of Physics, University of Chittagong, Chattogram, Bangladesh

**Arvee M. De Castro** Department of Earth and Space Sciences, Rizal Technological University, Mandaluyong City, Philippines

**Pranab Dewan** Research Laboratory for Computational Electromagnetics, Department of Physics, University of Chittagong, Chattogram, Bangladesh

**Rebecca Marie Dizon** Senior High School Department, Mapua University Muralla St. Intramuros, Manila, Philippines

**Zane Nikia C. Domingo** Department of Physics, Mapúa University, Manila, Philippines

**Iqra Ejaz** Small Satellite Technology and Research Lab, National Center of GIS and Space Application Institute of Space Technology, Islamabad, Pakistan

**Mohammad Rashed Iqbal Faruque** Space Science Centre (ANGKASA), Institute of Climate Change (IPI), Universiti Kebangsaan Malaysia (UKM), Bangi, Selangor D. E., Malaysia

**Akiko Fujimoto** Kyushu Institute of Technology, Fukuoka, Japan

**Kok Beng Gan** Department of Electrical, Electronic and Systems Engineering, Faculty of Engineering and Built Environment, Universiti Kebangsaan Malaysia, Bangi, Malaysia

**Mohammad Lutful Hakim** Pusat Sains Angkasa (ANGKASA), Institut Perubahan Iklim, Universiti Kebangsaan Malaysia, UKM Bangi, Selangor, Malaysia

**Nurul Shazana Abdul Hamid** Department of Applied Physics, Faculty of Science and Technology, Universiti Kebangsaan Malaysia, Bangi, Malaysia;
Space Science Centre (ANGKASA), Institute of Climate Change, Universiti Kebangsaan Malaysia, Bangi, Malaysia

**Z. S. Hamidi** Faculty of Applied Sciences, MARA Technology University, Shah Alam, Selangor, Malaysia

**Abu Hanif** Pusat Sains Angkasa (ANGKASA), Institut Perubahan Iklim, Universiti Kebangsaan Malaysia (UKM), UKM Bangi, Selangor, Malaysia

**Saif Hannan** Department of Electronic and Telecommunication Engineering, International Islamic University Chittagong (IIUC), Chattogram, Bangladesh

**Md. Ashraful Haque** Department of Electrical and Electronic Engineering, Universiti Teknologi PETRONAS, Perak, Malaysia

**Siti Fatimah Harun** Fakulti Kejuruteraan and Alam Bina, Universiti Kebangsaan Malaysia, Bangi, Malaysia

**Mohd Helmy Hashim** School of Electrical Engineering, College of Engineering, Universiti Teknologi MARA, Shah Alam, Selangor, Malaysia

**Moe Hayashi** Kyushu University, Fukuoka, Japan

**Chang Soon Tony Hii** Department of Electrical, Electronic and Systems Engineering, Faculty of Engineering and Built Environment, Universiti Kebangsaan Malaysia, Bangi, Malaysia

**Seikh Erfan Ahmed Himu** Department of Electrical and Electronic Engineering, International Islamic University Chittagong (IIUC), Chattogram, Bangladesh

**Ahasanul Hoque** Space Science Centre (ANGKSA), Universiti Kebangsaan Malaysia (UKM), Bangi, Selangor, Malaysia;
Institute of Climate Change, Universiti Kebangsaan Malaysia, Bangi, Malaysia

**Md. Inzamul Hoque** Department of Electronic and Telecommunication Engineering, International Islamic University Chittagong (IIUC), Chattogram, Bangladesh

**Md. Bellal Hossain** Space Science Centre, Institute of Climate Change, Universiti Kebangsaan Malaysia (UKM), Bangi, Selangor D.E., Malaysia

**Shawon Hossen** Electronic and Telecommunication Engineering (ETE), International Islamic University Chittagong, Kumira, Chattogram, Bangladesh

**Norlinah Mohamed Ibrahim** Department of Medicine, Faculty of Medicine, Neurology Unit, Universiti Kebangsaan Malaysia Medical Centre, Kuala Lumpur, Malaysia

**Nurul Hazrina Idris** Tropical Resource Mapping Research Group, Department of Geoinformation, Faculty of Built Environment and Survey, Universiti Teknologi Malaysia, Johor Bahru, Malaysia;
Geoscience and Digital Earth Centre (INSTeG), Research Institute for Sustainable Environment (RISE), Universiti Teknologi Malaysia, Johor Bahru, Malaysia

**A. N. Ishak** Malaysian Space Agency (MYSA), Kuala Lumpur, Malaysia

**Md. Rashedul Islam** Department of Electrical, Electronic and Systems Engineering, Universiti Kebangsaan Malaysia, Bangi, Malaysia;
Department of Electrical, Electronic and Communication Engineering, Pabna University of Science and Technology, Pabna, Bangladesh

**Md. Shabiul Islam** Faculty of Engineering (FOE), Multimedia University (MMU), Cyberjaya, Selangor, Malaysia

**Mohammad Tariqul Islam** Department of Electrical, Electronic and Systems Engineering, Faculty of Engineering and Built Environment, Universiti Kebangsaan Malaysia (UKM), Bangi, Selangor, Malaysia

**Sikder Sunbeam Islam** Department of Electrical and Electronic Engineering, International Islamic University Chittagong (IIUC), Chattogram, Bangladesh

**Mst Ishrat Jahan** Space Science Centre, Institute of Climate Change, Universiti Kebangsaan Malaysia UKM, Bangi, Selangor D.E., Malaysia

**Mohd Hasni Ja'afar** Department of Community Health, Faculty of Medicine, Universiti Kebangsaan Malaysia Medical Centre, Kuala Lumpur, Malaysia

**Mohamad Huzaimy Jusoh** School of Electrical Engineering, College of Engineering, Universiti Teknologi MARA, Shah Alam, Selangor, Malaysia

**Muhammad Amir khalil** Faculty of Engineering (FOE), Multimedia University (MMU), Cyberjaya, Selangor, Malaysia

**Nur Dalila Khirul Ashar** College of Engineering, Universiti Teknologi MARA Johor Branch, Masai, Johor, Malaysia

**C. C. Lau** Malaysian Space Agency (MYSA), Kuala Lumpur, Malaysia

**Lance Gabriel Licup** Senior High School Department, Mapúa University, Intramuros, Manila, Philippines

**Ernest P. Macalalad** Department of Physics, Mapúa University, Intramuros, Manila, Philippines

**Rehan Mahmood** Small Satellite Technology and Research Lab, National Center of GIS and Space Application Institute of Space Technology, Islamabad, Pakistan

**Mohd Fais Mansor** Department of Electrical, Electronic and Systems Engineering, Faculty of Engineering and Built Environment, Universiti Kebangsaan Malaysia, UKM Bangi, Selangor, Malaysia;
Fakulti Kejuruteraan and Alam Bina, Universiti Kebangsaan Malaysia, Bangi, Malaysia

**Mohd Nadzri Md Reba** Tropical Resource Mapping Research Group, Department of Geoinformation, Faculty of Built Environment and Survey, Universiti Teknologi Malaysia, Johor Bahru, Malaysia;
Geoscience and Digital Earth Centre (INSTeG), Research Institute for Sustainable Environment (RISE), Universiti Teknologi Malaysia, Johor Bahru, Malaysia

**Md. Mhedi Hasan** Department of Electrical, Electronic and Systems Engineering, Faculty of Engineering and Built Environment, Universiti Kebangsaan Malaysia (UKM), Bangi, Selangor, Malaysia;
Department of Information and Communication Technology (ICT), Faculty of Engineering, Comilla University, Cumilla, Bangladesh

**Md. Sharif Mia** Research Laboratory for Computational Electromagnetics, Department of Physics, University of Chittagong, Chattogram, Bangladesh

**Danial Ahmed Mir** Small Satellite Technology and Research Lab, National Center of GIS and Space Application Institute of Space Technology, Islamabad, Pakistan

**Norbahiah Misran** Microwave Technology Research Group, Department of Electrical, Electronic and Systems Engineering, Faculty of Engineering and Built Environment, Universiti Kebangsaan Malaysia (UKM), Bangi, Selangor, Malaysia;
Fakulti Kejuruteraan and Alam Bina, Universiti Kebangsaan Malaysia, Bangi, Malaysia

**Mohd Shawal Faizal Mohamad** Cardiology, Medical Department, Faculty of Medicine, Universiti Kebangsaan Malaysia Medical Centre, Kuala Lumpur, Malaysia

**Sulaiman Ali Mohammmad** Space Science Centre (ANGKASA), Institute of Climate Change (IPI), Universiti Kebangsaan Malaysia (UKM), Bangi, Selangor, Malaysia

**Aiffah Mohd Ali** School of Aerospace Engineering, Universiti Sains Malaysia (USM), Nibong Tebal, Pulau Pinang, Malaysia

**Zahira Mohd Radzi** Exploration and Space Science Division, Malaysian Space Agency (MYSA), Banting, Selangor, Malaysia

**C. Monstein** Istituto Ricerche Solari Aldo e Cele Daccò (IRSOL), Università Della Svizzera Italiana, Locarno, Switzerland

**Samantha Gevan Morales** Senior High School Department, Mapua University Muralla St. Intramuros, Manila, Philippines

**Muhammad Haikal Fayyadh Munadi** Department of Geoinformation, Faculty of Built Environment and Surveying, Universiti Teknologi Malaysia, Skudai, Johor Darul Takzim, Malaysia

**Ahmad Musa** Institut Perubahan Iklim (IPI), Universiti Kebangsaan Malaysia, UKM Bangi, Selangor, Malaysia

**Lin M. M. Myint** School of Engineering, King Mongkut's Institute of Technology Ladkrabang, Bangkok, Thailand

**Karthigesu Nagarajoo** Department of Transportation Engineering Technology, Faculty of Engineering Technology, Universiti Tun Hussein Onn, Batu Pahat, Johor, Malaysia

**Z. K. D. Nurlisman** Malaysian Space Agency (MYSA), Kuala Lumpur, Malaysia

**Dwi Andi Nurmantris** School of Applied Science, Telkom University, Bandung, Indonesia

**Wai Leong Pang** Faculty of Engineering, Multimedia University, Cyberjaya, Selangor, Malaysia

**Murman Dwi Prasetio** Fakultas Rekayasa Industri, Telkom University, Bandung, Jawa Barat, Indonesia

**Lyren Hope Pujida** Senior High School Department, Mapua University Muralla St. Intramuros, Manila, Philippines

**Hasanah Putri** School of Applied Science, Telkom University, Bandung, Indonesia

**Zahira Mohd Radzi** Malaysian Space Agency (MYSA), Banting, Malaysia

**Md. Asifur Rahman** Department of Electrical and Electronic Engineering, School of Science and Engineering, United International University, UIU, Dhaka, Bangladesh

**Md. Afzalur Rahman** Space Science Centre, Climate Change Institute, Universiti Kebangsaan Malaysia (UKM), Bangi, Malaysia;
Department of Electrical and Electronic Engineering, School of Science and Engineering, United International University, UIU, Dhaka, Bangladesh

**Nurul Huda Abd Rahman** School of Electrical Engineering, College of Engineering, Universiti Teknologi MARA, Shah Alam, Malaysia

**Amran Chowdhury Raihan** Department of Electrical and Electronic Engineering, International Islamic University Chittagong (IIUC), Chattogram, Bangladesh

**Md Raihan** Department of Electrical and Electronic Engineering, International Islamic University Chittagong (IIUC), Chattogram, Bangladesh

**Brelveenraj Kaur Rajwant Singh** School of Aerospace Engineering, Universiti Sains Malaysia (USM), Nibong Tebal, Pulau Pinang, Malaysia

**Setyawan Renaldy** School of Applied Science, Telkom University, Bandung, Indonesia

**Mardeni bin Roslee** Faculty of Engineering, Multimedia University, Cyberjaya, Selangor, Malaysia

**Dipon Saha** Department of Electrical and Electronic Engineering, Universiti Teknologi PETRONAS, Perak, Malaysia

**Norsuzlin Mohd Sahar** Space Science Centre, Institute of Climate Change (IPI), Universiti Kebangsaan Malaysia (UKM), Bangi, Selangor, Malaysia;
Pusat Sains Angkasa (ANGKASA), Institut Perubahan Iklim, Universiti Kebangsaan Malaysia, UKM Bangi, Selangor, Malaysia;
Space Science Centre (ANGKSA), Universiti Kebangsaan Malaysia (UKM), Bangi, Selangor, Malaysia

**Waseem Sajjad** Small Satellite Technology and Research Lab, National Center of GIS and Space Application Institute of Space Technology, Islamabad, Pakistan

**Nurul Izzati Salleh** Department of Electrical, Electronic and Systems Engineering, Faculty of Engineering and Built Environment, Universiti Kebangsaan Malaysia, Bangi, Malaysia

**Idahwati Sarudin** School of Physics, Universiti Sains Malaysia (USM), Pulau Pinang, Malaysia

**Arooj Shafique** Small Satellite Technology and Research Lab, National Center of GIS and Space Application Institute of Space Technology, Islamabad, Pakistan

**Raja Adibah Raja Halim Shah** Department of Applied Physics, Faculty of Science and Technology, Universiti Kebangsaan Malaysia, Bangi, Malaysia

**Usama Shakoor** Small Satellite Technology and Research Lab, National Center of GIS and Space Application Institute of Space Technology, Islamabad, Pakistan

**Mandeep Jit Singh** Department of Electrical, Electronic and Systems Engineering, Faculty of Engineering and Built Environment, Universiti Kebangsaan Malaysia, UKM, Bangi, Selangor, Malaysia

**Somkit Sophan** School of Engineering, King Mongkut's Institute of Technology Ladkrabang, Bangkok, Thailand

**Pornchai Supnithi** School of Engineering, King Mongkut's Institute of Technology Ladkrabang, Bangkok, Thailand

**Budi Syihabuddin** School of Electrical Engineering, Telkom University, Bandung, Indonesia

**Wai-Leong Teh** Space Science Centre, Institute of Climate Change, Universiti Kebangsaan Malaysia, Bangi, Malaysia

**Princess Tucio** Department of Earth and Space Sciences, Rizal Technological University, Mandaluyong City, Philippines

**Md. Kutub Uddin** Department of Electronic and Telecommunication Engineering, International Islamic University Chittagong (IIUC), Chattogram, Bangladesh

**Najeeb Ullah** Faculty of Engineering (F.O.E.), Multimedia University (MMU), Cyberjaya, Selangor, Malaysia

**Yousuf Ibn Habib Ullah** Electronic and Telecommunication Engineering (ETE), International Islamic University Chittagong, Kumira, Chattogram, Bangladesh

**R. Umar** East Coast Environmental Research Institute (ESERI), Universiti Sultan Zainal Abidin, Kuala Terengganu, Terengganu, Malaysia

**K. Wahid** Malaysian Space Agency (MYSA), Kuala Lumpur, Malaysia

**Yoshihide Yamada** Malaysia-Japan International Institute of Technology, Universiti Teknologi Malaysia, Kuala Lumpur, Malaysia

**Wong Hin Yong** Faculty of Engineering (FOE), Multimedia University (MMU), Cyberjaya, Selangor, Malaysia

**Akimasa Yoshikawa** International Research Center for Space and Planetary Environmental Science (i-SPES), Kyushu University, Fukuoka, Japan;
International Center for Space Weather Science and Education, Kyushu University, Fukuoka, Japan

**Huay Woon You** Pusat GENIUS@Pintar Negara, Universiti Kebangsaan Malaysia, Bangi, Malaysia

**Nasharuddin Zainal** Department of Electrical, Electronic and Systems Engineering, Faculty of Engineering and Built Environment, Universiti Kebangsaan Malaysia, Bangi, Malaysia

**Nurul Arbaatun Adawiah Zamanuri** Space Science Centre, Institute of Climate Change (IPI), Universiti Kebangsaan Malaysia (UKM), Bangi, Selangor, Malaysia

**Maheen Zulfiqar** Small Satellite Technology and Research Lab, National Center of GIS and Space Application Institute of Space Technology, Islamabad, Pakistan

# Chapter 1
# The Advanced Technology of Satellite Radar Altimetry for Wave Energy Resource Assessment in Terengganu, Malaysia

**Nurul Hazrina Idris and Muhammad Haikal Fayyadh Munadi**

**Abstract** The paper presents the assessment of wave energy resources along the coast of Terengganu using advanced Synthetic Aperture Radar (SAR) altimetry. The SAR altimetry data are expected to give a better accuracy when making observations near shore, compared to the previous low-resolution mode of altimeters (e.g. Jason series). The assessment is conducted along track 189 of the Sentinel 3A SAR altimeter from April 2017 to March 2018. The parameters of wave height, wind speed, wave period and energy power that are retrieved from the satellite are found well-agreed to the field observation data. The Terengganu coast is characterised by small wave with height of 0.5–2 m, and low wave period of 2–4 s. During the northeast monsoon, a high wave energy can be harvested (2–6 kW/m), meanwhile, during the other period, wave power is estimated to be low (< 2 kW/m). In conclusion, the Terengganu coast has a low wave climate. It is essential to have an effective and dependable technology that can maximise energy harvesting.

---

N. H. Idris (✉)
Tropical Resource Mapping Research Group, Faculty of Built Environment and Surveying, Universiti Teknologi Malaysia, 81310 Skudai, Johor Darul Takzim, Malaysia
e-mail: nurulhazrina@utm.my

Geoscience and Digital Earth Centre, Research Institute for Sustainability and Environment, Universiti Teknologi Malaysia, 81310 Skudai, Johor Darul Takzim, Malaysia

M. H. F. Munadi
Faculty of Built Environment and Surveying, Universiti Teknologi Malaysia, 81310 Skudai, Johor Darul Takzim, Malaysia

 
M. T. Islam et al. (eds.), *Proceedings of the 8th International Conference on Space Science and Communication*, Springer Proceedings in Physics 303,
https://doi.org/10.1007/978-981-97-0142-1_1

## 1.1 Introduction

In Malaysia, the primary sources of electricity generation for power resources are predominantly non-renewable, including coal, oil, and gas. The rise in Carbon Dioxide ($CO_2$) and greenhouse gas emissions resulting from energy sources has raised concerns. In addition to the depletion of fossil fuel reserves, there is a growing interest in the exploration of renewable energy (RE) resources as a potential solution. Renewable energy (RE) can be obtained by harnessing the kinetic energy generated by the motion of waves. According to [1], this energy source is both environmentally friendly and highly efficient, as it demonstrates significant energy potential across various geographical regions. Nevertheless, in comparison to other renewable energy sources like wind and tides, it can be observed that solar energy is relatively less developed.

The potential for wave renewable energy in Malaysia has been studied by a number of researchers. A research by [2] looked at the possibilities for wave energy throughout Peninsular Malaysia's eastern coast. According to the analysis obtained from oceanographic buoys, the average wave energy reaches a maximum of 6.5 kW/m, while exhibiting an energy storage capacity of approximately 18 MWh/m. The aforementioned data are corroborated by [2], who conducted a study revealing energy levels of ranging from 0.15 to 6.49 kW/m and an approximate capacity of wave storage up to 17.9 MWh/m. While both studies make valuable contributions to the comprehension of renewable energy in Malaysia, it is important to note that the data obtained from buoys have certain limitations in terms of its spatial and temporal coverage. Consequently, the current representation falls short in capturing the entirety of the designated region, specifically referring to the Malaysian seas.

In their study, [3] present a satellite altimetry method for evaluating wave energy in the Malaysian sea. The radar altimeter provides superior spatial and temporal data coverage compared to buoys. Based on their research, the researchers have determined that the mean wave energy and storage in the seas surrounding Malaysia range from 1.41 to 7.92 kW/m and 7.1 to 69.41 MWh/m, respectively. According to [4], the observation of wave energy during northeast monsoon varies from 8 to 20 kW/m; meanwhile during southwest monsoon, its range is from 4 to 5 kW/m. These findings suggest that a significant level of wave energy is harnessed during period of intense seasonal monsoon. According to a study conducted by [1], the waters of Terengganu and Sarawak have been identified as the most appropriate regions for the deployment of a wave energy converter. These areas exhibit an average potential energy ranging from 2.8 up to 8.6 kW/m.

In a previous study conducted by [4], the Terengganu coast was identified as a promising location for harnessing renewable energy from waves. The objective of this study is to conduct a comprehensive investigation into the potential of wave energy in the coastal region of Terengganu. In a previous study conducted by [4], the low-resolution mode of altimeters, such as Jason-2, was utilised. This particular mode is recognised for its inherent noise, particularly in close proximity to the shore, within a range of approximately 10 km [5]. This study focuses on the processing of the advanced radar altimeter obtained from the Synthetic Aperture Radar (SAR) mode. The objective is to estimate the wave significant height ($H_s$) over the region of Terengganu. The availability of wave data has significantly improved due to advancements in technology, resulting in enhanced accuracy. The Synthetic Aperture Radar (SAR) mode offers significant benefits in terms of reducing range noise and exhibits superior performance in proximity to land when compared to low-resolution mode altimeters.

## 1.2 Study Area and Data

The main focus area of the research is on the coastal water of Terengganu, specifically near Redang Island and Pulau Perhentian Besar (Fig. 1.1). The oceanic region is susceptible to seasonal monsoons, which span from late May to September (southwest monsoon) and November until April (northeast monsoon). According to [6], wind speeds normally do not exceed 7 m/s during the seasonal southwest monsoon. However, during the northeast monsoon, wind speeds are capable to go up to 15 m/s, particularly in east coast of Peninsular Malaysia.

The satellite altimetry data collected from the SAR mode along track 189 of the Sentinel 3A mission span from April 2017 to March 2018. The G-POD SARvatore portal is the platform through which they can be accessed. The data used in this study were obtained from the L2P product, specifically focusing on the parameters of significant wave height (SWH) and wind velocity. In order to verify the data, the parameters of wave height, wind speed, and wave period over Terengganu were acquired from the Volunteer Observing Ships (VOS) programme.

## 1.3 Methodology

The SAR altimetry data from Sentinel 3A satellite are obtained from European Space Agency Grid Processing on Demand (G-POD). The data preprocessing such as atmospheric and geophysical corrections is performed through the G-POD processor. The parameters of wind speed and wave height are obtained from 2017 to 2018.

Wave period can be calculated using an equation from [7],

$$\frac{U}{gT} = 0.048\left(\frac{U^2}{(gH_s)^{0.67}}\right), \tag{1.1}$$

where $g$ is gravitational acceleration, $U$ is wind speed value, and $H_s$ is SWH. In this study, the values of $U$ and $H_s$ are obtained from satellite radar altimeters.

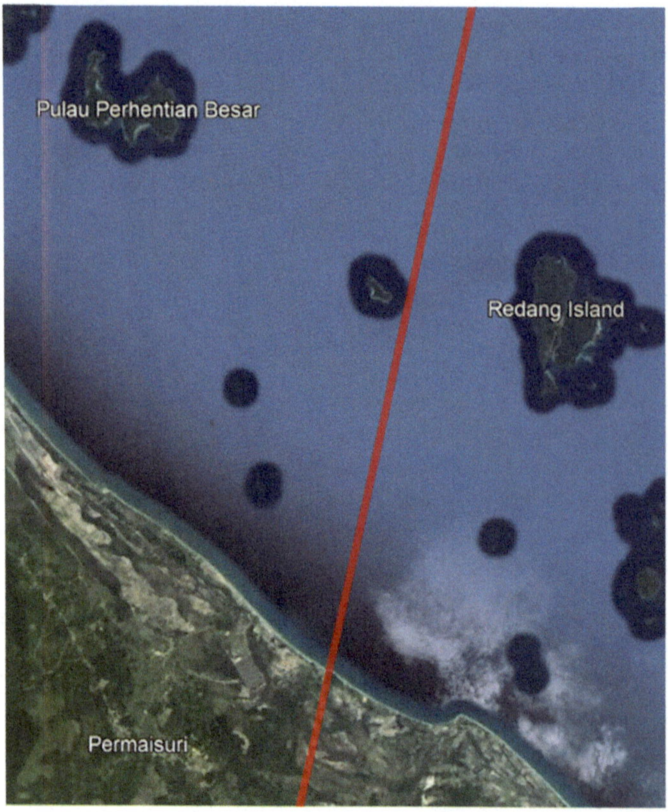


**Fig. 1.1** Study area over the Terengganu coast with a satellite track 189 (in red line) passing the area

The estimation of wave energy is derived using the following expression [3, 8]:

$$P = \frac{\rho g^2}{64\pi} H_s^2 T_e = 0.49 H_s^2 T_e, \tag{1.2}$$

where wave energy per unit of crest length (kW/m) is $P$, SWH (in unit m) $H_s$, wave period (in the unit s) is $T_e$, the density of seawater (assumed to be 1025 kg/m$^3$) is $\rho$, and $g$ is the gravity acceleration. Although Eq. 1.1 is practical for the deep ocean, it can also be applied to the coastal ocean around Malaysia [3].

**Table 1.1** RMSE and correlation of wave height, wind speed, wave period, and wave power

| Parameters | RMS error | Correlation |
|---|---|---|
| Wave height (m) | 0.55 | 0.84 |
| Wind speed (m/s) | 4.98 | 0.65 |
| Wave period (s) | 3.78 | 0.45 |
| Wave power (kW/m) | 3.40 | 0.46 |

## 1.4 Results and Analysis

### *1.4.1 Validation of Parameters*

The parameters of SWH, wind speed, wave period, and wave energy are compared with the data collected from the VOS. There are only nine samples which become available for comparison. Table 1.1 summarises the root mean square (RMS) error and correlation coefficient between both datasets.

Based on the results in Table 1.1, the RMS error and correlation of wave height and wind speed are reasonably acceptable. That is, the RMS error of wave height is 0.55 m and the correlation is 0.84, while the values of wind speed are 4.98 m/s and 0.65, respectively. The derived wave period and wave power are also well-agreed with the VOS datasets with correlation between 0.45 and 0.46 and RMS error of 3.78 s and 3.40 kW/m, respectively. The discrepancy is probably due to the difference in the spatiotemporal sampling of both datasets.

### *1.4.2 The Wave Energy Over Terengganu Coast*

Figure 1.2 represents the mean values of wave height, wind speed, wave period, and wave power from April 2017 to 2018 along track 189 near the coast of Terengganu. From April to October 2017, the value of wave height is relatively low ($\leq$ 1 m), and the value of wind speed is $\leq$ 4 m/s. However, this value was somewhat higher during the northeast monsoon season from November 2017 to March 2018. That is, the SWH is between 1 and 2 m, and the wind speed is 4–8 m/s. With the high value of wave height and wind speed during the northeast monsoon, the power of wave energy calculated during the period can reach up to 6 kW/m, particularly during December 2018. From December to March, the potential wave energy that can be harvested is between 2 and 6 kW/m.

The results from this study are consistent with the previous study by [9], where the wave energy is found higher during the northeast monsoon than in the other seasons. A study using a satellite altimeter by [3] identified that the potential wave energy of 1.41–7.92 kW/m can be harvested in Malaysia. In addition, [10] found that wave energy ranging from 0.15 to 6.49 kW/m can be harvested. According to [1], five potential locations have been identified as the best locations for wave energy

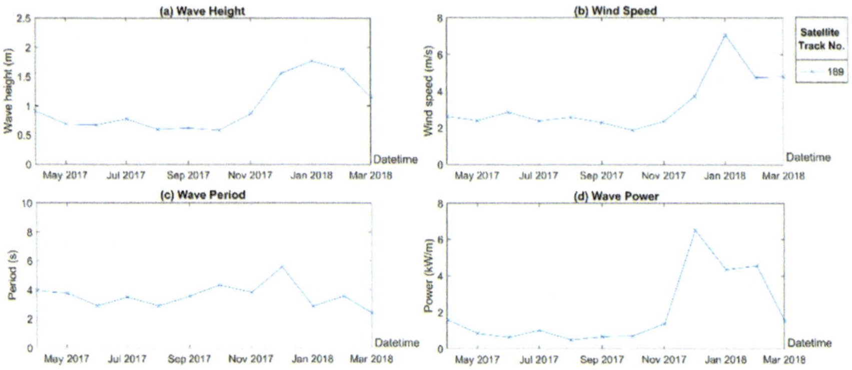


**Fig. 1.2** Mean profile of **a** wave height, **b** wind speed, **c** wave period, and **d** wave power from April 2017 to March 2018 from the Sentinel 3A track 189

generation in Terengganu water. They are near Kapas Island, Nyirih Island, Tenggol Island, and Kerteh Port.

## 1.5 Conclusion

In conclusion, the Terengganu coast has the potential for harvesting wave energy resources. The wind speed and wave height reach the peak during the northeast monsoon from December to March, thus enabling wave energy power of up to 6 kW/m during the period. An ongoing study is currently being conducted to analyse the SAR altimetry data for a longer period since 2015 when the SAR altimetry data from Sentinel 3A became available.

**Acknowledgements** We would like to acknowledge the European Space Agency (ESA) for providing the Sentinel 3A datasets, and the Department of Meteorological Malaysia for providing the VOS data. This research is supported by the Universiti Teknologi Malaysia UTM RA ICONIC Grant (Vot. Q.J130000.4352.09G73) and the Ministry of Higher Education Malaysia Fundamental Research Grant Scheme [FRGS/1/2019/WAB05/UTM/02/1].

## References

1. N.A.M. Nasir, K.N.A. Maulud, Wave power potential in Malaysian territorial waters. IOP Conf. Ser. Earth Environ. Sci. IOP Publishing (2016)
2. A. Muzathik, *Wave Energy Potential of Peninsular Malaysia* (2010)
3. O. Yaakob et al., Satellite-based wave data and wave energy resource assessment for South China Sea. Renew. Energy **88**, 359–371 (2016)
4. N.H. Idris, Wave energy resource assessment with improved satellite altimetry data over the Malaysian coastal sea. Arab. J. Geosci. **12**(15), 484 (2019)

5. N.H. Idris, S. Vignudelli, The technical progress and achievements of coastal altimetry waveform retracking over the Asia seas, in *Coastal Altimetry* (2023), pp. 21–40
6. A.A. Aziz, *Feasibility Study on Development of a Wind Turbine Energy Generation System for Community Requirements of Pulau Banggi Sabah.* A report from Mechanical Engineering Department, UTM, Malaysia, 2011
7. P.A. Hwang et al., A statistical comparison of wind speed, wave height, and wave period derived from satellite altimeters and ocean buoys in the Gulf of Mexico region. J. Geophys. Res. Oceans **103**(C5), 10451–10468 (1998)
8. L. Goddijn-Murphy et al., Wave energy resource assessment with AltiKa satellite altimetry: a case study at a wave energy site. Geophys. Res. Lett. **42**(13), 5452–5459 (2015)
9. A. Mirzaei, F. Tangang, L. Juneng, Wave energy potential along the east coast of Peninsular Malaysia. Energy **68**, 722–734 (2014)
10. A. Muzathik et al., Ocean wave energy along Terengganu coast of Malaysia, in *2nd International Conference on Production, Energy and Reliability (ICPER) 2010*, Universiti Teknologi Petronas, Malaysia, 2010

# Chapter 2
# Inland Waters Unveiled: Exploring Water Surface Elevation in Peninsular Malaysia Using Satellite Sentinel-3 SRAL Altimeter

**Mohd Adha Abdul Majid, Nurul Hazrina Idris, Mohd Nadzri Md Reba, and Muhammad Haikal Fayyadh Munadi**

**Abstract** Continuous monitoring of inland water resources is vital for sustainable management, especially during water scarcity events like El Niño in Malaysia. However, a research gap exists in studying small to medium-scale reservoirs using small footprint altimeters. This study proposes using the Satellite Sentinel-3 SRAL Altimeter (S3 SRAL) to retrieve accurate water surface elevation (WSE) time series. The methodology involves data processing with the SARvatore toolkit, WSE retrieval using proposed equations, data masking using lake boundaries, construction of WSE time series, and validation through comparison with S3 SRAL data and in situ gauge measurements. The primary objectives are to retrieve inland water surface elevation and validate data accuracy using regression, correlation coefficient ($R^2$), and root mean square error (RMSE) evaluation. Promising outcomes were obtained from research on small reservoirs in Peninsular Malaysia (< 150 km$^2$). S3 SRAL measurements showed strong agreement with gauged levels, indicating high correlation and low error rates. Temengor Reservoir (137.25 km$^2$) exhibited robust results ($R^2$ = 0.99, RMSE = 59.95 cm), while Chenderoh Reservoir (8.55 km$^2$) had slightly lower

M. A. Abdul Majid · M. H. F. Munadi
Department of Geoinformation, Faculty of Built Environment and Surveying, Universiti Teknologi Malaysia, 81310 Johor Bahru, Malaysia
e-mail: mohdadha@graduate.utm.my

M. H. F. Munadi
e-mail: mhfayyadh2@graduate.utm.my

N. H. Idris (✉) · M. N. Md Reba
Tropical Resource Mapping Research Group, Department of Geoinformation, Faculty of Built Environment and Survey, Universiti Teknologi Malaysia, 81310 Johor Bahru, Malaysia
e-mail: nurulhazrina@utm.my

Geoscience and Digital Earth Centre (INSTeG), Research Institute for Sustainable Environment (RISE), Universiti Teknologi Malaysia, 81310 Johor Bahru, Malaysia

M. N. Md Reba
e-mail: nadzri@utm.my

 
M. T. Islam et al. (eds.), *Proceedings of the 8th International Conference on Space Science and Communication*, Springer Proceedings in Physics 303,
https://doi.org/10.1007/978-981-97-0142-1_2

values ($R^2 = 0.74$, RMSE = 79.64 cm). The same methods were applied to the largest reservoir in Peninsular Malaysia, Kenyir Reservoir (326.38 $km^2$), highlighting the effectiveness of S3 SRAL. Historical S3 SRAL data since 2016 aided reservoir and lake management decisions and supported research on water level and climate change.

## 2.1 Introduction

Fresh water is a crucial resource globally, and its effective management is vital. In Malaysia, there are 103 dams managed by 28 operators, including notable organizations like Tenaga Nasional Berhad, the Department of Irrigation and Drainage, and the State Water Corporation, serving various purposes, including water supply, hydropower, drainage, irrigation, flood control, private use, agriculture, and leisure activities [1]. Water scarcity is a significant concern, particularly in Malaysia, where rapid development and population growth strain inland waters [2–5]. Challenges such as geopolitical factors, limited gauging of water levels, and restricted access to data hinder water resource monitoring and management [6, 7]. During the dry season, the impact of El Niño exacerbates water scarcity, emphasizing the need for effective solutions. Reports indicate greater interest and engagement from public participation and concern regarding water elevation data and its relevance to water resource management [8–12]. Experts suggest using satellite SRAL altimeters as an alternative approach for monitoring small to medium-sized inland water bodies due to their high spatial resolution capabilities [7, 13].

The recent technological advances in the Synthetic Aperture Radar Altimeter (SRAL) onboard the Sentinel-3 (S3) have made it possible to derive the inland surface water elevation, particularly in a small (less than 150 $km^2$) and complex reservoirs and lakes in Malaysia [10, 14–23]. The Synthetic Aperture Radar (SAR) mode that utilizes delay-Doppler modulation has higher resolution compared to the low rate mode (LRM) altimeter (such as Jason series), making it valuable for deriving water levels for this region [22, 24]. Regular calibration and validation of radar altimeters, combined with advancements in hardware design, processing algorithms, and the integration of SAR imagery and SRAL data, are vital for enhancing water level precision, facilitating accurate measurements of parameters across land, ice, and sea surfaces, and enabling comprehensive monitoring of inland water dynamics [25, 26]. Waveform retracking is a vital technique in SRAL for improving water level measurement accuracy. Retracking has algorithms used to measure geophysical quantities like water elevations, wave heights, and wind speeds from radar return waveforms. These algorithms analyze the waveforms to extract precise geophysical information. Three main categories of waveform retracking methods exist: Empirical Retracking, Physically Based Retracking, and Statistical Retracking. SAMOSA++ (SAR Altimetry MOde Studies and Applications), a notable physically based approach, enhances

SAR altimetry retracking methods by optimizing the detection of inland water bodies, rivers, and lakes through the incorporation of the Range-Integrated Power (RIP) waveform available in SAR data [13, 22].

This manuscript optimizes the utilization of the S3 SRAL altimeter, which offers finer along-track resolution ranging from a few meters to tens of meters, for accurate retrieval of water surface elevation (WSE) in small to medium-sized inland water bodies in Peninsular Malaysia. This approach is particularly important considering the limitations of conventional LRM pulse-limited altimeters, such as the Jason series, SARAL-AltiKa, Envisat, and ERS, which have lower spatial resolutions. Moreover, this study has the potential to make significant contributions to various Sustainable Development Goals (SDGs) outlined in the 2030 Agenda for Sustainable Development [27]. It aligns with SDG 6: Clean Water and Sanitation by enabling effective water resource management through improved retrieval of WSE. Additionally, the study supports SDG 11: Sustainable Cities and Communities by facilitating sustainable water resource management in urban areas. Lastly, it contributes to SDG 15: Life on Land by assisting in the protection of ecosystems and biodiversity through enhanced monitoring and understanding of inland water bodies. Therefore, this study aims to develop a virtual station using S3 SRAL to measure water surface elevation in Peninsular Malaysia. Its specific objectives are to derive time series of water surface elevation for reservoirs and lakes using S3 SRAL data and evaluate the accuracy of these measurements by comparing them with in situ water level gauge data. The accuracy assessment involves comparing the virtual stations to water gauge data, utilising linear regression ($R^2$) and Root Mean Square Error (RMSE) as evaluation metrics [14, 19, 28–32].

## 2.2 Materials

### *2.2.1 Study Area*

This study focuses on three main reservoirs, which are Chenderoh, Temengor, and Kenyir Reservoirs (see Fig. 2.1). These reservoirs were selected based on the availability of Satellite Sentinel-3A and S-3B SRAL Altimeter tracks, the presence of in situ water gauge data as a control, and the permission granted by TNB Genco. This allows for exploring inland water applications, particularly the water elevation monitoring. The Chenderoh Reservoir in Perak (8.55 $km^2$), the Temengor Reservoir in Perak (137.25 $km^2$), and the Kenyir Reservoir in Terengganu (326.38 $km^2$) are the study areas. The Kenyir Reservoir holds particular significance as one of the largest reservoirs in Peninsular Malaysia, with two available tracks from the S-3A and S-3B satellites to cover its expansive area.

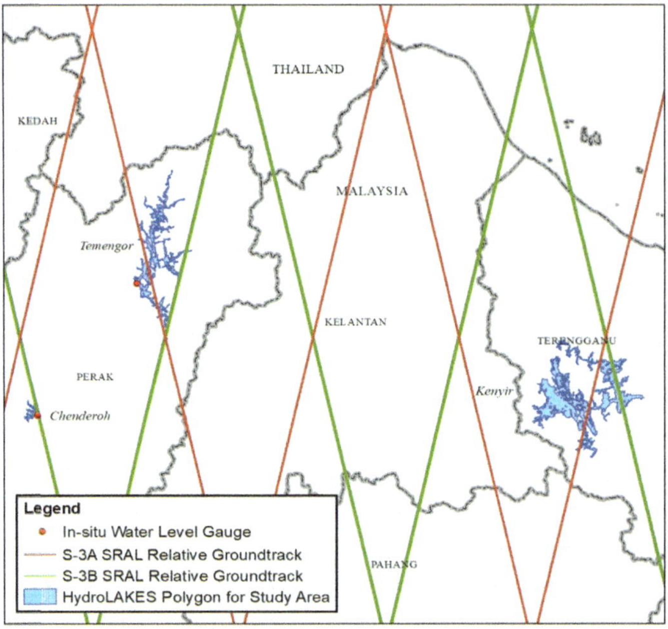


**Fig. 2.1** Study area includes Chenderoh, Temengor, and Kenyir Reservoirs, with the overlay of S-3A (red line) and S-3B (green line) footprints along with water level gauge

### *2.2.2 Data Acquisition*

**Sentinel-3A and Sentinel-3B Dataset**. The Sentinel-3 (S3) satellites, equipped with SRAL, are operated collaboratively by the European Space Agency (ESA) and the European Organization for the Exploitation of Meteorological Satellites (EUMETSAT). These missions involve low-Earth polar orbiting satellites orbiting at an average altitude of 815 km and following a 27-day repeat cycle. The first satellite, S-3A, was launched on 16 February 2016, followed by the second satellite, S-3B, on 25 April 2018. Both S-3A and S-3B have interleaved ground tracks, providing comprehensive and uniform coverage of the Earth's surface. This enhances the monitoring capabilities of the S3 missions. For Peninsular Malaysia, the SRAL data from S-3A and S-3B are obtained from 2016 to 2023 from the EarthConsole® platform owned by Progressive Systems [33, 34].

In this study, the S3 SRAL datasets from 2016 to 2023 within the designated study area were processed online using the EarthConsole® Parallel Processing Environment (P-PRO) service. Table 2.1 provides important information about the available S3 SRAL altimeter missions, including their relative orbit numbers and time frames. It serves as a quick reference for understanding the available missions, their associated reservoirs, and the corresponding relative orbit data and time frames used in this study.

**Table 2.1** Sentinel missions, relative orbit information, and date range for selected reservoirs in the study area

| Study area | Type of Sentinel mission | Relative orbit no. | Date range |
|---|---|---|---|
| Chenderoh Reservoir | Sentinel-3B | 053 | 16 January 2019–17 December 2022 |
| Temengor Reservoir | Sentinel-3A | 053 | 12 April 2016–3 January 2023 |
| Kenyir Reservoir | Sentinel-3A | 189 | 22 April 2016–13 January 2023 |
| | Sentinel-3B | 167 | 1 December 2018–12 December 2021 |

**In Situ Water Level Gauge Data**. The in situ water level gauge data for the Temengor and Chenderoh reservoirs are provided by TNB Generation Sdn. Bhd. (TNB Genco) and their location is shown in Fig. 2.1. The data over Chenderoh are available up to 2021; meanwhile, the data over Temengor are from 2016 through 2023. Unfortunately, no gauged data are available for the Kenyir reservoir.

**HydroLAKES Database**. The S3 SRAL data from S-3A and S-3B satellites underwent meticulous processing for this study. To ensure accuracy, the data were carefully selected based on lake boundary polygons obtained from the HydroLAKES database as shown in Fig. 2.1 [35]. This step effectively removed irrelevant and missing data values, resulting in a refined SRAL dataset. The lake boundary polygons have acted as masks, enabling the extraction of precise SRAL SAR altimetry readings for the lakes under investigation. This systematic approach guarantees the utilization of correct and pertinent data for further analysis. The procedure was carried out within the research region that includes the Temengor, Chenderoh, and Kenyir Reservoirs.

## 2.3 Methodology

This section describes the methodology of the study.

### *2.3.1 Data Processing*

In this study, the SRAL datasets were processed and downloaded using the SARvatore for Sentinel-3 processor service on the EarthConsole® P-PRO platform. This service converted all S3 SRAL altimeter datasets to Level-2 geophysical data in netCDF format. The selection parameters used in the study are presented in Table 2.2.

**Table 2.2** Parameter set in the EarthConsole® P-PRO service for this study

| Step | Sub-parameter | Input parameter |
|---|---|---|
| Processor | | SAR Versatile Altimetric Toolkit for Ocean Research and Exploitation (SARvatore for S3) |
| L1B step | Pick profile | Inland water (HPR) |
| | Pass direction | All |
| | Data posting rate | 80 Hz |
| | Range walk correction | N |
| | Hamming weighting window | Yes, apply it |
| | Exact beam forming | Approximated |
| | FFT zero-padding | N |
| | Radar receiving window size | 128 range bins |
| | Stack subset | 180 |
| | Antenna pattern compensation | N |
| | Dump SAR stack data in output | No |
| L2 step | Restrict the retracking on specific surfaces | Process only water |
| | Model | Multi-look |
| | PTR width alphap parameter | LUT |
| | SAMOSA Model Generation | SAM++ |
| | Dump RIP in output | Y |
| | Dump SAR echo waveforms in output | Y |
| | Choose the default tide model | FES2014b |
| | Choose the default mean sea surface model | DTU18 |
| ALES plus step | Append the ALES + SAR | Yes |

The chosen of some parameter input discussed as follows:

- Inland Water (HPR) Profile: Chosen to process Sentinel-3 data over inland water targets, as it provides accurate and precise measurements of inland water elevation [13].
- Hamming Weighting Window: Used to mitigate sidelobes' ambiguity during beam formation by suppressing spurious signals caused by sidelobes [13].
- SAM++ (SAMOSA Model Generation): Utilizes the Range-Integrated Power (RIP) waveform in addition to the standard SAR waveform to improve retracking accuracy. It also employs a Hamming Weighting Window and has demonstrated improved detection of inland water bodies compared to previous algorithms [13].
- Multi-looked Model: Divides SAR data into smaller sub-apertures for separate processing, resulting in higher-resolution images with reduced noise. It is robust to pitch mispointing direction [13].

### 2.3.2 *Retrieval of Water Surface Elevation (WSE)*

Four key variables must be taken into consideration to accurately estimate water elevation from the S3 SRAL for inland water applications [7, 36]. First, instrument corrections account for biases and errors inherent to the altimeter instrument, such as instrumental drift and calibration errors. Propagation errors form the second variable, arising as the altimeter pulse travels to and from the water surface. Atmospheric conditions can introduce errors due to factors like water vapor, rain, and disturbances, which are mitigated using correction models and algorithms. Surface correction, the third variable, considers factors related to the water surface, including tides, waves, and wind-induced roughness. Applying appropriate corrections helps to minimize the influence of these factors and improve the accuracy of water elevation measurements. The final variable, geophysical adjustment, accounts for gravitational attraction, Earth's rotation, and geoid variations to standardize water elevation within a consistent geodetic frame of reference. By addressing these correction factors, the accuracy and precision of water elevation data derived from the S3 SRAL can be significantly enhanced for inland water applications.

Equation 2.1 is applied to estimate the water surface elevation (WSE), considering important correction factors and closed water systems, as well as geoidal correction [7, 31].

$$\mathrm{WSE} = \left[H_{\mathrm{sat}} - \left(R + C_{\mathrm{iono}} + C_{\mathrm{dry}} + C_{\mathrm{wet}} + C_{\mathrm{st}} + C_{\mathrm{pt}}\right)\right] - \mathrm{Geoid}, \tag{2.1}$$

where

$H_{\mathrm{sat}}$ The precise altitude of a satellite obtained using a radio-positioning system in unit mete (m).

$R$ Range that refers to the vertical height or distance measured between the altimeter instrument and the water surface target at nadir in unit meter (m).

The altimeter calculates the range by measuring the time it takes for a signal to travel to the surface and back, using the speed of propagation.

$C_{iono}$ The corrections are due to a delay in pulse propagation through the ionosphere in unit meter (m).

$C_{dry}$ The corrections due to dry troposphere in unit meter (m).

$C_{wet}$ The corrections due to wet troposphere in unit meter (m).

$C_{st}$ The corrections due to solid earth tide in unit meter (m).

$C_{pt}$ The corrections due to polar tide in unit meter (m).

Geoid The geoid height in unit meter (m).

Geoidal correction, based on the Earth Gravitational Model 2008 (EGM2008), is applied in this study to ensure consistency between SRAL level-2 data and the geoid model utilized in the in situ gauge measurements. This correction aligns the water elevation measurements with the reference frame established by the gauge measurements.

### *2.3.3 Data Masking*

After completing the data correction process to obtain the WSE stage, the subsequent step involved applying data masking to derive a refined dataset specifically for the S3 SRAL. This process entailed removing all irrelevant and non-existent values by utilizing a mask that preserved only the data points intersecting with the reservoir or lake polygon within the study area, obtained from the HydroLAKES database. It is important to note that this spatial selection method was performed without applying any search distance, solely relying on the intersection with the reservoir or lake polygons within the study area.

### *2.3.4 The Construction of WSE Time Series*

The WSE for a specific date at the S3 SRAL virtual station is extracted by employing the central tendency method, which calculates the mean or average value, with no identification of data outliers. Subsequently, all datasets spanning from the initial date to the most recent one for the dedicated reservoir are grouped together to construct a WSE time series.

### *2.3.5 Data Validation*

A data validation process ensured accurate and reliable water level measurements. In this study, the agreement between measurements obtained from S3 SRAL and ground-based gauges was assessed using several technical techniques. These techniques included the utilization of linear regression, calculation of the correlation coefficient ($R^2$), and evaluation of the root mean square error (RMSE) using Eq. 2.2.

$$\text{RMSE} = \sqrt{\sum (P_i - O_i)^2 / n}. \tag{2.2}$$

In the formula, $\sum$ represents the sum, $P_i$ denotes the predicted value for the $i$th observation in the dataset (e.g., S3 SRAL measurements), $O_i$ represents the observed value for the $i$th observation in the dataset (e.g., in situ gauges reading), and $n$ is the sample size.

The methodology use in this study has limitations regarding several factor which can discuss as follows. Firstly, the method focuses solely on the Sentinel-3 Synthetic Aperture Radar (SAR) satellite altimeter, which has unique orbit characteristics and sensor parameters. Consequently, the results may not be comprehensive or directly applicable to other SAR systems, and the absence of hydrological models may limit the capture of all complexities in water level dynamics. Secondly, the method exclusively utilizes the SAR Versatile Altimetric Toolkit for Ocean Research and Exploitation (SARvatore) technique, omitting other techniques such as the delay-Doppler algorithm [37], fully focused SAR (FF-SAR) altimetry [38], and precision inland surface altimetry (PISA) [29]. Therefore, the extent to which the data capture local variations in water elevation using these other techniques remains unclear. Thirdly, the paper does not address the limitations inherent to SAR altimetry itself, such as the potential influence of atmospheric conditions on measurement accuracy. Fourthly, the study does not consider the impact of other factors, such as wind, temperature, and humidity, on the study area of all reservoirs. These factors can influence the accuracy of the estimated water elevation and may introduce uncertainties in the results. Fifth, the method solely utilizes data from one specific station, Chenderoh, and Temengor, with no in situ data available for Kenyir Reservoir for further analysis. This limited sampling may not be fully representative of the entire reservoir study area and could introduce potential spatial variability that is not captured in the analysis. Lastly, the study's temporal coverage for the reservoirs (2016–2023 for Temengor and Kenyir Reservoirs and 2019–2023 for Chenderoh Reservoir) may not be sufficient to capture long-term trends and changes in water elevation across all reservoirs. Therefore, although this study presents advancements in SAR altimetry, it is essential to recognize these limitations, which highlight the need for further research and development.

## 2.4 Results

### *2.4.1 WSE Time Series*

This section provides valuable insights into water elevation variations and trends for the Chenderoh, Temengor, and Kenyir Reservoirs, demonstrating the effectiveness of S3 SRAL in capturing and analyzing these dynamics. Figure 2.2 shows the WSE time series derived from S3 SRAL for the Chenderoh Reservoir. S-3B data collection has started in 2019 for this area (see Table 2.1). The Chenderoh Reservoir, with a size of 8.55 $km^2$ and a complex shape, fluctuates between 59.0 and 61.16 m. Despite limited in situ gauge data in 2021, there is a noticeable correlation between gauge data and S-3B SRAL time series data.

Moving to Fig. 2.3, we present the WSE time series derived from S3 SRAL for the Temengor Reservoir. The Temengor Reservoir covers an area of 137.25 $km^2$ and shows a relatively stable trend compared to in situ gauge measurements. The WSE fluctuates between 232.24 and 267.19 m.

Figure 2.4 displays the WSE time series derived from S3 SRAL for the Kenyir Reservoir. As the largest reservoir with an area of 326.38 $km^2$, the Kenyir Reservoir exhibits distinct patterns and variations in water surface elevation. There is no comparison to in situ gauge measurements, and the data solely rely on S-3A and S-3B SRAL data using the same method. The WSE fluctuates between 112.07 and 188.62 m from April 2016 to January 2023.

Table 2.3 portrays the dataset retrieval from S3 SRAL, revealing distinct patterns in WSE changes among the Chenderoh, Temengor, and Kenyir Reservoirs. Chenderoh Reservoir exhibits relatively stable WSE with minor fluctuations, lacking a clear trend of consistent increase or decrease. In contrast, both Temengor and Kenyir Reservoirs display noticeable fluctuations in WSE, with Temengor experiencing a significant increase in 2020 and Kenyir encountering a noteworthy drop in 2016.

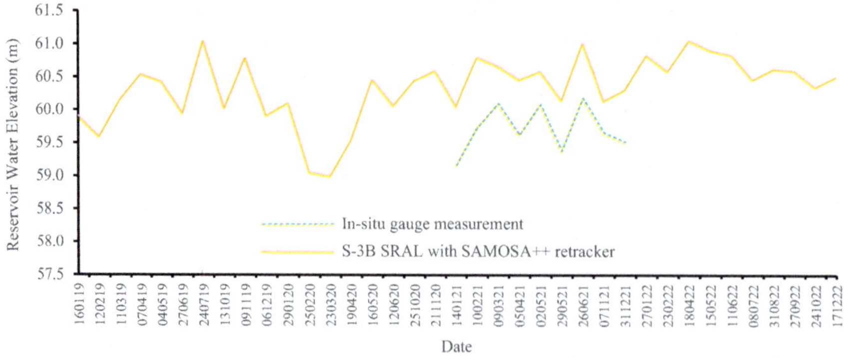


**Fig. 2.2** WSE time series for Chenderoh Reservoir: S-3B SRAL versus SAMOSA++ retracker (2019–2022) and in situ gauge measurements (2021)

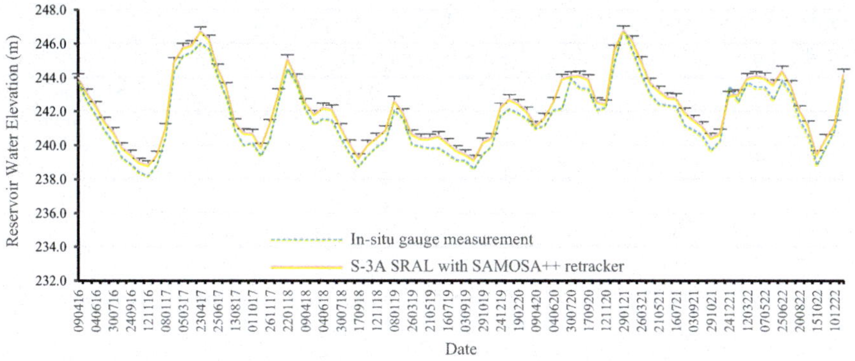


**Fig. 2.3** WSE time series for Temengor Reservoir (2016–2023) with comparison: S-3A SRAL with SAMOSA++ retrackers versus in situ gauge measurements

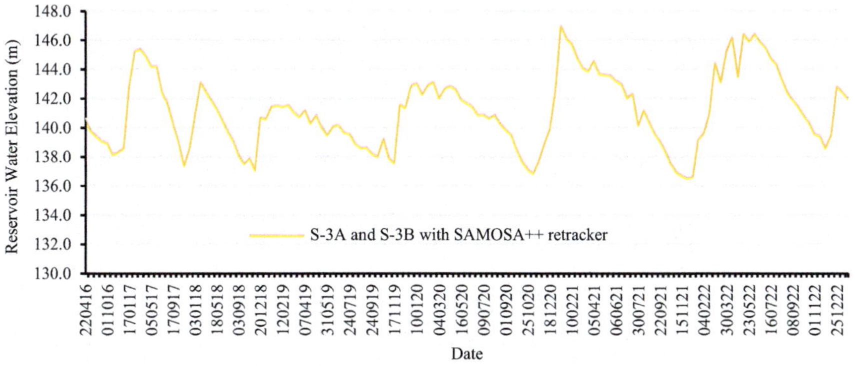


**Fig. 2.4** WSE time series for Kenyir Reservoir (2016–2023) with combination of S-3A and S-3B SRAL with SAMOSA++ retrackers

Overall, no consistent trend in WSE is observed across the dataset, but Kenyir Reservoir demonstrates higher variability compared to the other reservoirs.

### 2.4.2 *Validation Result*

Figure 2.5 shows the linear regression analysis results for the Chenderoh Reservoir in 2021. The graph indicates a positive correlation with an $R^2$ value of 0.7424 and an RMSE of 79.64 cm, representing the degree of fit between S-3B SRAL data and in situ gauge measurements for Chenderoh Reservoir. Moving to Fig. 2.6, we analyze the linear regression analysis for the Temengor Reservoir from 2016 to 2023. The graph reveals a nearly perfect positive correlation with an $R^2$ value of 0.9864 and

**Table 2.3** WSE statistics retrieval from S3 SRAL for the study area

| Study area | Year | WSE readings (m) | | | |
|---|---|---|---|---|---|
| | | Min | Max | Average | Standard deviation |
| Chenderoh Reservoir | 2019 | 59.60 | 61.06 | 60.33 | 0.47 |
| | 2020 | 59.00 | 60.72 | 59.90 | 0.63 |
| | 2021 | 60.06 | 61.16 | 60.48 | 0.33 |
| | 2022 | 60.35 | 61.07 | 60.66 | 0.21 |
| Temengor Reservoir | 2016 | 238.68 | 244.38 | 241.04 | 1.79 |
| | 2017 | 238.91 | 246.94 | 243.37 | 2.31 |
| | 2018 | 238.76 | 245.28 | 241.75 | 1.74 |
| | 2019 | 232.44 | 243.81 | 240.56 | 1.03 |
| | 2020 | 232.24 | 267.19 | 242.82 | 1.68 |
| | 2021 | 240.02 | 248.12 | 243.21 | 2.07 |
| | 2022 | 234.58 | 245.21 | 242.71 | 1.67 |
| | 2023 (January) | 244.04 | 244.54 | 244.30 | 0.12 |
| Kenyir Reservoir | 2016 | 128.51 | 141.30 | 139.13 | 1.07 |
| | 2017 | 135.17 | 147.14 | 142.57 | 2.73 |
| | 2018 | 131.27 | 143.64 | 140.25 | 1.96 |
| | 2019 | 116.86 | 188.62 | 140.05 | 2.92 |
| | 2020 | 112.07 | 144.10 | 140.89 | 2.39 |
| | 2021 | 131.49 | 147.49 | 141.66 | 3.29 |
| | 2022 | 132.90 | 147.40 | 142.89 | 2.62 |
| | 2023 (January) | 138.49 | 143.34 | 142.31 | 0.79 |

an RMSE of 59.95 cm, indicating a strong agreement and high correlation between S-3A SRAL measurements and in situ gauge data for the Temengor Reservoir.

Overall, these findings demonstrate the accuracy of S3 SRAL measurements in estimating water elevation. The high $R^2$ values and low RMSE values for both Chenderoh and Temengor Reservoirs validate the agreement between S-3A and S-3B SRAL data and in situ gauge measurements, establishing the reliability of S3 SRAL data in capturing water elevation dynamics. This methodology can be applied to other small to medium-sized reservoirs and lakes in Peninsular Malaysia, with Kenyir Reservoir serving as an example in this study.

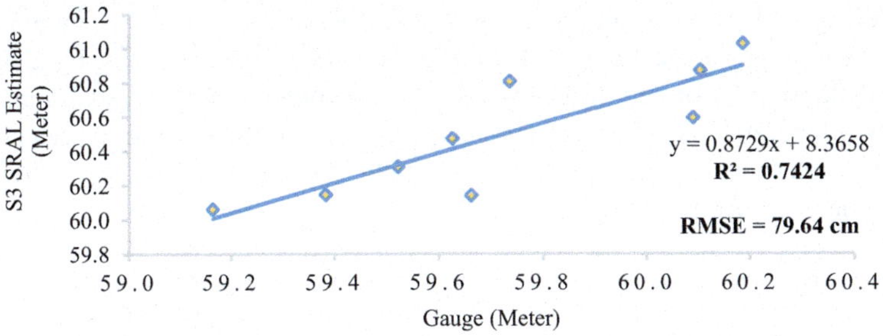


**Fig. 2.5** Comparison of water elevation estimates from satellite S-3B SRAL altimeters and in situ gauge measurements using scatter plots: a case study of Chenderoh Reservoir. Analysis of $R^2$ and RMSE for linear regression based on the 2021 dataset

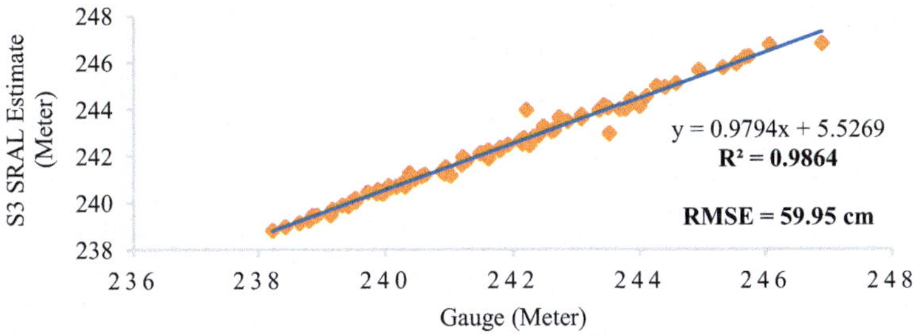


**Fig. 2.6** Comparison of water elevation estimates from satellite S-3A SRAL altimeters and in situ gauge measurements using scatter plots: a case study of Temengor Reservoir. Analysis of $R^2$ and RMSE for linear regression based on the 2016–2023 dataset

## 2.5 Conclusion and Recommendations

In conclusion, this study successfully used S3 SRAL to derive water surface elevation time series for reservoirs and lakes. It provided valuable insights into water level variations and trends for Chenderoh, Temengor, and Kenyir Reservoirs, demonstrating the effectiveness of S3 SRAL in capturing and analyzing these dynamics. The strong agreement between S-3A and S-3B SRAL data and in situ gauge measurements, indicated by high $R^2$ values and low RMSE values, confirmed the accuracy and reliability of S3 SRAL data for Chenderoh and Temengor Reservoirs. The proposed methodology can be applied to other small to medium-sized reservoirs and lakes in Peninsular Malaysia, using Kenyir Reservoir as an example. Historical S3 SRAL data since 2016 have been instrumental in supporting reservoir and lake management decisions and facilitating research on water levels and climate change.

Expanding the analysis to include other inland water types like rivers and wetlands throughout the country, including Sabah and Sarawak, as well as other Southeast Asian countries, is recommended. Stakeholders in water resource management and research institutions are encouraged to continue utilizing S3 SRAL data for ongoing monitoring and analysis of water surface elevation. Applying this methodology to a wider range of locations will improve our understanding of water dynamics and enable more informed decision-making in sustainable water resource management. Collaboration and data sharing among Southeast Asian countries can enhance regional water resource management and enable comparative studies on the impact of climate change on water levels.

**Acknowledgements** We would like to thank the ESA Network of Resource Initiative for providing the Sentinel-3A and 3B data through the SARvatore for Sentinel-3 on the EarthConsole® Parallel Processing Environment (P-PRO) platform, the Malaysian Space Agency, the Ministry of Science and Technology, and the TNB Generation Sdn. Bhd. (TNB Genco) for providing the in situ data.

## References

1. NRECC, *List of Dams Built by Various Agencies According to the Purpose of Their Construction*, ed. by NRECC (2018)
2. M.M. Hanafiah et al., Assessing water scarcity in Malaysia: a case study of rice production. Desalin. Water Treat. **149**, 274–287 (2019)
3. V. Shan, S. Singh, A. Haritash, Water crisis in the Asian countries: status and future trends, in *Resilience, Response, and Risk in Water Systems: Shifting Management and Natural Forcings Paradigms* (2020), pp. 173–194
4. G.A. Tularam, P. Marchisella, Water scarcity in Asia and its long-term water and border security implications for Australia, in *Securing Water and Wastewater Systems: Global Experiences* (2014), pp. 189–211
5. W.K. Mun, *Dam Water Level* (2012)
6. D. Zhao, *Generating Water Level Time Series from Satellite Altimetry Measurements for Inland Applications* (Institute of Geodesy, University of Stuttgart, Stuttgart, 2018), p. 58
7. S. Ghosh et al., The potential applications of satellite altimetry with SARAL/AltiKa for Indian inland waters. Proc. Natl. Acad. Sci. India Sect. A Phys. Sci. (2017)
8. N.K. Biswas et al., Towards a global reservoir assessment tool for predicting hydrologic impacts and operating patterns of existing and planned reservoirs. Environ. Modell. Softw. **140** (2021)
9. M.F. Ahmed, M.B. Mokhtar, L. Alam, Factors influencing people's willingness to participate in sustainable water resources management in Malaysia. J. Hydrol. Reg. Stud. **31** (2020)
10. N. Adebisi et al., Advances in estimating sea level rise: a review of tide gauge, satellite altimetry and spatial data science approaches. Ocean Coast. Manag. **208** (2021)
11. K. Anderson, M.J. Westoby, M.R. James, Low-budget topographic surveying comes of age: structure from motion photogrammetry in geography and the geosciences. Prog. Phys. Geogr. **43**, 163–173 (2019)
12. A.Y. Sun, B.R. Scanlon, How can big data and machine learning benefit environment and water management: a survey of methods, applications, and future directions. Environ. Res. Lett. **14** (2019)
13. S. Dinardo, *Techniques and Applications for Satellite SAR Altimetry Over Water, Land and Ice* (Department of Civil and Environmental Engineering, Technische Universität Darmstadt, Darmstadt, 2020), p. 164
14. Q. Gao et al., Analysis of retrackers' performances and water level retrieval over the Ebro river basin using Sentinel-3. Remote Sens. **11**(6), 718 (2019)
15. F. Ehlers et al., Validity assessment of SAMOSA retracking for fully-focused SAR altimeter waveforms. Adv. Space Res. **71**, 1377–1396 (2023)

16. R. Abileah, S. Vignudelli, A. Scozzari, A completely remote sensing approach to monitoring reservoirs water volume. Int. Water Technol. J. **1**, 63–77 (2011)
17. N.H. Idris, S. Vignudelli, X. Deng, Analysis of Sentinel-3A Synthetic Aperture Radar (SAR) altimetry waveforms over the Southeast Asia Region, in *IGARSS 2020–2020 IEEE International Geoscience and Remote Sensing Symposium* (IEEE, 2020)
18. T. Moreau et al., High-performance altimeter Doppler processing for measuring sea level height under varying sea state conditions. Adv. Space Res. **67**(6), 1870–1886 (2021)
19. K. Nielsen, O.B. Andersen, H. Ranndal, Validation of Sentinel-3A based lake level over US and Canada. Remote Sens. **12**(17), 2835 (2020)
20. Y. Le Roy et al., SRAL, a radar altimeter designed to measure a wide range of surface types, in *2009 IEEE International Geoscience and Remote Sensing Symposium* (IEEE, 2009)
21. J.-F. Crétaux, C. Birkett, Lake studies from satellite radar altimetry. C.R. Geosci. **338**(14–15), 1098–1112 (2006)
22. P. Bonnefond et al., Calibrating the SAR SSH of Sentinel-3A and CryoSat-2 over the Corsica facilities. Remote Sens. **10**(1), 92 (2018)
23. H. Eldardiry et al., Success stories of satellite radar altimeter applications. Bull. Am. Meteor. Soc. **103**(1), E33–E53 (2022)
24. R. Abileah et al., Coherent ranging with Envisat radar altimeter: a new perspective in analyzing altimeter data using Doppler processing. Remote Sens. Environ. **139**, 271–276 (2013)
25. Y.-L.S. Tsai et al., Monitoring large-scale inland water dynamics by fusing Sentinel-1 SAR and Sentinel-3 altimetry data and by analyzing causal effects of snowmelt. Remote Sens. **12**(23), 3896 (2020)
26. G.D. Quartly et al., An overview of requirements, procedures and current advances in the calibration/validation of radar altimeters. Remote Sens. **13**(1), 125 (2021)
27. UN, The 17 goals (2023). https://sdgs.un.org/goals. Accessed 16 May 2023
28. L. Jiang et al., Evaluation of Sentinel-3 SRAL SAR altimetry over Chinese rivers. Remote Sens. Environ. **237**, 111546 (2020)
29. R. Abileah, S. Vignudelli, Precise inland surface altimetry (PISA) with nadir specular echoes from Sentinel-3: algorithm and performance assessment. Remote Sens. Environ. **264**, 112580 (2021)
30. N. Xu et al., Global estimation and assessment of monthly lake/reservoir water level changes using ICESat-2 ATL13 products. Remote Sens. **13**(14), 2744 (2021)
31. S. Shu et al., Analysis of Sentinel-3 SAR altimetry waveform retracking algorithms for deriving temporally consistent water levels over ice-covered lakes. Remote Sens. Environ. **239**, 111643 (2020)
32. Y. Liu, H. Yue, Estimating the fluctuation of Lake Hulun, China, during 1975–2015 from satellite altimetry data. Environ. Monit. Assess. **189**, 1–16 (2017)
33. Systems, P. EarthConsole® (2023). https://ui-ppro.earthconsole.eu/. Accessed 16 May 2023
34. ESA, Copernicus Open Access Hub (2023). https://scihub.copernicus.eu/dhus/#/home. Accessed 16 May 2023
35. M.L. Messager et al., Estimating the volume and age of water stored in global lakes using a geo-statistical approach. Nat. Commun. **7**(1), 13603 (2016)
36. B. Iijima et al., Automated daily process for global ionospheric total electron content maps and satellite ocean altimeter ionospheric calibration based on global positioning system data. J. Atmos. Sol. Terr. Phys. **61**(16), 1205–1218 (1999)
37. R.K. Raney, The delay/Doppler radar altimeter. IEEE Trans. Geosci. Remote Sens. **36**(5), 1578–1588 (1998)
38. A. Egido, W.H. Smith, Fully focused SAR altimetry: theory and applications. IEEE Trans. Geosci. Remote Sens. **55**(1), 392–406 (2016)

# Chapter 3
# Design Optimization of UWB Microstrip Antenna Based on Fractal Koch Structure for C-Band and X-Band Satellite Communication System

**Setyawan Renaldy, Radial Anwar, and Budi Syihabuddin**

**Abstract** Satellite is still one of among the most utilized mediums in communication system. This paper presents an optimized antenna design, suitable to be used as a feed in pair with a parabolic dish reflector. Fractal Koch patch structure in combination with pyramidal ground plane structure is analyzed and implemented. Based on measurement result, the proposed design can operate at 3.4–11.4 GHz with gain of about 3.9 dBi at the center of the band. The measured radiation pattern shows a bi-directional coverage, which is agreed with the simulation. These results support the possibility to employ the proposed design as an antenna feed in C-band and X-band satellite communication system.

## 3.1 Introduction

Size miniaturization has been one of the most explored topics in the field of antenna microstrip design. Many techniques have been introduced, including utilization of slots in the patch or ground plane [1, 2], meandering [3], capacitive and/or inductive load insertion [4, 5], and by combining two or more techniques [6]. Another method that has been employed for miniaturization is by adopting fractal structure [7–12]. However, fractal structure is not used only for miniaturization. Instead, it is also used to achieve multiband operating frequencies [13–16] or to obtain an ultra-wideband (UWB) characteristic [11, 17, 18].

---

S. Renaldy · R. Anwar (✉)
School of Applied Science, Telkom University, Bandung, Indonesia
e-mail: radialanwar@tass.telkomuniversity.ac.id

B. Syihabuddin
School of Electrical Engineering, Telkom University, Bandung, Indonesia

 
M. T. Islam et al. (eds.), *Proceedings of the 8th International Conference on Space Science and Communication*, Springer Proceedings in Physics 303,
https://doi.org/10.1007/978-981-97-0142-1_3

This paper presents fractal Koch structure utilization in a microstrip antenna, in combination with pyramid shape ground plane design to achieve UWB operating frequency, with bandwidth more than 100%, for satellite application. The proposed design can operate at C- and X-band frequency, which is much higher compared to the Minkowski fractal in [11] that operates at UHF band. Moreover, it does not have any notch or rejected band that occurred in the designs explained in [17, 18]. Thus, the antenna in this paper provides a wider utilization option, including as feeder for a satellite dish antenna that operates at C-band, X-band, or both, and for 5.8 GHz ISM band application.

## 3.2 Methodology

The design process can be divided into five stages of optimization. All stages were initiated solely to obtain the widest operating frequency. First stage is by comparing the zeroth iteration of the fractal Koch structure with its first iteration. In this stage, the full ground plane is utilized, as shown in Fig. 3.1. Second stage was optimization of the first iteration structure by defecting the ground plane, in this case only by removing some of the ground plane area but preserving the rectangular shape (Fig. 3.2). Up to this stage, none of the designs provides 100% bandwidth or higher.

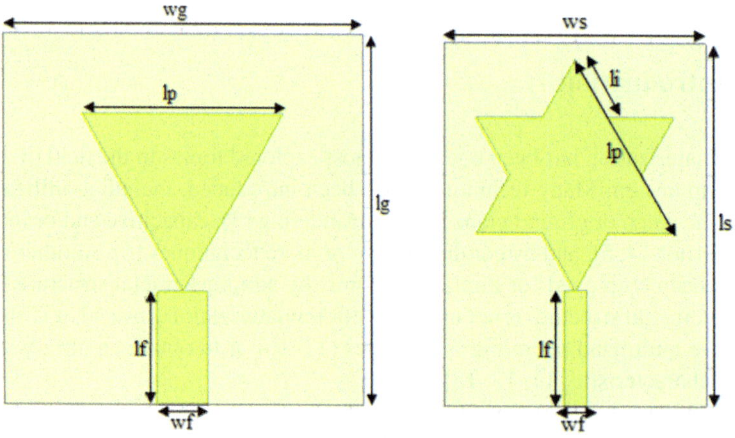


**Fig. 3.1** First-stage optimization. Left: zeroth iteration. Right: first iteration

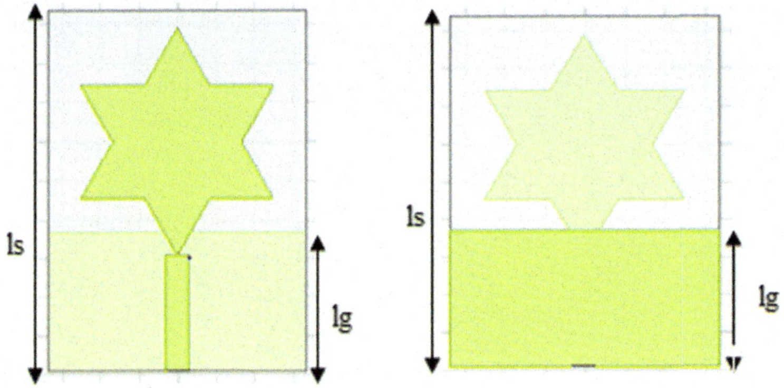


**Fig. 3.2** Second-stage optimization. Left: front side (patch). Right: back side (ground plane)

The third stage is comparing three different ground plane structures, the sawtooth, half circle, and the pyramid structure (Fig. 3.3). Simulation showed that the pyramid structure provides the widest bandwidth compared to the other two, and hence, the optimization is continued by using this ground plane structure, which is considered as the fourth stage. Last stage is by comparing the first and second fractal Koch iteration to ensure of which geometry gives the widest bandwidth, as shown in Fig. 3.4.

After the optimized design was achieved, the proposed design was fabricated and measured in terms of return loss. The radiation pattern and gain are also measured, but only at the middle of the frequency band.

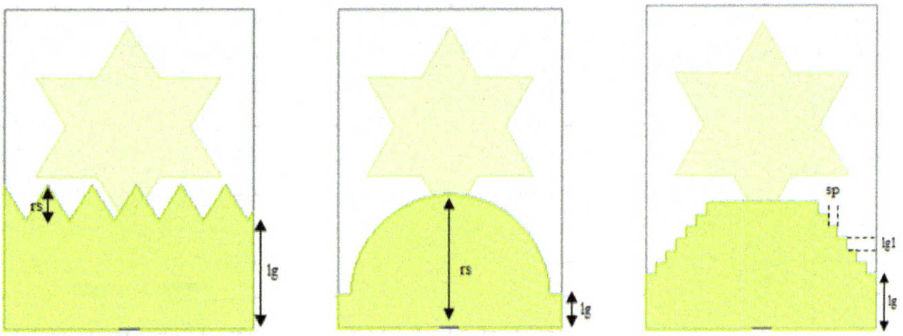


**Fig. 3.3** Third-stage optimization. Left: sawtooth. Middle: half circle. Right: pyramid

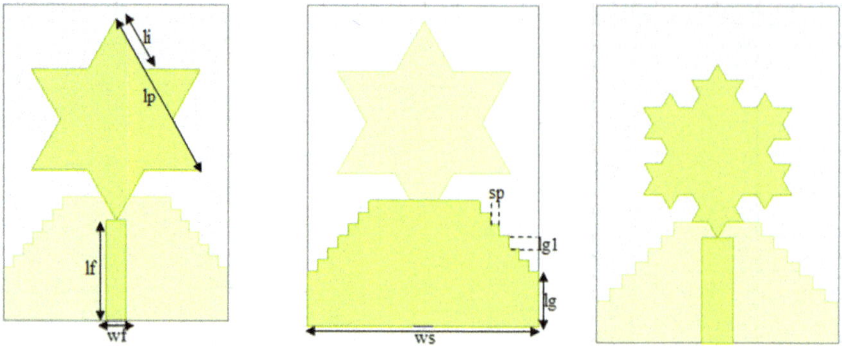


**Fig. 3.4** Fourth- and fifth-stage optimizations. Left: fourth optimization parameters (front side). Middle: fourth optimization parameters (back side). Right: second iteration

## 3.3 Result and Discussion

Four ground plane geometries have been simulated to see which one inherits the widest operating frequency. Each geometry went through a series of optimization process. The optimized result from each geometry is then compared, as shown in Fig. 3.5. It can be seen that the pyramid structure provides the widest bandwidth, reaching up to 107.4%. However, the obtained operating frequency only covers the C-band for satellite application. This initial design with the pyramid ground plane is further optimized by changing its geometry parameter, as listed in Table 3.1.

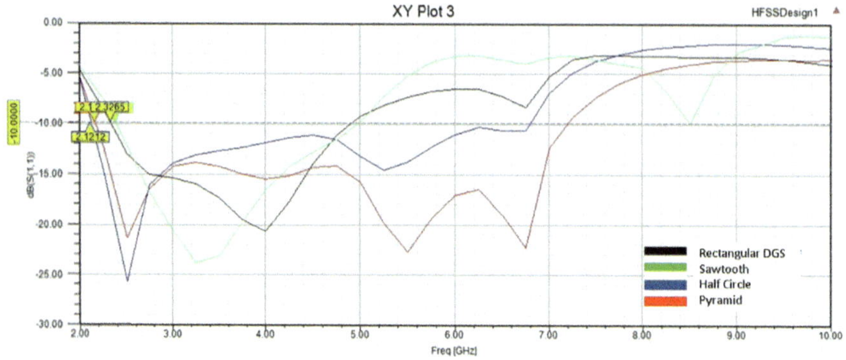


**Fig. 3.5** Return loss comparison of four ground plane geometries

**Table 3.1** Optimization with pyramid ground plane structure

| No | Lp (mm) | Li (mm) | Lf (mm) | Lg (mm) | Lgl (mm) | Sp (mm) | Bw (GHz) | Bw (%) |
|---|---|---|---|---|---|---|---|---|
| 1 | 9.093 | 3.031 | 7 | 1 | 1 | 1.25 | 4.656–6.186 | 28.22 |
| 2 | 14.289 | 4.763 | 10 | 4 | 1 | 1.25 | 3.31–4.188 | 23.42 |
| 3 | 29.228 | 9.743 | 13 | 7 | 1 | 1.25 | 10.831–12.88 | 17 |
| 4 | 9.093 | 3.031 | 7 | 1 | 1.25 | 2.5 | 4.659–9.978 | 73 |
| 5 | 14.289 | 4.763 | 10 | 4 | 1.25 | 2.5 | 3.336–13.211 | 119.36 |
| 6 | 29.228 | 9.743 | 13 | 7 | 1.25 | 2.5 | 2.610–8.4 | 105.84 |
| 7 | 9.093 | 3.031 | 7 | 1 | 1.5 | 3.75 | 0 | 0 |
| 8 | 14.289 | 4.763 | 10 | 4 | 1.5 | 3.75 | 3.208–4.443 | 32.28 |
| 9 | 29.228 | 9.743 | 13 | 7 | 1.5 | 3.75 | 2.585–6.886 | 90.83 |

From Table 3.1, design on the fifth optimization obtains the widest operating frequency, as well as covering the C-band and X-band on satellite application. Hence, this design was fabricated and measured. Figure 3.6 shows the fabricated antenna, while the geometry parameters are listed in Table 3.2.

Return loss comparison of the simulation and measurement result is shown in Fig. 3.7. Measurement result gives a slight narrower bandwidth, of about 108.943% (3.405–11.415 GHz). Nevertheless, this band still can cover the C-band and X-band satellite operating frequency. The proposed antenna was also measured for its radiation pattern. Measurement result shows a good agreement with simulation, that the proposed antenna inherits bi-directional coverage, as depicted in Fig. 3.8. The obtained beamwidth is of about 40°, with a slight smaller value on measurement

**Fig. 3.6** Fabricated antenna. Left: front side. Right: back side

**Table 3.2** Final design geometry

| Variable | Length (mm) |
|---|---|
| Lp | 14.289 |
| Li | 4.763 |
| Lf | 10 |
| Wf | 3 |
| Ws | 23.4 |
| Ls | 27.8 |
| Wg | 23.4 |
| Lg | 4 |
| Lg1 | 1.25 |
| Sp | 2.5 |

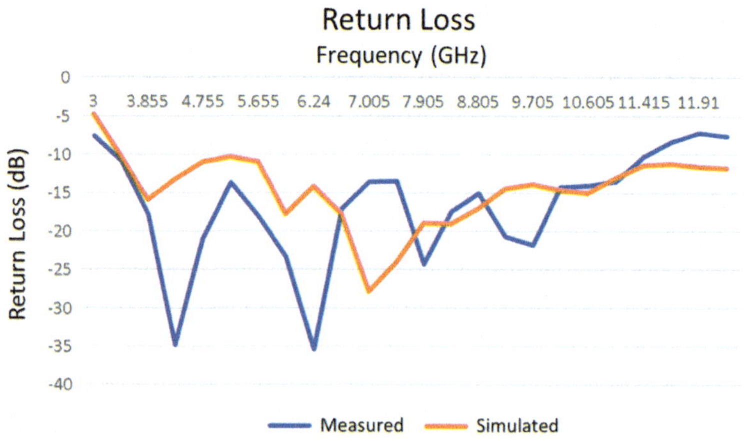


**Fig. 3.7** Return loss comparison

result. Gain of the proposed antenna was also simulated (3.12 dBi) and measured (3.94 dBi) at 6.25 GHz which is the center frequency of the band. Some differences between all simulation and measurement data have occurred probably due to unideal measurement location, soldering, and the utilization of connector which are not part of simulation. To provide a broader perspective, Table 3.3 list characteristic of antennas explained in some references compared to the proposed design in this paper.

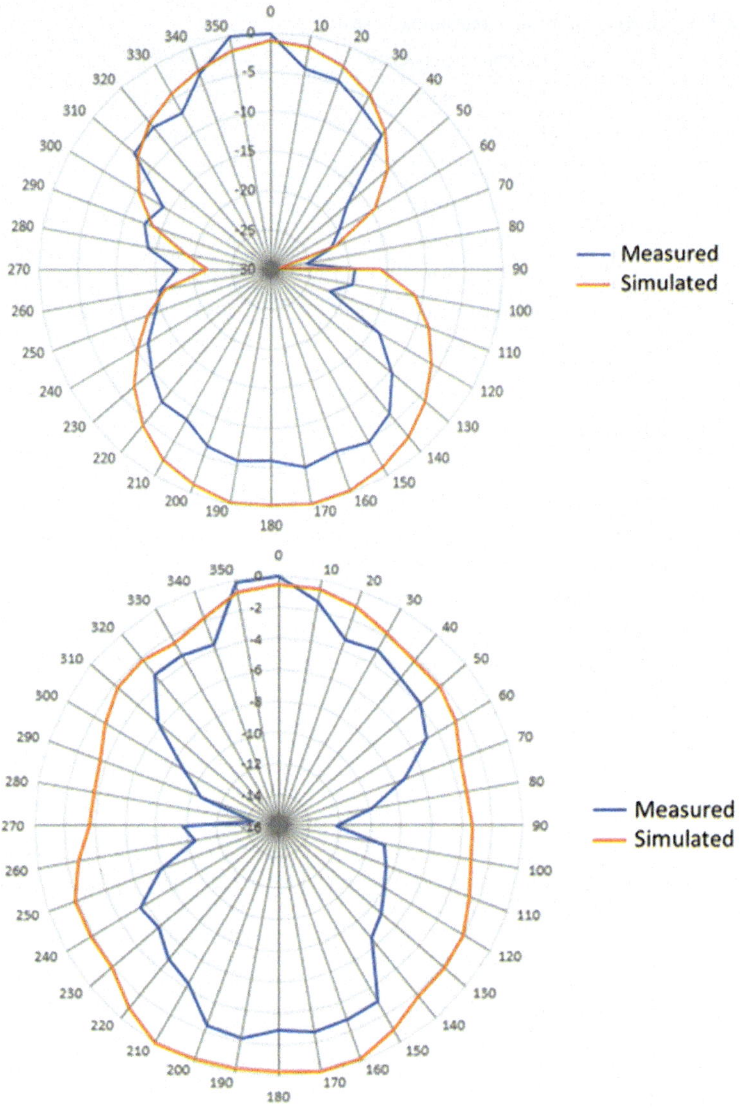


**Fig. 3.8** Radiation pattern of the proposed antenna. Top: azimuth. Bottom: elevation

**Table 3.3** Comparison of result from similar research

| References | Operating frequency | Bandwidth |
|---|---|---|
| [11] | UHF, L-band, S-band | UWB, 124% |
| [13] | Ka | Multiband |
| [14] | S-band, C-band, X-band | Multiband |
| [15] | C-band, X-band | Multiband |
| [16] | C-band, X-band | Multiband |
| [17] | S-band, C-band, X-band | UWB with rejected band |
| [18] | S-band, C-band, X-band | UWB with rejected band |
| Proposed design | C-band, X-band | UWB, 119% |

## 3.4 Conclusion

Design of a miniaturized UWB microstrip antenna has been presented in this paper. Bandwidth higher than 100% has been achieved by combining fractal Koch patch structure with pyramid shape on the ground plane. Simulations give a 119.36% bandwidth, while measurement provides 108.943%, which cover the C-band and X-band satellite operating frequency. With bi-directional radiation pattern, the proposed antenna is suitable to be paired with a parabolic dish reflector for satellite communication application.

## References

1. R. Azadegan, K. Sarabandi, A novel approach for miniaturization of slot antennas. IEEE Trans. Antennas Propag. **51**(3), 421–429 (2003)
2. L. Ammai, R. Anwar, D.A. Nurmantris, Analysis on multi rings defected ground structure for microstrip antenna miniaturization, in *2017 International Conference on Engineering Technology and Technopreneurship (ICE2T)* (Kuala Lumpur, Malaysia, 2017), pp. 1–4
3. V. Agus Sria Warman, R. Anwar, Y. Wahyu, Development of meandered microstrip antenna for RFID application. Int. J. Appl. Inf. Technol. 68–73 (2018)
4. P.L. Chi, R. Waterhouse, T. Itoh, Antenna miniaturization using slow wave enhancement factor from loaded transmission line models. IEEE Trans. Antennas Propag. **59**(1), 48–57 (2011)
5. C.R. Rowell, R.D. Murch, A capacitively loaded PIFA for compact mobile telephone handsets. IEEE Trans. Antennas Propag. **45**(5), 837–842 (1997)
6. N. Behdad, K. Sarabandi, Slot antenna miniaturization using distributed inductive loading, in *Proceedings of IEEE Antennas and Propagation Society International Symposium*, vol. 1 (2003), pp. 308–311
7. V.G. Sawant, V. Aruna, A. Iyer, P. Diksha, S. Deepthi, Design of high gain fractal antenna. Int. J. Res. Anal. Rev. **6**(1), 22–28 (2019)
8. M.M. Rahman, M.R. Islam, T.M. Faisal, A compact design and analysis of a fractal microstrip antenna for ultra-wideband applications. Am. J. Eng. Res. **3**(10), 1–11 (2019)
9. B. Kanthamma, P.V. Sridevi, Design and simulation of horizontally placed H-shaped fractal antenna. Int. J. Innov. Technol. Explor. Eng. **9**(1), 33–38 (2019)
10. M. Zulfin, A.H. Rambe, B. Budi, Design and analysis microstrip dipole using fractal Koch for 433 MHz applications. IOP Conf. Ser. Mater. Sci. Eng. **309**, 012047 (2018)

11. F. Wang, F. Bin, Q. Sun, J. Fan, F. Liang, X. Xiao, A novel UHF Minkowski fractal antenna for partial discharge detection. Microw. Opt. Technol. Lett. **59**, 1812–1819 (2017)
12. D. Tumakov, D. Chikrin, P. Kokunin, Miniaturization of a Koch-type fractal antenna for Wi-Fi applications. Fractal Fract. **4**(25) (2020)
13. M.R. Jena, B.B. Mangaraj, R. Pathak, An improved compact & multiband fractal antenna using the Koch curve geometry. Wireless Mob. Technol. **2**(1), 1–6 (2014)
14. G. Ramalakshmi, P. Mallikarjuna Rao, A novel metamaterial inspired 2nd iteration Koch fractal antenna for Wi-Fi, WLAN, C band and X band wireless communications. J. Phys. Conf. Ser. **2062**, 012004 (2021)
15. M. Gupta, V. Mathur, Koch boundary on the square patch microstrip antenna for ultra wideband applications. Alex. Eng. J. **57**(3), 2113–2122 (2018)
16. N. Sharma, V. Sharma, S.S. Bhatia, A novel hybrid fractal antenna for wireless applications. Prog. Electromagn. Res. M **73**, 25–35 (2018)
17. S.K. Terlapu, P.S.R. Chowdary, C. Jaya, V.V.S.S. Sameer Chakravarthy, S.C. Satpathy, On the design of fractal UWB wide-slot antenna with notch band characteristics, in *Microelectronics, Electromagnetics and Telecommunications*, ed. by J. Anguera, S. Satapathy, V. Bhateja, K. Sunitha. Lecture Notes in Electrical Engineering, vol. 471 (Springer, Singapore, 2018)
18. R.A. Lituma-Guartan, J.B. Benavides-Aucapiña, D.F. Poveda-Pulla, L.F. Guerrero-Vasquez, P.A. Chasi-Pesántez, A novel hybrid fractal antenna design for ultra-wideband application, in *2018 IEEE 10th Latin-American Conference on Communications (LATINCOM)* (Guadalajara, Mexico, 2018), pp. 1–5

# Chapter 4
# Wideband Monostatic/Bistatic RCS Reduction Using Optimized 1-Bit Coding Metasurface

**Md. Bellal Hossain, Mohammad Rashed Iqbal Faruque, and Sabirin Abdullah**

**Abstract** The 1-bit coding metasurface is employed in this paper to demonstrate a technique for decreasing radar cross-section (RCS) in both monostatic and bistatic levels that has been optimized using a binary particle swarm optimization (BPSO) algorithm. The configuration of the design comprises two unit cells, specifically a square and a circular ring, represented as elements '0' and '1'. The reflection phase between these unit cells exhibits a phase difference of approximately $180^\circ \pm 37^\circ$. This design operates within a frequency range of 22.5–30 GHz and achieves an RCS reduction of less than − 10 dB. The design of the coding element utilizes a cost-effective F4B substrate with a thickness of 1.5 mm. The BPSO algorithm, in conjunction with array theory, is employed to determine the optimal positioning of unit cells within the subarray to achieve wideband RCS reduction. Compared to the conventional checkerboard metasurface, the monostatic RCS reduction achieved by the optimized coding metasurface exhibits a wider bandwidth. Furthermore, the performance of the optimized metasurface is evaluated in a bistatic scenario at 27.5 GHz, demonstrating significant RCS reduction when compared to an equivalently sized metallic plate and checkerboard pattern. The simulation results indicate that the optimized coding metasurface presents a feasible solution for achieving significant RCS reduction in both monostatic and bistatic scenarios.

## 4.1 Introduction

Metamaterials, in contrast to naturally occurring materials composed of microscopic particles such as atoms and molecules, are artificial structures composed of subwavelength composite elements arranged in periodic or nonperiodic patterns. These structures possess exceptional capabilities to control electromagnetic waves based on

Md. B. Hossain · M. R. I. Faruque (✉) · S. Abdullah
Space Science Centre, Institute of Climate Change, Universiti Kebangsaan Malaysia (UKM), 43600, Bangi, Selangor D.E., Malaysia
e-mail: rashed@ukm.edu.my

 
M. T. Islam et al. (eds.), *Proceedings of the 8th International Conference on Space Science and Communication*, Springer Proceedings in Physics 303,
https://doi.org/10.1007/978-981-97-0142-1_4

specific needs, resulting in a wide array of extraordinary physical phenomena and practical applications [1, 2]. Metasurfaces have garnered considerable attention for their captivating applications, serving as the two-dimensional (2D) counterpart to metamaterials. These applications include notable applications such as invisibility cloak [3], electromagnetic (EM) wave absorbers [4, 5], stealth of target [6], antenna design [7], SAR reduction [8], sensors [9], satellite communications [10, 11], and so on. In a recent development, Giovampaola and Engheta introduced the concept of digital metamaterials as an innovative method for controlling the electromagnetic (EM) characteristics of metasurface unit cells using digital bits [12]. Building upon this concept, Cui et al. further advanced the concept of digital metamaterials by introducing coding-based metasurfaces [13]. Coding metasurfaces have garnered significant attention, with numerous designs proposed to achieve wideband radar cross-section (RCS) reduction [14]. The manipulation of electromagnetic waves using different coding sequences in metasurfaces is of great importance for achieving RCS reduction, as it is recognized as one of the most critical applications in this field. A basic and straightforward form of coding metasurface is the 1-bit coding metasurface, which involves dividing the metasurface elements into '0' and '1' states. The '0' element is designed to exhibit a phase response of 0°, whereas the '1' element has a phase response of 180° [15]. The technique employed to reduce radar cross-section (RCS) is based on the cancelation of opposite phases. This approach entails maintaining a phase difference of $180° \pm 37°$ within a designated frequency range [16]. Nevertheless, the impact of both monostatic and bistatic scenarios is not considered by these approaches.

This study explores the new perspectives offered by 1-bit coding metasurfaces to effectively decrease radar cross-section (RCS) in both monostatic and bistatic directions within the microwave frequency range spanning from 22.5 to 30 GHz. Moreover, the optimized 1-bit pattern exhibits significant effectiveness in attenuating bistatic radar cross-sections (RCS), as confirmed by the observed 2D and 3D scattering patterns at frequencies of 23, 26, and 27.5 GHz.

## 4.2 Optimized Coding Metasurface Pattern Design

The optimized coding metasurface is built using a combination of square and circular ring structures. The ring structure exhibits remarkable stability, remaining unaffected by variations in incident angle and polarization. The unit cell of the coding element comprises two layers, with a F4B substrate sandwiched between two copper layers. The F4B substrate has a dielectric constant of 2.65 and a tangent loss of 0.001.

The substrate thickness is 1.5 mm. The upper layer is designed with a square ring-shaped structure representing the '0' element, while the '1' element is depicted as a circular ring, as depicted in Fig. 4.1. The substrate length ($N$) and width ($M$), length ($T$) and width ($S$) of the square ring, the inner arm of the circular ring ($P$), and outer arm ($Q$) are 5 mm, 1.0 mm, 1.6 mm, and 2.3 mm, respectively. The reflection amplitude and phase characteristics of the elements are obtained through simulations using the frequency-domain solver of CST Microwave Studio, with unit cell boundary conditions applied. The simulations are conducted using a Floquet port configuration, where the unit cells define the boundary conditions in the *x*- and *y*-directions, and a Floquet port is established in the *z*-direction. The reflection phases of both components and their phase difference with reflection amplitude are depicted in Fig. 4.2. To achieve a reduction of $-$ 10 dB in RCS, it is necessary for the reflection phase difference between the two elements to vary between 143 and 217°. Ideally, the desired scenario involves finding two elements with a phase difference of $180° \pm 37°$ throughout the wide frequency range, while maintaining equal reflection amplitudes. Specifically, in the frequency range of 22.5–30 GHz, the phase difference exhibits variations of 143 and 217°, with exact values of 180° observed at 26 GHz.

Once the design of the coding metasurface elements '0' and '1' is complete, they are positioned on a two-dimensional surface, allowing the coding metasurface to control the behavior of electromagnetic waves. The proposed coding metasurface is constructed using an $8 \times 8$ subarray configuration, with overall dimensions of $40 \times 40$ mm$^2$. To optimize the arrangement of the metasurface, a combination of array theory and the binary particle swarm optimization (BPSO) algorithm is utilized. A MATLAB code is developed to implement the BPSO algorithm and facilitate the optimization process. The input for the BPSO algorithm consists of the amplitude and phase values of the two elements at the central frequency of 26 GHz. By executing 400 iterations of the BPSO algorithm, the desired arrangement of coding subarrays is determined. This arrangement is utilized to construct the optimized structure, ensuring the achievement of the lowest possible fitness value. Figure 4.3a shows the fitness function curve generated through numerical optimization using a MATLAB program. The curve demonstrates a rapid decrease during the initial phase

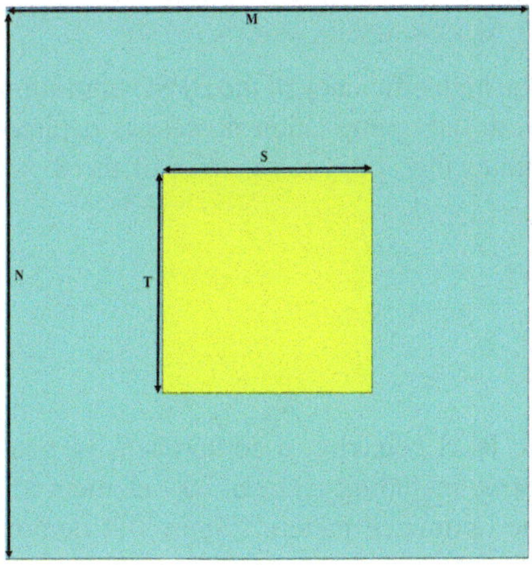


(a)

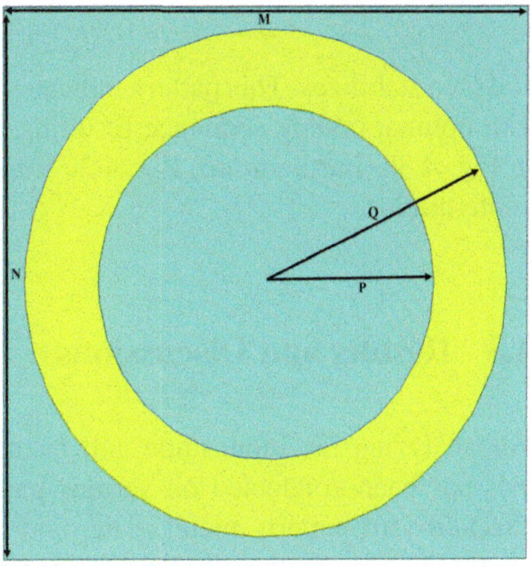


(b)

**Fig. 4.1** Schematic diagram of **a** element '0' view and **b** element '1' view

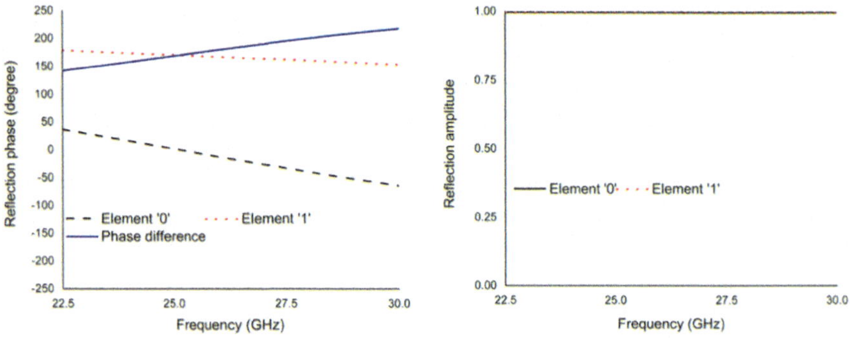


(a) (b)

**Fig. 4.2** Scenario of **a** reflection phase and **b** reflection amplitude

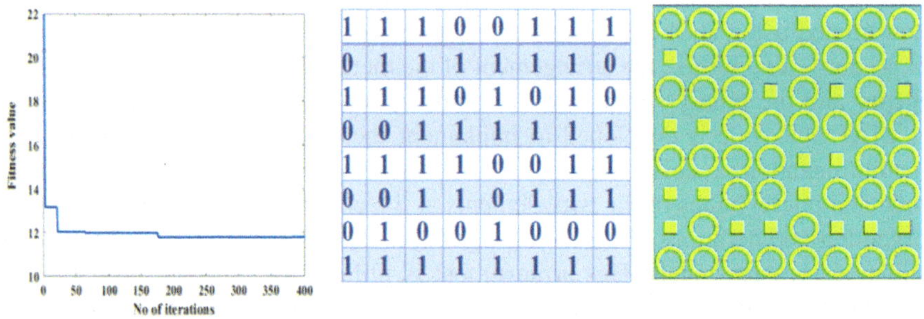


(a) (b) (c)

**Fig. 4.3** **a** Fitness value versus no. of iterations using BPSO algorithm, **b** optimized coding sequence and **c** optimized metasurface

and then stabilizes. This pattern indicates the high efficiency of the BPSO algorithm. The optimal coding sequence resulting from the optimization process is depicted in Fig. 4.3b. Furthermore, Fig. 4.3c illustrates the optimized pattern of the 8 × 8 configuration.

## 4.3 Results and Discussions

For analyzing the monostatic and bistatic RCS behavior, a comprehensive analysis has been conducted on various patterns, including element '0', element '1', checkerboard pattern, metal plate, and the optimized pattern. Figure 4.4 demonstrates the comparative reduction scenarios in the simulated monostatic RCS across these different patterns. The optimized pattern consistently achieves RCS reduction of less than − 10 dB throughout the frequency range of 24–30 GHz for normal incident angles. In contrast, the other sequences exhibited similar scattering patterns without notable RCS reduction. The exceptional performance of the optimized structure can be attributed to its enhanced ability to manipulate electromagnetic waves and the significant impact of the arrangement of coding elements on the scattering pattern. Notably, the optimized structure recorded the highest RCS reduction value of – 21 $dBm^2$ at 27.5 GHz.

Figures 4.5 and 4.6 provide visual representations of the scattering patterns in two-dimensional (2D) and three-dimensional (3D) formats for the metal plate, checkerboard pattern, and optimized pattern. These representations are observed under normal incidence at frequencies of 23, 26, and 27.5 GHz. The purpose of these figures is to analyze the influence of the metasurface and the phase cancelation mechanism by examining both 2D and 3D scenarios. Figure 4.5c demonstrates a significant reduction in backscattering for the optimized pattern, as opposed to the backscattering observed in the metal plate and checkerboard pattern at 27.5 GHz. These observations indicate that the optimized structure offers a wider range of scattered energies, causing electromagnetic waves to diffuse instead of reflecting directly back. Additionally, a slight decrease in backscattering of the main lobe is observed for the optimized pattern at 23 and 26 GHz, as shown in Fig. 4.5a, b.

Figure 4.6a, b show that both the metal plate and checkerboard pattern display similar specular reflections. Throughout all frequencies, these patterns exhibit normal reflections with four backscattered lobes in symmetrical directions, resulting from phase diffusion. Consequently, these patterns demonstrate a high radar cross-section (RCS). In contrast, the optimized pattern exhibits the ability to evenly redistribute scattered waves in multiple directions at 26 and 27.5 GHz. However, at 27.5 GHz, there is a significant decrease in energy in the normal direction of the metasurface compared to the frequencies of 23 and 26 GHz, as indicated in Fig. 4.6c. This outcome suggests that the optimized pattern effectively scatters the incident wave, facilitating phase cancelation across a broad frequency range. As a result, a substantial reduction

**Fig. 4.4** Simulation scenario of RCS reduction at different patterns

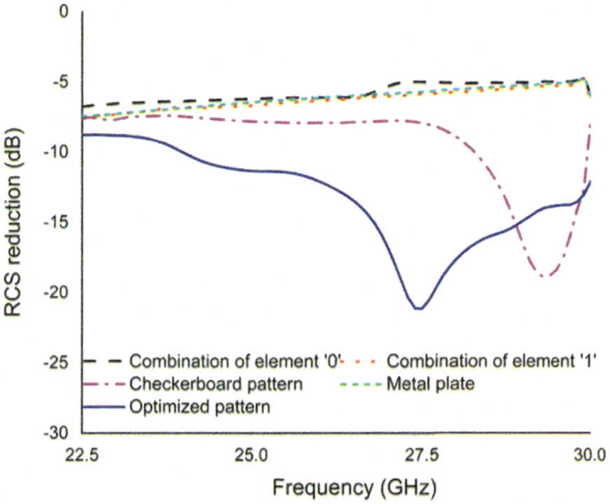

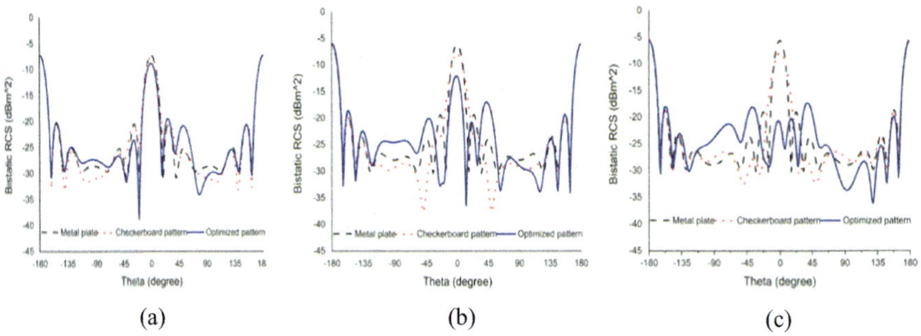


**Fig. 4.5** Two-dimensional scattering pattern of metal plate, checkerboard, and optimized pattern at **a** 23 GHz, **b** 26 GHz and **c** 27.5 GHz

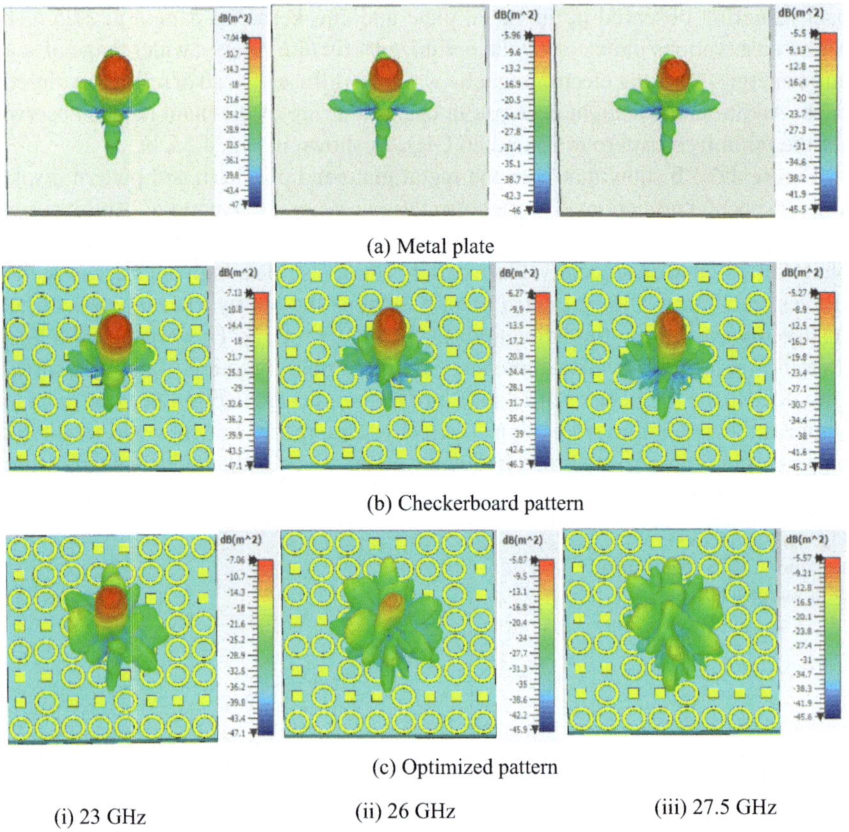


**Fig. 4.6** Three-dimensional scattering pattern at 23, 26, and 27.5 GHz of **a** metal plate, **b** checkerboard pattern and **c** optimized pattern

in radar cross-section (RCS) is achieved. In conclusion, the comparison demonstrates that the metal plate and checkerboard pattern configuration divide the energy in the normal direction into four beams. In contrast, the optimized pattern disperses the energy into multiple beams, leading to a reduction not only in the normal energy but also in the energy of the side lobes. Consequently, the optimized structure proves to be an improved and optimized pattern in comparison to the checkerboard pattern and metal plate.

## 4.4 Conclusion

In this paper, radar cross-section (RCS) reduction has been successfully achieved using an optimized coding pattern obtained through the application of a BPSO algorithm. The coding pattern consists of unit cells arranged in a combination of square and circular ring structures representing '0' and '1' elements, respectively. The phase difference between these unit cells varies from 1430 to 2170. Through extensive simulations, the results demonstrate the superiority of the proposed optimized coding pattern over both the checkerboard pattern and metal plate. The optimized coding pattern exhibits remarkable performance in achieving low monostatic and bistatic RCS, spanning a wide bandwidth from 24 to 30 GHz, while utilizing a more efficient BPSO optimization algorithm.

**Acknowledgements** The authors acknowledge to the research Universiti Grant, Universiti Kebangsaan Malaysia, Geran Translasi, UKM-TR2022-05 conduct the research work.

## References

1. V.G. Veselago, The electrodynamics of materials with simultaneously negative values of ε and μ. Sov. Phys. **10**, 509–514 (1968)
2. D.R. Smith, J.B. Pendry, M.C.K. Wiltshire, Metamaterials and negative refractive index. Science **305**, 788–792 (2004)
3. S.S. Islam, M.M. Hasan, M.R.I. Faruque, A new metamaterial-based wideband rectangular invisibility cloak. Appl. Phys. A **124**, 1–6 (2018)
4. M.B. Hossain, M.R.I. Faruque, M.T. Islam, Double elliptical resonator based quadruple band metamaterial absorber for EMI shielding applications in microwave regime. Alex. Eng. J. **69**, 193–206 (2023)
5. M.B. Hossain, M.R.I. Faruque, M.T. Islam, M. Singh, M. Jusoh, Triple band microwave metamaterial absorber based on double E-shaped symmetric split ring resonators for EMI shielding and stealth applications. J. Mater. Res. Technol. **18**, 1653–1668 (2022)
6. G. Zhao, S. Bi, M. Niu, Y. Cui, A zero refraction metamaterial and its application in electromagnetic stealth cloak. Mater. Today Commun. **21**, 100603 (2019)
7. Md.M. Hasan, M.T. Islam, S.K. Rahim, T. Alam, H. Rmili, A. Alzamil, Md.S. Islam, M.S. Soliman, A compact mu-near-zero metamaterial integrated wideband high-gain MIMO antenna for 5G new radio applications. Materials **16**(4), 1751 (2023)

8. M.I. Hossain, M.R.I. Faruque, M.T. Islam, Analysis on the effect of the distances and inclination angles between human head and mobile phone on SAR. Prog. Biophys. Mol. Biol. **119**, 103–110 (2015)
9. A.M. Siddiky, M.R.I. Faruque, M.T. Islam, S. Abdullah, M.U. Khandaker, K.S. Al-Mugren, Dual-square-split-ring-enclosed microstrip-based sensor for noninvasive label-free detection. Materials **15**(21), 7688 (2022)
10. M. Bellal Hossain, M. Rashed Iqbal Faruque, A.S. Alshammari, M. Tariqul Islam, Wide bandwidth enriched symmetric hexagonal split ring resonator based triple band negative permittivity metamaterial for satellite and Wi-Fi applications. Results Phys. **37**, 105511 (2022)
11. M.B. Hossain, M.R.I. Faruque, M.T. Islam, M.U. Khandaker, N. Tamam, A. Sulieman, Modified coptic cross shaped split-ring resonator based negative permittivity metamaterial for quad band satellite applications with high effective medium ratio. Materials **15**(9), 3389 (2022)
12. C.D. Giovampaola, N. Engheta, Digital metamaterials. Nat. Mater. **13**, 1115–1121 (2014)
13. T.J. Cui, M.Q. Qi, X. Wan, J. Zhao, Q. Cheng, Coding metamaterials, digital metamaterials and programmable metamaterials. Light Sci Appl. **3**(10), e218 (2014)
14. T. Ramachandran, M.R.I. Faruque, M.T. Islam, M.U. Khandaker, K.S. Al-Mugren, Development of diverse coding metamaterial structure for radar cross section reduction applications. Sci. Rep. **12**, 10958 (2022)
15. M.K.T. Al-Nuaimi, Y. He, W. Hong, Design of 1-bit coding engineered reflectors for EM-wave shaping and RCS modifications. IEEE Access **6**, 75422–75428 (2018)
16. H. Sun, C. Gu, X. Chen, Z. Li, L. Liu, B. Xu, Z. Zhou, Broadband and broad-angle polarization-independent metasurface for radar cross section reduction. Sci. Rep. **7**(1), 40782 (2017)

# Chapter 5
# Overview and Preliminary Subsystem Design of ICUBE-N: The National CubeSat of Pakistan

**Rehan Mahmood, Maheen Zulfiqar, Danial Ahmed Mir, Arooj Shafique, Osama Ali, Iqra Ejaz, Usama Shakoor, and Waseem Sajjad**

**Abstract** The evolution of CubeSats makes space access easier for educational and research institutes which was associated with large space organizations, traditionally. However, the high cost of the space missions still limits the access to space for developing countries. The better approach is then to have a multi-university CubeSat with efficient resource and cost sharing while providing the learning benefits to maximum students. The Institute of Space Technology (IST) is the only educational entity in Pakistan which developed and launched a CubeSat; ICUBE-1 in 2013. Small Satellite Technology Research Lab established in IST decided to develop a 3U (Unit) CubeSat; ICUBE-N (National) in collaboration with various universities of Pakistan in 2021. Different universities in Pakistan were first educated about the space mission analysis and design which is followed by the conceptual subsystem design of various ICUBE-N subsystems. This paper describes the preliminary subsystem design, analysis, and management strategies of ICUBE-N. The lesson learned during this multi-university CubeSat is also discussed.

## 5.1 Introduction

The Institute of Space Technology (IST) started the development of ICUBE-1 [1], a low-cost small spacecraft back in 2009 that was successfully launched in 2013. It was primarily developed by the undergraduate students of IST under the guidance and mentorship of IST faculty. The main mission of ICUBE-1 was technology demonstration; however, ICUBE-1 also carried an imaging payload that had a low-resolution Complementary Metal Oxide Semiconductor (CMOS) camera responsible for taking images of Earth. A satellite ground station with VHF, UHF, and S-band communication was also developed in IST as a part of the ICUBE-1 project. Following the

R. Mahmood (✉) · M. Zulfiqar · D. A. Mir · A. Shafique · O. Ali · I. Ejaz · U. Shakoor · W. Sajjad
Small Satellite Technology and Research Lab, National Center of GIS and Space Application
Institute of Space Technology, Islamabad, Pakistan
e-mail: rehan.mahmood@ist.edu.pk



M. T. Islam et al. (eds.), *Proceedings of the 8th International Conference on Space Science and Communication*, Springer Proceedings in Physics 303,
https://doi.org/10.1007/978-981-97-0142-1_5

successful launch of ICUBE-1, IST has also participated in the development of SSS-2A 3U CubeSat as a part of the Asia Pacific Space Corporation Organization (APSCO) Small Student Satellite development project.

In 2018, Higher Education Commission of Pakistan established National Center of GIS and Space Application (NCGSA) consisting of various laboratories in different universities of Pakistan. CubeSat development lab in IST was also renamed as Small Satellite Technology Research Lab (SSTRL) and included in NCGSA. Considering the resources required for space missions, the primary objectives of SSTRL are (i) to develop a 3U CubeSat by involving various universities of Pakistan, (ii) enable remote access of IST satellite ground station for various universities of Pakistan. Considering the objectives of SSTRL, ICUBE-N project was conceived in 2021 and it was decided to develop a 3U CubeSat while distributing the design and development work among the shortlisted universities with aim to facilitate the students in development of nanosatellites.

Since IST is the only institute in Pakistan which is imparting education related to space technology, the overall ICUBE-N development strategy is divided into two parts: firstly, to educate the students about the space environment, mission planning, and secondly, to involve students in the prototype development of various subsystems of ICUBE-N. The secondary mission objective is to successfully collect payload data and communicate with the ground station. In this paper, the preliminary design and analysis of ICUBE-N are discussed. In Sect. 5.2, brief overview of various subsystems of ICUBE-N is presented, while Sect. 5.3 presents the ICUBE-N mission control center and ground station. The testing of ICUBE-N is discussed in Sect. 5.4 which is followed by the discussion in Sect. 5.5 and conclusion in Sect. 5.6.

## 5.2 ICUBE-N Subsystems

In this section, preliminary design and analysis of various subsystems of ICUBE-N using COTS components are discussed. The experience gained during the development of ICUBE-1 and SSS-2A was used while defining the top-level details of each subsystem. Subsystems' detail and summary are listed in Table 5.1.

**Table 5.1** Subsystems' summary for ICUBE-N

| Subsystem | Details |
|---|---|
| Communication subsystem | Uplink frequency: 145–146 MHz<br>Downlink frequency: 145–146 MHz |
| Attitude determination and control system | Three-axis stabilized coarse sun sensors, three-axis magnetometer reaction wheels, magnetorquers |
| Solar panel | Triple junction GaAs solar cells |
| Antenna | Half-wave dipole deployable antennas |

**Table 5.2** Properties of various materials

| Material | Yield strength (MPa) | Density (kg/m$^3$) | Machinability |
|---|---|---|---|
| Stainless steel | 790 | 7760 | Easy |
| Titanium | 900 | 4429 | Hard |
| Al-6061-T6 | 320 | 2700 | Easy |
| Al-7075-T6 | 340 | 2810 | Easy |
| Composites | 640 | 1000 | Hard |
| Composites | 640 | 1000 | Hard |

### *5.2.1 Structure*

ICUBE-N structural subsystem follows standard 3U CubeSat dimension. The structure is designed to withstand the launch environment. The internal subsystems are integrated within the CubeSat in a modular manner allowing easy assembly and testing [2]. Each of these aspects is discussed below:

**Material Selection**. The basic criteria for the material selection of any CubeSat are based on the material strength, weight, machinability, and cost [3]. Comparison for some of the materials according to their yield, density, and machinability is listed in Table 5.2 [4]. Aluminum 6061-T6 is selected for manufacturing of ICUBE-N due to ease of machinability, availability, and lower density.

**Structural Design and Assembly**. Structural design of ICUBE-N is based on stacking techniques for PCB mounting in addition to outer frame that can endure launch process and space environment [5, 6]. The structure consists of side frames providing structural strength, while PCBs are stacked using the small rib structure. Solar panels will be mounted on the side frames, while the antenna deployment system will be placed on the top plate. The structure for ICUBE-N along with its subsystem arrangement is shown in Fig. 5.1.

### *5.2.2 Onboard Computer (OBC)*

Architecture of ICUBE-N OBC is proposed considering several factors, e.g., system architecture, performance, reliability, power consumption, communication, testing and validation, etc. The performance of OBC is mainly dependent on its processing unit, external peripherals, and the communication interfaces [7]. Fault tolerance [8] is an important consideration in reliability which can be used to mitigate faults while keeping the system running by using redundancy schemes, i.e., error detection and correction (EDAC), latchup protection, and watchdog timer (WDT) [9].

Cold redundancy strategy has been adopted in design for protection against radiation effects in space by providing the switching from affected prime system to the

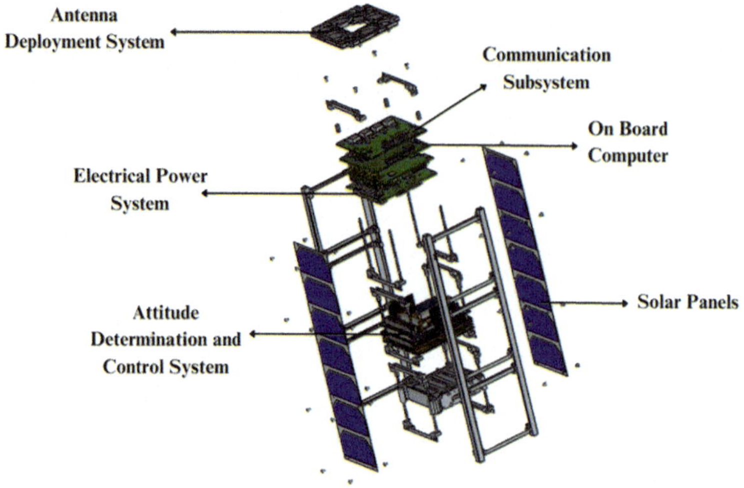


**Fig. 5.1** Exploded view of ICUBE-N

redundant system [10, 11]. Power switching between both microcontrollers (MCUs) is decided by Arbiter which continuously monitor the hard coded signal generated by the main microcontroller. Based on this signal, Arbiter decides which side will be the active one. A high-level block diagram for ICUBE-N OBC architecture is given in Fig. 5.2.

For OBC design, EFM32 Giant Gecko Series 1 (EFM32GG11) microcontroller with a Cortex-M4 processor is selected according to the requirements [12]. Selection of microcontroller is because of its flight heritage [13, 14], Low power consumption and availability of energy saving modes [15], Built-in CAN controller along with five different to shut down the idle peripherals to save energy.

### *5.2.3 Electrical Power Subsystem (EPS)*

The proposed top-level block diagram of ICUBE-N EPS is shown in Fig. 5.3. Seven SISP 32% GaAs SC-3GA-4 solar cells make up the majority of the solar panels. Four MPPT chargers, one for each axis, convert the energy collected by the solar panels. The power management and distribution (PMD) provide power to various load lines that may include different payloads. The energy storage is implemented on the separate PCB called a battery board. The battery board houses a battery assembled from four lithium-ion cells connected in 2P2S configuration. The digital control unit of EPS consists of PIC24FJ256GA110 controller which is a 16-bit general-purpose microcontroller that combines improved computational performance with a wide range of peripheral features. EPS health is collected with four 12-bit ADCS over SPI protocol.

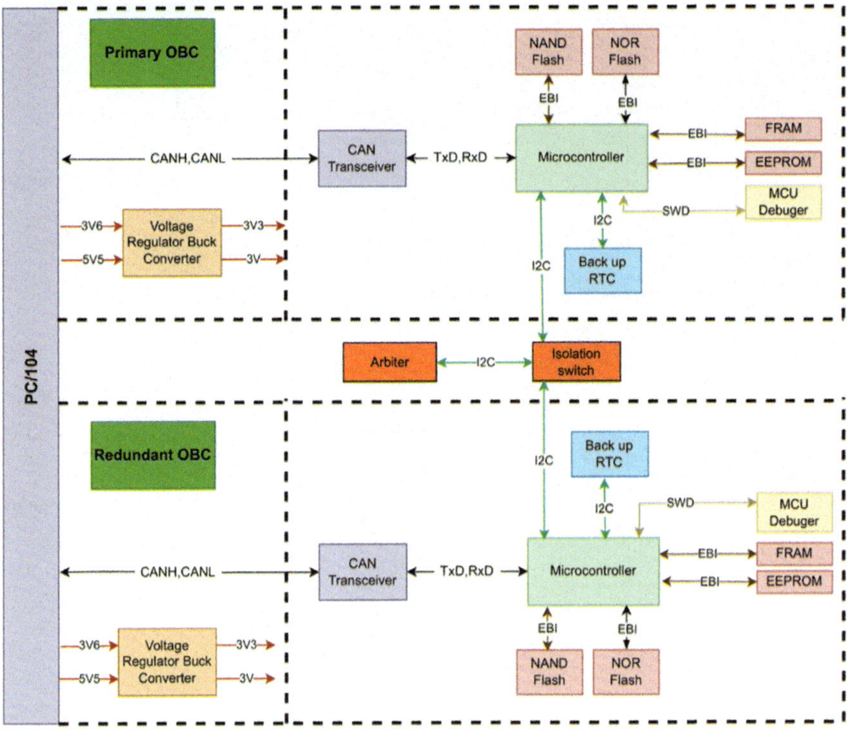


**Fig. 5.2** Block diagram of onboard computer architecture

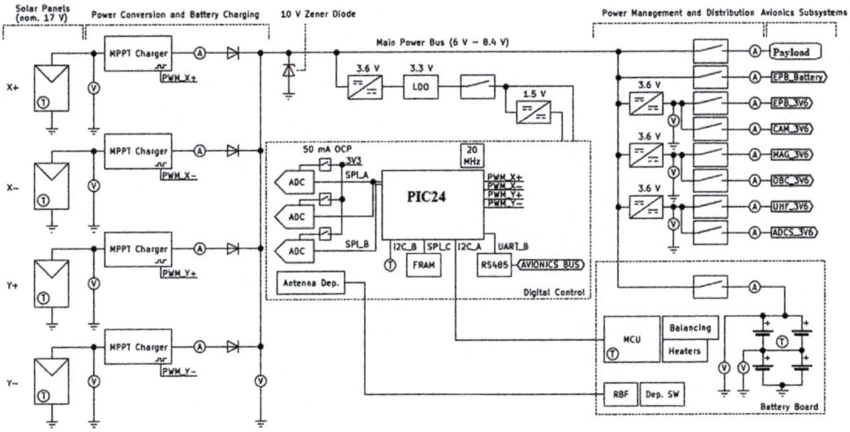


**Fig. 5.3** Top-level block diagram of proposed electrical power subsystem of ICUBE-N

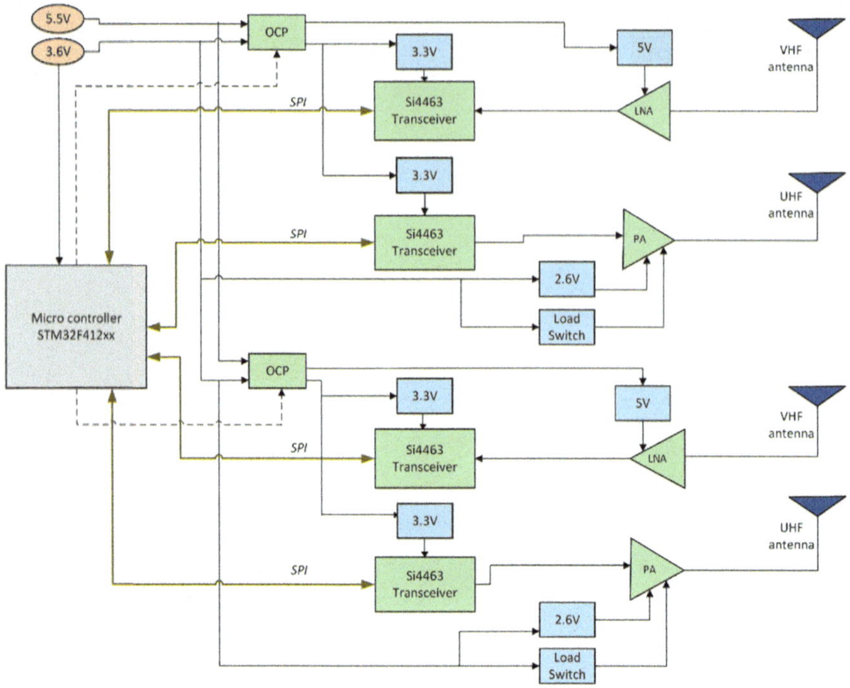


**Fig. 5.4** Top-level block diagram of communication system of ICUBE-N

### *5.2.4 Communication Subsystem*

An efficient communication system needs to have the features of high data rate, high bandwidth efficiency, less development cost, low power consumption with high signal quality [16]. The communication system of ICUBE-N will operate in VHF and UHF bands having uplink in VHF and downlink in UHF band.

The reason is being that many nanosatellites find it feasible to operate in UHF and VHF bands [17, 18]. Both the uplink and downlink are controlled by a single microcontroller. Both the links are managed by their own transceivers developing a full duplex topology. The hardware architecture of the uplink and downlink is depicted in the block diagram of communication subsystem in Fig. 5.4.

### *5.2.5 Attitude Determination and Control System (ADCS)*

The control system for any CubeSat consists of sensors, actuators, and a control algorithm embedded in a microcontroller [19, 20]. Since ICUBE-N will be launched in a Low Earth Sun-synchronous Orbit, the disturbance torques in this type of orbit are solar radiations, magnetic, gravity gradient, and atmospheric drag disturbances [19]. The worst-case disturbance torques are computed for ICUBE-N, see Table 5.3, with standard maximum mass and dimensions as given in [21] at an orbital altitude of 650 km.

**Attitude Determination**. In LEO the attitude determination of the CubeSat can be achieved using only magnetometer and sun sensors [19, 22]. Based on this, the attitude determination sensors for this CubeSat are coarse sun sensors and magnetometers. Ten-coarse sun sensor, SLCD-61N8, will compute the incident angle of the sunlight to the normal of their plane with accuracy less than 3°. Among these, PNI MicroMag3 has been chosen as the primary magnetometer because it contains high resolution per unit power and a digital interface [23]. Four magnetometers will be used improving the resolution of the system by the factor of 2 giving an accuracy of around 1° for the magnetic field in LEO [23]. The bias and residual magnetic field of other electronics in the CubeSat will be filtered out using a Kalman Filter (EF) or Extended Kalman Filter (EKF) [19]. Using literature study of already used MCU, it is concluded that Atmel controller series ATxmega256A3U is suggested because of its use in previous successful missions in CubeSat MOVE-II, SANDROID, HERMES and in a testing ADCS.

**Attitude Control**. ICUBE-N is three-axis stabilized with combination of three magnetorquers orthogonal to each other for CubeSat detumbling and three reaction wheels for pointing control [19]. Reaction wheels, consisting of the wheel and the motor, will have control over the three principal axes of ICUBE-N. Low-cost Faulhaber group BLDC motor is selected to be primary [24].

Using Table 5.3, aluminum 6061 reaction wheel moment of inertia, torque of Faulhaber 1509 B-flat, and formulae given in [19], a slew rate of around 3° can be achieved. Some key characteristics related to slew rate of reaction wheel are calculated using formulae given in [3, 19] and given in Table 5.4. Magnetorquers dump the reaction wheel momentum when they get saturated.

Work done in articles [25] inspired us to make in-house magnetic torque rods for ICUBE-N. While designing the coils for the magnetorquers, it is important to consider the material qualities, number of turns ($n$), and coil area ($A$) when doing the calculations [3]. "$A$" is dictated by structural considerations and "$n$" is the number of turns to be made.

**Table 5.3** Disturbance torques for ICUBE-N

| Disturbance | Torque (n Nm) |
|---|---|
| Solar radiation | 10.44 |
| Magnetic | 449.40 |
| Atmospheric drag | 16.23 |
| Gravity gradient | 56.00 |
| Total | 532.07 |

**Table 5.4** Slew maneuver summary

| Ramp-up time | 23.9 s |
|---|---|
| Angular acceleration | 0.0027 rad/s$^2$ |
| Angular velocity | 3.70 deg/s |
| Angular momentum | 0.002424 kg m$^2$ rad/s |
| Torque | $0.01 \times 10^{-3}$ |

## 5.3 ICUBE-N Ground Station

A satellite ground station was developed by IST to communication with amateur class satellite by communicating in VHF/UHF/S-band. Ground station consists of a Rotor Unit, (RC2800PRKX), CPU unit, Transceiver (1COM-910H), and High-speed packet radio controller (TNC31S). Yagi-Uda antennas (one VHF and one UHF) are mounted. Antennas are connected to the ground station via cables.

### *5.3.1 Mission Control Software*

The mission control software (MCS) is a Python-based application, providing a graphical way to interact and to monitor/control several parameters of ICUBE-N satellite. The main window GUI of MCS presents exactly six relevant subfunctionalities which can be independently accessed to monitor data. These subfunctionalities include ICUBE-N passes window, ICUBE-N Telemetry window, Track ICUBE-N window, ICUBE-N Payload data window. Figure 5.5 shows the main window of MCS.

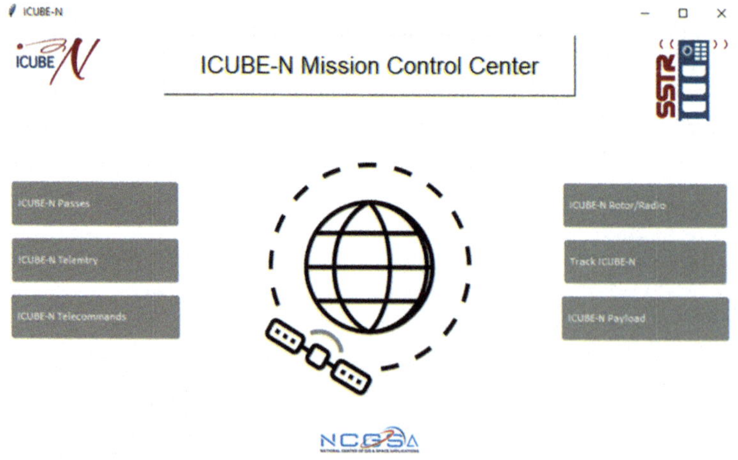


**Fig. 5.5** Mission control software main window

## 5.4 ICUBE-N Testing Plan

Before launching any satellite, it is necessary to test the overall system and the subsystems separately. There are different scenarios that can be defined using the flight records of previous CubeSats that can be used to verify the working and response of the system of ICUBE-N. Certain scenarios have been defined below along with the procedure of how the system should react to certain conditions. For easier understanding of testing procedure, the criteria to test the ADCS system have been divided into three categories: Sensor Testing, Actuator Testing, and Controller Testing [26].

### *5.4.1 ADCS Testing Plan*

It is important to calibrate and thoroughly test the ADCS before launch to ensure that it functions accurately and correctly. Using an air bearing to test the ADCS makes it easier to calibrate and test the system. The low-friction nature of the air bearing ensures that any external disturbances caused by friction or other factors are minimized leading to more accurate test results.

### *5.4.2 Structural Testing Plan*

Test plan for vibration testing of ICUBE-N shall be based on ASTM D3580-95 guidelines which will be conducted on all three axes with levels and frequencies determined by mission requirements and design considerations. The testing will be performed for 2–3 h per axis, and the ICUBE-N must remain functional and structurally intact during and after the vibration test.

## 5.5 Discussion

ICUBE-N is the second CubeSat being developed by any educational institute in Pakistan. Presented subsystems of Communication, power, OBC, ADCS comprise novel designs which are being developed from scratch. Preliminary designs of the systems are tested and are now implemented. As CubeSat development is a newer field in Pakistan, so it was our goal to involve maximum number of Pakistani students in the development process by conducting workshops and trainings.

## 5.6 Conclusion

In this paper, preliminary subsystem design and analysis of ICUBE-N are presented. Since the development of ICUBE-N is a multi-university project, it is expected that a significant number of students will be the beneficiary of the project. Furthermore, the educational value achieved by the students from this project will help these students in their professional life as well.

## References


1. R. Mahmood, K. Khurshid, A. Zafar, Q. Islam, ICUBE-1: first step towards developing an experimental pico-satellite at Institute of Space Technology. J. Space Technol. **1**, 5–10 (2011)
2. C. Cappelletti, S. Battistini, B. Malphrus, *CubeSat Handbook: From Mission Design to Operations* (Academic Press, 2020)
3. C. Cappelletti, S. Battistini, B.K. Malphrus, *CubeSat Handbook: From Mission Design to Operations* (Elsevier Science, 2020)
4. M. Ashby, *Material Property Data for Engineering Materials* (Engineering Department and Granta Design, 2016), p. 27
5. C. Güvenç, B. Topcu, C. Tola, Mechanical design and finite element analysis of a 3 unit CubeSat structure. Mach. Technol. Mater. **12**(5), 193–196 (2018)
6. E.E. Areda, J.R. Cordova-Alarcon, H. Masui, M. Cho, Development of innovative CubeSat platform for mass production. Appl. Sci. **12**(18), 9087 (2022)
7. E. Razzaghi, *Design and Qualification of On-Board Computer for Aalto-1 CubeSat* (Aalto University, School of Electrical Engineering, Finland, 2012)
8. T.I. Bajenescu, M.I. Bazu, *Reliability of Electronic Components: A Practical Guide to Electronic Systems Manufacturing* (Springer Science & Business Media, 2012)
9. L.T. Lumbwe, Development of an onboard computer (OBC) for a CubeSat. Ph.D. thesis, Cape Peninsula University of Technology, South Africa, 2013
10. K. Borschiov, *Generic On-Board-Computer Hardware and Software Development for Nanosatellite Applications* (York University, Canada, 2012)
11. S. Raje, A. Goel, S. Sharma, K. Aggarwal, D. Mantri, T. Kumar, Development of on board computer for a nanosatellite, in *68th International Astronautical Congress (IAC)* (International Astronautical Congress, Adelaide, Australia, 2017), pp. IAC-17-E2.4.7: 1–6
12. B.S. Rani, R. Santhosh, L.S. Prabhu, M. Federick, V. Kumar, S. Santhosh, A survey to select microcontroller for Sathyabama satellite's on board computer subsystem, in *Recent Advances in Space Technology Services and Climate Change 2010 (RSTS & CC-2010)* (IEEE, Chennai, 2010), pp. 134–137
13. S. Song, H. Kim, Y.-K. Chang, Design and implementation of 3U CubeSat platform architecture. Int. J. Aerosp. Eng. **2018**, 1–17 (2018)
14. P.J. Botma, The design and development of an ADCS OBC for a CubeSat. Ph.D. thesis, Stellenbosch University, Stellenbosch, 2011
15. Silicon Laboratories, *EFM32 Giant Gecko 11 Family Reference Manual*, 4 2016. Rev. 1.2 (2016)
16. C.O.O. Adeogun, Study on designing high data basebands for CubeSats' telecommunications systems. https://www.academia.edu/download/82047645/Study_20on_20Designing_20High_20Data_20Basebands_20_20for_20CubeSats_20Telecommunications_20Systems_1_.pdf. Accessed 13 Mar 2023
17. A. Kyrgiazos, B. Evans, P. Thompson, P.T. Mathiopoulos, S. Papaharalabos, A terabit/second satellite system for European broadband access: a feasibility study. Int. J. Satell. Commun. Network. **32**(2), 63–92 (2014)

18. J. Bouwmeester, J. Guo, Survey of worldwide pico- and nanosatellite missions, distributions and subsystem technology. Acta Astronaut. **67**(7–8), 854–862 (2010)
19. C.A. Kluever, *Space Flight Dynamics*. Aerospace Series (Wiley, 2018)
20. Guidance, Navigation, and Control | NASA. https://www.nasa.gov/smallsat-institute/sst-soa/guidance-navigation-and-control/. Accessed 13 Mar 2023
21. S. Loff, CubeSats Overview. http://www.nasa.gov/mission_pages/cubesats. Accessed 13 Mar 2023
22. J. Gerber, A 3-axis attitude control system hardware design for a CubeSat. Thesis, Stellenbosch University, Stellenbosch, South Africa, 2014
23. L. Regoli, M. Moldwin, J. Thoma, M. Pellioni, B. Bronner, Four-magnetometer board for CubeSat applications, in *32nd Annual AIAA/USU Conference on Small Satellites*. Session 3: Advanced Technologies I of SSC18-III-06 (2018), pp. 1–6
24. A. Gaude, V. Lappas, Design and structural analysis of a control moment gyroscope (CMG) actuator for CubeSats. Aerospace **7**(5), 55, 2–16 (2020)
25. J.C. Springmann, J.W. Cutler, Flight results of a low-cost attitude determination system. Acta Astronaut. **99**, 201–214 (2014)
26. A. Farhat, J. Ivase, Y. Lu, A. Snapp, *Attitude Determination and Control System for CubeSat* (Worcester Polytechnic University, Worcester, UK, 2013)

# Chapter 6
# Space Weather Study Through Analysis of Solar Radio Bursts Detected by a Single Station CALLISTO Spectrometer at the Space Technology Complex (STC), Banting, Malaysia

**C. C. Lau, Z. K. D. Nurlisman, K. Wahid, A. N. Ishak, Z. S. Hamidi, R. Umar, and C. Monstein**

**Abstract** The investigation of Solar Radio Bursts (SRBs) found by the CALLISTO (Compound Astronomical Low-cost Low-frequency Instrument for Spectroscopy and Transportable Observatory) station at the Space Technology Complex (STC), Banting, managed by the Malaysian Space Agency, is described in this paper. From 1 October 2021 to 31 December 2022, the SRBs were detected which encompassed the relocation of the pre-existing LPDA antenna and the substitution of the indoor LNA with an outdoor LNA (LNA2000). The e-CALLISTO website reported that the CALLISTO station at STC, Banting, Selangor, observed a total of 123 SRBs. There are 11 type II SRBs, 104 type III SRBs, 2 type IV SRBs, and 6 type V SRBs among them. It has been discovered that impulsive solar flares are linked to 15% of type III SRBs and 17% of type V SRBs. Coronal Mass Ejections (CMEs) are related for 63% of type II SRBs, 30% of type III SRBs, and 17% of type V SRBs. Furthermore, 37% of type II SRBs, 19% of type III SRBs, 33% of type V SRBs, and all type IV SRBs that have been analyzed are linked to solar flares and CMEs. Based on the analysis

C. C. Lau (✉) · Z. K. D. Nurlisman · K. Wahid · A. N. Ishak
Malaysian Space Agency (MYSA), Kuala Lumpur, Malaysia
e-mail: lau@mysa.gov.my

Z. S. Hamidi
Faculty of Applied Sciences, MARA Technology University, Shah Alam, Selangor, Malaysia

R. Umar
East Coast Environmental Research Institute (ESERI), Universiti Sultan Zainal Abidin, Kuala Terengganu, Terengganu, Malaysia

C. Monstein
Istituto Ricerche Solari Aldo e Cele Daccò (IRSOL), Università Della Svizzera Italiana, Locarno, Switzerland



M. T. Islam et al. (eds.), *Proceedings of the 8th International Conference on Space Science and Communication*, Springer Proceedings in Physics 303,
https://doi.org/10.1007/978-981-97-0142-1_6

conducted, the detection and analysis SRBs from ground observation can provide substantial information that can be utilized to identify and forecast space weather occurrences that may cause space weather hazards to our Earth and technology.

## 6.1 Introduction

Large-scale energetic events like Solar Flares and Coronal Mass Ejections (CMEs) can be generated by the Sun, which is the solar system's active celestial body. These events have the potential to trigger plasma oscillations, which may result in the emission of measurable radiation encompassing almost all of the electromagnetic spectrum, from high-energy gamma rays with energies in the hundreds of MeV to long-wavelength radio frequencies. A Solar Flare represents a sudden and intense outburst originating from the Sun's surface, emitting a highly concentrated burst of electromagnetic radiation. Solar flares occur due to the variability of the equilibrium of the coronal magnetic field within active regions with complex magnetic configurations. Energy stored in the active regions is rapidly converted into thermal, kinetic, and mechanical energies during the eruption. Meanwhile, a CME is a large and powerful explosion of plasma and magnetic fields that originates from the Sun's corona, its outermost atmosphere. This enormous event is believed to be the main trigger of geomagnetic disturbances, which can have a great impact on both ground- and space-based technological systems and components, that include the shutting down of power systems [1].

Typically, Solar Radio Bursts (SRBs) occur along with space weather events like solar flares and CMEs. These SRBs are commonly identified by the presence of baseline levels of radiation, overlaid with sudden bursts of activity. Investigating the fluctuations in radio signal attenuation resulting from ionospheric absorption is important, as major solar flares can give rise to a stream of highly energetic particles known as solar proton events. These occurrences may have an impact on the Earth's magnetosphere and ionosphere, having a particular negative effect on radio communication system performance and potentially a major effect on the local space weather conditions near the Earth [2]. Adverse space weather may directly affect satellites or have a severe impact on space-based technological systems, especially satellite communication systems and the global navigation satellite systems (GNSSs). Both astronauts and the flight crew may be at risk from the various radiations [3].

SRBs are solar flare-related radio waves that happen on the Sun's surface. Some of them can be detected by the ground-based antenna while they reach the Earth's surface.

SRBs typically originate from the same region of the solar atmosphere as solar flares, CMEs, and shocks, from which the Sun emits radio radiation into the interstellar medium [4]. The study of space weather conditions using affordable ground-based instrumentation on the Earth has created SRBs, a research topic that interests solar physicists nowadays.

Since the 1960s, SRBs have been classified into five main types below a few hundred MHz [5], which are SRBs Types I, II, III, IV, and V. SRBs are classified according to variations in the burst frequency and duration and morphologies of their spectroscopic shapes. Solar storms are related to type I SRBs, while CME occurrences are related to type II SRBs. Solar flares and the development of new sunspots are the causes of type III and type IV SRBs. Type V SRBs are extremely rare; however, they frequently arise following type III SRBs [6]. Each type of SRBs has its own connected activities due to the variety of processes associated with their production, which are explained in Table 6.1 and Fig. 6.1. The features of SRBs are shown in Table 6.1, and the diagram of SRBs is shown in Fig. 6.1 as a graph of frequency versus time. Solar phenomena, often in the form of plasma emissions from the solar corona, have the potential to penetrate Earth's atmosphere and perturb its magnetic field. This disruption can lead to disturbances in wireless communication, satellite functionality, interference with the Global Positioning System (GPS), and disruptions in various communication systems [1], damage and produce disruption surges on power grids, and raise health risk for astronauts. Therefore, it is important to understand the dynamics of SRBs, which is a key to understanding the space weather condition [7] and can be used as a tool in forecasting space weather.

In order to determine whether SRBs are associated with solar events, our goal is to evaluate SRBs data collected by the CALLISTO station at the Space Technology Complex in Banting, Selangor. The data provided in this study pertain to the observations made by CALLISTO Banting during its inaugural year of operation using the updated configuration, spanning from 1 October 2021 to 31 December 2022.

## 6.2 Observation and Methods

The Compound Astronomical Low-cost Low-frequency Instrument for Spectroscopy and Transportable Observatory, referred to as CALLISTO, was created by Christian Monstein, an engineer from ETH Zurich who had previously worked for the Institute of Astrophysics' Radio and Plasma Physics Group.

The CALLISTO system, which was designed to detect electromagnetic radiation at radio frequencies between 45 and 870 MHz, consists of an antenna, a low-noise amplifier (LNA), a spectrometer, control software, and logging software. Today, more than 200 CALLISTOs are operating simultaneously around the world, including in Malaysia. We gather the solar radio spectrum data 24/7 throughout the year online and keep it in a centralized database at the University of Applied Sciences Northwestern Switzerland (FHNW). A worldwide network of CALLISTO spectrometers, e-CALLISTO, observes the activities of the Sun in the radio region and provides valuable information on space weather. As a result, it is possible to monitor space weather conditions continuously on large scales, utilizing the spectrum that has been collected in this network [8].

**Table 6.1** The features of solar radio bursts (SRBs) (Credit to IPS Australia)

| Type | Characteristics | Duration | Frequency range | Associated phenomena |
|---|---|---|---|---|
| I | Short, narrow-bandwidth burst. Usually occur in large numbers with underlying continuum | Single burst: ~ 1 s<br>Storm: Hours–days | 80–200 MHz | Active regions, flares, eruptive prominences |
| II | Slow frequency drift bursts. Usually accompanied by a (usually stronger intensity) second harmonic | 3–30 min | Fundamental: 20–150 MHz | Flares, proton emission, magnetohydrodynamic shockwaves |
| III | Fast frequency drift bursts. Can occur singularly, in groups, or storm (often with underlying continuum). Can be accompanied by a second harmonic | Single burst: 1–3 s<br>Group: 1–5 min<br>Storm<br>Minutes–hours | 10 kHz to 1 GHz | Active regions, flares |
| IV | Stationary Type IV: Broadband continuum with fire structure<br>Moving Type IV: broadband, slow frequency drift, smooth continuum<br>Flare Continua: broadband, smooth continuum | Hours to days<br>30 min to 2 h<br>3–45 min | 20 MHz to 2 GHz<br>20–400 MHz<br>25–200 MHz | Flares, proton emission<br>Eruptive prominences, magnetohydrodynamic<br>Flares, proton emission |
| V | Smooth, short-lived continuum. Follows some type III bursts. Never occur in isolation | 1–3 min | 10–200 MHz | Same as type III bursts |

At the Space Technology Complex in Banting, Selangor, with coordinates of 2° 47′ 2.5″ N, 101° 30′ 24.8″ E, the CALLISTO system has been operationally located since 2016 [9]. It consists of four main components, which are the CLP5130-1 Log Periodic Dipole Antenna (LPDA), CALLISTO spectrometer, LNA2000 outdoor low-noise amplifier with a gain of more than 10 dB and a high signal-to-noise ratio (SNR), and control and logging software. Through the coaxial cable in the controlled building, a CLP5130-1 LPDA is connected to the LNA2000 and CALLISTO spectrometer. The output feeder is linked via an RS232 cable to a main computer for signal processing, control, monitoring, and maintenance. The observed spectrum is automatically saved in the frequency versus time (FIT) file format. CALLISTO's configuration diagram

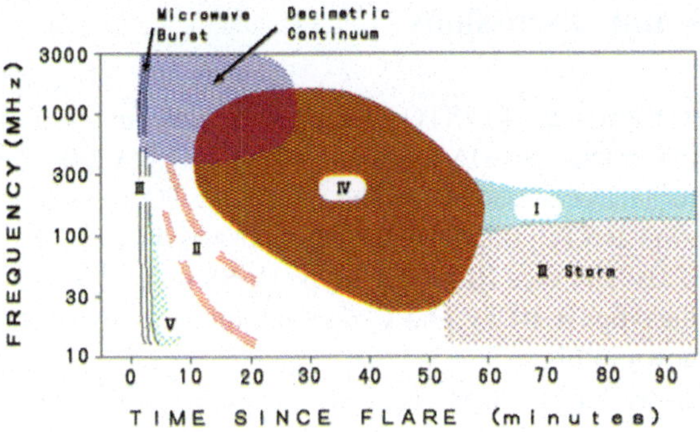


**Fig. 6.1** The diagram of SRBs as a graph of frequency versus time. *Source* Hiraiso/NICT website

is shown in Fig. 6.2. With effect from 1 October 2021, the antenna LPDA has been relocated within 5 m of the receiver, and the indoor LNA has been replaced with an outdoor LNA (LNA2000) with 20 dB gain to minimize signal loss. Over a period of almost 12 h, daily observations were conducted.

Based on SRBs detected by STC in Banting, Selangor, this study was conducted. Data were also obtained from the e-CALLISTO network, the Large Angle and Spectrometric Coronagraph (LASCO), the Solar Monitor, and the National Oceanic and Atmospheric Administration's Space Weather Prediction Center (SWPC).

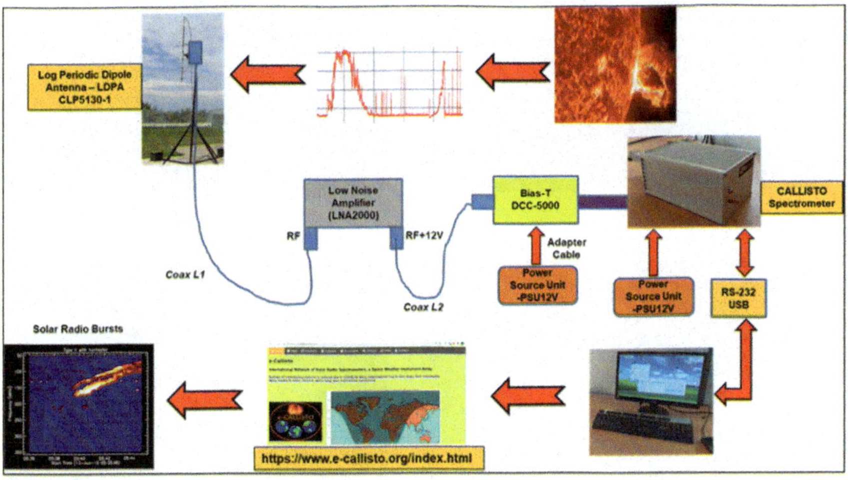


**Fig. 6.2** The CALLISTO system configuration at the space technology complex, Banting

## 6.3 Results and Discussions

Based on observations of CALLISTO Banting during the first year of operation with relocated existing antenna LPDA and outdoor LNA (LNA2000), which is from 1 October 2021 to 31 December 2022. The e-CALLISTO website reported that the CALLISTO station at STC, Banting, Selangor, observed a total of 123 SRBs. There are 11 type II SRBs, 104 type III SRBs, 2 type IV SRBs, and 6 type V SRBs among them. Some of the types II, III, IV, and V of solar radio emissions detected in Banting, Selangor, are shown in Fig. 6.3.

The GOES x-ray flares' onset and end times, flare classes, heliographic coordinates and group number of related active regions (Ars), and the time at which CMEs begin to form are among the burst parameters that are used to analyze the 123 observed bursts and their related flare properties. Table 6.2 displays the observed burst analyses.

According to the findings, 15% of type III SRBs and 17% of type IV SRBs are associated with impulsive solar flares. However, 63% of type II SRBs, 30% of type III SRBs, and 17% of type V SRBs are caused by CMEs. Additionally, CMEs and solar flares have been associated with 37% of type II SRBs, 19% of type III SRBs, and 33% of type V SRBs. All of the results have been compared with the findings of previously, associated research conducted by scientists at the University of Rwanda's

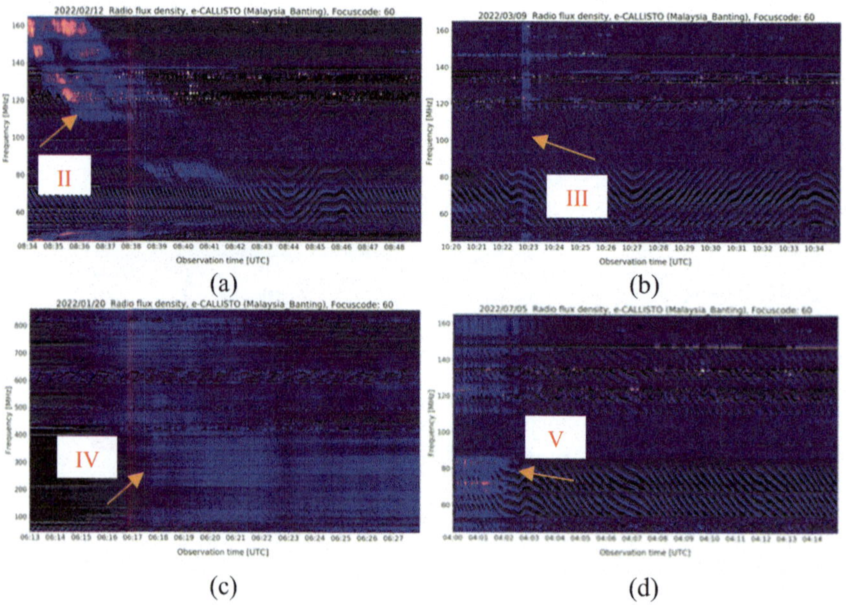


**Fig. 6.3.** Some of type II bursts on 12 February 2022 **a**, type III bursts on 9 March 2022 **b**, type IV bursts on 20 January 2022 **c**, and type V bursts on 5 July 2022 **d** that have been detected by CALLISTO at the space technology complex, Banting

**Table 6.2** Summary of SRBs detected by CALLISTO at the Space Technology Complex, Banting associated with Solar Flares and CMEs since 1 October 2021 until 31 December 2022

| Solar Activity | Type of SRBs detected | | | | | | | |
|---|---|---|---|---|---|---|---|---|
| | Quantity | | | | Percentage (%) | | | |
| | II | III | IV | V | II | III | IV | V |
| Flare associated | – | 16 | – | 1 | 0 | 15 | 0 | ~ 17 |
| CME associated | 7 | 31 | – | 1 | ~ 63 | ~ 30 | 0 | ~ 17 |
| Flare and CME associated | 4 | 20 | 2 | 2 | ~ 37 | ~ 19 | 100 | ~ 33 |
| No Flare and No CME associated | – | 37 | – | 2 | 0 | ~ 36 | 0 | ~ 33 |
| **Total** | **11** | **104** | **2** | **6** | | | | |

College of Education [8], who indicated that type II, III, and IV SRBs are related with solar flares and CMEs. In accordance with the findings of Hamidi et al., CMEs frequently coexist with type II SRBs [10].

Solar flares and CMEs happened a few minutes or hours before or after ground-based instrumentation like CALLISTO observed bursts. Although solar flares and CMEs are two unique solar activities, CMEs can occasionally occur before or after flare explosions. Large solar flares may occasionally be coupled with CMEs. During solar flares, the magnetic fields of the active region of the Sun rapidly transform into thermal, kinetic, and mechanical energy. The magnetic field of the Sun and high-energy particles are linked to both CMEs and solar flares. Both of these reactions generate SRBs and plasma emissions that can be detected by ground-based antennas as they reach the surface of the Earth [11]. Recently, solar physicists have conducted studies on the energy distribution of both solar activities. As a result of this work, it is possible to identify and predict space weather phenomena that could endanger our planet and technology by detecting and analyzing SRBs from ground observation.

## 6.4 Summary and Conclusion

We have detected 123 SRBs by the CALLISTO spectrometer located at the Space Technology Complex, Banting, Selangor, Malaysia, since 1 October 2021 until 31 December 2022. As a result, findings show that SRB types II, III, IV, and V are essential for evaluating space weather conditions and can be utilized to forecast space weather. Since Solar Cycle 25, all solar physicists are concerned about studying SRBs in the low-frequency radio region. There is a strong correlation between space weather hazards, including geomagnetic storms, solar flares, and CMEs observed. As a result, deep research into detecting SRBs utilizing the CALLISTO system could provide beneficial data and statistics relating to relevant solar activity and provide predictions about upcoming solar activities, which can minimize the impact of space weather on our technology, reorganizing the electrical grid, rerouting the northern and southern polar flight, and adjusting spacecraft.

**Acknowledgements** We appreciate the e-CALLISTO network, LASCO, Solar Monitor, and NOAA SWPC for making their data accessible online. Our sincere gratitude to the University Technology of MARA (UiTM) and the University of Sultan Zainal Abidin (UniSZA) for their advice and consultation in this CALLISTO research project.

## References

1. N.M. Ansor, Z.S. Hamidi, N.N.M. Shariff, An Automated system in detecting solar radio bursts type II and IV associated to multiple coronal mass ejections. J. Phys. Conf. Ser. **1411**(2019), 012015 (2019)
2. N.Z.M. Afandi, N.H. Sabri, R. Umar, C. Monstein, Burst-finder: burst recognition for e-CALLISTO spectra. Indian J. Phys. **94**, 947–957 (2020)
3. E.P. Carley, C. Baldovin, P. Benthem, M.M. Bisi, R.A. Fallows, P.T. Gallagher, M. Olberg, H. Rothkaehl, R. Vermeulen, N. Vilmer, D. Barnes, Radio observatories and instrumentation used in space weather science and operations. J. Space Weather Space Clim. **10**, 7 (2020)
4. S.N.A. Syed Zafar, R. Umar, N.H. Sabri, M.H. Jusoh, A.N. Dagang, A. Yoshikawa, Effects of solar flares and coronal mass ejections on earth's horizontal magnetic field and solar wind parameters during the minimum solar cycle 24. Mon. Not. R. Astron. Soc. **504**(3), 3812–3822 (2021)
5. J. Wild, S. Smerd, A. Weiss, Solar bursts. Ann. Rev. Astron. Astrophys. **291** (1963)
6. Z.S. Hamidi, N.N.M. Shariff, Evaluation of signal to noise ratio (SNR) of log periodic dipole antenna (LPDA), in *Business Engineering and Industrial Application Colloquium (BEIAC)*, (IEEE, 2013)
7. Z.S. Hamidi et al., Space weather: the significance of -CALLISTO (Malaysia) as one of contributor of solar radio bursts due to solar activity. Int. Lett. Chem. Phys. Astron. **26**, 37–44 (2014)
8. T. Ndacyayisenga, A.C. Umuhire, J. Uwamahoro, C. Monstein, Space weather study through analysis of solar radio bursts detected by a single station CALLISTO Spectromestudenter. Annal. Geophys. (2021)
9. M.M. Shafie, R. Umar, N.H. Sabri, Statistical analysis of radio frequency interference (RFI) caused by solar radiation during wet and dry. Malays. J. Fundam. Appl. Sci. **17**(4), 430–436 (2021)
10. Z. Hamidi, N.N.M. Shariff, C. Monstein, Characterization of selected solar radio bursts based on solar activity detected by e-CALLISTO (Malaysia). Int. Lett. Chem. Phys. Astron. **13**(2), 144–154 (2014)
11. N.H. Zainol et al., Effective data collection and analysis of solar radio bursts type II using automated CALLISTO network system. (IEEE, 2016)

# Chapter 7
# Mid-Latitude Ground Magnetic and Electric Field Responses Associated with SCW Development During Substorms

**Moe Hayashi, Akimasa Yoshikawa, and Akiko Fujimoto**

**Abstract** This study elucidates the development of the global three-dimensional current system from polar to equatorial regions during substorms. We extracted 50 cases from 1934 isolated substorms between 2006 and 2013. These 50 cases met two criteria: the electric field was calm for 1 h before the substorm, and clear electric field fluctuations were observed during substorm onset. We compared the 210 MM and mid-latitude magnetic field data to capture each event's distribution and temporal evolution of the substorm current wedge. Additionally, we captured the electric field response using the mid-latitude Paratunka station. The results revealed that the electric field amplitude was proportional to the SuperMAG aurora electrojet index magnitude (correlation coefficient = 0.67). In the central region of the substorm current wedge (SCW), predominantly westward electric fields were observed between the upward and downward field-aligned currents (FACs) (i.e., among 12 cases, wherein the observation points were at the center of the SCW, 10 showed westward electric fields). Moreover, a tendency for larger amplitudes of these westward electric fields was observed at the SCW center. This may reflect the projection of the primary potential formed by the charges injected into the upward and downward FAC regions on the ionosphere. Contrarily, a tendency for eastward electric fields was observed near the upward FAC of the SCW (i.e., among 24 cases, wherein the observation points were near or slightly inside the upward FAC, 14 showed eastward electric fields), which might be due to the twisting of the potential distribution associated with Hall polarization.

M. Hayashi (✉) · A. Yoshikawa
Kyushu University, 744 Motooka Nishi-Ku, Fukuoka, Japan
e-mail: hayashi.moe.045@s.kyushu-u.ac.jp

A. Fujimoto
Kyushu Institute of Technology, 680-4 Kawazu, Iizuka-Shi, Fukuoka, Japan



M. T. Islam et al. (eds.), *Proceedings of the 8th International Conference on Space Science and Communication*, Springer Proceedings in Physics 303,
https://doi.org/10.1007/978-981-97-0142-1_7

## 7.1 Introduction

Two primary ionospheric current systems develop during substorms: (1) The Region 1 type current (R1-current) and Region 2 type current (R2-current) in the polar ionosphere [1]. The R1-current varies in conjunction with the variation of the magnetospheric convection system and develops from the growth phase to the expansion phase of substorms. In contrast, the R2-current develops in correlation with the variation of the pressure gradient region in the inner magnetosphere during the substorm growth phase [1, 2]; (2) the substorm current wedge (SCW) current, which forms a current closure in the same direction as the R1-current and occurs during the onset of substorms [3]. R2-currents occurring at lower latitudes than SCW can cancel (shield) or exceed (over shield) the electric field created by R1-type currents [4, 5].

The magnetic field variations caused by FACs associated with the development of SCW are particularly prominent in the nighttime mid-latitude region, and their development has been modeled by McPherron et al. [3]. The spatiotemporal evolution of SCW structures, such as where they develop and how they evolve and decay spatially, is closely related to the development and decay of auroral storms [6]. However, a comprehensive understanding of this relationship is still lacking.

Furthermore, studies have been conducted on developing the global magnetosphere-ionosphere coupling (global M-I coupling) from the polar to mid-latitude regions. It has been revealed that during substorms, a circuit connecting the R2 FAC and the dayside equatorial current system (CEJ) develops [4]. The development of a circuit that connects the R2 FAC not only with the dayside but also with the nightside equatorial region (EEJ) has been demonstrated [7]. Moreover, the existence of a current circuit connecting the SCW with the dayside EEJ using Pi2 observations has been suggested [8]. Thus, although it has been observationally suggested that SCW development is related to R1/R2 ionospheric currents, no study has systematically examined the development of the current system from both the electric and magnetic fields.

The main purpose of this paper is to capture the spatiotemporal evolution of current structure associated with substorms and to clarify the magnetospheric-ionospheric response associated with the development of currents using magnetic and electric field observations. In this study, we compared ground magnetic field data from MAGDAS and SuperMAG with the McPherron model to capture the spatiotemporal development of SCW. Additionally, we analyzed the electromagnetic response by combining it with electric field observations using the FM-CW radar installed at PTK (Paratunka, Russia) by Kyushu University.

In Sect. 7.2, we describe the ground magnetic field data and electric field data used in this study; in Sect. 7.3, we present the results of event analysis for an isolated substorm that occurred on April 1, 2010; in Sect. 7.4, we report the results of three statistical analyses following the results from Sect. 7.3: (4.1) the relationship between SuperMAG aurora electrojet index (SML) variations and electric field variations; (4.2) the relationship between electric field direction and SCW; and (4.3) the relationship between SCW location and electric field magnitude; and in Sect. 7.5, we discuss and summarize the relationship between the magnetic field and electric field variations during substorms.

## 7.2 Data Set

We used magnetic field data to capture the spatiotemporal evolution of the SCW, and electric field data were used to analyze the electromagnetic response in the mid-latitude region.

### *7.2.1 Ground Magnetic Field Data*

Magnetic field data were used to capture the spatiotemporal evolution of the SCW structure. We utilized data from the 210 MM Magnetic Observation Network (including PTK) and seven mid-latitude observation points. Details of the magnetometers used in this study (i.e., latitude, longitude, and magnetic local time [MLT]) are shown in Table 7.1. The 210 MM is part of the MAGnetic Data Acquisition System/Circum-pan Pacific Magnetometer Network Data (MAGDAS/CPMN) [9, 10], details of which can be found on the International Space and Planetary Environment Research Center (i-SPES) at Kyushu University website (https://www.i-spes.kyushu-u.ac.jp/research.html#ttl01). The mid-latitude observation points used data available on the SuperMAG website (https://supermag.jhuapl.edu/mag/?fidelity=low&start=2001-0101T00%3A00%3A00.000Z&interval=23%3A59). These data were obtained from the INTERMAGNET, USGS, and GIMA observation networks, which have been standardized by SuperMAG [11].

### *7.2.2 Electric Field Data: FM-CW Radar Data*

We investigated the electromagnetic field response at mid-latitudes during substorms by comparing this electric field data with the magnetic field data.

The ionospheric electric field was derived from the Doppler shift observed by a frequency modulated continuous wave (FM-CW) high-frequency radar installed in PTK (magnetic latitude 45.8°, magnetic longitude 221.6°, $L = 2.05$, LT = UT + 10.5 h). Our radar is an improved version of the FM-CW ionosonde [12, 13]. The Doppler observations are based on the Fast Fourier Transform (FFT) technique, a variation of the technique to measure ocean waves [14].

The radar can measure the vertical drift velocity and its virtual height in the ionosphere. The observed Doppler frequency shift $\Delta f$ is defined by the following equation:

$$\Delta f = fo \times 2V^*/c \tag{7.1}$$

Here, $V^*$ represents the vertical drift velocity and *fo* is the transmission frequency. Thus, the vertical drift velocity in the ionosphere $V^*$ is given by the following equation:

$$V^* = c \times \Delta f/2fo \tag{7.2}$$

The electric field $E$ is defined by the product of the velocity $V$ and the magnetic field $B$, which are perpendicular to each other, as follows:

$$E = -V \times B \tag{7.3}$$

Among these velocities, only the component $V$ perpendicular to the ground is observable. If we consider the inclination angle θ of the magnetic field, $V$ can be expressed as $V = V\cos\theta$. Substituting this into Eq. (7.3), we obtain:

$$E = -\left(V^*/\cos\theta\right) \times B \tag{7.4}$$

The electric field $E$ was calculated by substituting the Doppler shift obtained from the FM-CW radar and the magnetic field at the ionospheric altitude calculated using the IGRF model into Eqs. (7.2) and (7.4), respectively. Notably, the electric field obtained by the FM-CW radar included only the east–west component.

## 7.3 Case Study

We analyzed an isolated substorm that occurred on April 1, 2010. To analyze the nightside magnetic and electric field responses during substorm occurrence, we utilized 1-min resolution magnetic field data of MAGDAS and SuperMAG (Table 7.1), 1-min resolution electric field data, and the SML provided by SuperMAG.

Figure 7.1 shows the variations in the SML (Fig. 1a) and electric field (Fig. 1b). The electric field had positive values corresponding to eastward electric fields and negative values corresponding to westward electric fields. The vertical dashed lines indicate the onset time, and the electric field slightly turned westward after the onset.

Figure 1c and d presents the variation in mid-latitude magnetic fields during this event. The magnetic stations are arranged in order of decreasing magnetic longitude, starting east of PTK. PTK, where the electric field was observed, is located between SHU and MMB. Focusing on the east–west component of the magnetic field (Fig. 1d), it decreased eastward of SIT and increased westward of T59. By comparing this magnetic field variation with the mid-latitude magnetic field variation model by McPherron et al. [3] based on the SCW, the influence of the downward FAC of the SCW was evident in the east of SIT, whereas the influence of the upward FAC of the SCW was observed in the west of T59 in the east–west component. Furthermore,

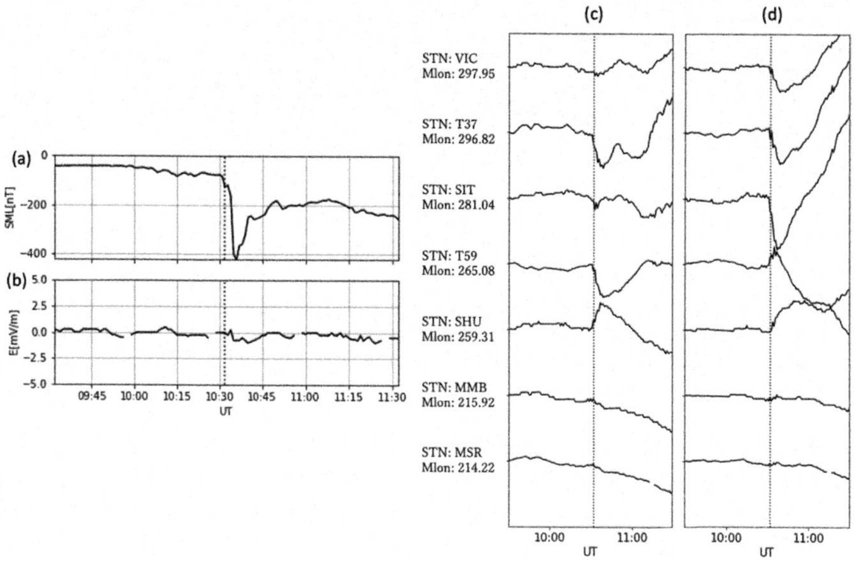


**Fig. 7.1** **a** SML and **b** electric field variation during an isolated substorm on April 1, 2010. The vertical dotted line indicates the onset. **c** north–south and **d** east–west components of magnetic field variations in the mid-latitude zone on the same day. They are sorted from high to low magnetic longitude

examining the north–south component of the magnetic field (Fig. 1c), the upward trend of the magnetic field at SHU indicated that SHU was located within the region of both upward and downward FACs. Therefore, PTK, situated west of SHU, was positioned near the upward current, which explains the observation of westward electric fields.

## 7.4 Statistical Analyses

We analyzed the isolated substorm events, specifically the 5716 events between September 21, 2006, and February 26, 2013, which were selected from the isolated substorm onset list using the SML [15]. We further filtered these events based on the position of PTK during 19–24 LT and 0–5 LT (LT = UT + 10.5 h). We extracted 1790 events for which electric field data were available for the 2-h period around the substorm onset. Furthermore, 50 events with no noise or data gaps for 1 h before or after the substorm onset and a clear electric field variation with the onset were selected.

Figure 7.2 presents an example of electric field variations, wherein the horizontal axis represents UT, the vertical axis represents the electric field $E$ [mV/m], and the black vertical lines indicate the substorm onset times. Variations in both eastward and

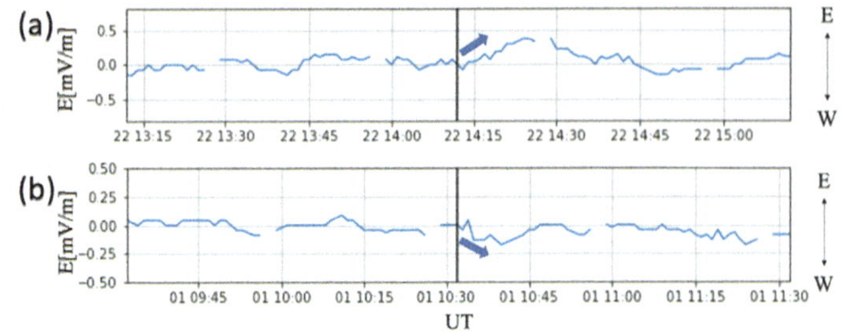


**Fig. 7.2** Example of electric field variation. **a** and **b** show the electric field variations observed on 2007/08/22 and 2010/04/01, respectively. The horizontal axis represents UT (date + time) and the vertical axis represents the electric field $E$ [mV/m]. The black vertical line indicates the onset

westward electric fields were observed, with 18 of the selected 50 events showing eastward variations and 23 showing westward variations. The other 9 events were defined as no change because the post-onset variation was the same or smaller than the pre-substorm variations.

### *7.4.1 Relationship Between SML Variations and Electric Field Variations*

To investigate the relationship between the intensity of auroral electrojet [AEJ]) and electric fields, we compared the variations (from onset values to maximum amplitude) of the SML (i.e., an index of the AEJ) and the electric field. Figure 7.3 shows the correlation between the SML and the electric field; the horizontal axis represents the amplitude of the SML and the vertical axis represents the amplitude of the electric field. Although some scatter was observed, a tendency for larger electric field variations was observed as the SML increased, indicating a good correlation between the intensity of AEJ and the intensity of the electric field. The light blue-dashed line in Fig. 7.3 represents the approximate linear fit, and the correlation coefficient between the SML intensity and electric field intensity was 0.67. This finding suggested a good correlation between the SML and the mid-latitude electric field and that the electric fields on the nightside during substorms were likely to have been generated by the AEJ and the associated SCW.

### *7.4.2 Relationship Between Electric Field Direction and SCW*

Using ground-based magnetic field data for all events, we determined the relative position of PTK with respect to the SCW and investigated the relationship between

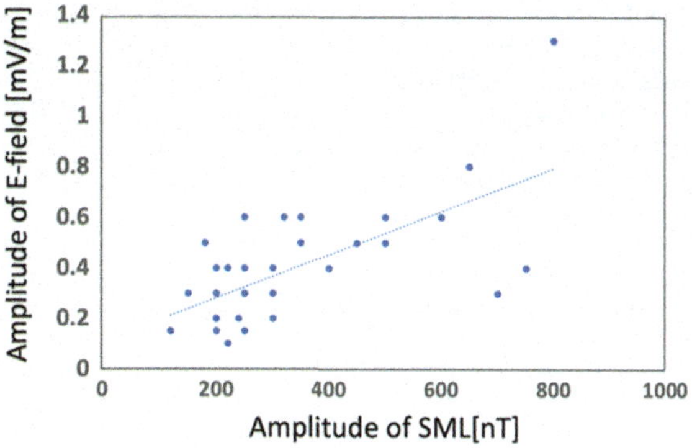


**Fig. 7.3** Correlation plot of SML amplitude and electric field amplitude. Each event's SML and electric field values are represented by a point, with an approximate straight line shown as a dotted line. The correlation coefficient between SML intensity and electric field intensity was 0.67

the electric field direction and the SCW. The relative position of PTK and the SCW was determined by the approach described in Sect. 7.2.

Figure 4a illustrates the relationship between the electric field direction and the position of PTK relative to the SCW. The horizontal axis represents the longitude spread of the SCW, and the areas surrounded by red-dotted lines indicate the location of upward and downward FACs. The vertical axis represents the number of events where the electric field was observed at that position, and the histogram colors represent the electric field direction. For example, in the upward region, the gray histogram shows a count of 6, indicating that when PTK was near the upward current, eastward electric fields were observed on six occasions.

When focusing on the SCW center, westward electric fields tended to be observed more frequently. This finding suggested that the ionospheric potential (Fig. 4b) formed at the base of the SCW extended to mid-latitudes, generating westward electric fields between the upward and downward FACs.

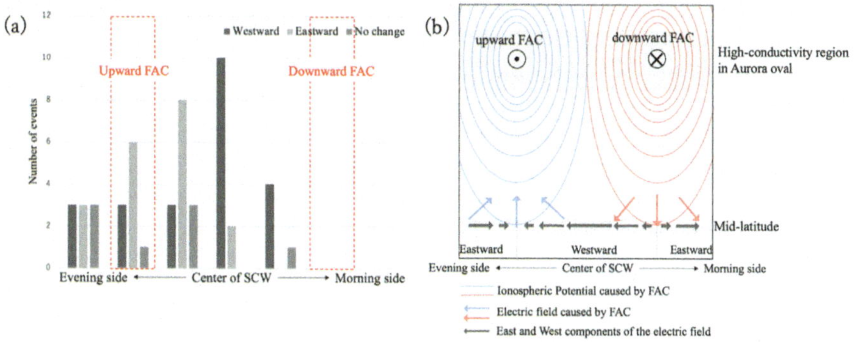


**Fig. 7.4** **a** Histogram of the counted number of electric field variation events at each position of PTK relative to the SCW. Colors indicate the direction of the electric field, and the red-dotted line indicates the positions of upward and downward FACs. **b** Diagram of primary potential at SCW foot; blue and red lines show potential at upward and downward FACs. Electric fields at mid-latitude are represented by blue and red arrows; the gray arrow indicates east–west component

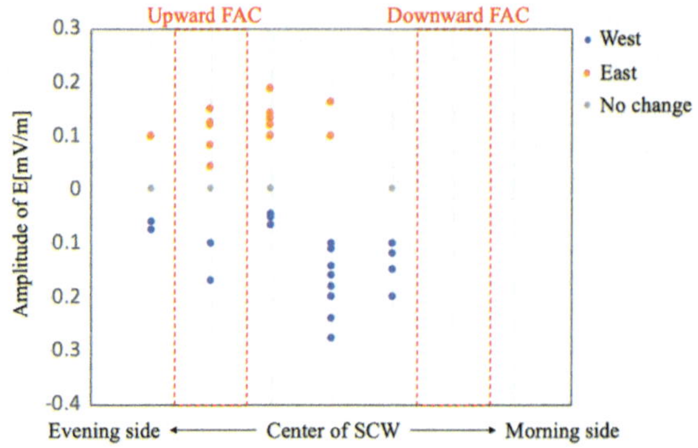


**Fig. 7.5** Magnitude of the electric field variation at the relative position of PTK to the SCW. Positive and negative values correspond to the eastward and westward electric fields, respectively, and the red-dotted lines indicate the positions of upward and downward FACs

Furthermore, when considering the upward FAC region in Fig. 4b, a higher occurrence of eastward electric fields was observed, as compared to Fig. 4b, which shows that it did not align with the direction of the electric field created by the primary potential. It indicates that it was a result of the distortion effect. Detailed interpretations of these trends are discussed in Chap. 5.

### *7.4.3 Relationship Between SCW Position and Electric Field Magnitude*

Figure 7.5 presents a diagram classifying the magnitude of the electric field for each event based on the position of the SCW. The horizontal axis represents the longitudinal spread of the SCW, and the vertical axis represents the variation in the electric field. Positive values indicate eastward electric fields and negative values indicate westward electric fields. In the central region of the SCW, an increase in the amplitude of the westward electric field was observed, and the electric field tended to decrease as it moved away from the center. This finding supported the idea that the primary potential generated at the base of the SCW excited a strong electric field at the center of the SCW, similar to the results presented in Sect. 4.2. In contrast, near the upward component, the eastward electric field tended to increase, which is believed to have been the effect of the Hall currents distorting the potential. Detailed interpretations of these trends are discussed in Chap. 5.

## 7.5 Discussion and Summary

Eastward and westward electric fields were observed on the nightside during substorm occurrences. Comparing the magnitude of these electric fields to the size of the SML showed that the magnitude of the electric fields was proportional to the size of the SML, suggesting a correlation between the electric fields and the AEJ. Regarding the direction of the electric fields, a predominance of westward electric fields was observed at the center of the SCW, and the magnitude of the westward electric fields tended to increase as they approached the center. This finding suggested that the electric fields created by the primary potential (Fig. 4b) formed at the base of the SCW extended to mid-latitudes.

In contrast, eastward electric fields were predominantly observed near the upward FAC, which may have reflected the potential structure's distortion due to the conductivity's inhomogeneity, as proposed in previous studies [16–19]. Figure 7.6 schematically represents the potential near the SCW, and the resulting electric fields based on the results of Yoshikawa et al. [19]. Figure 7.6a–c illustrate the process of inducing charges in the conductivity gradient region by the Hall effect, and Fig. 7.6d represents the distorted potential [19] and the electric fields created by this potential at mid-latitudes. The gray regions in Fig. 7.6a–c represent the high-conductivity regions near the upward and downward FACs, assuming low conductivity in the surrounding regions. First, the Hall currents (red and blue arrows in Fig. 7.6) formed at the roots of the FAC-induced charges in the boundary region (Fig. 7.6a). Next, Hall currents (red and blue arrows) were formed around the induced charges from (Fig. 7.6a), and these Hall currents induced charges (red $+$ and $-$) in the conductivity gradient region (Fig. 7.6b). The total induced charges in the conductivity gradient region, considering Fig. 7.6a and b, are represented in Fig. 7.6c. Figure 7.6d represents the potential distribution distorted by these effects, based on Yoshikawa et al. [19]. The solid black line is shown with blue and red arrows indicating the resulting electric fields at mid-latitudes, and the gray arrows represent the east–west components of the electric fields. By focusing on the east–west components of the electric fields, it can be explained that the electric fields shifted slightly toward the morning side from the fields shown in Fig. 7.4b and that eastward electric fields were formed below or slightly inside the upward FAC, resulting in the predominance of eastward electric fields near the upward FAC. Furthermore, Figs. 7.4a and 7.5 suggest that the distribution and magnitude of westward electric fields were not symmetrical but biased toward the region slightly below the downward FAC. This finding can also be explained by the potential shown in Fig. 7.6.

Among the analyzed electric fields, variations with time scales shorter than the fluctuations of the SML were observed. These variations may have been caused by induced electric fields resulting from magnetic field variations generated by the FACs rather than electric fields produced by the potential formed at the base of the current wedge. Nakahara et al. [20] investigated the variations in electric and magnetic fields on the ground during substorms and reported that the MLT distribution of the east–west component of the magnetic field induced by the FACs coincided with the northeast and southwest components of the electric field. This result suggested that the magnetic field components generated by the FACs excited induced electric fields on the ground. Similar induced electric fields are also likely to occur in the ionospheric electric field, generating variations on short time scales.

In this paper, we investigated the electric and magnetic field responses for 50 events in which the electric field clearly fluctuates during substorms. The results

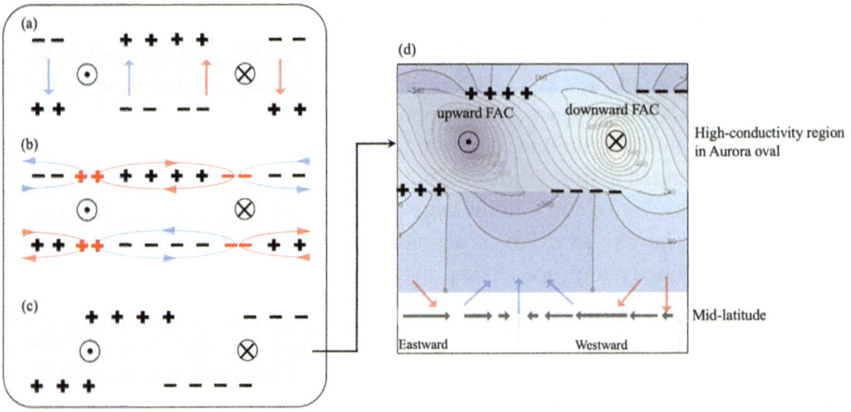


**Fig. 7.6** Schematic diagram of the potential due to the Hall effect. **a–c** The process of charge induction in the conductivity gradient region by the Hall effect. The gray region indicates the high conductivity of the auroral region. **d** Twisted electric potential distribution based on Yoshikawa et al. [19]. The mid-latitude electric field is indicated by blue/red arrows and the east–west component is indicated by gray arrows

showed that the electric field reflects the projection of the potential formed in the FAC region on the ionosphere. However, there are many events for which such clear fluctuations cannot be confirmed. Understanding the factors contributing to these differences will be critical for a comprehensive understanding of the global M-I coupling system.

**Acknowledgements** This work was supported by JSPS KAKENHI Grant Numbers JP22J21352 JP20H01961, JP21K03646, JP21H04518, JP22K21345. MAGDAS/CPMN magnetic data were provided by the principal investigator of MAGDAS/CPMN project (http://magdas.serc.kyushu-u.ac.jp/). We would like to thank the national institutes that support INTERMAGNET for promoting high standards of magnetic observatory practice (www.intermagnet.org).

## Appendix

This appendix shows the details of the magnetometers used in this study (i.e., latitude, longitude, and magnetic local time [MLT]). These 210 MM and mid-latitude magnetic field data were used to capture the spatiotemporal evolution of the SCW structure (Table 7.1).

**Table 7.1** List of the MAGDAS and SuperMAG magnetometer stations

| MAGDAS station | | | | | | | | | |
|---|---|---|---|---|---|---|---|---|---|
| Station | Geographic | | Geomagnetic | | Station | Geographic | | Geomagnetic | |
| | Lat | Long | Lat | Long | | Lat | Long | Lat | Long |
| CST | 68.5 | 179.2 | 63.60 | 231.49 | AMA | 28.17 | 129.33 | 18.83 | 200.00 |
| KTN | 75.94 | 137.71 | 66.76 | 197.88 | CBI | 27.15 | 142.3 | 18.80 | 212.18 |
| TIK | 71.59 | 128.78 | 62.05 | 193.97 | OKN | 26.75 | 128.22 | 17.35 | 199.06 |
| CHD | 70.62 | 147.89 | 62.26 | 207.33 | LNP | 25 | 121.17 | 15.27 | 192.56 |
| ZGN | 66.75 | 123.26 | 57.03 | 191.16 | HLN | 23.9 | 121.55 | 14.18 | 192.96 |
| ZYK | 65.75 | 150.78 | 57.77 | 211.67 | TGG | 17.66 | 121.76 | 7.96 | 193.43 |
| YAK | 62.02 | 129.72 | 52.59 | 196.76 | MUT | 14.37 | 121.02 | 4.64 | 192.85 |
| MGD | 59.97 | 150.86 | 52.12 | 213.82 | GUA | 13.58 | 144.87 | 5.55 | 216.08 |
| PTK | 52.94 | 158.25 | 46.01 | 222.00 | GAM | 13.58 | 144.87 | 66.76 | 216.08 |
| MSR | 44.37 | 142.27 | 35.91 | 209.98 | LGZ | 13.15 | 123.74 | 62.05 | 195.55 |
| RIK | 43.48 | 143.76 | 35.16 | 211.42 | CEB | 10.36 | 123.91 | 57.03 | 195.85 |
| ASB | 43.46 | 142.17 | 35.00 | 210.02 | YAP | 9.5 | 138.08 | 65.23 | 209.84 |
| PPI | 42.98 | 131.73 | 33.74 | 200.92 | CDO | 8.46 | 124.63 | 38.36 | 196.66 |
| BJI | 40.28 | 116.18 | 30.39 | 187.41 | MEL | 7.33 | 134.5 | 46.87 | 206.49 |
| TNO | 39.37 | 141.6 | 30.89 | 210.08 | KOR | 7.33 | 134.5 | 55.42 | 206.49 |
| ONW | 38.43 | 141.47 | 29.94 | 210.09 | DAV | 7 | 125.4 | 35.16 | 197.49 |
| ZAO | 38.1 | 140.53 | 29.53 | 209.29 | PON | 7 | 158.33 | 35.00 | 230.03 |
| OIS | 35.18 | 139.17 | 26.52 | 208.40 | POH | 7 | 158.33 | 33.74 | 230.03 |
| KOC | 33.5 | 133.5 | 24.41 | 203.41 | SBH | 6.02 | 116.07 | 29.94 | 188.25 |
| KUJ | 33.06 | 131.23 | 23.82 | 201.37 | RAN | 5.95 | 116.68 | 29.53 | 188.86 |
| KAG | 31.48 | 130.72 | 22.22 | 201.03 | MND | 1.44 | 124.84 | 21.93 | 197.22 |
| YMK | 31.2 | 130.62 | 21.93 | 200.96 | | | | | |
| *SuperMAG* | | | | | | | | | |
| VIC | 48.52 | 236.58 | 53.68 | 297.95 | SHU | 55.35 | 199.54 | 53.09 | 259.31 |
| T37 | 53.81 | 237.17 | 59.14 | 296.82 | MMB | 43.91 | 144.19 | 37.13 | 215.92 |
| SIT | 57.06 | 224.67 | 59.74 | 281.04 | MSR | 44.37 | 142.27 | 37.66 | 214.22 |
| T59 | 57.79 | 207.60 | 57.04 | 265.08 | | | | | |

## References


1. T. Iijima, T.A. Potemra, Field-aligned currents in the dayside cusp observed by Triad. J. Geophys. Res. (1896–1977) **81**, 5971 (1976)
2. T. Iijima, T.A. Potemra, Large-scale characteristics of field-aligned currents associated with substorms. J. Geophys. Res. Space Phys. **83**, 599 (1978)
3. R.L. McPherron, C.T. Russell, M.P. Aubry, Satellite studies of magnetospheric substorms on August 15, 1968. 9. Phenomenological model for substorms. J. Geophys. Res. **78**, 3131 (1973)

4. T. Kikuchi, H. Lühr, K. Schlegel, H. Tachihara, M. Shinohara, T.-I. Kitamura, Penetration of auroral electric fields to the equator during a substorm. J. Geophys. Res. **105**(23), 251–23, 261 (2000)
5. B. Veenadhari, T. Kikuchi, S. Kumar, R.S. Tulasi, D. Chakrabarty, Y. Ebihara, G.D. Reeves, Signatures of substorm related overshielding electric field at equatorial latitudes under steady southward IMF Bz during main phase of magnetic storm. Adv. Space Res. **64**(10), 1975–1988 (2019)
6. J.W. Gjerloev, R.A. Hoffman, J.B. Sigwarth, L.A. Frank, Statistical description of the bulge-type auroral substorm in the far ultraviolet. J. Geophys. Res. **112**, 07213 (2007)
7. K.K. Hashimoto, T. Kikuchi, I. Tomizawa, T. Nagatsuma, Substorm overshielding electric field at low latitude on the nightside as observed by the HF Doppler sounder and magnetometers. J. Geophys. Res. Space Phys. **122**(10) 851 (2017)
8. S. Imajo, A. Yoshikawa, T. Uozumi, S. Ohtani, A. Nakamizo, R. Marshall, B.M. Shevtsov, V.A. Akulichev, U. Sukhbaatar, A. Liedloff, K. Yumoto, Pi2 pulsations observed around the dawn terminator. J. Geophys. Res. Space Phys. **120**, 2088 (2015)
9. K. Yumoto, MAGDAS Group: MAGDAS project and its application for space weather, in *The Solar Influence on the Heliosphere and Earth's Environment: Recent Progress and Prospects*, ed. by N. Gopalswamy, A. Bhattacharyya (Quest, Mumbai, India 2006)
10. K. Yumoto, MAGDAS group: space weather activities at SERC for IHY: MAGDAS. Bull. Astronom. Soci. India **35**, 511–522 (2007).
11. J.W. Gjerloev, The SuperMAG data processing technique. J. Geophys. Res. **117**, A09213 (2012)
12. K. Nozaki, T. Kikuchi, A new multimode FM/CW ionosonde, memoirs of national institute of polar research. Special issue **47**, 217–224 (1987)
13. K. Nozaki, T. Kikuchi, Preliminary results of the multimode FM/CW ionosonde experiment. Proc. NIPR Symp. Upper Atmos. Phys. **1**, 204–209 (1988)
14. D.E. Barrick, *FM/CW radar signals and digital processing, NOAA Technical Report ERL 283-WPL 26* (NOAA, Boulder, Colo, 1973)
15. S. Ohtani, J. Gjerloev, Is the substorm current wedge an ensemble of wedgelets?: revisit to midlatitude positive bays. J. Geophys. Res. **125** (2020)
16. G. Atkinson, D. Hutchison, Effect of the day night ionospheric conductivity gradient on polar cap convective flow. J. Geophys. Res. **83**, 725 (1978)
17. A. Nakamizo, A. Yoshikawa, Deformation of ionospheric potential pattern by ionospheric Hall polarization. J. Geophys. Res. Space Phys. **124**, 7553 (2019)
18. R.A. Wolf, Effects of ionospheric conductivity on convective flow of plasma in the magnetosphere. J. Geophys. Res. **75**, 4677 (1970)
19. A. Yoshikawa, O. Amm, H. Vanhamäki, R. Fujii, Illustration of cowling channel coupling to the shear Alfven wave. J. Geophys. Res. Space Physics **118**, 6405 (2013)
20. M. Nakahara, A. Yoshikawa, T. Uozumi, A. Fujimoto, Electromagnetic induction responses to geomagnetic disturbances at low-and-mid-latitudes. In *International Conference on Space Weather and Satellite Application* 7th-9th August (2018)

# Chapter 8
# Ion Velocity Distributions in a Small-Scale Magnetic Flux Rope: A Case Study

**Wai-Leong Teh**

**Abstract** Magnetic flux rope is of a helical magnetic field configuration generated by magnetic reconnection. A magnetic flux rope is characterized by a bipolar magnetic field component normal to a boundary layer and an increase of magnetic field strength at the reversal of the bipolar field. Ion behaviors have rarely been investigated in a small-scale magnetic flux rope due to a lack of high-time-resolution ion plasma measurements. In this paper, a case study is presented using Magnetospheric Multiscale observations of a small-scale flux rope during a magnetopause crossing from magnetosphere to magnetosheath. The diameter of the flux rope was ~ 236 km, corresponding to ~ 5.4 ion inertial lengths. In the negative normal field region, crescent-shaped ion velocity distributions were formed in a plane perpendicular to magnetic field, but their crescent distributions were not similar. By contrast, different-shaped velocity distributions were found in the positive normal field region. Moreover, the velocity distribution map at the flux rope center was different than others. Although the ion motions were different within the flux rope, they all had a positive $\mathbf{E} \times \mathbf{B}$ drift.

## 8.1 Introduction

It has been shown that magnetic reconnection is a fundamental plasma process where magnetic energy is rapidly converted into thermal and kinetic energy in a tiny diffusion region. Reconnection signatures have been observed in various plasma environments in space, e.g., magnetopause, magnetotail, and solar wind. Magnetic flux rope is a by-product of reconnection, consisting of a helical magnetic field structure with a field strength enhancement at its center [1–3]. Simulation and observational results indicate that energetic particles can be produced inside a small-scale flux rope [4–6].

W.-L. Teh (✉)
Space Science Centre, Institute of Climate Change, Universiti Kebangsaan Malaysia, 43600 Bangi, Malaysia
e-mail: wteh@ukm.edu.my



M. T. Islam et al. (eds.), *Proceedings of the 8th International Conference on Space Science and Communication*, Springer Proceedings in Physics 303,
https://doi.org/10.1007/978-981-97-0142-1_8

Ion velocity distributions have been studied near the reconnection $X$-line observed by the Geotail spacecraft in the magnetotail, where high-energy (>20 keV) ions had a semicircular pattern in a plane perpendicular to magnetic field [7]. Such ion feature was resulted from meandering motions in a thin current sheet. From THEMIS observations of a tailward reconnection outflow, a crescent-shaped thermal ion distribution was found in the perpendicular plane [8].

Magnetospheric Multiscale (MMS) mission was designed to provide electromagnetic field and plasma measurements at high time resolutions for studying magnetic reconnection in a small-scale diffusion region [9]. The MMS mission has four identical spacecraft with the same scientific instruments, which are formed as a tetrahedron configuration with inter-spacecraft separations varying from 10 to 400 km, depending on the mission phases [10].

The ion plasma measurements by the MMS spacecraft provide an opportunity to study ion behaviors in a small-scale magnetic flux rope, which has rarely been investigated. In this paper, a case study is presented using MMS measurements of three-dimensional (3-D) ion velocity distributions. For different parts of a flux rope, a 2-D slice through a 3-D ion velocity distribution is generated for comparison.

## 8.2 MMS Observations

In this study, the magnetic field and plasma data from fluxgate magnetometer (FGM) [11] and fast plasma investigation (FPI) [12] instruments were obtained at https://lasp.colorado.edu/mms/sdc/public/. Figure 8.1 shows the 2-day (10–11 January 2016) orbit of the four MMS spacecraft in the ecliptic plane. In geocentric solar ecliptic (GSE) coordinates, the $x$ axis is from Earth toward Sun, and the $z$ axis is perpendicular to the ecliptic plane of the Earth (positive = north), and $y = z \times x$ (positive = dusk). At ~ 09:00:58 UT, on 10 January 2016, the MMS tetrahedron made an inbound magnetopause crossing from magnetosphere to magnetosheath, around (8.9, – 2.6, – 0.8) $R_E$ in GSE coordinates. The average spacecraft separation was ~ 40 km during the magnetopause crossing.

Figure 8.2 shows omni-directional ion and electron energy spectrograms, magnetic field, and ion plasma velocity, in the time interval between 09:01:02 UT and 09:01:08 UT. The data were based on the MMS1 spacecraft. In panel (c), a magnetic flux rope is identified, in which a bipolar pattern is shown in $B_x$ and the magnetic field strength is enhanced at the reversal of the bipolar $B_x$ around ~ 09:01:5.2 UT. This flux rope was previously studied for the evolution of a magnetic flux rope [13]. The duration of the flux rope was ~ 2 s, starting from ~ 09:01:04 UT to ~ 09:01:06 UT. Teh et al. (2017b) estimated that the diameter of the flux rope was ~ 236 km, corresponding to ~ 5.4 ion inertial lengths (a magnetosheath density of 27 $cm^{-3}$). Figure 8.2d shows that the ion velocity $V_x$ was decreased inside the flux rope, while the $V_y$ and $V_z$ were increased near the edge of the positive $B_x$ of the flux rope.

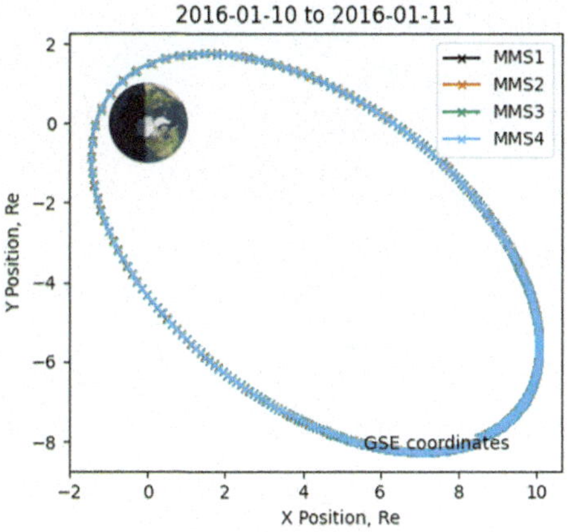


**Fig. 8.1** MMS spacecraft orbit in the GSE $x$–$y$ plane during 10–11 January 2016. The $R_E$ is the Earth's radius

## 8.3 Ion Velocity Distributions of a Flux Rope

The 3-D electron and ion velocity distributions were measured at 30 ms and 150 ms, respectively. The measured energy range for ions and electrons is from 10 eV to 30 keV. Figures 8.3, 8.4, and 8.5 show cuts perpendicular to magnetic field through 3-D ion velocity distributions, for the negative $B_x$ of the flux rope (Figs. 8.3 and 8.4) and the center of the flux rope (Fig. 8.5). The ion energy range in the maps was limited to 1–30 keV, and the velocity distribution maps were generated in order at times ~ 09:01:4.5 UT, ~ 09:01:4.8 UT, and ~ 09:01:5.2 UT. The $x$ axis points toward $(b \times v) \times b$ and the $y$ axis is directed along $b \times v$. Here $b$ and $v$ are unit vectors of magnetic field and ion plasma velocity. The 2-D slices were generated by Python codes [14], which can be accessed on GitHub at https://github.com/spedas/pyspedas.

One can find that the ion velocity distribution coded in red was different at the three different locations of the flux rope. At ~ 09:01:4.5 UT, a crescent-shaped ion distribution was observed, and its center was off the $+V_{V\,\mathrm{perp}B}$ axis (Fig. 8.3), while a narrow one was found at time ~ 09:01:4.8 UT and was centered along the $+V_{V\,\mathrm{perp}B}$ axis (Fig. 8.4). By contrast, one finds an irregular ion distribution at the center of the flux rope.

For the positive $B_x$ region of the flux rope, the ion velocity distribution maps were shown at times ~ 09:01:5.6 UT and ~ 09:01:5.8 UT (Fig. 8.6). Clearly, the two velocity distribution maps were different from those seen in the negative $B_x$ region and at the center of the flux rope.

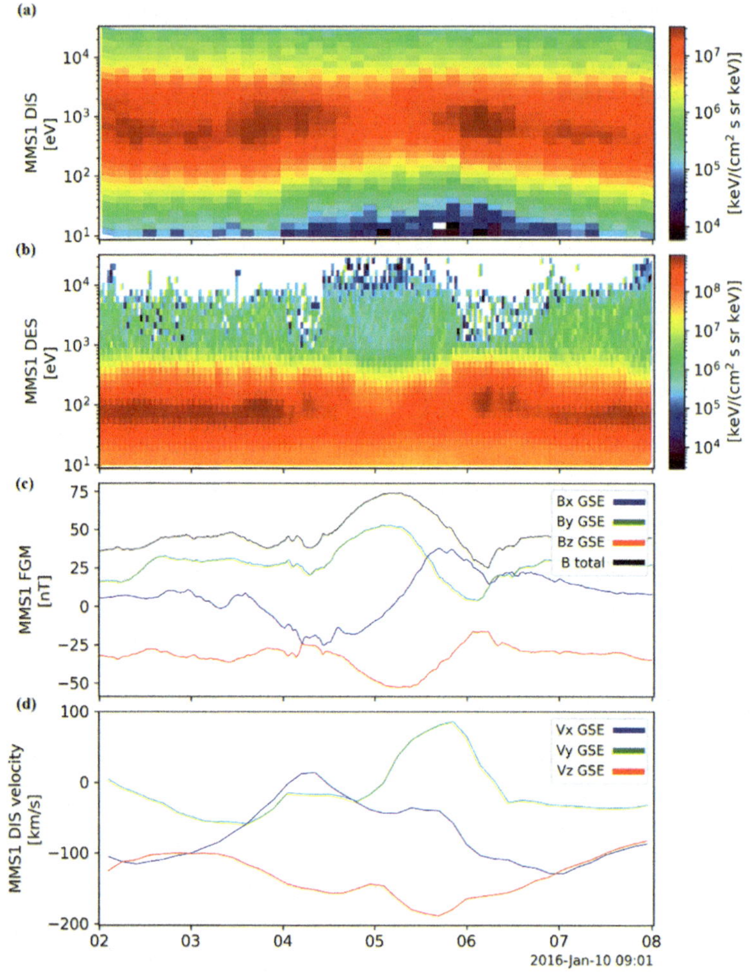


**Fig. 8.2** Magnetic flux rope profile

## 8.4 Summary and Discussion

Ion behaviors in a small-scale magnetic flux rope have been examined using MMS measurements of 3-D ion velocity distribution at 150 ms time resolution. A 2-D slice through a 3-D velocity distribution for different parts of the flux rope was generated for comparison. In the negative $B_x$ region of the flux rope, crescent-shaped ion distributions were formed in the plane perpendicular to the magnetic field, but their crescents were not similar. By contrast, different-shaped distributions were seen in the positive $B_x$ region of the flux rope. Moreover, at the flux rope center the velocity distribution map was different from those seen in the negative and positive $B_x$ regions of the flux rope.

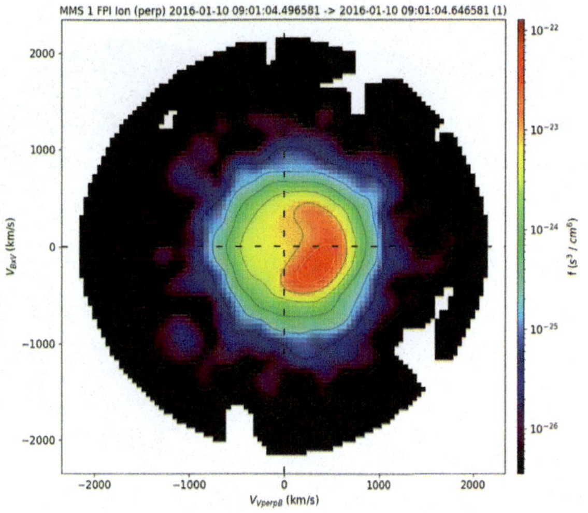


**Fig. 8.3** Ion velocity distribution in the perpendicular plane, in the negative $B_x$ region of the flux rope at time ~ 09:01:4.5 UT

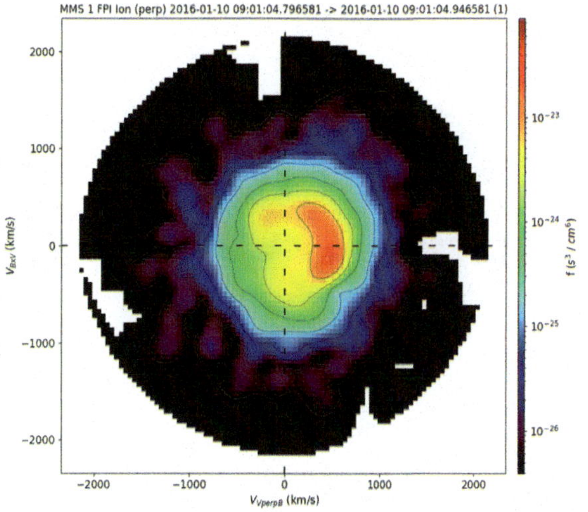


**Fig. 8.4** The same figure format as Fig. 8.3, in the negative $B_x$ region of the flux rope at time ~ 09:01:4.8 UT

The crescent-shaped ion populations in the negative $B_x$ region of the flux rope are resulted from the meandering motions. Such ion populations are similar to those discovered in the tailward reconnection outflow in the magnetotail [8]. It is interesting that the meandering ion motion was observed in the small-scale magnetic flux rope. The 2-D distribution maps indicate that the ion motions were different within the flux rope, but they all had a positive $V_{V\text{perp}B}$ due to the $\mathbf{E} \times \mathbf{B}$ drift.

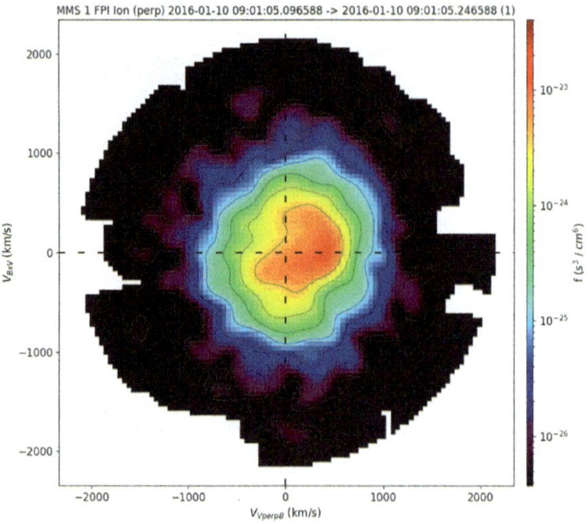


**Fig. 8.5** The same figure format as Fig. 8.3, at the center of the flux rope at time ~ 09:01:5.2 UT

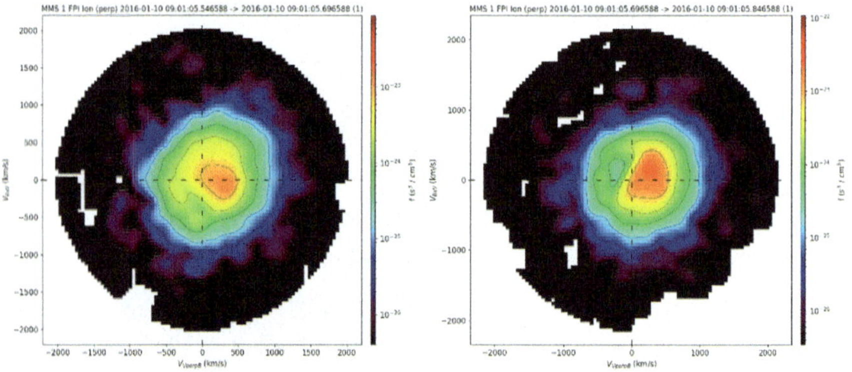


**Fig. 8.6** The same figure format as Fig. 8.3, in the positive $B_x$ region of the flux rope at times ~ 09:01:5.6 UT and ~ 09:01:5.8 UT

**Acknowledgements** This work was funded by the Universiti Kebangsaan Malaysia grant (GP-2021-K020730).

## References

1. W.-L. Teh, R. Nakamura, H. Karimabadi, W. Baumjohann, T.L. Zhang, Correlation of core field polarity of magnetotail flux ropes with the IMF $B_y$: Reconnection guide field dependency. J. Geophys. Res. Space Phys. **119**, 2933–2944 (2014)

2. W.-L. Teh, R.E. Denton, B.U.Ö. Sonnerup, C. Pollock, MMS observations of oblique small-scale magnetopause flux ropes near the ion diffusion region during weak guide-field reconnection. Geophys. Res. Lett. **44**, 6517–6524 (2017)
3. W.-L. Teh, T.K.M. Nakamura, R. Nakamura, T. Umeda, Oblique ion-scale magnetotail flux ropes generated by secondary tearing modes. J. Geophys. Res. Space Phys. **123**, 8122–8130 (2018)
4. J.F. Drake, M. Swisdak, K.M. Schoeffler, B.N. Rogers, S. Kobayashi, Formation of secondary islands during magnetic reconnection. Geophys. Res. Lett. **33**(13) (2006)
5. M. Oka, T.-D. Phan, S. Krucker, M. Fujimoto, I. Shinohara, Electron acceleration by multi-island coalescence. Astrophys. J. **714**, 915 (2010)
6. S. Lu, A.V. Artemyev, V. Angelopoulos, P.L. Pritchett, Energetic electron acceleration by ion-scale magnetic islands in turbulent magnetic reconnection: particle-in-cell simulations and ARTEMIS observations. Astrophys. J. **896**, 105 (2020)
7. T. Nagai, I. Shinohara, S. Zenitani, Ion acceleration processes in magnetic reconnection: Geotail observations in the magnetotail. J. Geophys. Res. Space Phys. **120**, 1766–1783 (2015)
8. A. Runov, M. Grandin, M. Palmroth, M. Battarbee, U. Ganse, H. Hietala et al., Ion distribution functions in magnetotail reconnection: global hybrid-Vlasov simulation results. Ann. Geophys. **39**, 599–612 (2021)
9. J.L. Burch, T.E. Moore, R.B. Torbert, B.L. Giles, Magnetospheric multiscale overview and science objectives. Space Sci. Rev. **199**, 5–21 (2016)
10. S.A. Fuselier, W.S. Lewis, C. Schiff, R. Ergun, J.L. Burch, S.M. Petrinec, K.J. Trattner, Space Sci. Rev. **199**, 77–103 (2016)
11. C.T. Russell, B.J. Anderson, W. Baumjohann, K.R. Bromund, D. Dearborn, D. Fischer et al., The magnetospheric multiscale magnetometers. Space Sci. Rev. **199**, 189–256 (2016)
12. C. Pollock, T.E. Moore, A. Jacques, J.L. Burch, U. Gliese, Y. Saito et al., Fast plasma investigation for magnetospheric multiscale. Space Sci. Rev. **199**, 331–406 (2016)
13. W.-L. Teh, T.K.M. Nakamura, R. Nakamura, W. Baumjohann, C.T. Russell, C. Pollock et al., Evolution of a typical ion-scale magnetic flux rope caused by thermal pressure enhancement. J. Geophys. Res. Space Phys. **122**, 2040–2050 (2017)
14. E.W. Grimes, B. Harter, N. Hatzigeorgiu, A. Drozdov, J.W. Lewis, V. Angelopoulos et al., The space physics environment data analysis system in Python. Front. Astron. Space Sci. **9**, 1020815 (2022)

# Chapter 9
# Geomagnetic Field Effects Over the Philippines During Strong Solar Flares in April 2022

**Zane Nikia C. Domingo, Ernest P. Macalalad, and Akimasa Yoshikawa**

**Abstract** During solar storms, conductivities in Earth's ionospheric regions become more enhanced. This phenomenon produces effects on Earth's current systems to which emissions of solar radiation largely contribute. Such radiations are analyzed as short timescale variations of the X-ray and EUV fluxes. This was observed for the two X- and three M-class solar flares on 17, 20, and 21 April 2022. All of which are geomagnetically quiet days with nominal Kp, Hp, and Dst indices. Magnetic responses were observed from magnetometers in the Philippines, located in the low-latitude (MUT) and near-equatorial (DAV) stations. These coexisting variations observed with Earth's magnetic field are referred to as magnetic crochets. Measurements also showed that solar flare effects (Sfe) are more pronounced near the equator, where high conductivity generates more variations in Earth's magnetic fields. Yet, the largest variations were not consistently observed for the strongest flares. At MUT, manifest variations in $H_{sfe}$ (horizontal) and $Z_{sfe}$ (downward) were aligned with the X1.1 and M1.89 flares on day 17. Variations are generally homogeneous on days 20 and 21. Meanwhile, the largest variations at DAV were found during the M9.7 flare on day 21. It is inferred that increases in $H_{sfe}$ remain greater than in $Z_{sfe}$ at both stations. Fluctuations in $H_{sfe}$ are nonetheless more prolonged and heightened at DAV station.

Z. N. C. Domingo · E. P. Macalalad (✉)
Department of Physics, Mapúa University, Manila, Philippines
e-mail: epmacalalad@mapua.edu.ph

A. Yoshikawa
International Center for Space Weather Science and Education, Kyushu University, Fukuoka, Japan

 
M. T. Islam et al. (eds.), *Proceedings of the 8th International Conference on Space Science and Communication*, Springer Proceedings in Physics 303,
https://doi.org/10.1007/978-981-97-0142-1_9

## 9.1 Introduction

Solar flares are explosive bursts of electromagnetic waves that originate at active regions of the Sun. The intensification of this ionizing radiation manifests at shorter wavelengths where the energies are higher [1]. Since satellite and navigation systems largely rely on the electron densities, finding the relation between a solar flare and the amount of radiation absorption can be extended to examining short-lived effects induced by the interplanetary magnetic field to Earth's own magnetic field. Such effects include magnetic crochets or solar flare effects (Sfe) which are short-lived geomagnetic disturbances mostly confined to Earth's sunlit hemisphere. These can be observed through ground-based detectors and seen on magnetometer data. Its characterization originated from the morphology of this phenomenon, capturing the "hook" structure reflected on magnetograms [2]. In this paper, the terms magnetic crochet and Sfe will be used interchangeably.

Surges of X-ray and extreme-ultraviolet (EUV) radiation from the Sun directly cause the enhancement in the upper atmospheric ionization [3]. These ionization enhancements alter the ionospheric currents in the E-region which can be associated with the magnetic crochets [4]. It was also suggested that aside from the changes in the D-region conductivities subsequently indicating the changes in the lower ionosphere, these may also be giving rise to the magnetic crochets [5].

The magnetic fields produced by the Earth's core generate electric currents that significantly contribute to the Earth's magnetic field. Among the known contributors to the geomagnetic field accounting for only a fraction of the variations on tens of nano Tesla (nT) are electric currents that flow in the ionosphere [6]. These currents with which the magnetic crochets are associated become present with the Sq (solar-quiet) currents. These are regular geomagnetic field variations for which most ionospheric currents and secondary currents from the Earth's interior are responsible [6]. While both current systems are generated by the dynamo action at which atmospheric winds behave, each was reported in earlier studies to be primarily existing in separate regions. The main discrepancy was on the ionospheric region within which they are restricted, with the Sfe ionospheric currents downwardly extending into the D-region and the Sq, under normal conditions, limited within the E-region [7]. It was, however, estimated that there is an equal contribution from the electric currents in the D- and E-regions to the Sfe current [8]. This shows that the geomagnetic effects are attributed to the current systems governing each region.

When solar wind-driven geomagnetic disturbances are absent, Sq currents become dominantly visible [4]. As a background current, Sq increases in its amplitudes because of external geomagnetic disturbances. It was pointed out by scientists that there's a similarity between the Sfe and preflare Sq variations in terms of magnitude and direction [9]. These enhancements in the background Sq must then be the attributed manifestations of the crochet arising from the increased ionization. Moreover, the responses of the equatorial electrojet (EEJ) to solar flares have also been investigated by researchers and the effects of its presence on the Earth's geomagnetic field. EEJ is referred to as a narrow band of daytime eastward electric current in the atmosphere along the dip equator [10]. Some researchers have been focusing on the Sq-EEJ relationship and employing an approach to quantify the exchange of magnetic signals between stations that are located in different latitudes. This shows that Sq and EEJ are coupled current systems. It was also suggested that the effect of the eastward ionospheric current with Sq combined with the (eastward) EEJ and (westward) counter EEJ shows the increase and decrease, respectively, in the horizontal geomagnetic field [11]. It was also found that the EEJ enhancement that would be induced by the solar flare is essentially proportional to the pre-existing EEJ [12].

In this paper, the geomagnetic effects of a series of solar flares on April 22 are investigated. There are two X-class flares, on April 17 and 20, and three M-class flares, on April 17, 20, and 21. Since we're building to a solar maximum in 2025, the Sun's activities can still become more frequent and intense. The geomagnetic signature of crochets is also examined in the magnetometer data from low-latitude and equatorial stations in the Philippines.

## 9.2 Data and Measurements

### *9.2.1 Kp, Dst, Hp30, and Hp60 Indices*

The strength of geomagnetic disturbance in Earth's magnetic field is measured using three indices. The Kp index provided the 3-hourly observation of the global geomagnetic field. On the other hand, a higher temporal resolution for the geomagnetic indices, a new group of indices collectively called Hpo is used. The "H" stands for half-hourly for 30 min (*Hp30*) or hourly (*Hp60*), "p" for planetary, and "o" indicates open-endedness [13]. Hpo indices are designed in a similar manner as the derivation of Kp but overcoming the latter's limited 3-h observation and upper limit at 9 [13]. Although the results in this paper do not involve any strong geomagnetic activity, thereby no longer demonstrating the upper limit resolution, the H index (for H30 and H60), from which the Hpo is derived, is akin to the K index for Kp. Normally, evaluation of the indices makes use of conversion tables between magnitude of geomagnetic disturbance and integer values. This is done for H as it was for K using a similar frequency distribution [13]. The limits of H are multiplied by a factor that is experimentally determined to achieve compatibility with the magnitude of the geomagnetic disturbance being studied [13]. This derivation is no longer discussed in this paper. Kp, Hp30, and Hp60 indices were obtained from GFZ German Research Centre for Geosciences (https://kp.gfz-potsdam.de/en/data). Moreover, the hourly measurement of the Dst was obtained from NASA/GSFC OMNIWeb Interface (https://omniweb.gsfc.nasa.gov/form/dx1.html) This would measure the strength of the ring current around Earth's magnetic field and indicate the emergence of any magnetic storm.

**Table 9.1** Classification of flares during April 2022

| Day | Start (UT) | Max (UT) | End (UT) | Max (LT) | Region | Class |
|---|---|---|---|---|---|---|
| 17 April 2022 | 02:00 | 02:11 | 02:23 | 10:11 | 2993 | M1.89 |
| 17 April 2022 | 03:17 | 03:34 | 03:51 | 11:34 | 2994 | X1.1 |
| 20 April 2022 | 01:20 | 01:36 | 01:44 | 09:36 | 2992 | M7.29 |
| 20 April 2022 | 03:41 | 03:57 | 04:04 | 11:57 | 2992 | X2.25 |
| 21 April 2022 | 01:47 | 01:59 | 02:05 | 09:59 | 2993 | M9.7 |

### *9.2.2 Solar Data (X-ray, EUV Flux)*

The characterizations of the solar flares being studied are presented in Table 9.1. M-class flares are observed to have preceded the major X-class flare events on days 17 and 20. Variations in the solar X-ray and EUV fluxes are also investigated. These short-lived flux variations can aid in probing into the magnetic reconnection process since these are closely related to the impulsive releases of energy during solar flares [14]. This would also show the relation of the strength of a flare to the largest variation in the geomagnetic field.

The solar X-ray and EUV flux data used in observing the behaviors of the X-class and M-class flares were obtained from the Geostationary Operational Environmental Satellite (GOES) archives, accessible at https://www.ngdc.noaa.gov/stp/satellite/goes-r.html. Level 2 data of 1-min average fluxes were selected for observing the peak fluxes of the flare events. The EUV solar irradiances chosen are of wavelengths 25.6 and 30.4 nm. This will aid in other studies that investigate the effects of radiation absorption on satellite signals and radio frequencies.

### *9.2.3 Magnetometer Data*

The construction of ground-based magnetometer networks allowed the monitoring and measuring of variations in Earth's magnetic field [15–18]. These reflect the magnetosphere-ionosphere responses to space disturbances primarily induced by solar activities. The magnetometer data used in this study were extracted from MAGDAS-9 detectors located at Muntinlupa and Davao, Philippines (see Fig. 9.1). Data from six stations in the country (Tuguegarao, Muntinlupa, Legazpi, Cebu, Cagayan de Oro, and Davao) and in Taiwan (Hualien) were requested but only two stations (Muntinlupa and Davao) had complete data. It was assumed that among the six, only Muntinlupa and Davao stations were operational. Muntinlupa station is located at 14.37 °N, 121.02 °E with a dip latitude of 7.32°, while Davao station is at 7 °N, 125.4 °E with a dip latitude of – 0.24°. The latitudinal difference between the said stations offers a favorable opportunity to examine the strengths of the magnetic responses based on the two locations.

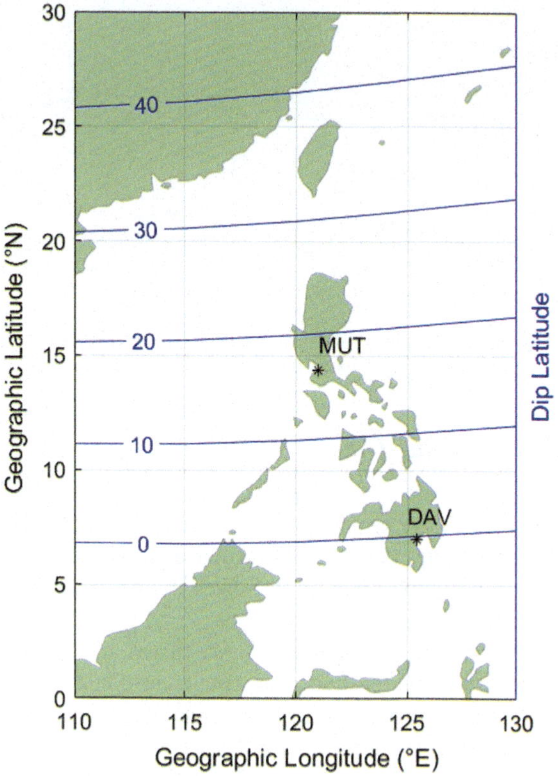


**Fig. 9.1** Map over the Philippine region illustrating the low-latitude and near-equatorial stations, along with the geomagnetic equator

## 9.3 Observations and Discussions

### *9.3.1 Kp, Dst, Hpo Indices*

The flares on days 17, 20, and 21 occurred on geomagnetically quiet days with Kp $\leq$ 3. As there was no need to extend the days that succeeded the flare events, Dst measurements confirmed the non-emergence of geomagnetic storms during the observed magnetic field variations. On day 17, the Dst minimum of −11nT occurred at 4 UT. The Dst values on day 20 remain relatively maintained until reaching a maximum of 5nT at 4 UT. On day 21, the Dst maximum of 14nT occurred at 2 UT, which is near the maximum time of the flare (Fig. 9.2).

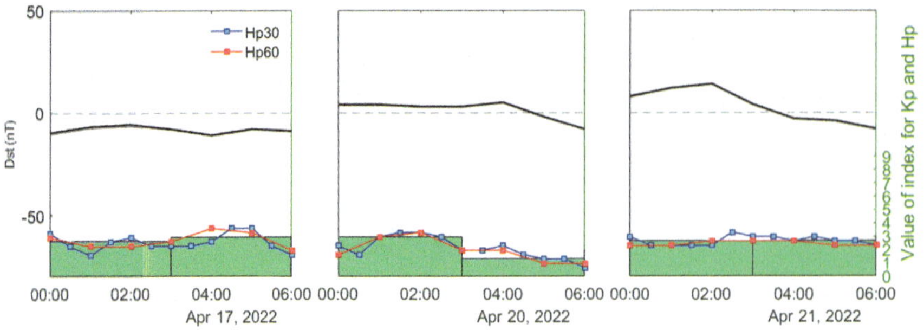


**Fig. 9.2** Geomagnetic indices at the interval 00:00–06:00 UT. The Kp and Hp are superimposed with their combined axis on the right, while the Dst has its axis on the left

## 9.3.2 Solar Data

The successive graphs in Fig. 9.3 describe the flare events in terms of the solar and magnetometer data. The two solar data include X-ray and EUV flux. These are in order from top to bottom, followed by the magnetometer data from MUT and DAV stations.

As seen in Fig. 9.3, the X-ray fluxes are illustrated in the first row. The X1.1 flare primarily reached a maximum at 03:34 UT. Prior to this, a weaker M1.89 flare was recorded at 02:11 UT. In between peak fluxes, there clearly is no retained elevation. If examined separately, the pre-ascent of the amplitude for the weaker event is more apparent compared to the abrupt rise of amplitude toward the stronger event. Conversely, the declining amplitude is more pronounced and prolonged for the main event.

The X2.25 flare on April 20, with its maximum at 03:57 UT, was preceded by the M7.29 flare, with its maximum at 01:36 UT. There also appears to be an intermedial peak, momentarily rising only to return to its relatively steady amplitude. Eventually, this abruptly rises toward the X2.25 flare at maximum. As soon as the amplitude decreases again, there is no longer an overall escalating behavior. The narrower peak of the X2.25 suggests a shorter-lived X-ray flux increase compared to the M7.29.

The M9.7 flare on April 21 induced an elevation with its X-ray flux that remained after reaching the maximum at 01:59 UT. Within the interval of six hours, neither preceding nor succeeding M-class flares were observed. Among the three M-class flares observed, the M9.7 is the strongest to be recorded.

The electromagnetic radiation in which solar energy travels exists as wavelengths and frequencies of different ranges. The absorption of EUV photons takes place in Earth's upper atmosphere. With so much thermal energy dominating the thermosphere (80–600 km) emitted by EUV radiation, ionization is driven more. Presented in the second row of Fig. 9.3 are the EUV fluxes observed for wavelengths 25.6 and 30.4 nm, dominated by spectral lines from He II. These increased significantly only for X1.1 on day 17. Conversely, there was a small increase in both wavelengths for M7.29. The succeeding increase only occurred in the 25.6 nm for X2.25. This was

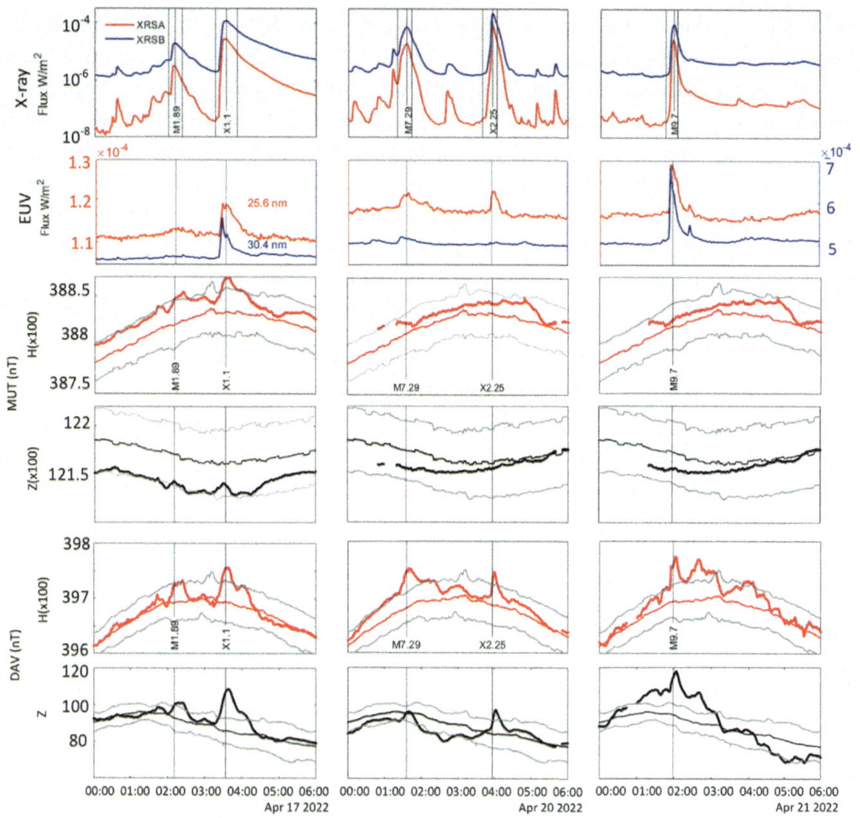


**Fig. 9.3** From top to bottom, included are the two solar data and magnetometer data at the interval 00:00–06:00 UT. Illustrated in the first row are the consecutive vertical black lines representing the start, maximum peak, and end of the flares. The daily magnetic variations (thicker red for *H* and black for *Z*), average quiet days (same color but of thinner lines), and standard deviation (gray) are superimposed for MUT and DAV

beyond the time at which the corresponding X-ray flux was of peak value. Moreover, the EUV fluxes elevated at higher intensities during M9.7 on day 21.

### *9.3.3 Magnetometer (H and Z Components)*

As seen in the last four rows in Fig. 9.3, the daily variations for the magnetometer data are represented by the thicker lines for horizontal (red) and downward (black) components. Whereas the average quiet days are represented by thinner lines of similar colors, the standard deviations are uniformly represented by the thin gray lines. The average of the magnetometer data for the month of April was obtained to procure the quiet days that should show the general behavior of the magnetic field. The standard deviation was obtained to quantify digression in the amplitude. Superimposing the average quiet days, the deviation, and the daily magnetometer data showcases how much enhancement there is between geomagnetic components. Rescaling some components provided an easier comparison between the variations in different components. The augmentation in (horizontal) $H$ and (downward) $Z$ transpired shortly after the time at which the flares are at maximum. Without rescaling, $H_{sfe}$ is initially in order of $10^4$nT for both stations and $Z_{sfe}$ is in $10^4$nT only for MUT station. Whereas non-rescaled components are in tens to hundreds nT. Around this time, the magnetic declination in Davao is around $-1.16°$ and Muntinlupa is $-2.51°$. Due to the small value of the declination, observations are assumed insignificant, hence these are no longer included and analyzed.

**MUT Station.** There are mostly no abrupt increases or decreases in the amplitudes in the low-latitude region. There is also a generally prolonged depression in $Z_{sfe}$ on days 17, 20, and 21. While this behavior is more homogeneous on days 20 and 21, apparent enhancements are only present on day 17. Difference from $H_{mean}$ was recorded to be 35.52 nT at 03:38 UT, shortly after X1.1. A longer delay was observed for the largest $Z_{sfe}$ at 03:51 UT. With the same flare intensity as the horizontal component, $Z_{sfe}$ deviated by a smaller amount of 32.80 nT.

**DAV Station.** Detections from DAV station show more pronounced effects in $H_{sfe}$, similar to low-latitude magnetic responses. This had the largest variation on day 21 during the M9.7 flare. An $H_{sfe}$ maximum of 397.7 nT was reached at 02:03 UT, deviating by 79.33 nT. This also revealed a prolonged response time, transpiring beyond the time at which the flare is at maximum. The largest $Z_{sfe}$ variation also occurred during the strongest M-class flare. An amplitude of 117.6 nT was reached at 02:04 UT with a corresponding variation of 23.68 nT from the mean. On all days, the fluctuations in the downward and vertical components are manifest within the 6-h interval.

At MUT, the largest magnetic variations correspond to the X1.1 flare on day 17. While at DAV, these are on day 21 during the M9.7 flare, which approaches the X-class intensity. This exhibits that even the strongest flare in a series doesn't induce extreme magnetic field variations owing to other factors. The solar zenith angle which is dictated by the season and the time of the day also may account for varying strength of the radiation. The duration is also different for all flares, hence this may have an effect on the magnetic field exposure time from the solar radiation.

## 9.4 Conclusions

In this study, an analysis of the flare events during the solar cycle 25 in April 2022 is presented. Geomagnetic conditions remain quiet during these flares. This indicates that the coexisting observations made on the crochets from the magnetic variations do not arise from geomagnetic disturbances. Solar radiation in different wavelengths is also shown to describe the flares that are in phase with the fluctuations observed in Earth's magnetic field. These magnetic variations were obtained from the low-latitude (MUT) and near-equatorial (DAV) stations in the Philippines. At MUT, magnetic variations were the most significant on day 17 during the M1.89 and X1.1 flares. None of the $H_{\mathrm{sfe}}$ and $Z_{\mathrm{sfe}}$ during the stronger flares on days 20 and 21 fluctuated as much. Meanwhile, augmentations in the magnetic fields are obvious on days 17, 20, and 21 at DAV. The highest variation was on day 21 during the M9.7 flare. This demonstrates that concurrent enhancements with solar radiation do not necessarily show the largest measurements for the largest magnetic variations. Also, the enhancements mostly occurring at DAV show that there is an increase in ionospheric conductivities near the geomagnetic equator.

**Recommendations**

The flares examined in this paper were only among the strongest during April 2022. Magnetic variations during relatively weaker flares can be covered for future comparison with others of greater intensity. The corresponding increase in conductivities may also be further investigated by looking into the governing current systems. Magnetic responses may be extended to studying the other magnetic field components while also comparing different dates during which the flare of same intensity occurred.

**Acknowledgements** The authors would like to express their gratitude to the International Research Center for Space and Planetary Environmental Sciences (i-SPES), Kyushu University for providing the magnetometer data under the MAGDAS/CPMN project.

## References

1. J.J. Curto, S. Marsal, E. Blanch, D. Altadill, Analysis of the solar flare effects of 6 sep-tember 2017 in the ionosphere and in the earth's magnetic field using spherical El-ementary current systems. Space Weather **16**, 1709–1720 (2018). https://doi.org/10.1029/2018SW001927
2. J.J. Curto, Geomagnetic solar flare effects: a review. J. Space Weather Space Climate **10** (2020)
3. L. Qian, A.G. Burns, S.C. Solomon, P.C. Chamberlin, Solar flare impacts on iono-spheric electrodyamics. Geophys. Res. Lett. **39** (2012). https://doi.org/10.1029/2012GL051102
4. N.M.N. Annadurai, N.S.A. Hamid, Y. Yamazaki, A. Yoshikawa, Investigation of Unu-sual solar flare effect on the global ionospheric current system. J. Geophys. Res. Space Phys. **123**, 8599–8609 (2018). https://doi.org/10.1029/2018JA025601
5. B. Gavrilov, A.N. Lyakhov, Y. Poklad et al., Geophysical effects of solar flare on 6 September 2017. SPIE-Intl. Soc. Optical Eng. 29 (2018)
6. Y. Yamazaki, A. Maute, Sq and EEJ—a review on the daily variation of the ge-omagnetic field caused by ionospheric dynamo currents. Space Sci. Rev. **206**, 299–405 (2017)
7. A.D. Richmond, S.V. Venkateswaran, Geomagnetic crochets and associated iono-spheric current systems. Radio Sci. **6**, 139–164 (1971). https://doi.org/10.1029/RS006i002p00139
8. H. Volland, J. Taubenheim, On the ionospheric current system of the geomagnetic solar flare effect (s.f.e.). Pergamon Press Ltd. (1958)
9. Y. Yamazaki, K. Yumoto, A. Yoshikawa et al., Characteristics of counter-Sq SFE (SFE) at the dip equator CPMN stations. J. Geophys. Res. Space Phys. **114** (2009). https://doi.org/10.1029/2009JA014124

10. S. Chapman, The equatorial electrojet as detected from the abnormal electric current distribution above Huaneayo, Peru, and elsewhere (1951)
11. R.G. Rastogi, H. Chandra, K. Yumoto, Unique examples of solar flare effects in ge-omagnetic H field during partial counter electrojet along CPMN longitude sec-tor. Earth Planets Space **65**, 1027–1040 (2013). https://doi.org/10.5047/eps.2013.04.004
12. M.A. Abdu, P.A.B. Nogueira, J.R. Souza et al., Equatorial electrojet responses to in-tense solar flares under geomagnetic disturbance time electric fields. J. Geophys. Res. Space. Phys. **122**, 3570–3585 (2017). https://doi.org/10.1002/2016JA023667
13. Y. Yamazaki, J. Matzka, C. Stolle et al., Geomagnetic activity index hpo. Geophys. Res. Lett. **49** (2022). https://doi.org/10.1029/2022GL098860
14. W. Lu, S.-M. Liu, N. Zong-Jun, Statistical properties of radio flux densities of solar flares (2020)
15. K. Yumoto, The STEP 210° Magnetic Meridian Network Project (1996)
16. K. Yumoto, C. Group, Characteristics of Pi 2 magnetic pulsations observed at the CPMN stations: a review of the STEP results (2001)
17. K. Yumoto, MAGDAS project and its application for space weather (2006)
18. K. Yumoto, Space weather activities at SERC for IHY: MAGDAS. Bull. Astron. Soc. India 511–522 (2007)

# Chapter 10
# The Variations of Precipitable Water Vapor and Rainfall in Puerto Princesa and Legazpi City During the Philippine Southwest Monsoon Season from 2013 to 2022

**Julia Marie Cabral, Rebecca Marie Dizon, Samantha Gevan Morales, Lyren Hope Pujida, and Ernest P. Macalalad**

**Abstract** Precipitable Water Vapor (PWV), the measure of the entire atmospheric water vapor carried in a vertical air column, can be used to represent measurements of water vapor (Mina in Mongabay 2021). The variations and correlation between PWV and rainfall over Puerto Princesa, Palawan, and Legazpi City, Albay, during the Philippine Southwest Monsoon (SWM) Season from 2013 to 2022 are presented through the use of radiosonde databases of both PWV and Rainfall, as extracted from Integrated Global Radiosonde Archives (IGRA) and Global Surface Summary of the Day (GSOD). It was interpreted by Microsoft Excel and MATLAB. A time series analysis with the Lomb-Scargle periodogram, Mann–Kendall test, and One-Way ANOVA graph was used to examine the temporal variations and analyze the correlations of the data extracted from respective sources. The results demonstrated an intriguing trend between the two stations and their locations. In the ten-year course of the study, Legazpi provided a higher peak of PWV and rainfall than Puerto Princesa, despite being the opposite course of SWM winds. PWV and Rainfall usually peak during the start of the SWM Season (June) and fall during the end of it (October). Lomb-Scargle presented apparent periodicities of 4.4 and 6.7 years for Puerto Princesa and Legazpi, while the Mann–Kendall test showed a downward trend for PWV and an upward trend for Rainfall. No significant difference exists between the variables presented, and the stations have a strong correlation, according to the One-Way ANOVA Graph. This study can be a reference for future studies of the Philippines' shifting climate.

J. M. Cabral · R. M. Dizon · S. G. Morales · L. H. Pujida
Senior High School Department, Mapua University Muralla St. Intramuros, 1002 Manila, Philippines

E. P. Macalalad (✉)
Department of Physics, Mapúa University, Muralla St, Intramuros, Manila, Philippines
e-mail: epmacalalad@mapua.edu.ph



M. T. Islam et al. (eds.), *Proceedings of the 8th International Conference on Space Science and Communication*, Springer Proceedings in Physics 303,
https://doi.org/10.1007/978-981-97-0142-1_10

## 10.1 Introduction

The Philippines is one of the countries highly vulnerable to natural disasters. About 74% of the country's total population is harmed in various courses of action by natural catastrophes [1]. Furthermore, 60% of the total land area of the Philippines is susceptible to these natural disasters, as the country is situated within the Pacific Ocean and the Ring of Fire, both hotspots of tropical storms and monsoons [1].

In the context of rainfall, the Philippines regularly experiences a climate substantially affected by rainfall driven by continental scales, such as the El Niño–Southern Oscillation (ENSO), mesoscale systems, and monsoons spanning annually [2]. Monsoons in the Philippines are categorized into two types: Northeast Monsoon (NWM) and Southwest Monsoon (SWM) [3]. The Southwest Monsoon, or habagat, refers to the warm, moist winds that stipulate heavy rainfall emerging from the southwest Indian Ocean from late April to early October [4].

The presence of such rain and wind systems brought by SWMs heavily affects the Philippines [5, 6]. Since the amount of rain varies annually, its patterns, if there may be any, are deemed relevant to mitigate its effects. Numerous studies have been conducted to analyze these patterns using various climatological instruments, such as the Global Satellite Mapping of Precipitation (GSMaP) and Global Precipitation Measurement (GPM) products [7], the global navigation satellite system (GNSS) [8], and radiosonde [9]. Radiosonde is a balloon-like device used to measure several atmospheric variables such as temperature, relative humidity, pressure, and precipitable water vapor (PWV). Meanwhile, PWV refers to the measure of the entire atmospheric water vapor carried in a vertical air column of a unit area from the Earth's surface to the upper atmosphere. On that note, other studies have explored PWV in investigating short [10] and long-period trends [5]. However, PWV studies within the SWM Season remain underexplored.

Therefore, this study solely focused on the analysis of variations of PWV and rainfall using time series analysis over the Philippines, specifically Puerto Princesa City, Palawan, and Legazpi City, Albay, Philippines, from 2013 to 2022, a comparison between the two cities was investigated to verify if the values between the two cities presented by the analysis had significant differences. Without other parameters like Northeast Monsoon (NWM), characteristics will not be included alongside pollutants, lightning, surface freezing, evapotranspiration, fire activities, and solar radiation. Other impacts of other natural disasters, such as tropical cyclones, on the SWMs and other factors with which an SWM is associated as it travels from different areas of responsibility. PWV data analyzed for the study was obtained from the IGRA, provided by NOAA's National Centers for Environmental Information. The data set investigated was limited from 2013 to 2022 and two weather stations, Puerto Princesa and Legazpi City. It was extracted in Microsoft Excel and arranged chronologically according to the date and time of the recorded PWV. Other variables recorded by the radiosonde, such as relative humidity and pressure, were not considered. Meanwhile, the rainfall data needed was acquired from the GSOD provided by US National Centers for Environmental Information. Limitations to the year and location of the data sets are the same as of the PWV. It was extracted in Microsoft Excel and arranged chronologically according to the date and time of the recorded precipitation and temperature. Other variables recorded, such as wind speed and dew points, were not considered.

## 10.2 Methods

PWV data analyzed for the study was obtained from IGRA, a radiosonde archive. In contrast, the rainfall data needed was acquired from the GSOD, a collection of daily weather data. Both PWV and rainfall data were archived and collected by NOAA. The data sets investigated were limited to 10 years, from 2013 to 2022, and two weather stations, Puerto Princesa and Legazpi City. A time series analysis was employed to determine the periodicity. Due to the data gaps in the PWV, the Lomb-Scargle Periodogram was used. Unlike conventional methods, which rely on a series of linear equations, the Lomb-Scargle periodogram presents an analytical solution to unevenly distributed data [11]. It would aid in figuring out how frequently PWV peaks would appear throughout the station's 10-year data set.

## 10.3 Results and Discussion

### *10.3.1 Monthly PWV Graphs of Puerto Princesa and Legazpi from 2013 to 2022*

Figure 10.1 shows the monthly graphs of Puerto Princesa and Legazpi's PWV from 2013 to 2022 during the SWM Season (June–October). Puerto Princesa peaks the highest during the month of June, the beginning of the SWM Season, with 54.7 mm in June 2019; however, it varies between July, September, and October as well. The lowest peak of the monthly PWV average is during September, with only 46.6 mm. On the other hand, Legazpi often peaks during July 2014, with its highest being 57.1 mm. An interesting behavior in the SWM winds also led August 2016 to have a peak of 56.5 mm. Furthermore, October presented a drastic drop in PWV average during October 2018 with 46.5 mm, making it the station's lowest peak from 2013 to 2022. The beginning and end of the SWM Season are habitually the time when PWV peaks and drops, as observed in the graphs below.

The results demonstrate that the PWV data in Legazpi station is consistently higher than that of Puerto Princesa, despite being located on the opposite side of the natural route of SWM winds. This finding is consistent with a previous study that examined PWV occurrence and variability during the pre-monsoon period, which investigated how PWV behaves during the SWM Season, considering its moisture,

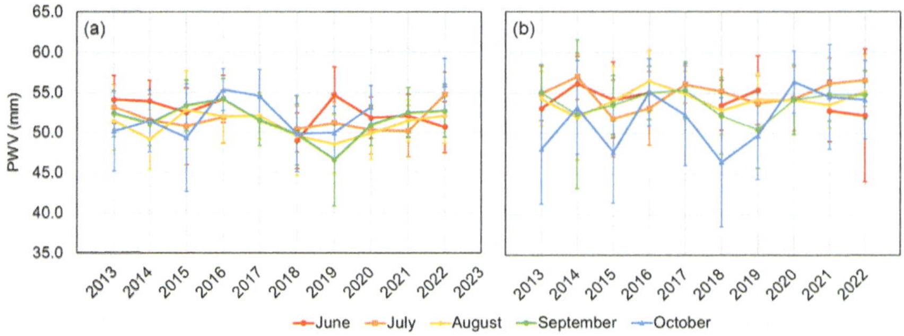


**Fig. 10.1** Monthly PWV of **a** Puerto Princesa and **b** Legazpi from 2013 to 2022

climate, and weather conditions [10]. The correlation between the two station locations has an impact on the comprehensiveness of humidity observations in both radiosonde stations.

The higher PWV values observed in Legazpi compared to Puerto Princesa, despite their contrasting positions in relation to the SWM winds, can be attributed to various factors. One possible explanation is the geographical location and topography of the regions. Legazpi City, located in Albay, is situated on the eastern coast of the Philippines, which faces the Pacific Ocean. This coastal location exposes Legazpi to the direct influence of warm and moist air masses coming from the southwest Indian Ocean during the SWM Season [5]. In contrast, Puerto Princesa City, located in Palawan, is located farther west and may be shielded to some extent from the direct impact of the SWM winds by land masses. This indicates that there is a relationship between PWV and local topography [12]. Regions with significant topographical features, such as coastal areas, can experience enhanced PWV values due to the complex interaction of air masses with the terrain [13]. This could explain why Legazpi, being situated on the coast, exhibits higher PWV values compared to Puerto Princesa, which may have less topographical influence [12].

### *10.3.2 Annual PWV Graphs of Puerto Princesa and Legazpi from 2013 to 2022*

As observed in the graphs presented in Fig. 10.2, the annual PWV mean variation is measured using error bars by the standard deviation of each year. The annual PWV mean variation of Puerto Princesa peaked in 2022 with a 53.2 mm value, while its lowest is the year 2018 with a 49.7 mm value. To further explain the graph, it illustrated that throughout the ten-year course analysis, 2014, 2015, 2018, 2019, 2020, and 2021 have indicated that there are low data sets of PWV with only annual variations ranging only from 49 to 51 mm, while 2013, 2016, 2017, and 2022 has shown concentrated data with PWV ranging from 52 to 53 mm. On the other hand,

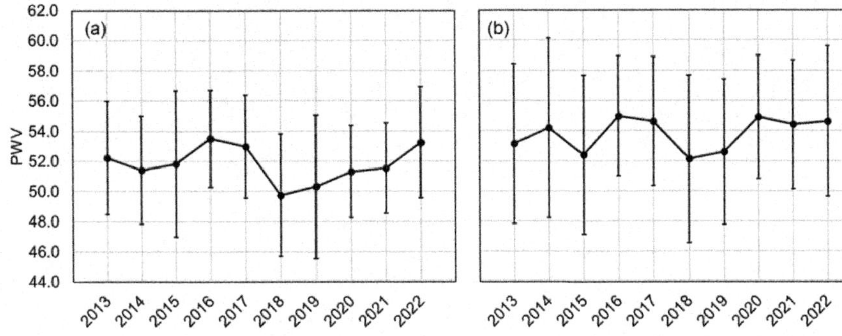


**Fig. 10.2** Annual PWV of **a** Puerto Princesa and **b** Legazpi from 2013 to 2022

as shown in the graphs below, Legazpi had an interesting, consistent PWV mean variation with 2013 having 53.2 mm; 2014, 2017, 2020–2022 having 54 mm; 2015, 2018, and 2019 having 52 mm making it the city's lowest peak and having the highest peak in 2016 with 55.0 mm. The movement between these two cities further proves the influence of the station's respective area, its climate, and weather conditions during the SWM Season.

A similar study conducted in 2020 utilized radiosonde data to analyze PWV over the Philippines [14]. The study presented a graph illustrating the annual rise and fall of PWV values, simplifying the analysis by condensing the data to an average. The erratic rise observed in Puerto Princesa during specific years suggests its susceptibility to weather conditions associated with the SWM. In contrast, Legazpi displays a more stable average, although with a larger margin of error. Puerto Princesa demonstrates a higher average PWV value compared to Legazpi, with a difference of two millimeters, while Puerto Princesa also holds the lowest average at 49.7 mm, and Legazpi maintains an average of 50 mm from 2013 to 2022.

The annual PWV mean variations observed in Puerto Princesa and Legazpi highlight the interplay between climate, weather conditions, and geographical factors influencing PWV values. The fluctuations in PWV values from year to year reflect the dynamic nature of the SWM Season and its impact on moisture patterns. The study about the impact of large-scale atmospheric circulation patterns, such as the Madden–Julian Oscillation (MJO), on PWV variability in the Philippines during the SWM Season [15]. The study found that the MJO, a tropical disturbance that affects atmospheric convection and moisture transport, can modulate PWV values, leading to interannual variations. This suggests that the year-to-year changes in PWV observed in Puerto Princesa and Legazpi may be influenced by the modulation of large-scale atmospheric patterns.

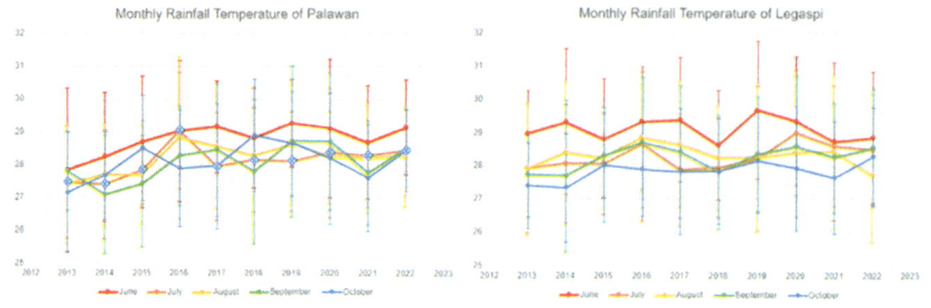


**Fig. 10.3** Monthly temperature of Puerto Princesa and Legazpi from 2013 to 2022

## *10.3.3 Monthly Temperature and Precipitation Graphs of Puerto Princesa and Legazpi from 2013 to 2022*

Figure 10.3 presents the monthly graphs of temperature for Puerto Princesa and Legazpi from 2013 to 2022 during the SWM Season (June-October). Both stations exhibit a similar pattern, with the highest temperature peaks occurring in June 2019 at 29.3 °C for Puerto Princesa and 29.7 °C for Legazpi. The lowest temperature peaks were observed in September 2014 at 27.1 °C for Puerto Princesa and October 2014 at 27.3 °C for Legazpi. The error bars presented also showed a wide range of complexity as each is relatively close to one another.

## *10.3.4 Monthly Rainfall of Legazpi and Puerto Princesa from 2013 to 2022*

Figure 10.4 displays the monthly graphs of rainfall precipitation for Puerto Princesa and Legazpi from 2013 to 2022 during the SWM Season. It reveals an interesting behavior where Puerto Princesa consistently exhibits higher rainfall values compared to Legazpi. In Puerto Princesa, the highest peak occurred in July 2013 at 16.3 mm, followed by a significant drop in subsequent years. Only in October 2017 and 2021 did Puerto Princesa experience another peak, reaching 14.7 mm and 14.1 mm, respectively. In contrast, Legazpi shows more consistent data, with rainfall precipitation ranging from 11.5 mm to 14.0 mm from 2013 to 2022.

The observed differences in rainfall patterns between Puerto Princesa and Legazpi can be attributed to various factors, including geographical location and local weather conditions. Puerto Princesa, located in Palawan, lies in the natural path of the SWM winds and is known for its higher humidity levels and increased precipitation, particularly during the SWM Season. The monthly temperature and rainfall graphs for both Legazpi and Puerto Princesa demonstrate the significant influence of local humidity, weather conditions, geographical placement, and overall climate on the observed variations and the rise and fall of the data sets.

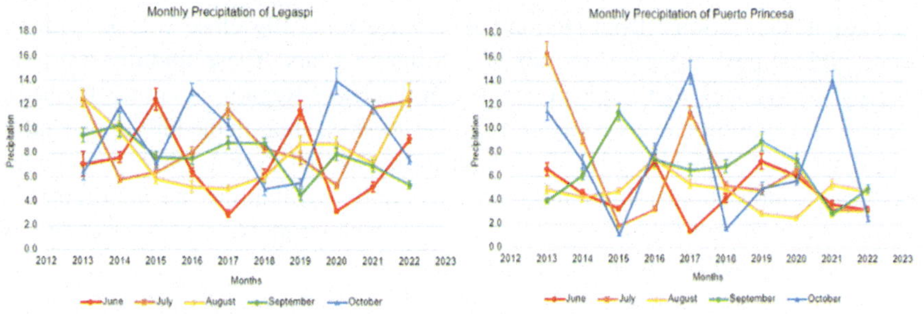


**Fig. 10.4** Monthly precipitation of Puerto Princesa and Legazpi from 2013 to 2022

Spatial and temporal distribution of rainfall in the Philippines during the SWM Season has been identified by previous studies. There is an influence of geographical factors, such as land-sea breeze convergence zones and topographical features, on rainfall patterns [15]. The location of Palawan and proximity to the South China Sea and the Sulu Sea contribute to its higher rainfall values, as it is more exposed to moisture-laden air masses brought by the SWM winds.

### *10.3.5 Annual Rainfall of Legazpi and Puerto Princesa from 2013 to 2022*

Figure 10.5 depicts the annual mean rainfall variation for Legazpi and Puerto Princesa from 2013 to 2022, with error bars as the indication of the standard deviation for each year. The graph illustrates that the annual mean rainfall values for both Legazpi and Puerto Princesa are relatively close to each other, with the same highest peak of 28.7 mm occurring in different years. Legazpi reached its peak in 2016 and 2020 at 28.7 mm, while Puerto Princesa achieved the same peak in 2019. The standard deviations for each year also exhibit proximity, as shown in the graph.

The results regarding the annual mean rainfall temperature demonstrate similar rainfall patterns between Legazpi and Puerto Princesa, despite their distinct

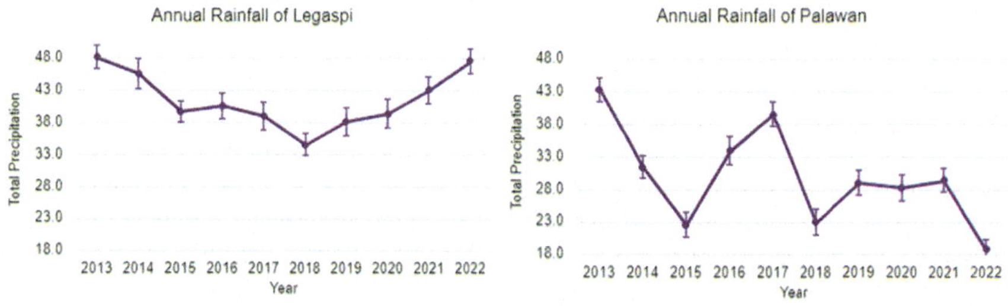


**Fig. 10.5** Annual rainfall of Puerto Princesa and Legazpi from 2013 to 2022

geographical locations. Both stations experience relatively close values, with the same highest peak at 28.7 mm, albeit in different years. These findings are noteworthy, considering that Legazpi, located in the Bicol Region, is known for lower rainfall and humidity levels compared to other western areas in the Philippines influenced by the SWM [12, 13]. The graph indicates that both Legazpi and Puerto Princesa exhibit comparable annual mean rainfall, highlighting the importance of variations in rainfall frequency and intensity, which strongly impact the economy, food security, livelihoods, and political prosperity [4].

The similar annual mean rainfall values observed between Legazpi and Puerto Princesa can be attributed to the influence of the SWM and other large-scale atmospheric circulation patterns affecting rainfall distribution in the Philippines. There is a spatial and temporal distribution of rainfall in the Philippines during the SWM Season. The Intertropical Convergence Zone (ITCZ) and other atmospheric features influenced the rainfall variability. The ITCZ, a band of low pressure near the equator, plays a significant role in controlling rainfall patterns during the SWM. This could explain the similar annual mean rainfall values observed between Legazpi and Puerto Princesa, as they are both affected by the influence of the ITCZ and the SWM [16].

### *10.3.6 Lomb-Scargle Power Spectrum and Fit Periodicity Plot of Legazpi and Puerto Princesa*

Figure 10.6 displays the Lomb-Scargle periodogram results for PWV over Puerto Princesa, Palawan, and Legazpi City, Albay. The analysis revealed apparent periodicities of 4.4 and 6.7 years for Puerto Princesa and Legazpi, respectively. These dominant periods and frequencies were identified using the periodogram, providing insights into the patterns observed during the ten-year period and potential future data sets for Legazpi and Puerto Princesa during the SWM Season. It should be noted that these periodicities may still change as additional years are incorporated into the data set.

The utilization of the Lomb-Scargle periodogram in this study aligns with a global-scale analysis of precipitable water vapor characteristics. This broader study involved over 192 stations and investigated annual, semi-annual, and seasonal periods, validating the efficacy of the Lomb-Scargle method [15]. PWV trends and variations in Laoag, Legazpi, Mactan, and Puerto Princesa from 2012 to 2019 utilized radiosonde data and statistical methods, including the Lomb-Scargle periodogram and Mann–Kendall test [17]. The results aligned with the Lomb-Scargle periodogram findings in this analysis, revealing a downward trend in PWV for Legazpi and Puerto Princesa and detecting periodicities of 4.4 and 6.7 years for Puerto Princesa and Legazpi, respectively. The study also confirmed the seasonal patterns of PWV, with higher levels during the wet months and variations during the dry months. The observed correlation between PWV, surface temperature, and relative humidity supported the

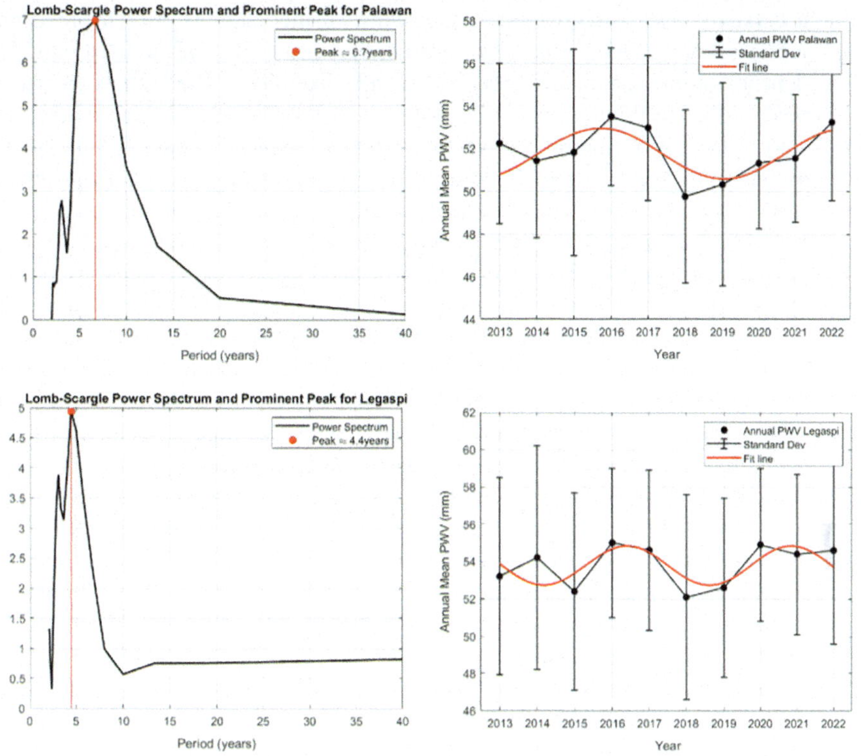


**Fig. 10.6** Lomb-Scargle periodogram (col 1) and fit periodicity plot (col 2) of Puerto Princesa, Palawan (row 1) and Legazpi (row 2)

understanding that higher temperatures and humidity levels contribute to increased PWV [17].

## 10.4 Conclusion and Recommendations

The comparative study between Puerto Princesa City, Palawan, and Legazpi City, Albay, Philippines, showed the evident distinction between these two geographical locations and how the variations of PWV and Rainfall affect their places through extracted recurring data. This research study has officially accomplished its two specific objectives and demonstrated the following results with the researchers' recommendations:

(1) Using IGRA v2 data and GSOD CSV files, the researchers were able to obtain enough data resources and information. The researchers were initially going to use Python. Due to time restrictions, Microsoft Excel functions were more

accessible for the researchers to analyze the weather patterns given the missing IGRA v2 and GSOD data. Therefore, the program that will be used entirely depends on the researcher's preference and capabilities. The highest recorded peak in Puerto Princesa occurred in June 2019 at 54.7 mm, which marks the start of the SWM Season. Conversely, Legazpi frequently peaks in July 2014, with its highest at 57.1 mm. Compared to Legazpi, where the weather was stable from June to October from 2013 to 2022, the monthly PWV for Puerto Princesa has varied in value over time, showing that the region suffers more weather anomalies frequently. Both Puerto Princesa and Legazpi had their highest peaks in June 2019 at 29.3 °C for Palawan and 29.7 °C for Legazpi. This pattern can be seen in the rainfall and precipitation data collected between 2013 and 2022 during the Southwest Monsoon Season. But for Puerto Princesa and Legazpi, their lowest peaks were in September and October of 2014, respectively, at 27.1 °C and 27.3 °C. The researchers recommend comparing outcomes with alternative rainfall data sources, such as satellite precipitation estimates.

(2) As a result of apparent periodicities of 4.4 and 6.7 years through the Lomb-Scargle periodogram, respectively, the Annual PWV values for Puerto Princesa and Legazpi were provided. The researchers concluded that there is a decreasing trend in both cities. They recommend that future researchers apply alternative methods for detecting anomalies, such as those based on machine learning, to identify significant lows and highs.

(3) The researchers highly recommend future studies to scrutinize further the difference in meteorological patterns between Southwest Monsoon (SWM) and Northeast Monsoon (NWM) with close regard to the influence of ENSO-induced rainfall (El Niño and La Niña) on the periodicity, trends, and anomalies of the PWV and precipitation variations and patterns throughout the ten years that heavily impacts the weather climate in the two selected cities, Puerto Princesa and Legazpi City.

## References

1. R. Mina, Philippines looks to improve disaster preparedness with geospatial tech. Mongabay (2021)
2. OECD. Approaches in the Philippines to increased coherence in climate change adaptation and disaster risk reduction in common ground between the Paris agreement and the sendai framework: climate change adaptation and disaster risk reduction (Paris, OECD Publishing 2020)
3. J. Matsumoto, L.M.P. Olaguera, D. Nguyen-Le, H. Kubota, M.Q. Villafuerte II., Climatological seasonal changes of wind and rainfall in the Philippines. Int. J. Climatol. **40**(11), 4843–4857 (2020)
4. PAGASA Climate Advisories [Online]. Available: https://www.pagasa.dost.gov.ph/climate/climate-advisories
5. A.M.S. Desquitado, M.R.R. Perez, R.S.R. Puchero, E.P. Macalalad, A climatological study of typhoons over the Philippine Area of Responsibility from 1989–2018. in *E3S Web of Conferences*, vol. 200 (2020)

6. L.M.P. Olaguera, J. Matsumoto, J.A. Manalo, The contribution of non-tropical cyclone vortices to the rainfall of the Philippines. Int. J. Climatol. **43**(4), 1871–1885 (2023)
7. G. Bagtasa, Assessment of tropical cyclone rainfall from GSMaP and GPM products and their application to analog forecasting in the Philippines. Atmosphere **13**(9), (2022)
8. E.P. Macalalad, R.V. Macalalad, Variation of GNSS-derived precipitable water vapor over Manila in 2017. in *2019 6th International Conference on Space Science and Communication (IconSpace)* (IEEE, 2019), pp. 235–238
9. A.L.S. Domingo, E.P. Macalalad, Multi-year precipitable water vapor observations in East Manila, Philippines using radiosonde and global navigation satellite system. in *E3S Web of Conferences*, vol. 199 (2020)
10. Y. Rivera, K. Capacete, S.K. Rodriguez, A.R. David, E. Macalalad, Variations of precipitable water vapor from 2012 to 2019 over the Philippines using radiosondes. IOP Conf. Ser.: Earth Environ. Sci. **880**(1), 012001, (2021)
11. J.T. VanderPlas, Understanding the lomb–scargle periodogram. Astrophys. J. Suppl. Ser. **236**(1), 16 (2018)
12. M.A. Dela Cruz, S. Nakamura, N. Hanasaki, J. Boulange, Integrated evaluation of changing water resources in an active ecotourism area: the case of Puerto Princesa City, Palawan, Philippines. Sustainability **13**(9), 4826 (2021)
13. H. Kubota, R. Shirooka, J. Matsumoto, E.O. Cayanan, F.D. Hilario, Tropical cyclone influence on the long-term variability of Philippine summer monsoon onset. Prog Earth Planet Sci **4**, 27 (2017)
14. S. J. Fletcher Chapter 14 - Observations in Data Assimilation for the Geosciences: From Theory to Application, Fort Collins, Elsevier, 599–626 (2017).
15. Q. Zhao, Y. Yao, W. Yao, Studies of precipitable water vapour characteristics on a global scale. Int. J. Remote Sens. **40**(1), 1–17 (2019)
16. J.C.H. Chiang, Y. Kushnir, A. Giannini, Deconstructing Atlantic Intertropical Convergence Zone variability: influence of the local cross-equatorial sea surface temperature gradient and remote forcing from the eastern equatorial Pacific. J. Geophys. Res. Atmosp. **107**(D1), ACL-3 (2002)
17. Y. Rivera, K.C. Capacete, S.K. Rodriguez, A.R. David, E. Macalalad, Variations of precipitable water vapor from 2012 to 2019 over the Philippines using radiosondes, IOP Conference Series: Earth and Environmental Science (2021)

# Chapter 11
# Rapid Intensifying Tropical Cyclones in the Philippine Area of Responsibility from the year 1991 to 2020

**Mark Francis James Bona, Uman Buñi, Vincent Cordova, Lance Gabriel Licup, and Ernest P. Macalalad**

**Abstract** Rapidly Intensifying Tropical Cyclones are one of the most destructive natural disasters, especially if it rapidly intensifies while nearing landfall point. The researchers are looking for the trends, and periodicities of Rapidly Intensifying Tropical Cyclones from 1991 to 2020 inside the Philippine Area of Responsibility (PAR). For the trends, the researchers have used the Mann–Kendall test, while for the periodicities, the researchers have used the Lomb-Scargle periodogram. On average, around 18.7 tropical cyclones have been named by the Philippine Atmospheric Geophysical and Astronomical Services Administration yearly from 1991 to 2020, with 2.2 of those rapidly intensifying (30 kt increase in 24 h) tropical cyclones. Rapid intensifying tropical cyclones pose threats to Filipino people, especially those who are exposed to tropical cyclones. There were no significant changes in the trends, while there were approximately 21.4 years of gap in the average peak of rapidly intensifying tropical cyclones in terms of periodicities. From 1991 to 2020, September to October recorded the highest number of tropical cyclones that rapidly intensified.

## 11.1 Introduction

Tropical cyclones are considered as rapidly rotating low-pressure areas over the warm ocean. The favorable environment has likely contributed to the formation of typhoons with the sea surface temperature as high as 30 °C [1]. The region where maximum intensification takes place in the Western North Pacific can be found from 130°E to 150°E, while the initial disturbances latitude is found to be 15°–20°N in

M. F. J. Bona · U. Buñi · V. Cordova · L. G. Licup
Senior High School Department, Mapúa University, Muralla St, Intramuros, Manila, Philippines
e-mail: epmacalalad@mapua.edu.ph

E. P. Macalalad (✉)
Department of Physics, Mapúa University, Muralla St, Intramuros, Manila, Philippines
e-mail: epmacalalad@mapua.edu.ph



M. T. Islam et al. (eds.), *Proceedings of the 8th International Conference on Space Science and Communication*, Springer Proceedings in Physics 303,
https://doi.org/10.1007/978-981-97-0142-1_11

August while 5°–10° in winter [2]. Over the 95th percentile of the 24 h wind speed changes from 1989 to 2000 resulted in 30 knots increase. This led to the definition of Rapid Intensifying Tropical Cyclones (RITC) [3].

The mean number of tropical cyclones that enter the Philippine Area of Responsibility (PAR) annually is 20, with 9 of them making landfall on the location usually found on the southern part Luzon Island and the eastern part of the Visayas [4]. Landfall is declared when the eye or the tropical cyclone's center hits a piece of land.

The researchers focused on the tropical cyclones that crossed the PAR from 1991 to 2020. Other tropical cyclones that formed outside PAR or outside the study period will not be considered. The researchers will use the JMA best track set to determine the tropical cyclones that enter the PAR; its intensity changes to determine if the tropical cyclones rapidly intensify during the said study period. Due to the scope of our study, tropical cyclones that rapidly intensified outside the area of responsibility, though they entered the PAR, will be ignored. Additionally, only typhoons would be considered due to the Japan Meteorological Agency not providing tropical cyclone strength during tropical depression or the category below.

When it comes to rapid intensification in the Northwestern Pacific, tropical cyclones can have more than one rapid intensification within their lifespan. On average, five rapid intensifications occur each year. And the strong rapid intensification episodes have increased by 0.25 events $yr^{-1}$. However, its long-term trends remain insignificant [5]. One of the studies conducted due to rapid intensification was Typhoon Hagibis. It was noted that it moves toward the highest Tropical Cyclone Heat Potential (TCHP). Despite the increasing TCHP, there was no impact when it came to the tropical cyclone track [6].

This research studies the tropical cyclone that rapidly intensifies inside the Philippine Area of Responsibility. Tropical cyclones that rapidly intensify, especially those that make landfall, threaten a human's life. More studies are needed to have more understanding when it comes to rapid intensification.

## 11.2 Methodology

The best track data set obtained through the JMA website contains the tropical cyclone names, latitude, longitude, pressure, time, and intensity in terms of knots. The researchers only considered the tropical cyclones that move toward PAR (Fig. 11.1) either rapidly intensifies or not.

The Japan Meteorological Agency (JMA) uses a 10-min average to determine the maximum sustained winds. Other information used includes the center position, central pressure, accuracy of the center position, maximum wind gust, wind radii of 30 and 50 knots, wind direction and speed, and a forecast cone. Tropical cyclone information is issued when a low storm intensity or higher exists in the coverage area of tropical cyclone is expected to evolve into tropical storm intensity within the next 24 h.

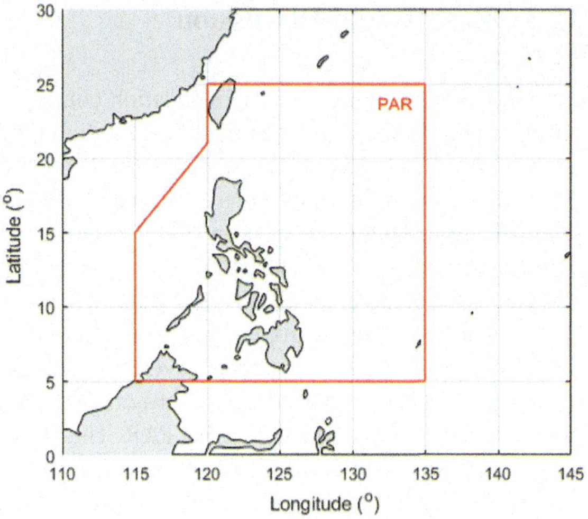


**Fig. 11.1** The Philippine area of responsibility (PAR)

Tropical cyclones that remain outside the PAR are not considered. The researchers used a 10-min wind scale prescribed by the World Meteorological Organization for the intensity changes. With the information provided, tropical cyclones from 1991 to 2020 are collected, and the tropical cyclone intensity and position are utilized during the intensification phase. The researchers used the JMA data to see a tropical cyclone's position and intensity. This information would be processed so that the researchers can determine if a tropical cyclone rapidly intensified or not and if it entered the PAR. The researchers determined if the tropical cyclone rapidly intensifies by calculating its intensity changes within 24 h. Rapid intensification is determined when the tropical cyclone has intensified by at least 30 knots within 24 h.

The researchers determined the number of storms that rapidly intensify inside the PAR. PAR is considered the smallest and innermost boundary closest to the Philippine Islands in which tropical cyclones inside the area warrant the issuance of a Severe Weather Bulletin. It determined the percentages of tropical cyclones that rapidly intensify versus those that entered the PAR but did not rapidly intensify. The Mann–Kendall (MK) test is used to determine the overall trend of the data set. It determines if the variable of interest has a monotonic upward or decreasing trend across time. The term "monotonic upward (downward) trend" refers to a variable that consistently increases (decreases) with time, while the trend may or may not be linear. The Lomb-Scargle periodogram is used to calculate the periodicity of the occurrence of RI typhoons that enter PAR [6]. Lomb-Scargle periodogram provides an analytical solution compared to standard procedures that require a set of linear equations [7, 8]. It is used for period and frequency analysis of time series. The two problems of the periodogram are that it cannot account for measurement errors and statistical fluctuations, which are then solved by Gilliland and Baliunas [9] and Irwin et al. [10]. However, this was fixed by introducing weighted sums, then the mean of the data was subtracted for the analysis [8].

## 11.3 Results and Discussion

Figure 11.2 shows the rapid intensification count yearly from 1991 to 2020, while the red curved line shows the best-fit periodic curve, and the blue line shows the linear trend. 2006 was when the most rapid intensification occurred, while 1997, 2002, 2009, 2018, and 2019 were the years wherein there was no recorded rapid intensification. Though it has been slightly increasing, the Mann–Kendall test shows an insignificant and inconclusive trend. The apparent periodicity is 21.4 years peak. Though Desquitado et al. (2020) state that there was around 12-year dominant periodicity when it comes to tropical cyclone count [11], there are more various effects that affect the intensity of a tropical cyclone, such as ENSO status in which tropical cyclones do tend to form more eastward during El Niño years [11]. The periodicity pertains to the number of RIs entering PAR. This means that based on the current data set, it takes about 21 years for PAR to have the most number of RIs in a year. Thus, it is recommended that additional years be introduced prior and/or after the current data set to check if this trend will hold or not. Compared to La Niña, it could be speculated that the typhoons could rapidly intensify outside the PAR before entering [11]. The active phase of Category 4 and 5 tropical cyclones is assassinated with the eastward shift of tropical cyclone formation locations, which corresponds to the warming of sea surface temperature in the tropical region of Central and Eastern Pacific [12]. On average, 2.27 out of 18.73 tropical cyclones have rapidly intensified inside the PAR annually. Regarding the interquartile range, the values range from − 2.75 to 7.25, of which 2 is the median, with no anomalies being found. Overall, there were five years where no rapid intensification occurred, four years where one tropical cyclone rapidly intensified, nine years where two tropical cyclones rapidly intensified, five years where three tropical cyclone rapidly intensified, five years where four tropical cyclone rapidly intensified, and one occurrence where 5 and 6 tropical cyclones have rapidly intensified. Song et al. (2020) study results also yield the same result as this study in terms of the trend, which results in a slight increase but an insignificant trend [15]. An insignificant trend pertains that there have been no significant changes in terms of the trends. As such, it is not considered a sign from a statistical standpoint [13, 14].

Figure 11.3 shows the percentage of RITC relative to the total number of tropical cyclones that entered PAR. It shows that the highest percentage (38%) of tropical cyclones that entered PAR is in 2007, involving 13 tropical cyclones, with 5 of them rapidly intensifying. In comparison, the most amount in terms of the number of typhoons that entered PAR was the year 2004, which had a 16% in terms of tropical cyclones that rapidly intensified [11]. The linear graph for the rapid intensification percentages yields a similar result to the count. The slope is now at positive 0.0825; however, the Mann–Kendall test yields an inconclusive result. However, the apparent periodicity slightly increased to 23.1 years in terms of percentages. The researchers can speculate that the number of tropical cyclones that rapidly intensifies is independent of the percentages of the tropical cyclones.

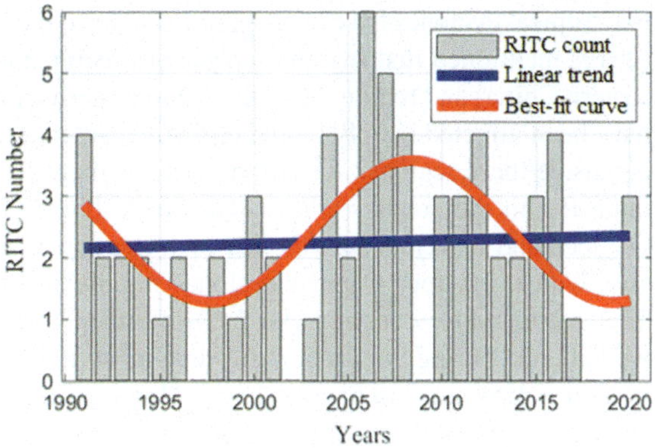


**Fig. 11.2** Annual count of RITC, linear trend, and 12-year period best-fit curve of rapid intensification count from 1991 to 2020

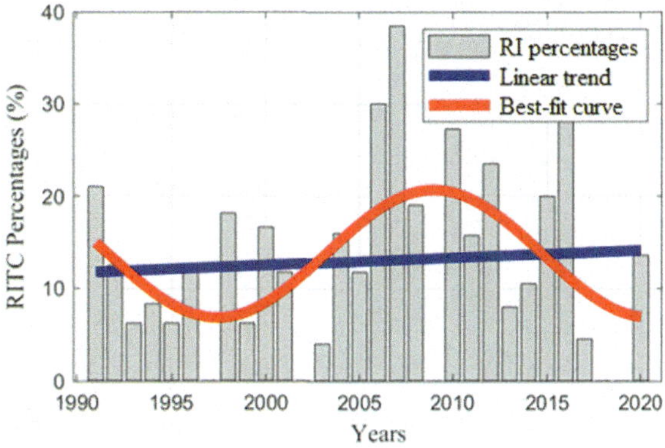


**Fig. 11.3** Annual percentage of RITC, linear trend, and 12-year period best-fit curve of rapid intensification count from 1991 to 2020

Rapid intensifying tropical cyclone tracks for different seasons are shown in Fig. 11.4. Figure 11.4a shows that the season of Dec–Feb has the smallest number of tropical cyclones that rapidly intensified inside the PAR. Tropical cyclones that rapidly intensified during this season are seen to be during December. These tropical cyclones generally moved west to the west-northwest, making landfall in the eastern portion of the Philippines. It supports Gracía-Herrera et al. (2007) where February is part of those months wherein minimum incidence of typhoons occur [16], which can be related to this study where there were no recorded tropical cyclones that rapidly intensified inside the PAR during January to March. As shown in Fig. 11.4b, the season of March, April May, characterized as the Summer Season [11], has six rapid intensifying tropical cyclones with the majority of those occurring during May. The tropical cyclones have a recurving track during this month. Other tropical cyclones that did not recurve have caused significant impacts on land by making close approaches and landfall. During June, July, and August (Fig. 11.4c) wherein most of this season is known to be the Southwest Monsoon season, 26 recorded tropical cyclones have rapidly intensified during the said period of which July has the most recorded at 12, followed by August, which is at nine and lastly, June in which five tropical cyclones have been recorded to have rapidly intensified. Yang and Chen (2019) mentioned that region 1 (Japan, Korean peninsula, and the coast of Northeastern China) and region 2 (Vietnam and eastern Chinese coasts including Taiwan) have been affected mainly by the rapid intensification. While the Philippines which is outside these regions, it is still not safe as there are tropical cyclones that rapidly intensify hit the country [17]. As shown in Fig. 11.4d, the months of September, October, and November are the seasons where most tropical cyclones rapidly intensify, with thirty-four tropical cyclones rapidly intensifying, with fifteen each for September and October and four for November. Yang and Chen (2019) study is also well aligned during September, in which most tropical cyclones that rapidly intensified track move toward regions 1 or 2. However, the Philippines is now the most hit during October and November compared to regions 1 and 2 [17]. Meanwhile, Holliday and Thompson (1979), the peak frequency of significant intensification occurs between September and October [18].

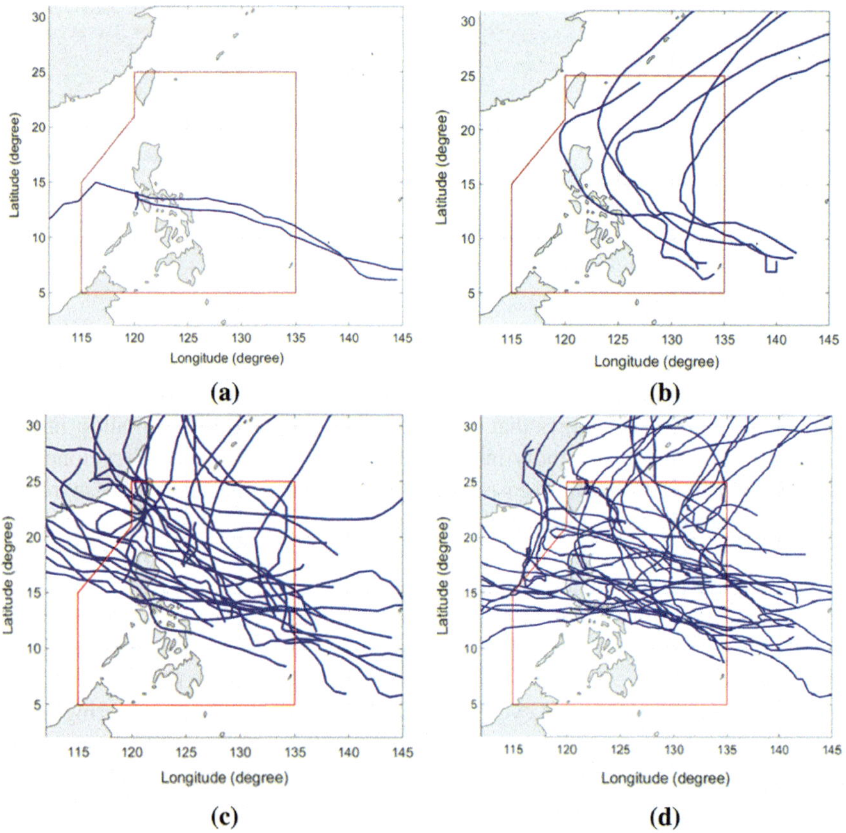


**Fig. 11.4** Rapid intensifying tropical cyclone tracks for **a** Dec–Jan–Feb, **b** Mar–Apr–May, **c** Jun–Jul–Aug, and **d** Sep–Oct–Nov for 1991–2020

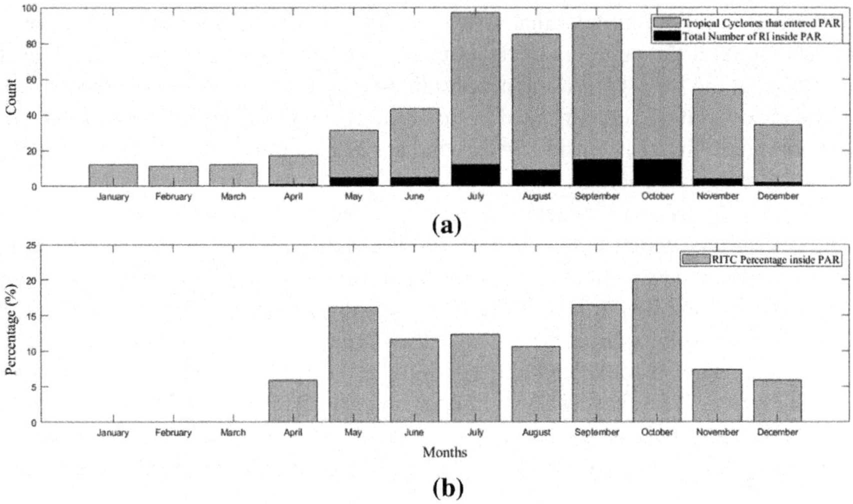


**Fig. 11.5.** **a** Monthly count of total and RI tropical cyclones, and **b** percentage of RI tropical cyclones that entered PAR

It can be seen from Fig. 11.5 that during September and October, it can be found that there was the highest tropical cyclone amount that rapidly intensified inside the PAR. The months of July, September, and October do have the top 3 highest average number of rapid intensifications inside PAR. It is also somewhat similar to Holliday and Thompson study in which, in their study stated that July to November has the highest amount of rapid intensification [17]. However, there was a bit of contrast in November in which there were fewer frequencies in terms of the number of tropical cyclones that rapidly intensified inside PAR. It could mean that those who rapidly intensified during November can be found outside the PAR since Holliday and Thompson study covers the whole Northwestern Pacific [18] in contrast to this study, which covers inside the PAR only.

## 11.4 Summary and Recommendation

In summary, the researchers found that 18.7 tropical cyclones entered the PAR each year from 1991 to 2020 have been named by PAGASA. Of those tropical cyclones, around 2.27 of them rapidly intensify based on the Japan Meteorological Agency Best Track Set. This was different from an estimated five rapid intensification from a 1979 to 2018 study [18]. However, the scope of that study covers the whole Northwestern Pacific, while this study covers tropical cyclones that rapidly intensified inside the Philippine Area of Responsibility. The Mann–Kendall test showed an inconclusive trend, and for the Lomb-Scargle periodogram, it was determined that there was an approximate 21.4 years peak while the rapid intensification percentage had 23.1-year periodicities. It is a reminder that even though this peaks at 21.4 years or 23.1 years depending on if the rapid intensification count or percentage is being tackled, even if you are at the lowest peak, the people should not let their guards down since even at the lowest values, some tropical cyclones do still rapidly intensify and those who did not rapidly intensify inside PAR; wind speed is not the only risk factor when it comes to a tropical cyclone.

The researchers recommend that future researchers continue to research these natural disasters, which would further help the communities and government respond to these such scenarios. Additionally, the researchers also recommend trying the Joint Typhoon Warning Center Best Track Set to see if there are any massive differences since they are using a 1-min wind scale. They are not using the Koba scale in terms of Dvorak values. It could lead to massive differences in rapid intensification count, especially after it finishes its eyewall replacement cycle. The researchers would like to ask future researchers about the impact of rapid intensification in terms of damages, whether agricultural, commercial, etc., to provide additional knowledge for people.

**Acknowledgements** The authors would like to acknowledge the Japan Meteorological Agency for the tropical cycle data set used in this study.

## References

1. L. Oey, Y. Lin, The influence of environments on the intensity change of Typhoon Soudelor. Atmosphere **12**(2), 1–15 (2021)
2. W. Gray, Global View of the origin of tropical disturbances and storms. Mon. Weather Rev. **96**(10), 669–700 (1968)
3. J. Kaplan, M. Demaria, J. Knaff, Large scale characteristics of rapidly intensifying tropical cyclones in the North Atlantic Basin. Weather Forecast. **18**(6), 1093–1108 (2003)
4. J. Lapidez, J. Tablazon, L. Dasallas, L. Gonzalo, K. Cabacaba, M. Ramos, J. Suarez, J. Santiago, A. Lagmay, V. Malano, Identification of storm surge vulnerable areas in the Philippines through the simulation of Typhoon Haiyan-induced storm surge levels over historical tracks. Nat. Hazards Earth Syst. Sci. **15**(7), 1473–1481 (2015)
5. J. Song, Y. Duan, P.J. Klotzbach, Increasing trend in rapid intensification magnitude of tropical cyclones over the western North Pacific. Environ. Res. Lett. **15**(8)
6. A. Wada, J. Chan, Increasing TCHP in the Western North Pacific and its influence on the intensity of FAXAI and HAGIBIS in. SOLA **17A**, 29–32 (2019)
7. Y. Rivera, K. Capacete, S. Rodriguez, A. David, E. Macalalad, Variations of precipitable water vapor from 2012 to 2019 over the Philippines using radiosondes, in *IOP Conference Series: Earth and Environmental Science*, vol. 880 (2021)
8. M. Zechmeister, M. Kürster, The generalised lomb-scargle peridogram a new formalism for the floating-mean and Keplerian peridograms. Astron. Astrophys. **496**(2), 577–584 (2008)
9. R. Gilliland, S. Baliunas, Objective characterization of stellar activity cycles. I. methods and solar cycle analyses. Astrophys. J. **314**(2), 766 (1987)
10. A. Irwin, B. Campbell, C. Morbey, G. Walker, S. Yang, Long-period radial-velocity variations of Arcturus. Publ. Astron. Soc. Pac. **101**, 147–159 (1989)
11. A. Desquitado, M. Perez, R. Puchero, E. Macalalad, A climatological study of typhoons over the Philippine area of responsibility from 1989–2018, in *E3S Web of Conferences*, vol. 200 (2020)

12. P. Hsu, C. Ho, S. Liang, N. Kuo, Impacts of two types of El Niño and La Niña events on typhoon activity. Adv. Meteorol. **2013** (2013)
13. P. Hsu, P. Chu, H. Murakami, X. Zhao, An abrupt decrease in the late-season typhoon activity over the Western North Pacific. J. Clim. **27**(11), 4296–4312 (2014)
14. J. Song, Y. Duan, P. Klotzbach, Increasing trend in rapid intensification magnitude of tropical cyclones over the western North Pacific. Environ. Res. Lett. **15** (2020)
15. X. Wang, C. Wang, L. Zhang, X. Wang, Multidecadal variability of tropical cyclone rapid intensification in the Western North Pacific. J. Clim. **28**(9), 3806–3820 (2015)
16. R. García-Herrera, P. Ribera, E. Hernández, L. Gimeno, Northwest Pacific typhoons documented by the Philippine Jesuits 1566–1900. Climate Dyn. **112** (2007)
17. J. Yang, M. Chen, Landfalls of tropical cyclones with rapid intensification in the Western North Pacific. Nat. Hazards Earth Syst. Sci. Discuss. (2019)
18. C. Holliday, A. Thompson, Climatological characteristics of rapidly intensifying typhoons. Mon. Weather Rev. **107**(8), 1022–1034 (1979)

# Chapter 12
# Mean Solar Quiet Profiling of Geomagnetic Field H-Component for Langkawi Observatory in Solar Cycle-24

**Mohd Helmy Hashim, Mohamad Huzaimy Jusoh, Nurul Shazana Abdul Hamid, Nur Dalila Khirul Ashar, Zahira Mohd Radzi, and Akimasa Yoshikawa**

**Abstract** It is well known that the geomagnetic field exhibits consistent diurnal, seasonal, and solar cycle fluctuations. In this paper, the seasonal variations of the quiet day (Sq) of the geomagnetic field in the recent solar cycle were obtained to characterise the profiling behaviour of the geomagnetic field using the horizontal component (*H-component*) at Langkawi. Long-term data analysis was limited to Magnetic Data Acquisition System (MAGDAS) data from 2011, 2012, 2013, 2016, and 2018. It was discovered that the seasonal variations of the geomagnetic field component during Sq demonstrated Langkawi as an equatorial station where the horizontal component of the geomagnetic field, MSq(H), peaked during the equinox for most of the year. The value is comparable to the predicted value by the model of the International Geomagnetic Reference Field (IGRF) at 41,794 nT. This information

M. H. Hashim · M. H. Jusoh (✉)
School of Electrical Engineering, College of Engineering, Universiti Teknologi MARA, Shah Alam, Selangor, Malaysia
e-mail: huzaimy@uitm.edu.my

N. S. Abdul Hamid
Department of Applied Physics, Faculty of Science and Technology, Universiti Kebangsaan Malaysia, Bangi, Selangor, Malaysia

N. D. Khirul Ashar
Faculty of Engineering, Universiti Putra Malaysia, Serdang, Selangor, Malaysia

College of Engineering, Universiti Teknologi MARA Johor Branch, Masai, Johor, Malaysia

Z. Mohd Radzi
Exploration and Space Science Division, Malaysian Space Agency (MYSA), Banting, Selangor, Malaysia

A. Yoshikawa
International Research Center for Space and Planetary Environmental Science (i-SPES), Kyushu University, Fukuoka, Japan

 
M. T. Islam et al. (eds.), *Proceedings of the 8th International Conference on Space Science and Communication*, Springer Proceedings in Physics 303,
https://doi.org/10.1007/978-981-97-0142-1_12

can be used to improve the accuracy of geomagnetic models and predictions, as well as gain a better understanding of the regular variations of Sq in Langkawi, one of equatorial observatories in Southeast Asia.

## 12.1 Introduction

Geomagnetic studies have emerged as an important field of research, providing a comprehensive understanding of the Earth's magnetic field and its interactions with external factors. The geomagnetic field, which extends from the depths of the Earth's core to outer space, acts as a shield against the influences of a variety of natural phenomena and technological applications. Low solar activity, also known as "solar quiet" or "Sq", provides an opportunity to analyse the geomagnetic field in its natural state. The study of the Sq geomagnetic field has aroused considerable interest and has been the subject of extensive investigation throughout prior solar cycles. Examining the geomagnetic field during quiet solar conditions permits us to identify its intrinsic behaviour and the fundamental physical mechanisms that govern its variations. This knowledge is extremely useful as a baseline for comparison in research into the effects of solar activity on the Earth's geomagnetic field.

To study this phenomenon, researchers have analysed geomagnetic data from observatories around the world, taking into account different geographic and geomagnetic locations. Since the pioneering work of Chapman and Stagg [1] and subsequent investigations by Hibberd [2], the comprehension of the changes in the Sq geomagnetic field has evolved. Pedatella et al. [3] investigated how daily geomagnetic variations are primarily caused by external factors, in particular the currents flowing over the Earth's atmosphere. In this case, ionospheric, plasma density, and solar radiation are considered as indicators of solar variations as described by Joselyn [4], which affect the Sq variations. This could be related to the first observation obtained by Matsushita and Maeda in 1965 [5], in which the properties of the Sq field demonstrated a substantial reliance on season, hemisphere, and longitude. According to the Lloyd season classification [6], there are three distinct seasons associated with the summer or June solstice (May, June, July, and August), the winter or December solstice (November, December, January, and February), and the equinox (September, October, March, and April) [6]. In 2020, Madeeha Talha et al. [7] used Sq field data from the South Asia region to characterise the local geomagnetic field based on seasons.

### *12.1.1 Solar Cycle-24*

The most recently observed solar cycle-24 was one of the weakest ever recorded, which makes it particularly important for this study because it has been reported to be mild [8]. Moreover, the solar cycle was preceded by an exceptionally calm and prolonged solar minimum [9].

### *12.1.2 Langkawi Observatory (LKW)—Equatorial Study*

Langkawi Observatory (LKW) is one of the geomagnetic observers located close to the equatorial region in the Southeast Asian region under the Magnetic Data Acquisition System (MAGDAS) magnetometer chain network, which monitors the daily geomagnetic field near the deep geomagnetic equatorial region [10] (Fig. 12.1). Its location is also close to the geographic (6.30°N, 99.78°E) and magnetic (– 2.32°N, 172.29°E) [11] equator, which is the region where the magnetic field lines are approximately horizontal, *H*-component, and became one of the most significant locations to study the characteristics of the interaction of solar radiation and low-latitude Sun-Earth coupling. The equatorial region significant to Equatorial Electrojet (EEJ) and its unique ionospheric features give a compelling setting to study the variation of the geomagnetic field, especially during the solar quiet, Sq. Therefore, this study aims to explore, characterise, and analyse the geomagnetic field variation of the LKW equatorial Southeast Asian region during Sq in a specific season and to illuminate its implications for space weather and geophysical processes during solar cycle-24.

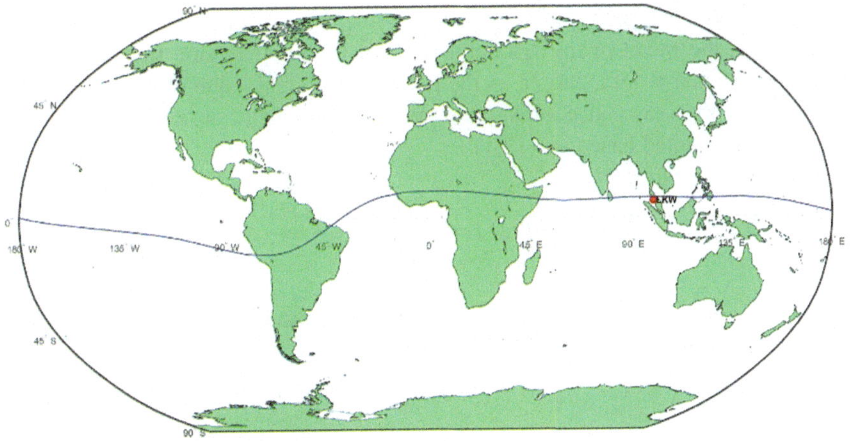


**Fig. 12.1** Location of LKW station

## 12.2 Methodology

### 12.2.1 Solar Cycle Period Determination

The annual sunspot number (SSN) collected from the OMNI online data centre was plotted to determine the solar cycle-24 inclination period. The phases of the solar cycle were then identified throughout the plotted SSN period.

### 12.2.2 Computing the Mean Solar Quiet Day Variations

In this study, geomagnetic field data with a resolution of one second for the *H-component* were used to investigate geomagnetic variation. Due to the fact that the *H-component* contributes significantly to the total geomagnetic field in low-latitude regions, preliminary studies have mainly investigated the geomagnetic variations with a focus on the *H-component* [12]. Data from three seasons comprising the December solstice, equinox, and June solstice in solar cycle-24 were considered for analysis according to the Lloyd season classification (see Fig. 12.2). Five quiet days (QD) for each month were identified and selected according to the World Data Center for Geomagnetism (WDC). The five days were confirmed so that the Magnetic Activity Indices ($K_p$) $\leq 3$ as suggested by Hamid et al. [13], thereby the Eqs. (12.1) and (12.2) were used to calculate the Sq(H) variation.

$$M_{\text{sqm}}(H) = \frac{\sum_{d=1}^{5} H_d}{5} \tag{12.1}$$

$M_{\text{sqm}}(H)$ is the mean of the $H$-components of the geomagnetic field for the five quietest days in a month, and $H_d$ is the geomagnetic field value of each day. To get the mean of the seasonal geomagnetic field $H$-component, an average of all four months is calculated based on (12.1).

$$M\text{Sq}\,(H) = \frac{\sum_{d=1}^{4} M_{\text{sqm}}(H)}{4} \tag{12.2}$$

$M\text{Sq}(H)$ is the mean of the $H$-components of the geomagnetic field for a season.

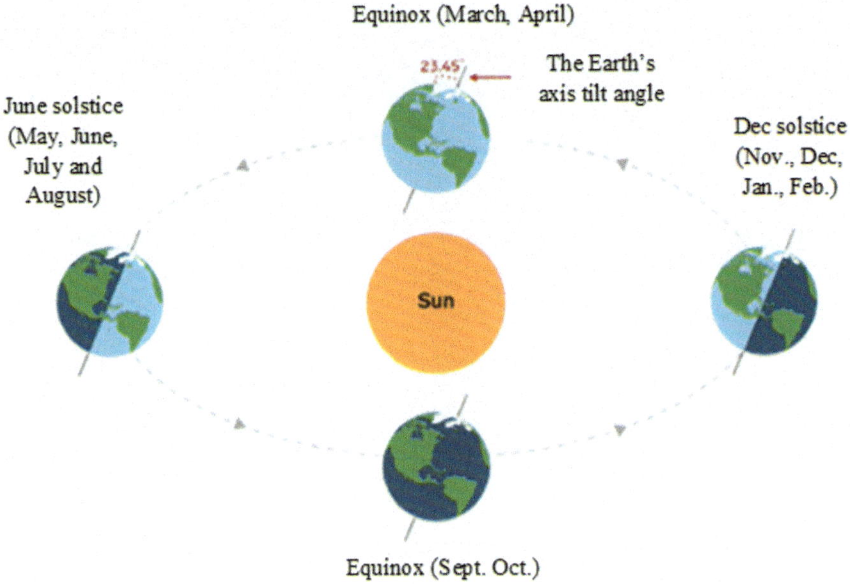


**Fig. 12.2** Earth's equinoxes and solstices based on the Lloyd seasonal classification

### *12.2.3 International Geomagnetic Reference Field—A Comparison*

The International Geomagnetic Reference Field (IGRF) is a set of spherical harmonic coefficients, that provided input for a mathematical model to define the large-scale, time-varying of the Earth's magnetic field between epochs 1900 A.D. and the present. It has been used by the scientific community to study the core of the Earth's field, induction of electromagnetic, lithosphere local magnetic anomaly, space weather, and satellite orientation applications. The IGRF is regularly revised every five years [14]. For this study, the latest version from IGRF, Magnetic Chart (2020.0) for geomagnetic *H-component*, was obtained from WDC [15] and used to compare and validate the MSq(H) over the solar cycle-24 at the LKW station.

## 12.3 Result and Discussion

### *12.3.1 Solar Cycle-24*

Figure 12.3 depicts the annual SSN. It shows that solar cycle-24 began to incline in 2009, reached its first peak in 2011, and finally reached its last peak in 2014. It continued to decline until it reached its deep solar cycle in 2019 (which is considered

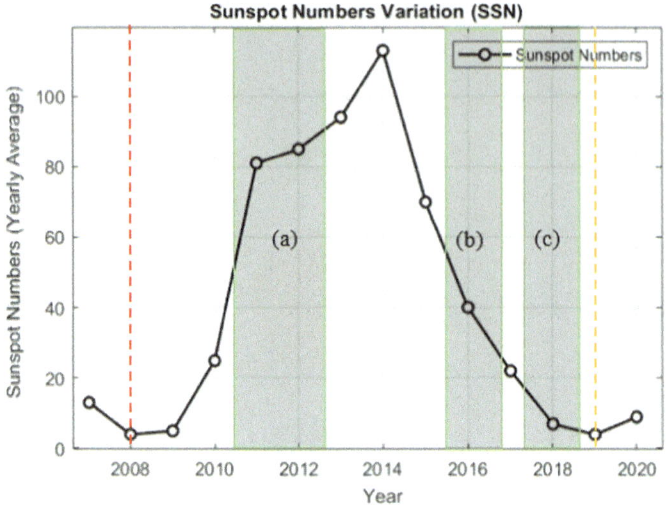


**Fig. 12.3** Sunspot number (SSN) annual average for solar cycle-24

the low point), marking the end of solar cycle-24. The shaded area indicates the data covered in this paper during inclination phase (a), declination phase (b), and minimum phase (c), while the red and yellow dashed lines show the beginning (2008) and the end (2019) phases of solar cycle-24 respectively.

## 12.3.2 Mean Geomagnetic Field Variation MSq(H) During Solar Cycle-24

Seasonal Dependence at LKW. MSq(H) variations for the LKW station during solar cycle-24 were calculated. Despite having limited raw statistical data for the years 2008, 2009, 2010, 2014, 2017, and 2019, it was possible to obtain the average for each season in 2011, 2012, 2013, 2016, and 2018 (Fig. 12.4). For almost every year, LKW station showed the behaviour of equatorial observatories (i.e. MSq(H) equinox maximum). A higher MSq(H) during the equinox corresponds to the conductivities in the ionospheric region that can be observed, which is characterised by the peak value during the equinox as the LKW station is located in the equatorial region.

For the year 2011, analysis shows that the highest peak of the average geomagnetic field occurred during the equinox, approximately at 41,580 nT. On the other hand, the lowest peak occurred during the June Solstice approximately at 41,540 nT. The central peak was observed during the December Solstice, measuring around 41,560 nT. The results are comparable to the year 2012, when the highest peak of the average geomagnetic field was observed during the equinox, with a value of about 41,580 nT. The central peak was observed again during the December Solstice,

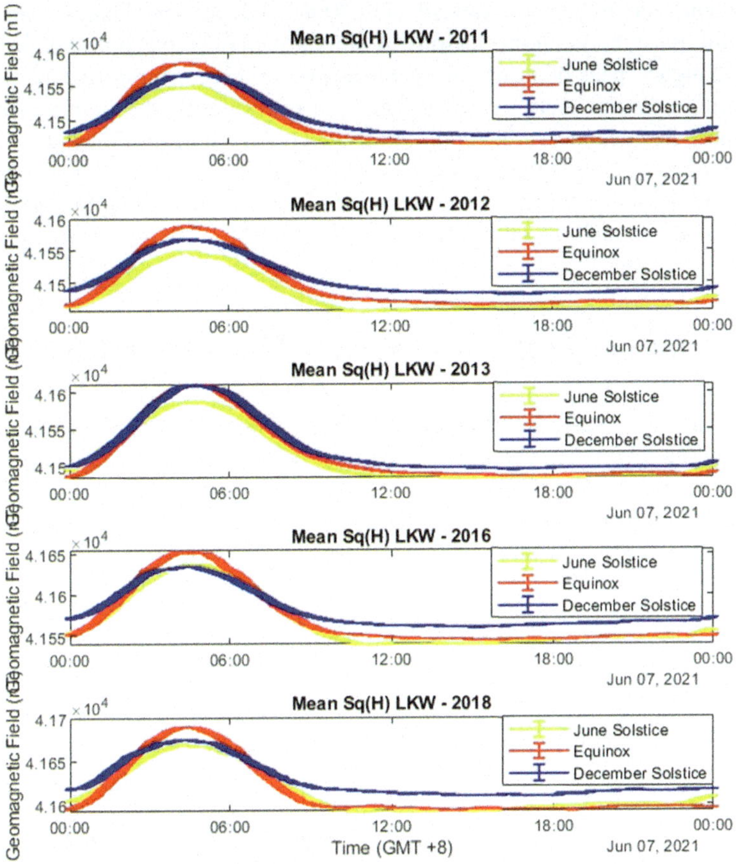


**Fig. 12.4** Annual seasonal MSq(H) of LKW in solar cycle-24

measuring about 41,560 nT, while the lowest peak occurred during the December Solstice, at around 41,540 nT.

Unlike the previous years, in 2013, the highest peak of the average geomagnetic field was observed during the December Solstice, measuring about 41,610 nT, and the middle peak was observed during the Equinox, measuring around 41,600 nT. Although there was just a marginally significant variation between the two seasons, it is a very significant finding to be highlighted here that shows deviations from expected behaviour for equatorial stations. The geomagnetic field during the Equinox was lower than it was during the December Solstice. The lowest peak occurred during the June Solstice at about 41,580 nT.

In 2016, the finding shows the average geomagnetic field peaked at around 41,650 nT during the equinox, which is the pattern back to show its behaviour as an equatorial station. The peaks during the June Solstice and the December Solstice were both found to be approximately 41,630 nT, showing that there was no significant difference between the seasons. A similar pattern emerged in the year 2018, with the highest peak of the average geomagnetic field occurring during the equinox, measuring approximately 41,690 nT, and the peaks during the June Solstice and December Solstice measuring approximately 41,670 nT, again showing no significant difference between the two seasons. This result is in good agreement with those in the earlier studies in which maximum MSq(H) was observed during the equinox at the equatorial and low-latitude regions [16–18].

### 12.3.3 International Geomagnetic Reference Field (IGRF)

As shown in Fig. 12.4, the variation of annual seasonal MSq(H) for LKW in solar cycle-24 (based on data obtained for the respective years) showed an approximate value of 41,400–41,690 nT. Based on IGRF (version 2020.0) from Fig. 12.5, Malaysia is located in the range of 41,794 nT for the *H-component* of the geomagnetic field, confirming the MSq(H) values obtained were in a significant range.

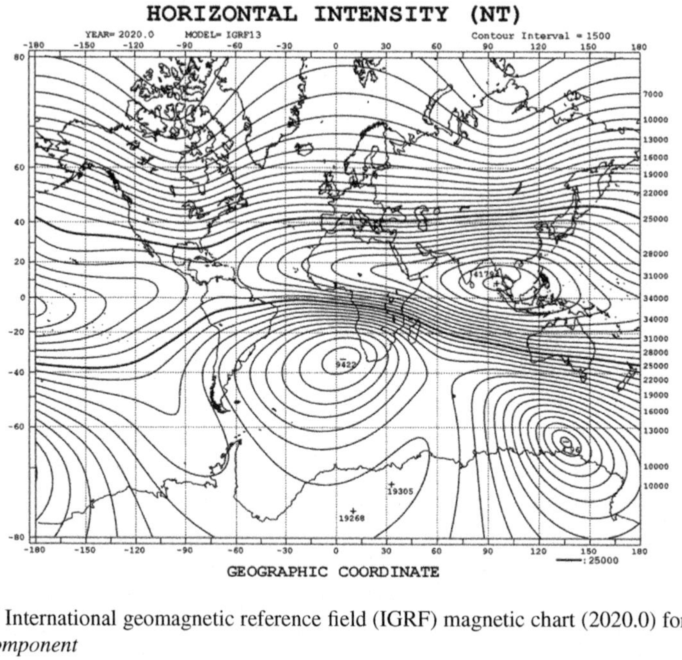


**Fig. 12.5** International geomagnetic reference field (IGRF) magnetic chart (2020.0) for geomagnetic *H-component*

## 12.4 Conclusion

In conclusion, this study discovered a significant relationship between the geomagnetic MSq(H) field and the solar cycle. The seasonal geomagnetic field variation during Sq exhibited by the LKW observatory presents its behaviour as an equatorial station when, almost all of the year, the MSq(H) shows the highest value during the equinox. LKW observatory Sq(H) is approximately at the value of the International Geomagnetic Reference Field (IGRF), 41,794 nT.

**Acknowledgements** The authors are highly grateful to the MAGDAS CPMN group and the Malaysia Space Agency (MYSA) for providing data. The authors are also extremely grateful to Jabatan Perkhidmatan Awam (JPA) for their financial support. Special thanks to the College of Engineering, Universiti Teknologi MARA (UiTM), the members of the Centre for Satellite Communication and Universiti Kebangsaan Malaysia (UKM) for the opportunity to participate in this project.

## References

1. S. Ch, On the variability of the quiet-day diurnal magnetic variation at Eskdalemuir and Greenwich. Proc. R. Soc. London. Ser. A, Contain. Pap. a Math. Phys. Char. **123**(791), 27–53 (1929)
2. F. Hibberd, Day-to-day variability of the sq geomagnetic field variation. Aust. J. Phys. **34**(1), 81 (1981)
3. N.M. Pedatella, J.M. Forbes, A.D. Richmond, Seasonal and longitudinal variations of the solar quiet (Sq) current system during solar minimum determined by CHAMP satellite magnetic field observations. J. Geophys. Res. Sp. Phys. **116**(A4), 4317 (2011)
4. J.A. Joselyn, Geomagnetic quiet day selection (1989)
5. S. Matsushita, H. Maeda, On the geomagnetic lunar daily variation field. J. Geophys. Res. **70**(11), 2559–2578 (1965)
6. E.O. Elemo, M.O. Ehigiator, R. Ehigiator-Irughe, Seasonal variations of the vertical total electron content (VTEC) of the ionosphere at the GNSS cor station (SEERL) UNIBEN and three other cors stations in Nigeria. Niger. J. Technol. **37**(2), 286 (2018)
7. M. Talha, G. Murtaza, J.L. Rasson, N. Ahmed, M. Peerzada, Sq(H) field variations at Sonmiani geomagnetic observatory, Pakistan for solar cycle 24. Adv. Sp. Res. **67**(1), 66–74 (2021)
8. N. Gopalswamy, S. Akiyama, S. Yashiro, H. Xie, P. Makela, G. Michalek, The mild space weather in solar cycle **24** (2015). [Online]. Available: http://arxiv.org/abs/1508.01603
9. S. Basu, The peculiar solar cycle 24-where do we stand?. J. Phys. Conf. Ser. **440**(1), 12001 (2013)
10. Law, Communication, Engineering, Sensing, and Science. ASM Sci. Malaysia Space **12**(2), (2019)
11. W. Nur et al., Comparison of EEJ longitudinal variation from satellite and ground measurements over different solar activity levels (2021)
12. R.G. Rastogi, K.N. Iyer, Quiet day variation of geomagnetic H-field at low latitudes. J. Geomagn. Geoelectr. **28**(6), 461–479 (1976)
13. N.S.A. Hamid et al., Relationship between the equatorial electrojet and global sq currents at the dip equator region. Earth, Planets Sp. **66**(1), (2014)
14. P. Alken et al., International geomagnetic reference field: the thirteenth generation. Earth, Planets Sp. **73**(1), (2021)

15. P. Alken, International geomagnetic reference field IGRF-13. WDC Solid Earth Geophysics (2019). [Online]. Available: https://www.ngdc.noaa.gov/IAGA/vmod/igrf.html
16. Y. Yamazaki, K. Häusler, J.A. Wild, Day-to-day variability of midlatitude ionospheric currents due to magnetospheric and lower atmospheric forcing. J. Geophys. Res. Sp. Phys. Day **121**(7), 7067–7086 (2016)
17. T.N. Obiekezie, F.N. Okeke, Solar quiet day ionospheric source current in the West African region. J. Adv. Res. **4**(3), 303–305 (2013)
18. F.N. Okeke, C.A. Onwumechili, B.A. Rabiu, Day-to-day variability of geomagnetic hourly amplitudes at low latitudes. Geophys. J. Int. **134**(2), 484–500 (1998)

# Chapter 13
# A Compact Multiband Microstrip Antenna Design for 5G IOT and Satellite Communication Applications

**Md. Afzalur Rahman, Samir Salem Al-Bawri, Mohammad Tariqul Islam, Mandeep Jit Singh, Dipon Saha, and Md. Ashraful Haque**


**Abstract** In this study, a microstrip patch antenna was designed to enable complete coverage of satellite communication. The antenna is capable of covering almost four bands, including S, C, Ka, and Q/V bands. The bandwidth achieved for these bands is relatively large, with the maximum bandwidth of 15.6 GHz achieved at the Q/V band and 11.4 GHz achieved at the Ka band. In the microwave range, the S and C bands have a bandwidth of 2.1 GHz, which is considerable. The antenna achieves a maximum gain of 9.1 dBi at the Q/V band. Moreover, the efficiency of the antenna is remarkably high, with almost 99% efficiency achieved at the Ka band and above 90% efficiency achieved for other bands. The proposed antenna also has a good radiation pattern, and it is considered a suitable contender for the satellite communication system and 5G IOT applications.



Md. A. Rahman · S. S. Al-Bawri (✉)
Space Science Centre, Climate Change Institute, Universiti Kebangsaan Malaysia (UKM), 43600 Bangi, Malaysia
e-mail: samir@ukm.edu.my; s.albawri@gmail.com

M. T. Islam · M. J. Singh
Department of Electrical, Electronic and Systems Engineering, Faculty of Engineering and Built Environment, Universiti Kebangsaan Malaysia, UKM, 43600 Bangi, Selangor, Malaysia

D. Saha · Md. A. Haque
Department of Electrical and Electronic Engineering, Universiti Teknologi PETRONAS, Seri Iskandar, 32610 Perak, Malaysia





M. T. Islam et al. (eds.), *Proceedings of the 8th International Conference on Space Science and Communication*, Springer Proceedings in Physics 303,
https://doi.org/10.1007/978-981-97-0142-1_13

## 13.1 Introduction

Satellite communication is crucial for various space applications like Earth observation, deep space exploration, and interplanetary communication. As satellite communication systems advance, there is a growing need for antennas that are smaller in size, more resistant to interference, and capable of multiple functions [1]. Microstrip antennas are becoming increasingly popular in space communication due to their advantages, such as low profile, lightweight, and easy integration on spacecraft. Microstrip antennas have gained widespread acceptance and are increasingly used in various applications such as mobile communication, satellite communication, aviation, radar, and missile systems, replacing other types of antennas [2]. Multiband antennas are widely used in wireless communication and are essential in modern systems. Achieving multiband operation in microstrip antennas for space communication is a topic of great interest as it enables efficient utilization of limited frequency bands in space.

In [1], authors have designed a multiband antenna which is covered only L and S bands, and moreover the size of the antenna is comparatively large. In [3–6], authors have presented some antennas in S, C, or X bands for satellite communication. The main drawback of these antennas is narrow bandwidth. In some paper, authors have applied the MIMO technique for sub-6 GHz frequency bands, which are considered as S or C bands [7–10]. There are a lot of antennas for satellite communication presented only by Ku-band in [11–13]. The size of these antennas is comparatively large, and the bandwidth of these antennas is narrow. In [14], A hexagonal shaped antenna was presented for Millimeter-Wave application, which covered Ka band. The authors designed an antenna, which cover Q/V band for satellite communication in [15].

In this study, a microstrip patch antenna has been presented, which has achieved multiband operation for satellite communication. There are almost four bands covered by this antenna with a large bandwidth. The frequency bands utilized for different satellite services are as follows: S band, operating within the range of 2–4 GHz, is designated for Mobile Satellite Service (MSS). C band, spanning from 4 to 8 GHz, is allocated for Fixed Satellite Service (FSS). Ka band, operating within the range of 26–40 GHz, is also employed for Broadcast Satellite Service (BSS), as stated in reference [16]. The other band is Q/V band (40–60) GHz which is working for Earth observation satellites according to reference [17]. These bands are used in commercial applications in satellite communications [16]. On the other hand, S and Ka bands also work for 5G IOT applications, as stated in [18]. This study explored recent advancements and research trends in microstrip antennas for satellite communication with a focus on achieving multiband operation for satellite communication applications and 5G IOT applications.

## 13.2 Design Methodology

This paper proposes a microstrip patch antenna for a satellite communication system. The dimensional front and back view of the antenna are presented in Fig. 13.1. There is a patch connected with two stubs in the front, and in the back using a partial ground with a metal plate of a perfect electric conductor for increasing the bands. Rogers RT5880(lossy) is used as a substrate with a thickness of 1.6 mm. The total size of the antenna is $15 \times 19 \times 1.6\ \text{mm}^3$. The initial antenna was designed using the following Eqs. (13.1–13.5) [19].

Patch width is calculated by using Eq. (13.1).

$$W = \frac{c_o}{2f_r}\sqrt{\frac{2}{\epsilon_r + 1}} \tag{13.1}$$

Patch length is calculated by using Eqs. (13.2–13.5).

$$L = L_{\text{eff}} - 2\Delta L \tag{13.2}$$

$$L_{\text{eff}} = \frac{c_o}{2f_r\sqrt{\varepsilon_{\text{reff}}}} \tag{13.3}$$

$$\Delta L = 0.412\frac{\left(\frac{W}{h} + 0.264\right)(\varepsilon_{\text{reff}} + 0.3)}{\left(\frac{W}{h} + 0.813\right)(\varepsilon_{\text{reff}} - 0.3)} \tag{13.4}$$

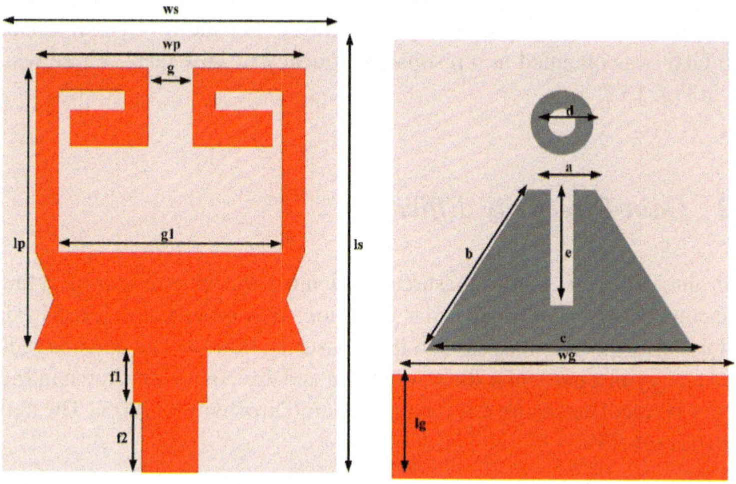


**Fig. 13.1** Front and back view. Where ws = 15, ls = 19, wp = 12, lp = 12.2, $g = 2$, $g1 = 10$, $f1 = 2.3$, $f2 = 3$ lg = 4.5, wg = 15, $a = 3$, $b = 8.32$, $c = 12$, $d = 2.8$, $e = 5$ (in mm)

$$\varepsilon_{\text{reff}} = \frac{\varepsilon_r + 1}{2} + \frac{\varepsilon_r - 1}{2}\left(1 + \frac{12h}{W}\right)^{-0.5} \tag{13.5}$$

where $c_o$ = Electromegnetic wave velocity

$f_{\text{r}}$ = Lowest resonent frequency

$\in_{\text{r}}$ = Dielectric constant of substrate

$L_{\text{eff}}$ = Effective length

$\Delta L$ = Length extension

$\varepsilon_{\text{reff}}$ = Effective dielectric constant.

## 13.3 Result Analysis

### *13.3.1 Reflection Coefficient and Bandwidth*

Reflection coefficient measures reflected energy in a transmission system and is expressed in dB. Bandwidth in antennas refers to the range of frequencies over which an antenna can transmit/receive signals effectively and is measured in Hz. The proposed antenna achieved good reflection coefficient and bandwidth for all six bands. At 3.36 GHz, a reflection coefficient of – 32.8 dB was achieved. The antenna also achieved reflection coefficient of – 34 dB, – 25 dB, and – 25 dB at 29 GHz, 36 GHz, and 46.7 GHz, respectively. For S and C bands, a bandwidth of 2.1 GHz was achieved at 3.36 GHz. At 29 GHz and 36 GHz resonant frequencies for the Ka band, a bandwidth of 11.4 GHz was achieved. For the Q/V band, a bandwidth of 15.6 GHz was obtained at a resonant frequency of 46.7 GHz. These results are shown in Fig. 13.2.

### *13.3.2 Gain-Directivity, Efficiency*

Gain, in antenna, refers to the measure of how much the antenna amplifies the power of an incoming signal in a particular direction. It is usually measured in decibels (dBi). Directivity in antenna refers to the measuring how well an antenna radiates or receives energy in a particular direction. Gain and directivity are shown in Fig. 13.3. Efficiency is calculated using this formula, (Gain/Directivity) * 100%. The maximum efficiency is got 99%.

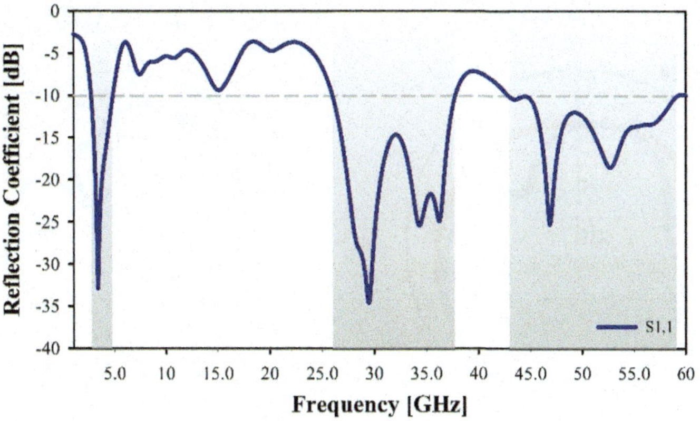


**Fig. 13.2** Reflection coefficient of proposed antenna

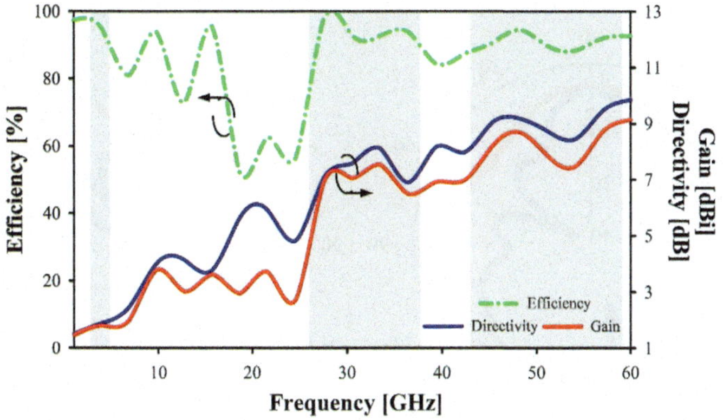


**Fig. 13.3** Gain-directivity and efficiency curve of proposed antenna

### *13.3.3 2D Radiation Pattern*

The radiation pattern of the proposed antenna is shown in Fig. 13.4, indicating the normalized E-Field pattern at resonance frequencies of 3.36, 29, 36, and 46.7 GHz for azimuth angles of phi = 0 and 90. The E-Field pattern at 3.36 GHz has a lobe magnitude of 13.8 dBV/m at phi 0 and 14.4 dBV/m at phi 90. Similarly, at 29 GHz, the magnitude is 15.2 dBV/m at phi 0 and 19.4 dBV/m at phi 90. At 36 GHz, the magnitude is 19.2 dBV/m at phi 0 and 20.4 dBV/m at phi 90. Finally, at 46.7 GHz, the magnitude is 13.4 dBV/m at phi 0 and 22.8 dBV/m at phi 90. These results indicate that the proposed antenna radiation performs exceptionally well, making it suitable for satellite communication applications.

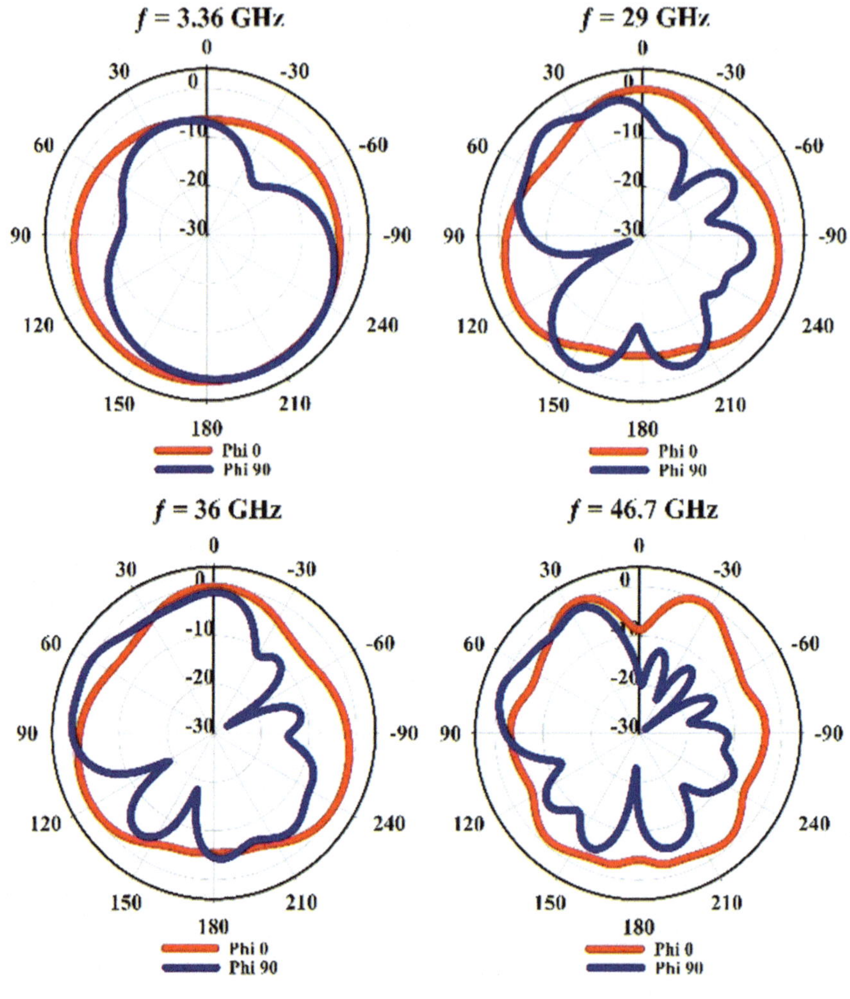


**Fig. 13.4** 2D radiation pattern for $\varphi = 0$ and $\varphi = 90$ in E-Field

## 13.4 Conclusion

A microstrip patch antenna design has been proposed in this research paper, utilizing the CST Studio Suite Software. The antenna demonstrates a compact structure with a low profile and high space utilization, providing coverage for five multiple bands for satellite communication. The bandwidth for each band is large, enabling high data rates and providing a suitable gain. The directional characteristics of the antenna are also commendable. The proposed design covers the S, C, Ka, and Q/V bands, making it a highly suitable antenna for satellite communication applications. Overall, the proposed microstrip patch antenna design exhibits excellent performance and represents a significant advancement in the field of satellite communication and 5G IOT applications.

**Acknowledgements** This work was supported by Geran Universiti Penyelidikan (GUP), Universiti Kebangsaan Malaysia (UKM), Selangor, Malaysia under grant number GUP-2022-075.

## References

1. C. Zheng, D. Li, X. Yu, P. Chen, A multi-band common-aperture phased array antenna for satellite communication, in *2022 International Conference on Microwave and Millimeter Wave Technology (ICMMT)* (IEEE, Harbin, China, 2022), pp. 1–3. https://doi.org/10.1109/ICMMT55580.2022.10022687
2. M. Sekhar, S.N. Bhavanam, P. Siddaiah, Triple frequency circular patch antenna, in *2014 IEEE International Conference on Computational Intelligence and Computing Research* (IEEE, 2014), pp. 1–3
3. S.S. Al-Bawri, M.T. Islam, M.S. Islam, M.J. Singh, H. Alsaif, Massive metamaterial system-loaded MIMO antenna array for 5G base stations. Sci. Rep. **12**, 14311 (2022)
4. G. Jegan, Multi band microstrip patch antenna for satellite communication, in *Recent Advances in Space Technology Services and Climate Change 2010 (RSTS & CC-2010).* (IEEE, 2010), pp. 153–156
5. P.P. Kumar, R. Avula, C.R. Naidu, N. Santhosi, K.S. Madhuri, L.L. Sri, Space filled hybrid multiband antenna for S–C–X band applications, in *2021 5th International Conference on Computer, Communication and Signal Processing (ICCCSP).* (IEEE, 2021), pp. 131–136
6. S.S. Al-Bawri, M.T. Islam, M.J. Singh, E. Alyan, M. Jusoh, T. Sabapathy, S. Padmanathan, K. Hossain, Broadband sub-6 GHz slot-based MIMO antenna for 5G NR bands mobile applications. J. Phys. Conf. Ser. (IOP Publishing, 2021), p. 012038
7. S. Padmanathan, A.A. Al-Hadi, A.M. Elshirkasi, S.S. Al-Bawri, M.T. Islam, T. Sabapathy, M. Jusoh, P. Akkaraekthalin, P.J. Soh, Compact multiband reconfigurable MIMO antenna for sub-6GHz 5G mobile terminal. IEEE Access **10**, 60241–60252 (2022)
8. S.S. Al-Bawri, M.S. Islam, K.S.B. Sahaq, M.S. Marai, M. Jusoh, T. Sabapathy, S. Padmanathan, M.T. Islam, Multilayer base station antenna at 3.5 GHz for future 5G indoor systems, in *2019 First International Conference of Intelligent Computing and Engineering (ICOICE)* (IEEE, 2019), pp. 1–4
9. M. Rahman, M.S. Islam, M.T. Islam, S.S. Al-Bawri, W.H. Yong, Metamaterial-based compact antenna with defected ground structure for 5G and beyond. Comput Mater Contin. **71**, 2383–2399 (2022)
10. M.A. Haque, M.A. Zakariya, N.S.S. Singh, M.A. Rahman, L.C. Paul, Parametric study of a dual-band quasi-Yagi antenna for LTE application. Bull. Electr. Eng. Inform. **12**, 1513–1522 (2023)
11. M.A. Haque, N. Sarker, N.S. Sawaran Singh, M.A. Rahman, M.N. Hasan, M. Islam, M.A. Zakariya, L.C. Paul, A.H. Sharker, G.E.M. Abro, Dual band antenna design and prediction of resonance frequency using machine learning approaches. Appl. Sci. **12**, 10505 (2022)
12. B. Wahiba, B. Nassima, O. Souad, R.M. Amin, The multi band Ku antenna design for space applications, in *2022 2nd International Conference of Smart Systems and Emerging Technologies (SMARTTECH)* (IEEE, 2022), pp. 226–231
13. M.H. Novak, J.L. Volakis, Ultrawideband antennas for multiband satellite communications at UHF–Ku frequencies. IEEE Trans. Antennas Propag. **63**, 1334–1341 (2015)

14. S.S. Al-Bawri, M.T. Islam, T. Shabbir, G. Muhammad, M.S. Islam, H.Y. Wong, Hexagonal shaped near zero index (NZI) metamaterial based MIMO antenna for millimeter-wave application. IEEE Access **8**, 181003–181013 (2020)
15. J. Ladkani, N. Agarwal, J. Thakkar, A. Jain, D.K. Singh, Space qualified Q/V band receiver for high throughput communication satellites, in *2022 IEEE Microwaves, Antennas, and Propagation Conference (MAPCON)* (IEEE, 2022), pp. 259–263
16. B.R. Elbert, The satellite communication applications handbook. Artech House (2004)
17. F. Cuervo, J. Ebert, M. Schmidt, P.-D. Arapoglou, Q-Band LEO earth observation data downlink: radiowave propagation and system performance. IEEE Access **9**, 165611–165617 (2021). https://doi.org/10.1109/ACCESS.2021.3133390
18. N. Hussain, N. Kim, Integrated microwave and mm-wave MIMO antenna module with 360° pattern diversity for 5G internet of things. IEEE Internet Things J. **9**, 24777–24789 (2022)
19. M.A.K. Jilani, M.A. Hossain, F.Y. Lamia, M.A. Rahman, A.H. Sharker, M.R. Ahsan, Design of 2* 1 patch array antenna for 5G communications systems using mm-wave frequency band, in *2022 IEEE 12th Annual Computing and Communication Workshop and Conference (CCWC)* (IEEE, 2022), pp 0857–0862

# Chapter 14
# Development of Multiband Antenna for Satellite Applications

**Nurul Arbaatun Adawiah Zamanuri, Amira Farhana Azhar, Khairayu Badron, and Norsuzlin Mohd Sahar**

**Abstract** Satellite technology has progressed from the experimental stage to become sophisticated and powerful in a very short period of time. The use of artificial satellites to provide connectivity between various sites on Earth is known as satellite applications. Many studies have been conducted in order to increase the quality of satellite applications. One of the fundamental goals of achieving a high-quality satellite application is to achieve a very high gain. The tremendous gain of single band antennas is well recognized. However, a single band antenna can only use one frequency band at a time. On the other hand, multiband antenna offers the advantages of a single band antenna across multiple bands although they do not always provide the same gain as a single band antenna. As a result, this project is being carried out in order to design multiband antenna that will subsequently enhance the quality of satellite applications. In this paper, a multiband microstrip patch antenna is designed, simulated, fabricated and tested to operate for satellite applications resonating at 9 and 12 GHz. The performance of the simulated and fabricated antenna is assessed in terms of fundamental antenna characteristics. From the analysis and experimental results, it can be seen that the proposed antenna exhibits s11 of − 35.366 dB and − 32.269 dB at 9 GHz and 12 GHz, respectively. It also exhibits a good directivity with value of 8.0855 dBi at 9 GHz and 7.4503 dBi at 12 GHz. The gain is considered acceptable with 7.356 dB at 9 GHz and 6.1065 dB at 12 GHz. Based on the measured results, the fabricated multiband microstrip antenna shows an acceptable s11 value of − 22.13 dB at 9 GHz and − 21.043 dB at 12 GHz.

---

N. A. A. Zamanuri (✉) · N. M. Sahar (✉)
Space Science Centre, Institute of Climate Change (IPI), Universiti Kebangsaan Malaysia (UKM), UKM, 43600 Bangi, Selangor, Malaysia
e-mail: p119274@siswa.ukm.edu.my

N. M. Sahar
e-mail: norsuzlin@ukm.edu.my

A. F. Azhar · K. Badron
Faculty of Engineering, Jalan Gombak, International Islamic University Malaysia(IIUM), Kuala Lumpur, Selangor, Malaysia

 
M. T. Islam et al. (eds.), *Proceedings of the 8th International Conference on Space Science and Communication*, Springer Proceedings in Physics 303,
https://doi.org/10.1007/978-981-97-0142-1_14

## 14.1 Introduction

Satellites are now used for various purposes, with communication and sensing being the most common application. These applications are divided into channels or frequency ranges that antennas will operate in. In this matter, multiband antennas have been of huge assistance to satellite applications, hence enhancing the quality of the overall satellite system. A multiband antenna is specifically designed to operate in multiband frequencies. Its design incorporates different active elements or sections that function in distinct frequency band. The ability to operate in multiple frequency bands at the same time has been the key reason why multiband antennas are adopted for many applications. Additionally, multiband antennas are cost-effective and easier to mount compared to single band antennas. In recent years, researchers have conducted numerous studies focus on how to construct a suitable antenna for satellite development. Owing to that, microstrip patch antennas are claimed to be the finest choice for wireless communication systems due to their unique features [1]. These features include low cost, lightweight design, ease of fabrication, and robustness, allowing them to be easily mounted on any solid surface of a device. Ankita and Verma [2] presented a research study on a microstrip patch antenna with a coaxial feed. This particular microstrip patch antenna demonstrates operation at multiple frequencies, specifically in the X and Ku-band frequencies. However, the antenna's bandwidth is very narrow. The K-band microstrip patch antenna was introduced by Chakma et al. [3]. This antenna is designed to operate in a single frequency band and exhibits low return loss. However, the author did not include any comparisons with fabricated results in their study. Similarly, in the paper by Al Janabi et al. [4], the proposed design also demonstrates low return loss, with values of − 32.98 dB for the Ku-band and 55.029 dB for the Ka-band. However, like the previous study, the authors did not provide any comparisons with fabricated results. Furthermore, in the study by Sharma et al. [5], the authors present a high gain microstrip patch antenna. However, they did not specifically mention the return loss, which is considered a crucial parameter for assessing the effectiveness of an antenna. Microstrip patch antennas, on the other hand, are claimed to be finest choice for wireless communication system due to their unique features [6]. This antenna can be easily integrated into a variety of equipment, including satellites, mobile phones, and even planes [7].

In this paper, a multiband microstrip patch antenna is introduced. The antenna exhibits resonance frequencies at 9 and 12 GHz, allowing for operation in multiple frequency bands. The uniqueness of this design lies in its ability to exhibit multiband frequencies while maintaining good directivity. Furthermore, the simulated measurements are compared with the actual measured results to assess its effectiveness. The aforementioned uniqueness of the proposed design indicates its potential to enhance satellite quality. In this article, a more detailed discussion of the processes involved in designing and modifying the antenna size to achieve the desired outcome will be delved into. Moving forward to the "Results" section, the simulation results obtained using CST-MWS software will be described, as well as the measured results of the fabricated antenna design will be presented. Lastly, in the "Conclusion" section, the performance of the multiband microstrip patch antenna will be summarized, and concluding remarks will be provided.

## 14.2 Methodology

### *14.2.1 Design of Single Band Microstrip Patch Antenna*

The microstrip patch antenna was originally designed and simulated using Computer Simulation Technology (CST) Microwave Studio, a dedicated microwave tool. Figure 14.1 shows the proposed design of the microstrip patch antenna. The antenna design was based on a benchmark paper. For the substrate material, Rogers RT/duroid 5880 was utilized, with a dielectric constant of $\varepsilon r = 2.2$, a loss tangent of $\delta = 0.0009$, and a thickness of $t = 1.6$. In addition, the microstrip patch antenna underwent further modification and optimization through the implementation of a slot configuration, as depicted in Fig. 14.2. The objective was to enhance the reflection coefficient and resonance frequency of the antenna. The design parameters for the microstrip patch antenna are detailed in Table 14.1. The slot modification was performed by conducting a parameter analysis, allowing for fine-tuning and optimization of the antenna's performance. Table 14.1 compares the initial parameters of the microstrip antenna before any optimization and after optimization.

Figure 14.3 provides a front view illustration of the microstrip patch antenna after the optimization process. This optimized design aims to improve the antenna's performance. Additionally, Table 14.2 summarizes all the dimensions associated with the optimized microstrip patch antenna. The table provides a comprehensive overview of the specific dimensions that have been modified and optimized during the design process.

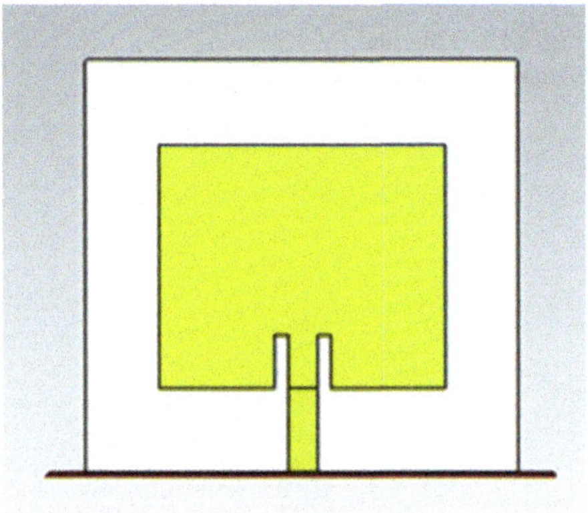

**Fig. 14.1** The microstrip patch antenna without optimization

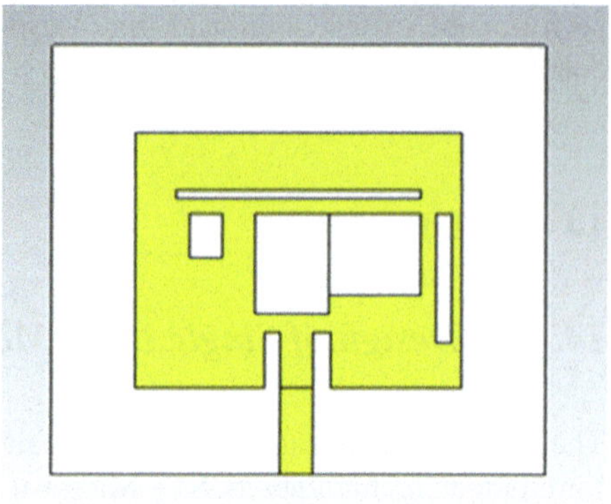

**Fig. 14.2** The microstrip patch antenna with optimization and slot configurations

**Table 14.1** Parameters of the proposed antenna

| Parameters | Microstrip patch antenna without optimization | Microstrip patch antenna with optimization |
|---|---|---|
| Resonant frequency | 9 GHz | 9 and 12 GHz |
| Width of patch | 21.96 mm | 19.9 mm |
| Length of patch | 17.79 mm | 17.79 mm |
| Dielectric material | RT/duroid 5880 | RT/duroid 5880 |
| Dielectric constant | 2.2 | 2.2 |
| Dielectric height | 1.60 mm | 1.60 mm |
| Dielectric length | 30 mm | 30 mm |
| Dielectric width | 30 mm | 30 mm |
| Ground length | 30 mm | 30 mm |
| Ground width | 30 mm | 30 mm |
| Width of feedline | 2 mm | 2 mm |
| Length of feedline | 6.105 mm | 6.105 mm |

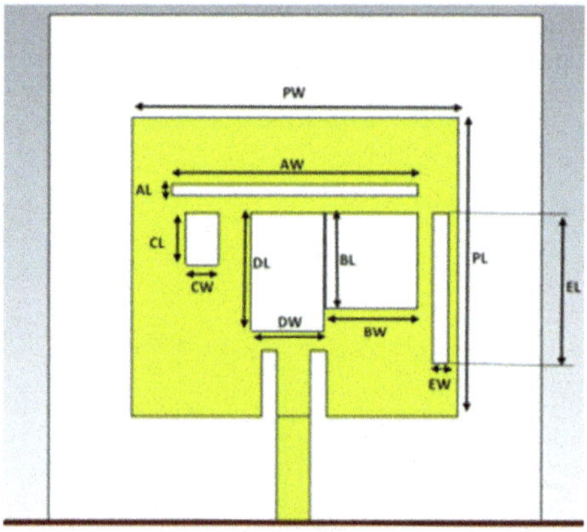


**Fig. 14.3** Labelled parameter of design antenna

**Table 14.2** Parameter dimension of microstrip patch antenna with optimization

| Parameter | Dimension (mm) | Parameter | Font size and style |
|---|---|---|---|
| PW | 19.90 | BL | 5.70 |
| PL | 17.79 | CW | 2.00 |
| AW | 15.00 | CL | 3.10 |
| AL | 0.655 | DW | 4.47 |
| BW | 11.10 | DL | 7.00 |
| EW | 1.00 | EL | 9.00 |

## 14.3 Result and Analysis

In this section, we present the results and analysis of the antenna's performance in terms of various parameters including reflection coefficient (S11), gain, radiation pattern, directivity, and VSWR. We also conduct an analysis comparing different substrate materials based on their reflection coefficients. Furthermore, we compare the simulation results with the measured results to evaluate the accuracy and reliability of the simulation model.

### *14.3.1 Analysis of Reflection Coefficient ($S_{11}$) of Proposed Microstrip Patch Antenna*

The simulated results of the microstrip patch antenna using CST Microwave Studio Software are presented in this section. Figure 14.4 illustrates the resonant frequency of the microstrip patch antenna. To achieve the optimal results, a parameter analysis was performed, wherein one parameter was varied while keeping others constant. This analysis provided insights into the effects of different parameters on the microstrip an. Figure 14.4 specifically displays the simulated return loss of the multiband microstrip antenna, indicating the effectiveness of the design in achieving desired frequency characteristics.

### *14.3.2 Analysis of Gain, Directivity and VSWR*

In order to assess antenna's performance, the performance of the antenna, an analysis was conducted on parameters such as gain, directivity and VSWR. Figure 14.5a presents the results of the gain analysis for the proposed microstrip patch antenna. The antenna exhibits a gain of 7.356 dB at a frequency of 9 GHz, and a gain of 6.1065 dB at a frequency of 12 GHz. These values indicate the amplification capability of the antenna at the respective frequencies. Figure 14.5b indicates the directivity of radiation pattern of antenna in a specified direction. At 9 GHz, the microstrip antenna

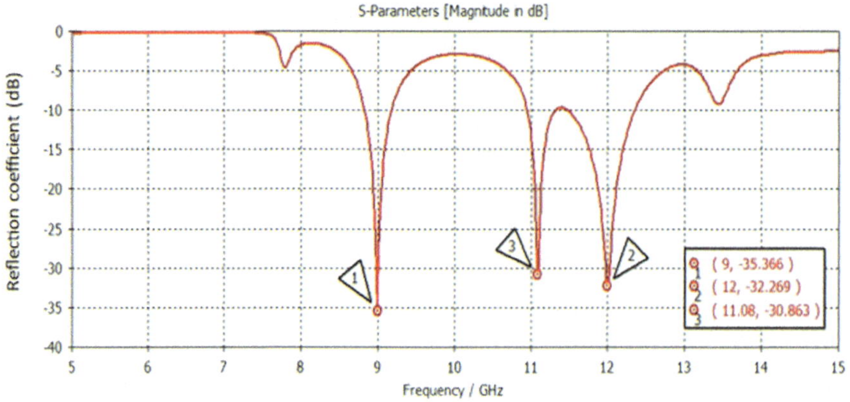


**Fig. 14.4** The reflection coefficient of proposed microstrip patch antenna with optimization

radiated main lobe directivity at 8.0855 dBi, whereas, at 12 GHz, it is observed in Fig. 14.5b that the antenna has a directivity value of 7.4503 dBi. Furthermore, Fig. 14.5c depicts the voltage standing wave ratio (VSWR). Summarization of the antenna performance can be referred in Table 14.3.

### *14.3.3 Measurement of Antenna*

To assess the antenna's performance in real-world applications, the proposed design was fabricated and tested using a Programmable Vector Analyzer (PNA). The objective of this testing was to validate the simulated results, particularly regarding the return loss. The simulation data were compared with the data obtained from the PNA measurements. Figure 14.6a depicts the front and back view of fabricated microstrip patch antenna, while Fig. 14.6b indicates the procedure when doing the measurement. As shown in Fig. 14.7, the fabricated antenna design demonstrated a reflection coefficient of – 22.13 dB at a frequency of 9.077 GHz and – 21.70 dB at a frequency of 12.043 GHz, compared to the simulated results which had reflection coefficients of – 35.366 dB and – 32.269 dB at frequencies of 9.077 GHz and 12.043 GHz, respectively. These results indicate that the fabricated antenna closely aligns with the simulated performance, confirming the accuracy and reliability of the design. Although there are variations between the measurement outcomes and the simulation findings, these differences are unavoidable due to errors introduced during fabrication and factors inherent in the measurement environment.

Table 14.4 provides a summary of recent work on alternative antenna design techniques in the field of microstrip patch antennas. Upon reviewing the table, it is evident that the proposed design stands out with its low return loss and high gain directivity, surpassing other approaches. Additionally, a comparison between simulated and measured results has been conducted to assess the effectiveness of the

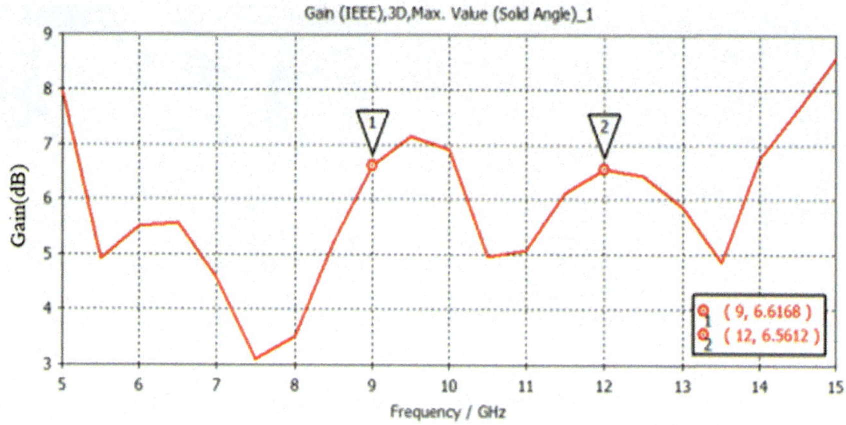


**(a)**

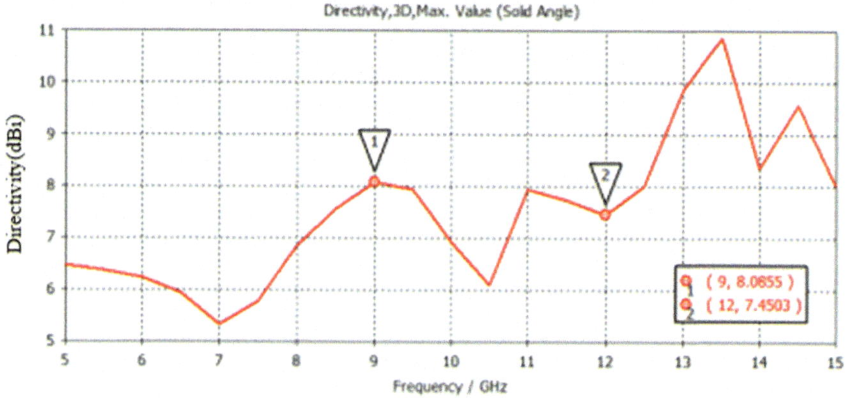


**(b)**

**Fig. 14.5** Analysis of **a** gain **b** directivity

**Table 14.3** Simulated results of the microstrip patch antenna with slot configuration in terms of gain (dB), directivity (dBi) and VSWR

| Parameters | 9 GHz | 12 GHz |
|---|---|---|
| Gain (dB) | 7.356 | 6.1065 |
| Directivity (dBi) | 8.0855 | 7.4503 |
| VSWR | 1.0347 | 1.0499 |

proposed design in real-world scenarios. This analysis helps ascertain the practical performance and viability of the proposed design in practical applications.

(a) (b)

**Fig. 14.6** Experiment setup **a** front and back view of fabricated antenna **b** measurement setup using VNA

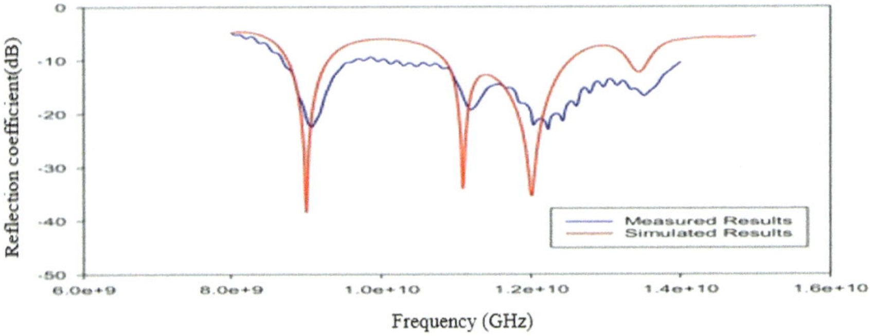


**Fig. 14.7** Comparison for simulated and measured reflection coefficient

**Table 14.4** Comparative analysis of different antenna designs in microstrip patch antenna for satellite applications

| References | Resonance band | Advantage | Disadvantage |
|---|---|---|---|
| Ankita and Verma [2] | X and Ku-band | Wider bandwidth and better impedance matching | Antenna bandwidth is very narrow |
| Chakma et al. [3] | K-band | Low return loss (− 41.013 and 37.408 dB) | No fabrication results |
| Al-Janabi et al. [4] | Ku and Ka-band | Low return loss (− 32.99 and 55.029 dB) | No fabrication results |
| Sharma et al. [5] | C-band | High gain (5, 7.6, 7 dB) | Return loss not specifically mentioned |
| Proposed | X-band | • Low return loss (− 35.37 and − 32.27 dB) and high gain directivity<br>• Make comparison between simulated and measured results | – |

## 14.4 Conclusion

In conclusion, a multiband microstrip patch antenna has been developed for satellite applications operating at 9 and 12 GHz. The performance of the antenna including reflection coefficient, gain, directivity, radiation pattern and VSWR has been thoroughly investigated and discussed. The antenna was first designed and was found not to work at desired frequency. The antenna was then further modified with optimization of antenna parameter and slot configuration in order to increase the quality of the antenna to work at desired resonance frequency. Furthermore, the fabricated antenna was tested to validate its functionality. Based on the result discussed, it can be summarized that the proposed multiband microstrip patch antenna can be an alternative choice for satellite applications.

## References

1. M.T. Islam, N.U. Anika, M.S. Mahmud, M.H. Islam, M.R.H. Abir, M.H. Chy, Design of a microstrip patch antenna for both C-band and X-band applications. in *2nd International Conference on Advanced Information and Communication Technology (ICAICT)*, pp. 7–10 (2020)
2. A. Sood, P. Verma, Design of dual band microstrip patch antenna for satellite communication and radar applications (2016)
3. N. Chakma, M.M. Farhad, A.J. Islam, S. Chakraborty, M.S. Bin Nesar, M.A. Muktadir, Performance analysis of a microstrip patch antenna for satellite communications at K band. in *2018 21st International Conference of Computer and Information Technology (ICCIT)*, pp. 1–4 (2018)
4. F. Al-Janabi, M.J. Singh, A.P.S. Pharwaha, Development of microstrip antenna for satellite application at Ku/Ka band. J. Commun. 118–125 (2021)
5. S.B. Sharma, A. Ugle, K. Parikh, A novel u-slot aperture coupled annular-ring microstrip patch antenna for multiband GNSS applications. in *14th European Conference on Antennas and Propagation (EuCAP)*, pp. 1–3 (2020)
6. P.L. Vijayvergiya, R.K. Panigrahi, Single-layer single-patch dual band antenna for satellite applications. IET Microwaves, Antennas Propag. **11**(5), 664–669. https://doi.org/10.1049/iet-map.2016.0393 (2017)
7. R. Zhi, M. Han, J. Bai, W. Wu, G. Liu, Miniature multiband antenna for WLAN and X-BAND satellite communication applications. Prog. Electromagnet. Res. Lett. **75**, 13–18. https://doi.org/10.2528/pierl18021805 (2018)

# Chapter 15
# U Coupled Rotationally Repeated SRR Shaped Metamaterial Absorber for 5G and Multiband Applications

**Md. Kutub Uddin, Shawon Hossen, Saif Hannan, Badariah Bais, and Mandeep Jit Singh**

**Abstract** We suggest developing and analyzing a rotationally symmetric metamaterial absorber (MMA) in this paper for wide-angle, polarization-insensitive wireless systems. The MMA, which consists of a single-layered electric ring resonator (ERR) with a four-fold rotational symmetry form, is employed to achieve the required absorption. Between 3.5 and 15 GHz, four strong absorption peaks have been noticed. The absorption may be more excellent in the microwave spectrum at 3.5, 5.28, 7.8, and 13.82 GHz, respectively, as high as 99%, 98%, 92%, and 95%. Numerical analysis and electromagnetic responses characterize the wide-angle and polarization insensitivity to co- and cross-polar waves. The unit cell is subwavelength compact because its thickness and size are 0.02λ and 0.22λ at the lowest working frequency. With a mean of 99% absorptivity at 3.5, 5.28, 7.8, and 13.82 GHz, as well as conventional and oblique incidence up to 80° and 180°, accordingly, the absorber showed similar results for cross-polar waves. The absorber can be used for practical amateur 5G, aeronautical radio navigation services, ground-based tactical radar systems, etc.

## 15.1 Introduction

There has been a lot of interest in electromagnetic (EM) metamaterials over the past decade due to their double negative features [1]. It is feasible to achieve desirable resonances at desired frequencies by adjusting the surface electrical response depending

Md. K. Uddin · S. Hossen · S. Hannan
Electronic and Telecommunication Engineering (ETE), International Islamic University Chittagong, Kumira, Chattogram 4318, Bangladesh

B. Bais · M. J. Singh (✉)
Department of Electrical, Electronic and Systems Engineering, Faculty of Engineering and Built Environment, Universiti Kebangsaan Malaysia, 43600 UKM Bangi, Selangor, Malaysia
e-mail: mandeep@ukm.edu.my

 
M. T. Islam et al. (eds.), *Proceedings of the 8th International Conference on Space Science and Communication*, Springer Proceedings in Physics 303,
https://doi.org/10.1007/978-981-97-0142-1_15

on the applied EM wave. Broadband absorptions at operational frequencies have been the subject of extensive study [2–4].

However, they have all grouped parts into the conducting patch and used a multilayer dielectric substrate [5, 6]. These MM absorbers were produced using a commercial or specially designed substrate. While these absorbers have a large effective frequency range and are polarization-insensitive at any incident angle, they are still useless for applications that require a narrower bandwidth, such as sensing, mutual coupling between antennas in a multiband system, beamforming interference between antennas, and so on.

When electromagnetic waves are transmitted through a material, the material is said to be MMA if its permittivity or permeability value is negative [7]. Split rings and their variants frequently maintain these MMAs [8–11]. Numerous designs of split-ring resonators or complementary split-ring resonators have been proposed for use in metamaterial-assisted applications. The split-ring resonator is a commonly used method for generating metamaterial features and absorbers, as shown by previous research [12, 13]. Split rings and variants have, therefore, been utilized to improve the growth of patches [14, 15].

Today's sophisticated antenna technologies use MIMO antennas because they strongly emphasize variables like miniaturization size, low cost, high transfer rates, and broad bandwidth [16–18]. Multiple input multiple outputs (MIMO) devices transmit and receive signals from various angles using antennas. Perfect multi-band absorbers can be employed in antenna arrays to improve beamforming efficiency, reduce sidelobes, and limit mutual contact between the elements [19, 20]. An ideal MM absorber does not change its absorption characteristics under any of the following conditions: normal incidence, oblique incidence, co-polarization, cross-polarization, or rotation [21–23]. Even with wideband absorption, most absorbers produced with a stacked substrate and lumped components inserted into the metallic patch or ground have a bound indent angle range [24, 25]. For these absorbers, the insensitivity to cross-polarization presents a significant challenge. A non-negative result of the refractive index is produced because it is theoretically unattainable to entirely absorb an item with SNG (single negative) or DNG (double negative) measurements of permittivity and permeability [26–28] over the entire absorption band. For this reason, the efficacy of these absorbers in terms of RCS reduction and stealth performance is debatable, as they cannot provide absolute absorption throughout the full spectrum.

This study proposes a simple technique for manufacturing a suitable multiband MM absorber (insensitive to any co- or cross-polarization) in the S, C, and X band microwave systems. The suggested MM absorber used a hybrid CSRR structure, defected ground, and other elements to produce the required absorptions. The great degree of uniformity in the single cell of the designed absorber makes it insensitive to the incidence angles and polarization directions of the incident waveform. This proposed work achieved 99%, 98%, 92%, and 95% at 3.5, 5.28, 7.8, and 13.82 GHz, respectively. Two peaks had absorptivity values of 99 and 98%. The patch structure is suggested to be split-spot symmetric or rotationally symmetric for the ideal MMA to absorb the EM wave entirely [23, 28, 29]. The suggested metamaterial is independent of incidence angle and polarization and exhibits four unique resonance frequencies, each with its optimum absorptions and metamaterial properties. Considering this, 5G and multiband wireless applications are compatible with the suggested absorber.

## 15.2 Unit Cell Design

The requirements for the metamaterial absorber elements, including size, the perfect arrangement for insensitivity to cross-polarization and rotation, and the simple manufacture of the resonator structure, were achieved during the development of the unit cell.

A FR4 substrate plate with a thickness of 1.6 mm and a dielectric constant of 4.4 was used to create the absorber. A 10 mm square measurement substrate was employed to get the required resonance frequency. For the hybrid resonator and the grounding element, annealed copper with a 0.035 mm depth was used on both ends of the substrate. The suggested unit cell has a resonator made up of four square rings. At its base, the substrate is metal-ground. Between the square rings is another square ring, and in the center is another square ring. On rectangle rings, the split gap is just 0.04 mm as opposed to the SSRR's 0.2 mm split spacing. In Table 15.1, unit cell segment and split gap sizes are shown. The metamaterial that makes up the absorber has been designed and modeled using CST based on the frequency integration technique (FIT). Using Eq. (15.1) [23], we have determined the absorption coefficient $A(\omega)$.

$$A(\omega) = 1 - |S_{11}(\omega)|^2 - |S_{21}(\omega)|^2 = 1 - R - T \tag{15.1}$$

where $S_{11}$ (and its subscripts, 1 and 2), respectively, and $S_{21}$ (and its subscripts, 1 and 2) are used for caused reflection and propagation, since the backplane of the absorber layout is created entirely of metal (see Fig. 15.1), its transmittance is now zero ($|S_{21}(\omega)|^2 = 0$). In addition, the impedances of the periodic structure and the grounded lossy substrate are combined and normalized to free space impedance as the total normalized complex impedance ($z = \text{re}(z) + \text{im}(z) \times j$) [30, 31]. To

**Table 15.1** Design considerations for the particular cell

| Parameters | Parameters value (mm) |
|---|---|
| $L_1$ | 10 |
| $L_2$ | 4.25 |
| $L_3$ | 3.25 |
| $L_4$ | 2.25 |
| $U_1$ | 4 |
| $U_2$ | 0.80 |
| r | 1 |

achieve complete absorption, the impedance must be perfectly matched to free space, requiring a reflection coefficient of zero and real and imaginary components of the overall normalized impedance that are close to unity and zero, respectively. Step-by-step designs are shown in Fig. 15.1. $S_{11}$ and $S_{21}$ reflection coefficients [32, 33] can also express the normalized impedance:

$$z = \pm\sqrt{\frac{(1+S_{11})^2 - S_{21}^2}{(1-S_{11})^2 - S_{21}^2}} \tag{15.2}$$

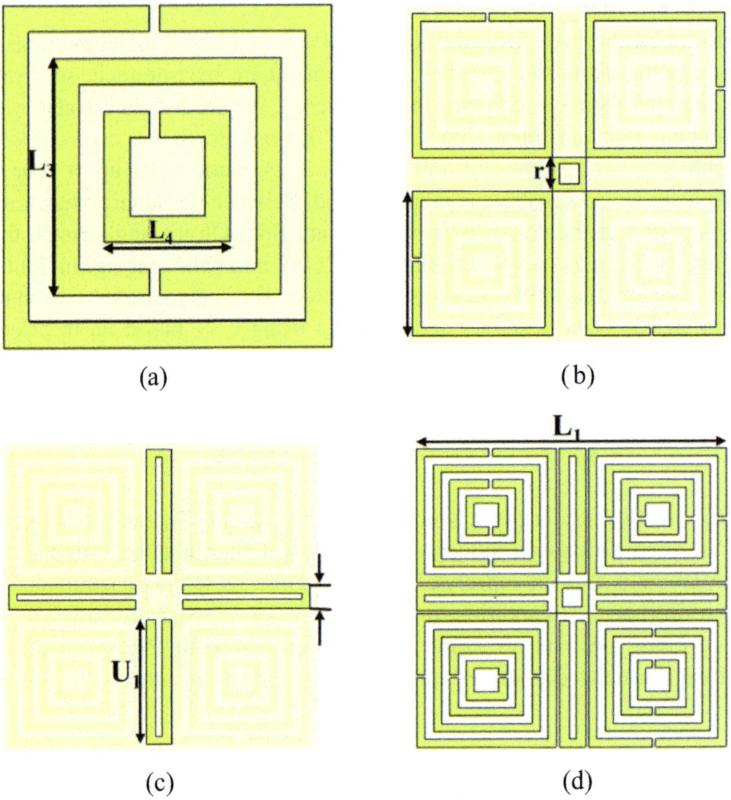


**Fig. 15.1** Geometric view of metamaterial absorber **a** unit cell **b** four outside rings and one little inside ring **c** U shaped resonator **d** overall structure

## 15.3 Theoretical Concept and Result Analysis

The S-Parameters template provided by the CST post-processing module was used to extract material properties [34]. This method gives the permittivity, permeability, and index of refraction of the desired unit cell. A dependable retrieval method is used to retrieve effective parameters in CST. Furthermore, the values of relative permittivity ($\varepsilon_r$) and relative permeability ($\mu_r$) may be determined and represented by the Eqs. (15.3) and (15.4).

$$\varepsilon_r = \frac{c}{j\pi f d} \times \frac{(1 - V_1)}{(1 + V_1)} \tag{15.3}$$

$$\mu_r = \frac{c}{j\pi f d} \times \frac{(1 - V_2)}{(1 + V_2)} \tag{15.4}$$

Here d is the substrate's depth, and c is the speed of light.

A higher frequency region's synthetic magnetism property with negative permittivity provides the refractive index, which describes how an electromagnetic wave is propagated through a material.

$$\text{Refractive index}: \eta = \frac{1}{kd}\{[[e^{\text{inkd}})]'' + 2m\pi] - i[\ln(e^{\text{inkd}})]'\} \tag{15.5}$$

Permittivity ($\varepsilon$), permeability ($\mu$), and refractive index ($\eta$) were the outputs of the CST numerical simulation (see Figs. 15.2 and 15.3) that the transmission coefficient of four resonating frequencies is less than − 10 dB. These frequencies are 3.5, 5.28, 7.8, and 13.82 GHz. The S, C, and Ku bands are included in these spectra. The data were obtained with the robust retrieval strategy within the context of the CST and NRW methodology, implemented in the MATLAB software.

At a lower resonance frequency (3.58 GHz), the values of permeability and permittivity are negative. The outcome of this phenomenon is the manifestation of Single Negative (SNG) metamaterial characteristics inside the lower frequency spectrum. However, as the frequency reaches 5.28 GHz, the material exhibits negative permeability and positive permittivity. As a result, we might call this behavior Single Negative (SNG) behavior. Besides, at 7.8 GHz, it shows DNG behavior. The Drude function may also be used as an explanation for the negative permittivity:

$$\varepsilon(\omega) = 1 - \frac{\omega_p^2}{\omega(\omega + i\Gamma)} \tag{15.6}$$

The $|S_{11}|$ minima and each of the resonances have this pattern. The refractive index graph in Fig. 15.3 shows an almost zero value regarding the resonance frequencies. Bandwidth may be increased using the ENG property [35], while the near-zero refractive index property can be used to improve the gain and directivity of an antenna [36].

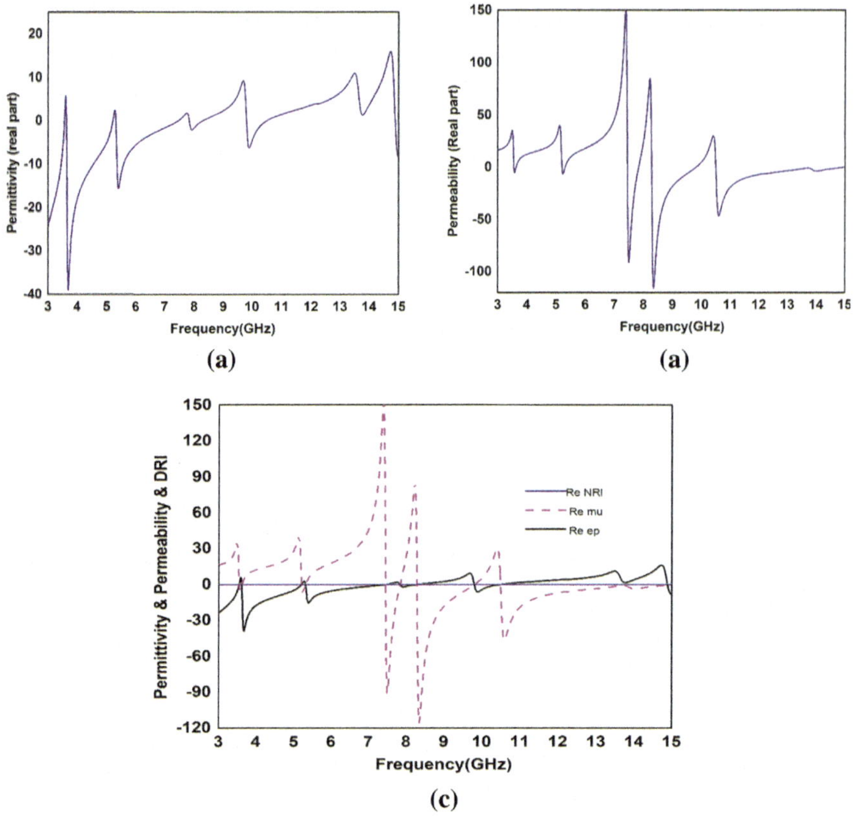


**Fig. 15.2** **a** Permittivity (real) **b** permeability (real) **c** Permittivity, permeability, and DRI

**Fig. 15.3** Refractive index (DRI)

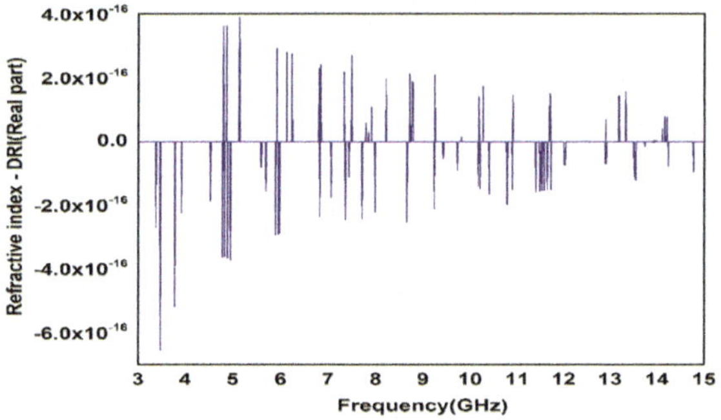

Fig. 15.4 Absorption

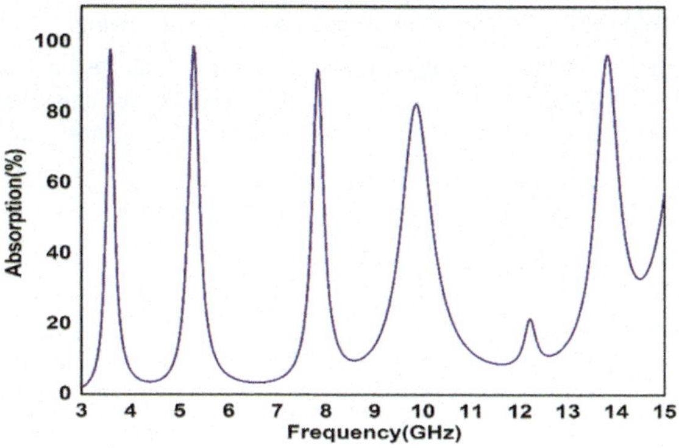


This absorber has achieved 99%, 98%,92%, and 95% at 3.5, 5.28, 7.8, and 13.82 GHz, respectively, as shown in Fig. 15.4.

Figure 15.5a shows the unit cell's small square ring; its four rectangular rings have been identified to have the most surface currents. Rings are responsible for the resonating in the S-band at 3.58 GHz. The inductive action of the metal strips causes the mutual coupling of rectangular rings. This explains why such a strong current flows around the rings' edge. At 5.28 GHz, the surface current is distributed more evenly around the ring, as seen in Fig. 15.5b. The current concentration increases with frequency, as seen in Fig. 15.5c, with a peak at 7.27 GHz across four SSRRs. The current concentration is steadily reducing at the 13.5 GHz rings depicted in Fig. 15.5d. It is also worth noting that the current comes largely from the center ring. When electrical charges travel, a magnetic field is created [37], according to Ampere's law. Therefore, the magnetic field strength in Fig. 15.5's H field rises when charge flow rises.

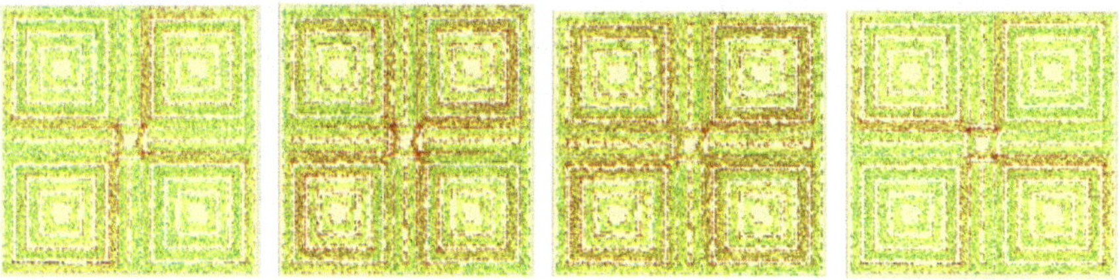

(a) (b) (c) (d)

Fig. 15.5 Dispersion of surface current **a** 3.5 GHz **b** 5.28 GHz **c** 7.8 GHz **d**13.82 GHz

Table 15.2 Existing work compared to proposed work

| Ref | Absorption peaks | Absorption (%) | Unit cell rotational symmetry | Dimension of the unit cell (mm) | Maximum angle of absorption (degree) | Co- & cross-polar absorption | Effective medium ratio (EMR) |
|---|---|---|---|---|---|---|---|
| [38] | 1.54, 3.5, 10.4 | Not mentioned | No | 11 × 11 × 1.6 | Not mentioned | No | 7.8, and 2.63 |
| [39] | 3.5 | 98.5 | No | 10 × 10 × 1.6 | up to 60° | No | Not mentioned |
| [40] | 3.5,5.9 | 99.25 99.4 | No | 17 × 17 × 1.6 | Not mentioned | No | Not mentioned |
| [41] | 2.4, 5 | 2.4, 5 | No | 40 × 40 × 11 | Not mentioned | No | 3.40 |
| **Our work** | **3.5, 5.28, 7.27 and 13.5** | **99, 98, 92 and 95** | **Yes** | **10 × 10 × 1.6** | **Theta = 80° Phi = 180°** | **Yes** | **8.57** |

Bold is used to highlight the proposed work

## 15.4 Comparison with Recent Work

Table 15.2 lists the present MMA's patch design, size, substrate materials, frequency of operation, absorption, and metamaterial properties with a thorough comparison. The paper [38] shows an SNG multiband absorber with a maximum EMR value of 7.8. Therefore, article [39] suggests a metamaterial absorber with 98.5% absorption. In addition, the paper [40] shows a metamaterial absorber with 99% absorptivity, but the size is not compact. However, a flexible microwave absorber with 80% absorption and large size is shown in the paper [41].

The constructed MMA has many potential uses, including in 5G, aeronautical radio navigation services, land-based tactical radar systems, etc. The recommended MMA improved current MMAs and achieved great EMR values and higher absorption spectrum frequencies in a tiny size, making it an attractive replacement for the C and Ku bands.

## 15.5 Conclusion

This paper presents a unique polarization-insensitive rotationally symmetric metamaterial absorber. Due to its resistance to polarization variations and consistent performance when exposed to electromagnetic waves at non-perpendicular angles, this technology has promising potential for integration into everyday wireless environments. The proposed metamaterial absorber (MMA) is insensitive to incident angles regarding normal (up to 80°) and oblique (up to 180°) incidents. With wireless S, C, and Ku band equipment, the suggested MMA can be utilized for various purposes, including 5G mobile service, aeronautical radio navigation services, land-based tactical radar systems, etc.

**Funding Statement** The authors acknowledge the Fundamental Research Grant Scheme (FRGS), grant number FRGS/1/2022/TK07/UKM/01/1 funded by the Ministry of Higher Education (MOHE), Malaysia.

## References

1. S. Hannan, M.T. Islam, M.S. Soliman, M.R.I. Faruque, N. Misran, M.S. Islam (2022) A co-polarization-insensitive metamaterial absorber for 5G n78 mobile devices at 3.5 GHz to reduce the specific absorption rate. Sci. Rep. **12**(1), 11193
2. M.S. Hasan et al., A symmetric plus-shape resonator based dual band perfect metamaterial absorber for Ku band wireless applications: Opt. Mater. (Amst) **143**, 114224 (2023)
3. S. Hannan, M.T. Islam, M.R.I. Faruque, M.E.H. Chowdhury, F. Musharavati, Angle-insensitive co-polarized metamaterial absorber based on equivalent circuit analysis for dual band WiFi applications. Sci. Rep. **11**(1), 13791 (2021)
4. M.K. Uddin et al, Split-ring enclosed K-shaped rotational symmetric metamaterial absorber for multiband wireless applications, in *2023 International Conference on Next-Generation Computing, IoT and Machine Learning (NCIM)* (IEEE, 2023), pp 1–6
5. H. Chen, W.-B. Lu, Z.-G. Liu, J. Zhang, A.-Q. Zhang, B. Wu, Experimental demonstration of microwave absorber using large-area multilayer graphene-based frequency selective surface. IEEE Trans. Microw. Theory Tech. **66**(8), 3807–3816 (2018)
6. S. Hossen, M.K. Uddin, A. Gafur, S. Zahidur Rashid, S. Hannan, A.Z.M. Imran, Design and analysis of an antenna using ring slotted rectangular reflector with double substrate for 5G mm-wave application, in *2023 International Conference on Next-Generation Computing, IoT and Machine Learning (NCIM)* (IEEE, 2023), pp 1–6
7. S.E. Mendhe, Y.P. Kosta, Metamaterial properties and applications. Int. J. Inf. Technol. Knowl. Manage. **4**(1), 85–89 (2011)
8. K. Diest, Numerical methods for metamaterial design. (Springer, 2013)
9. R. Azim, M.T. Islam, A.T. Mobashsher, Design of a dual band-notch UWB slot antenna by means of simple parasitic slits. IEEE Antennas Wirel. Propag. Lett. **12**, 1412–1415 (2013)
10. M.S. Alam, N. Misran, B. Yatim, M.T. Islam, Development of electromagnetic band gap structures in the perspective of microstrip antenna design. Int. J. Antennas Propag. **2013** (2013)
11. T.J. Sheng, M.S. Islam, N. Misran, M.H. Baharuddin, H. Arshad, M.R. Islam, M.E.H. Chowdhury, H. Rmili, M.T. Islam, An internet of things based smart waste management system using LoRa and tensorflow deep learning model. IEEE Access **8**, 148793–148811 (2020)
12. C.M. Watts, X. Liu, W.J. Padilla, Metamaterial electromagnetic wave absorbers. Adv. Mater. **24**(23), OP98-OP120 (2012)
13. M. Samsuzzaman, M.T. Islam, A semicircular shaped super wideband patch antenna with high bandwidth dimension ratio. Microw. Opt. Technol. Lett. **57**(2), 445–452 (2015)
14. S. Hannan, M.T. Islam, S.H.A. Almalki, M.R.I. Faruque, M.S.M, M.S. Islam, Rotational symmetry engineered, polarization and incident angle-insensitive, perfect metamaterial absorber for X and Ku band wireless applications. Sci. Rep. **12**(1), 3740 (2022)
15. A. Rahman, M.T. Islam, M.J. Singh, S. Kibria, M. Akhtaruzzaman, Electromagnetic performances analysis of an ultra-wideband and flexible material antenna in microwave breast imaging: to implement a wearable medical bra. Sci. Rep. **6**(1), 1–11 (2016)

16. S. Hassan Ghadeer et al., An innovative fractal monopole MIMO antenna for modern 5G applications. AEU-Int. J. Electr. Commun. **159**, 154480 (2023)
17. M.T. Islam, M. Cho, M. Samsuzzaman, S. Kibria, Compact antenna for small satellite applications [Antenna Applications Corner]. IEEE Antennas Propag. Magaz. **57**(2), 30–36 (2015)
18. M.Z. Mahmud, M.T. Islam, N. Misran, S. Kibria, M. Samsuzzaman, Microwave imaging for breast tumor detection using uniplanar AMC based CPW-fed microstrip antenna. IEEE Access **6**, 44763–44775 (2018)
19. D. Ashar, A.K.S. yadav, M. Sharma, Design and simulation of compact MIMO antenna for the 5G communication in C-band. J. Phys. Conf. Ser. **2312**(1), 012032 (2022)
20. M. Habib Ullah, M.T. Islam, J.S. Mandeep, N. Misran, A new double L-shaped multiband patch antenna on a polymer resin material substrate. Appl. Phys. A **110**, 199–205 (2013)
21. R.M.H. Bilal, M.A. Baqir, P.K. Choudhury, M.M. Ali, A.A. Rahim, W. Kamal, Polarization-insensitive multiband metamaterial absorber operating in the 5G spectrum. Optik (Stuttg) **216**, 164958 (2020)
22. J.J. Tiang, M.T. Islam, N. Misran, J.S. Mandeep, Slot loaded circular microstrip antenna with meandered slits. J. Electromagn. Waves Appl. **25**(13), 1851–1862 (2011)
23. M. Rokunuzzaman, M. Samsuzzaman, M.T. Islam, Unidirectional wideband 3-D antenna for human head-imaging application. IEEE Antennas Wirel. Propag. Lett. **16**, 169–172 (2016)
24. S. Hannan, M.T. Islam, M.R.I. Faruque, H. Rmili, Polarization-independent perfect metamaterial absorber for C, X and Ku band applications. J. Market. Res. **15**, 3722–3732 (2021)
25. R. Azim, M.T. Islam, N. Misran, S.W. Cheung, Y. Yamada, Planar UWB antenna with multi-slotted ground plane. Microw. Opt. Technol. Lett. **53**(5), 966–968. (2011)
26. S. Hannan et al., A filling-factor engineered, perfect metamaterial absorber for multiple applications at frequencies set by IEEE in C and X bands. J. Market. Res. **19**, 934–946 (2022)
27. S. Hannan, M.T. Islam, N.M. Sahar, K. Mat, M.E. Chowdhury, H. Rmili, Modified-segmented split-ring based polarization and angle-insensitive multiband metamaterial absorber for X, Ku and K band applications. IEEE Access **8**, 144051–144063 (2020)
28. S. Hannan, M.T. Islam, S.H. Almalki, M.R.I. Faruque, M.S. Islam, Rotational symmetry engineered, polarization and incident angle-insensitive, perfect metamaterial absorber for X and Ku band wireless applications. Sci. Rep. **12**(1), 3740 (2022)
29. M. Shahidul Islam, M.T. Islam, A.F. Almutairi, G.K. Beng, N. Misran, N. Amin, Monitoring of the human body signal through the Internet of Things (IoT) based LoRa wireless network system. Appl. Sci. **9**(9), 1884 (2019)
30. S. Hannan, M.T. Islam, A. Hoque, M.J. Singh, A.F. Almutairi, Design of a novel double negative metamaterial absorber atom for Ku and K band applications. Electronics **8**(8), 853 (2019)
31. N.I. Landy, S. Sajuyigbe, J.J. Mock, D.R. Smith, W.J. Padilla, Perfect metamaterial absorber. Phys. Rev. Lett. **100**(20), 207402 (2008)
32. R. Azim, M.T. Islam, J.S. Mandeep, A.T. Mobashsher, A planar circular ring ultra-wideband antenna with dual band-notched characteristics. J. Electromagn. Waves Appl. **26**, 14–15 (2012)
33. F. Costa, S. Genovesi, A. Monorchio, G. Manara, A circuit-based model for the interpretation of perfect metamaterial absorbers. IEEE Trans. Antennas Propag. **61**(3), 1201–1209 (2012)
34. M.T. Islam, M. Samsuzzaman, M.T. Islam, S. Kibria, M.J. Singh, A homogeneous breast phantom measurement system with an improved modified microwave imaging antenna sensor. Sensors **18**(9) (2018)
35. D.M. Pozar, Microwave engineering. John wiley & sons (2011)
36. M. Moniruzzaman et al., Gap coupled symmetric split ring resonator based near zero index ENG metamaterial for gain improvement of monopole antenna. Sci. Rep. **12**(1), 7406 (2022)
37. M.S. Wartak, K.L. Tsakmakidis, O. Hess, Introduction to metamaterials. Phys. Can. **67**(1), 30–34 (2011)
38. M.M. Hasan, M.R.I. Faruque, M.T. Islam, Beam steering of eye shape metamaterial design on dispersive media by FDTD method. Int. J. Numer. Model. Electron. Netw. Dev. Fields **31**(5), e2319 (2018)

39. N. Dhar, M.A. Rahman, M.A. Hossain, Design and exploration of functioning of a DZ shaped SNG multiband metamaterial for L-, S-, and X-bands applications. SN Appl. Sci. **2**(6), 1077 (2020)
40. P. Garg, P. Jain, Analysis of a novel metamaterial absorber using equivalent circuit model operating at 3.5 GHz, in *2020 IEEE International Conference on Computational Electromagnetics (ICCEM)* (IEEE, 2020), pp. 246–247
41. L. Vaz, V. Barros, A. Silva, M.W. Silva, V.P. Magri, T. Ferreira, Chipless RFID tag based on metamaterial absorbers, XL Simpósio Brasileiro de Telecomunicações e Processamento de Sinais (2022)

# Chapter 16
# Labyrinth Maze-Shaped Split Ring EM Metamaterial Absorber for C and X Band Application

**Md. Inzamul Hoque, Neyamul Alam, Md. Kutub Uddin, Saif Hannan, Norsuzlin Mohd Sahar, and Ahasanul Hoque**

**Abstract** This work aims to provide a maze-shaped split ring metamaterial absorber (MMA) with a high absorption rate. The unit cell's resonator's well-engineered split cut assists in obtaining this adequate absorption. This MMA is created using a 10 mm unit cell and an FR4 substrate (permittivity = 4.4). In this MMA, EM waves have been utilized in TEM mode at normal and oblique incidences of a maximum of 80° and 180°, respectively, and the same absorption of 93.70%, 93.32%, 99.96%, and 96.04% at four different resonance frequencies has been gained. In the C and X bands, resonance frequencies of 4.44, 6.24, 8.50, and 11.50 GHz, the design functions as a single negative (SNG) MM absorber. Radar absorption, wireless communication, microwave imaging, and radio astronomy applications are feasible with the suggested metamaterial.

## 16.1 Introduction

For a couple of years, metamaterial (MM) absorbers have become the immersing field for scientists and engineers because of a wide variety of applications in mechanical, optical, acoustic, and electromagnetic [1]. MMs have been shown in every technologically significant spectral area to date, ranging from radio [2, 3], microwave [4, 5],

Md. I. Hoque (✉) · N. Alam · Md. K. Uddin · S. Hannan
Department of Electronic and Telecommunication Engineering, International Islamic University Chittagong (IIUC), Chattogram 4318, Bangladesh
e-mail: inzamulhoqueete.iiuc@gmail.com

S. Hannan
e-mail: sh@iiuc.ac.bd

N. M. Sahar · A. Hoque
Space Science Centre (ANGKSA), Universiti Kebangsaan Malaysia (UKM), 43600 Bangi, Selangor, Malaysia
e-mail: ahasanul@ukm.edu.my



M. T. Islam et al. (eds.), *Proceedings of the 8th International Conference on Space Science and Communication*, Springer Proceedings in Physics 303,
https://doi.org/10.1007/978-981-97-0142-1_16

mm-wave, THz, MIR, and NIR [6, 7] to the near optical [8]. They also help to make antennas smaller so that devices can be modeled that are both small and effective [9].

Metamaterials develop into structures that show negative permeability and permittivity [10] or else when electromagnetic waves pass through them. Both show the refractive index to a negative value [11]. Due to some structural insensitivity to cross-polarization, practically all absorbers are not ideal [12]. For this reason, some common types of resonators, such as split ring resonators [13], complementary split ring resonators, and coupled non-coupled SRRs, are utilized in structural design [1]. These absorbers are capable of highly efficient, narrowband EM wave absorption at some specific frequencies [14].

This paper proposes a simple approach to design a metamaterial absorber in a single substrate layered, rotational, and used maze-shaped split ring resonator as structural design. In the proposed metamaterial absorber, a commercially available FR4 substrate has been used which has a dielectric constant of 4.3 and is 1.6 mm thick. Some perfectly measured split cut variation is given to have an effect on electromagnetic (EM) waves. This specifically designed absorber shows four separate resonance frequencies (4.44, 6.24, 8.50, 11.50 GHz) with maximum absorptions of 99.96%. And when EM waves are applied, both normal and oblique incidence has no effect on the absorber. So here, Radar systems may utilize the 4.4 GHz frequency to minimize object cross-section (RCS). Wireless communication systems can make use of the 6.24 GHz frequency, particularly for enhancing signal reception and lowering interference. Microwave imaging systems can take advantage of 8.54 GHz, and radio astronomy applications can take advantage of 11.50 GHz.

## 16.2 Design of the Unit Cell and Theory of Absorption

The labyrinth maze-shaped split ring structural designed absorber is displayed in Fig. 16.1. Three layers combine to form the absorber, the top layer consists of an annealed copper (thickness 0.035) resonator where a combination of square split ring structure has been designed, the middle layer is a dielectric substrate FR4 ($\varepsilon_1 = 4.3$, $\tan\delta = 0.025$), and a thin layer of copper at the bottom completely covers the bottom surface.

The geometric dimensions of the proposed unit cell are optimized through calculation and simulation repeatedly. The computerized simulation was performed with the use of Computer Simulation Technology (CST) software. The unit cell is a specific size of 10 mm × 10 mm. In this design, there are eight square split ring resonators where we used a rotational split cut of 0.20 mm in every resonator. But only the resonator in the middle has no cut on it. There is a gap of 0.50 mm between the two resonators. The geometric parameters of the top layer resonators are also displayed in Fig. 16.1.

At various resonance frequencies, these split ring resonators operate simultaneously. In the absence of any transmission or reflection through the absorber, the

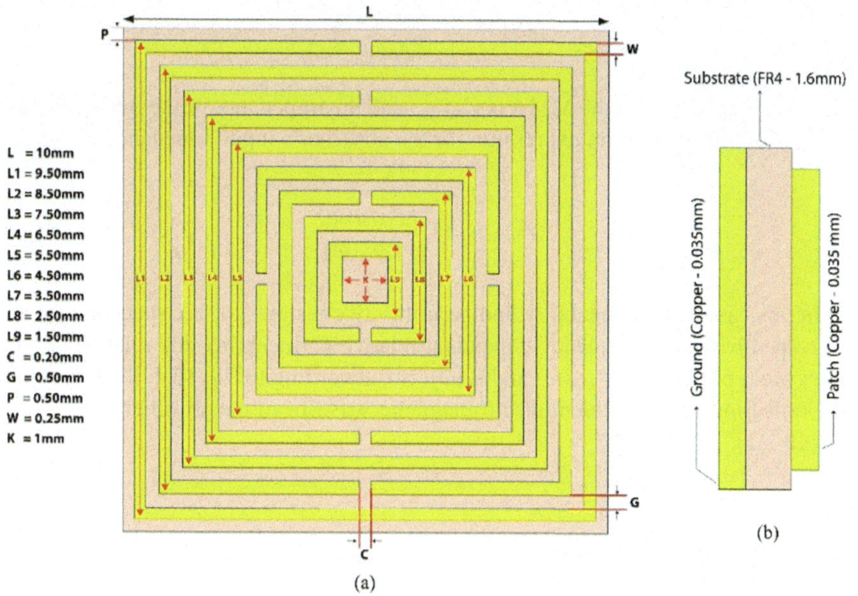


**Fig. 16.1** Unit cell with detailed dimensions (**a**) and side view (**b**)

absorber will perfectly absorb electromagnetic waves. There won't be much transmission through the unit cell because it has solid copper ground. As the absorption usual formula [15–17],

$$A = 1 - |S11|^2 - |S21|^2. \tag{16.1}$$

In this equation, absorption is $A$, reflection coefficient is $|S11|^2$, and transmission coefficient is $|S21|^2$.

On both surfaces of the MMA, two waveguide ports are normally used to calculate the $S$ parameters. The $S11$ and $S21$ scattering ($S$) parameters are explained as follows [18]:

$$S11 = \frac{\sqrt{\text{Power reflected from port 1}}}{\sqrt{\text{Power incident on port 1}}}, \tag{16.2}$$

$$S21 = \frac{\sqrt{\text{Power reflected from port 2}}}{\sqrt{\text{Power incident on port 1}}}. \tag{16.3}$$

The maximum absorption may be accomplished if both of the reflection and transmission coefficients become zero. When an EM wave encounters a normal incidence, the reflection coefficient can be obtained by [14]

$$|S11|^2 = \frac{Z_{\mathrm{mm}(\omega)-Z_0}}{Z_{\mathrm{mm}(\omega)+Z_0}}. \tag{16.4}$$

Here, the impedance of the MMA is $Z_{\mathrm{mm}(\omega)}$ and the impedance of the free space (air) is $Z_0$. The following is an expression for the absorber's input impedance $Z_{\mathrm{mm}(\omega)}$:

$$Z_{\mathrm{mm}(\omega)} = \sqrt{\frac{\mu_0 \mu_{\mathrm{r}}(\omega)}{\varepsilon_0 \varepsilon_{\mathrm{r}}(\omega)}}. \tag{16.5}$$

In this case, the permeability and permittivity constants of free space are $\mu_0$ and $\varepsilon_0$, respectively, the frequency-dependent relative permeability and relative permittivity are represented by $\mu_{\mathrm{r}}(\omega)$ and $\varepsilon_{\mathrm{r}}(\omega)$. So, according to Eq. 16.4, there won't be any reflection of the incident electromagnetic wave from the absorber if $Z_{\mathrm{mm}(\omega)} = Z_0$ [19–21].

## 16.3 Result Analysis

The suggested MM unit cell's absorption bandwidth was obtained by adding the relevant reflection and transmission coefficients where the frequency ranges around each resonance peak with a maximum value of – 10 dB [22]. In Fig. 16.2, which shows the results of simulations, the measured *S* parameters and absorption usually correlate with the mentioned resonance frequency. Here, four separate resonance frequencies are 4.44, 6.24, 8.50, and 11.50 GHz. This frequency range covers the C and X bands respectfully. As a result, it was reported that the suggested MMA was a strong candidate with a maximum of 93.7058, 93.32, 99.96, and 96.04% at the mentioned resonance frequencies.

The resonator was configured to 80 and 180° theta and phi polarizing angles of the incident EM waves because of the perfect symmetry cut in the resonator architecture. According to Fig. 16.3, the unit cell of the proposed architecture exhibits incident angle insensitive absorption effectiveness.

The optimum absorption rate must be shown when the relative permeability or permittivity value is near zero or negative [23]. Due to this, we identified a negative value for permittivity, which can be observed in Fig. 16.4 when we considered our resonance frequency.

The unit cell's impedance should be close to or equal to the free space impedance ($Z_0 \approx 377\ \Omega$) at the resonance frequency to enable maximal absorption. The impedance of the absorber unit cell is displayed in Fig. 16.5. Here, the unit cell's input impedance exhibits values close to $Z_0$ at resonance frequencies.

The current distribution in the surface is displayed in Fig. 16.6. The significant surface current flowing through the various split ring sections of the resonator is what causes the negative permittivity or permeability value. This causes an effective refractive index that is negative or very close to zero. Analyzing the active inductive

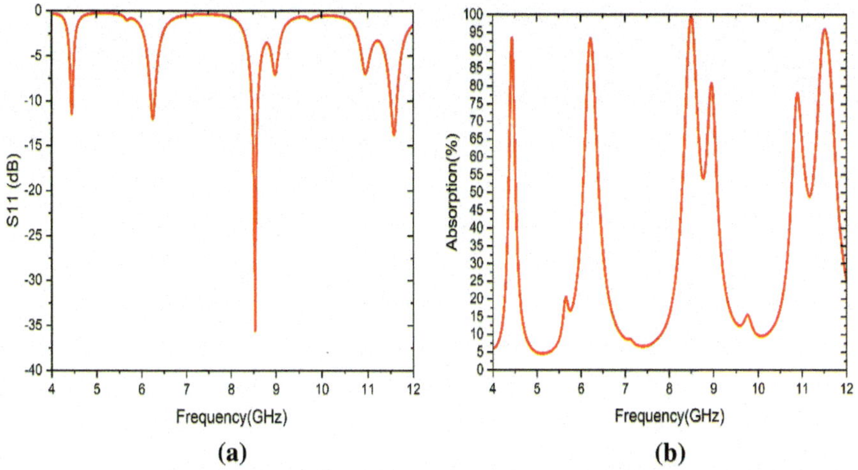


**Fig. 16.2** **a** $S11$ performance of the absorber for unit cell and **b** absorption of the proposed absorber

**Fig. 16.3** Incident angle (simulated) of normal (theta polarization) and oblique (phi polarization) incidence in various EM waves

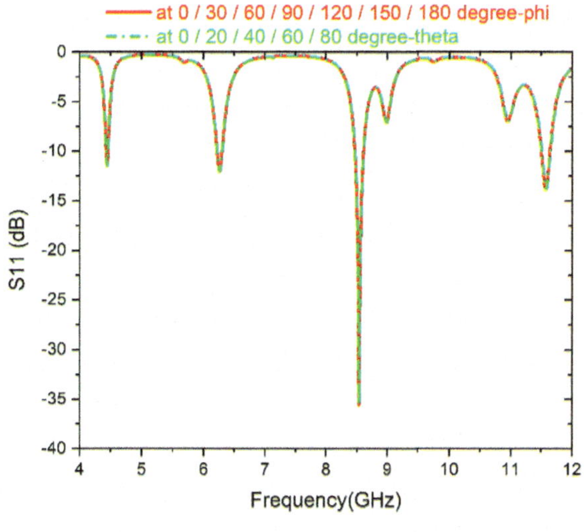


component at the resonance frequency is made possible by the surface current distribution. The capacitive part of that resonance frequency is represented by the copper cut that separates the following copper segments.

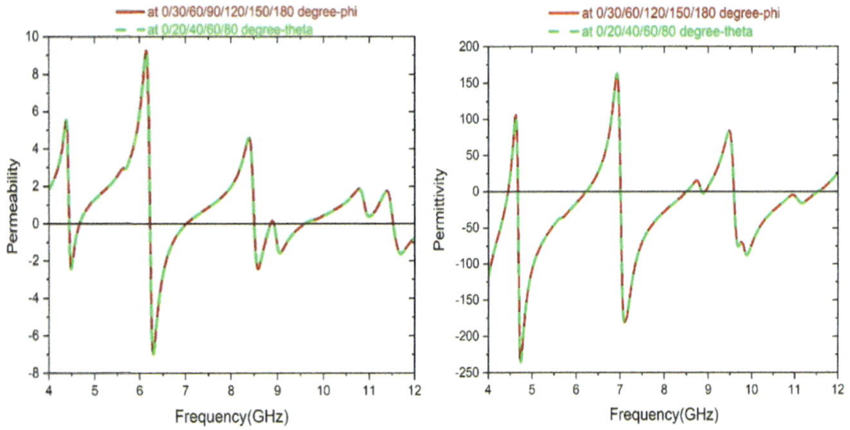


**Fig. 16.4** Simulated and measured **a** permeability and **b** permittivity

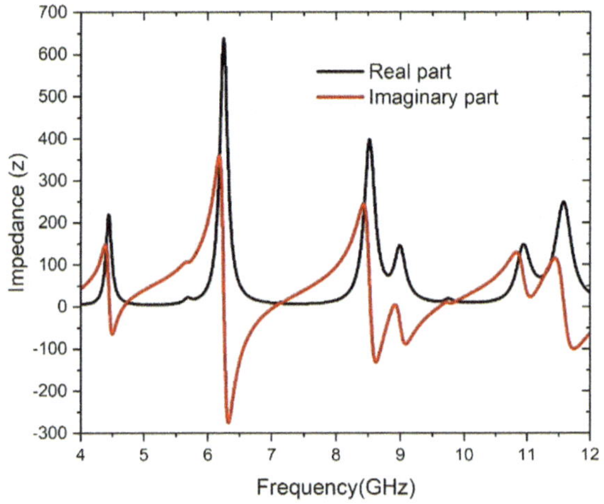


**Fig. 16.5** Unit cell's impedance when various frequencies are applied

## 16.4 Discussion

A comparison of some elements, such as unit cell measurement, the range of resonance frequencies, maximum absorption, and one of the important parts, which is SNG or DNG, has been introduced in Table 16.1 to demonstrate the added value of the proposed metamaterial absorber (MMA) over other similar absorbers that already exist. Table 16.1 clearly shows that our suggested absorber is excellent in comparison due to its effective unit cell size, relatively easy resonator design, single substrate layer, a good amount of resonance frequency, and superior absorption. The absorber is a suitable metamaterial absorber partially due to this unique characteristic and single negative (SNG) permittivity value. Additionally, due to the

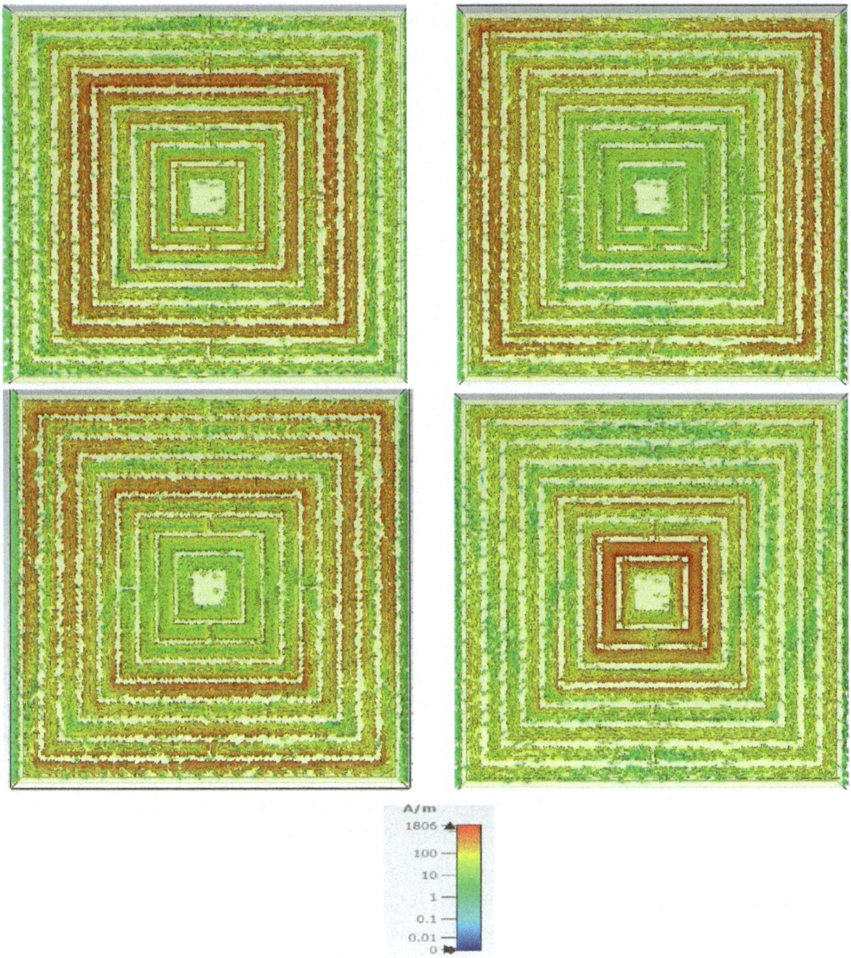


**Fig. 16.6** Surface current distribution of the unit cell

angle of incidence insensitivity, it makes an excellent absorber. Radar absorption (at 4.44 GHZ frequency), wireless communication (at 6.24 GHZ frequency), microwave imaging (at 8.5003 GHz frequency), and radio astronomy (at 11.507 GHz frequency) applications at the C and X bands are all possible with this absorber.

**Table 16.1** Proposed MMA is compared to relevant works

| References | Year | Unit cell size (mm) | SNG or DNG | Resonance frequency | Max absorption (%) | Max polarization angle insensitivity |
|---|---|---|---|---|---|---|
| [24] | 2019 | 24 | Not mentioned | 8.6, 10.2, 11.95 | 84 | Not mentioned |
| [25] | 2019 | 25 | Not mentioned | 5.2, 10.4 | 99 | Not mentioned |
| [26] | 2021 | 13.6 | Not mentioned | 4.5, 11.4 | 90 | 90° |
| [27] | 2021 | 11.5 | Not mentioned | 6.16, 9.22, 11.6 | 97 | 90° |
| [28] | 2022 | 16.5 | Not mentioned | 6.8, 8.5, 11.9 | 90 | 90° |
| Proposed work | 2023 | 10 | SNG | 4.44, 6.24, 8.5003, 11.507 | 99.96 | 180° |

## 16.5 Conclusion

In the following paper, a square maze-shaped split ring MMA is proposed for C and X bands. According to the insensitive incident angle of normal (theta polarization) and oblique (phi polarization) incidence in various EM waves, the suggested metamaterial absorber (MMA) was demonstrated to be a perfect metamaterial absorber (MMA). Any incident EM wave at these resonance frequencies will be adequately absorbed by the absorber regardless of the incident angle. The simulation result revealed four particular resonances at 4.44, 6.24, 8.50, and 11.50 GHz. It exhibits single negative (SNG) permittivity attributes, ensuring the metamaterial properties at the resonant frequencies. Additionally, it can effectively absorb up to a maximum of 99.96% of the applied electromagnetic (EM) waves. Due to the design's excellent performance, it may be an appropriate applicant for use in Radar absorption, wireless communication, microwave imaging, and radio astronomy, and numerous upcoming significant opportunities in C- and X-band.

**Acknowledgements** This work is supported by Ministry of Higher Education (MOHE) through Fundamental Research Grant Scheme (FRGS), code grant FRGS/1/2022/TK07/UKM/02/23.

## References

1. S. Hannan, M.T. Islam, M.R. Faruque, M.E. Chowdhury, F. Musharavati, Angle insensitive co-polarized metamaterial absorber based on equivalent circuit analysis for dual-band WiFi applications. Sci. Rep. **11**(1) (2021)
2. T.J. Sheng et al., An internet of things based smart waste management system using LoRa and TensorFlow deep learning model. IEEE Access **8**, 148793–148811 (2020)

3. M. Samsuzzaman, M.T. Islam, A semicircular shaped super wideband patch antenna with high bandwidth dimension ratio. Microw. Opt. Technol. Lett. **57**(2), 445–452 (2014)
4. M.T. Islam, M. Cho, M. Samsuzzaman, S. Kibria, Compact antenna for small satellite applications [antenna applications corner]. IEEE Antennas Propag. Mag. **57**(2), 30–36 (2015)
5. Md.Z. Mahmud, M.T. Islam, N. Misran, S. Kibria, Md. Samsuzzaman, Microwave imaging for breast tumor detection using uniplanar AMC based CPW-fed microstrip antenna. IEEE Access **6**, 44763–44775 (2018)
6. M. Shahidul Islam et al., Monitoring of the human body signals through the internet of things (IoT) based LoRa wireless network system. Appl. Sci. **9**(9), 1884 (2019)
7. Md. Rokunuzzaman, Md. Samsuzzaman, M.T. Islam, Unidirectional wideband 3-D antenna for human head-imaging application. IEEE Antennas Wirel. Propag. Lett. **16**, 169–172 (2017)
8. N.I. Landy, S. Sajuyigbe, J.J. Mock, D.R. Smith, W.J. Padilla, Perfect metamaterial absorber. Phys. Rev. Lett. **100**(20) (2008)
9. Md.A. Rahman, A. Noman, M.J. Uddin, S. Hannan, A. Musa, M.A. Rahman, Square split enclosed labyrinth maze shaped metamaterial absorber for hydrocarbon sensing application, in *2023 International Conference on Next-Generation Computing, IoT and Machine Learning (NCIM)*, Gazipur, Bangladesh, June 2023 (IEEE, 2023), pp. 1–6
10. R. Azim, M.T. Islam, J.S. Mandeep, A.T. Mobashsher, A planar circular ring ultra wideband antenna with dual band-notched characteristics. J. Electromagn. Waves Appl. **26**(14–15), 2022–2032 (2012)
11. S. Hannan, M.T. Islam, M.R. Faruque, H. Rmili, Polarization-independent perfect metamaterial absorber for C, X and Ku band applications. J. Mater. Res. Technol. **15**, 3722–3732 (2021)
12. J.J. Tiang, M.T. Islam, N. Misran, J.S. Mandeep, Slot loaded circular microstrip antenna with meandered slits. J. Electromagn. Waves Appl. **25**(13), 1851–1862 (2011)
13. M.H. Ullah, M.T. Islam, J.S. Mandeep, N. Misran, A new double L-shaped multiband patch antenna on a polymer resin material substrate. Appl. Phys. A **110**(1), 199–205 (2012)
14. S. Hannan, M.T. Islam, A.F. Almutairi, M.R. Faruque, Wide bandwidth angle- and polarization-insensitive symmetric metamaterial absorber for X and Ku band applications. Sci. Rep. **10**(1) (2020)
15. S. Hannan, M.T. Islam, M.S. Soliman, N.B. Mohd Sahar, M.S. Jit Singh, M.R. Faruque, A. Alzamil, A filling-factor engineered, perfect metamaterial absorber for multiple applications at frequencies set by IEEE in C and X bands. J. Mater. Res. Technol. **19**, 934–946 (2022)
16. R. Azim, M.T. Islam, A.T. Mobashsher, Design of a dual band-notch UWB slot antenna by means of simple parasitic slits. IEEE Antennas Wirel. Propag. Lett. **12**, 1412–1415 (2013)
17. Md.S. Alam, N. Misran, B. Yatim, M.T. Islam, Development of electromagnetic band gap structures in the perspective of microstrip antenna design. Int. J. Antennas Propag. **2013**, 1–22 (2013)
18. S. Hannan, M.T. Islam, S.H. Almalki, M.R. Faruque, M. Salaheldeen M, M.S. Islam, Rotational symmetry engineered, polarization and incident angle-insensitive, perfect metamaterial absorber for X and Ku band wireless applications. Sci. Rep. **12**(1) (2022)
19. A. Rahman, M.T. Islam, M.J. Singh, S. Kibria, Md. Akhtaruzzaman, Electromagnetic performances analysis of an ultra-wideband and flexible material antenna in microwave breast imaging: to implement a wearable medical bra. Sci. Rep. **6**(1) (2016)
20. R. Azim, M.T. Islam, N. Misran, S.W. Cheung, Y. Yamada, Planar UWB antenna with multi-slotted ground plane. Microw. Opt. Technol. Lett. **53**(5), 966–968 (2011)
21. S. Hossen, Md.K. Uddin, A. Gafur, S. Zahidur Rashid, S. Hannan, A.Z.M. Imran, Design and analysis of an antenna using ring slotted rectangular reflector with double substrate for 5G mmWave application, in *2023 International Conference on Next-Generation Computing, IoT and Machine Learning (NCIM)*, Gazipur, Bangladesh, June 2023 (IEEE, 2023), pp. 1–6
22. Md.K. Uddin et al., Split-ring enclosed K-shaped rotational symmetric metamaterial absorber for multiband wireless applications, in *2023 International Conference on Next-Generation Computing, IoT and Machine Learning (NCIM)*, Gazipur, Bangladesh, June 2023 (IEEE, 2023), pp. 1–6

23. A.R. Chowdhury, S. Hannan, M.K. Uddin, Md.E.H. Bhuiyan, M.I. Hoque, M.T. Islam, Gas stove burner shape with inductive tailed rotational symmetric metamaterial absorber for C and X band application, in *2023 International Conference on Next-Generation Computing, IoT and Machine Learning (NCIM)*, Gazipur, Bangladesh, June 2023 (IEEE, 2023), pp. 1–5
24. S. Jafari, M. Naderi, A. Hatami, F.B. Zarrabi, Microwave Jerusalem cross absorber by metamaterial split ring resonator load to obtain polarization independence with triple band application. AEU-Int. J. Electron. Commun. **101**, 138–144 (2019)
25. M. Chaluvadi, P.H. Rao, Wide-angle and polarization insensitive dual-band metamaterial absorber, in *2019 IEEE 5th Global Electromagnetic Compatibility Conference (GEMCCON)* (2019)
26. S. Kalraiya, R.K. Chaudhary, R.K. Gangwar, Polarization-independent triple band ultra-thin conformal metamaterial absorber for C- and X-frequency bands. AEU-Int. J. Electron. Commun. **135**, 153752 (2021)
27. D.T. Phan, T.K. Nguyen, N.H. Nguyen, D.T. Le, X.K. Bui, D.L. Vu, C.L. Truong, T.Q. Nguyen, Lightweight, ultra-wideband, and polarization-insensitive metamaterial absorber using a multi-layer dielectric structure for C- and X-band applications. Phys. Status Solidi (b) **258**(10), 2100175 (2021)
28. P. Ranjan, C. Barde, K. Roy, R. Sinha, S. Kumar, D. Das, Pixelated wideband metamaterial absorber for X-band applications, in *Lecture Notes in Electrical Engineering* (2022), pp. 553–562

# Chapter 17
# Octagonal Enclosed Octagonal-Shaped Metamaterial Absorber Based on $SiO_2$ for the Infrared Region

**Md Raihan, Seikh Erfan Ahmed Himu, Amran Chowdhury Raihan, Saif Hannan, Sikder Sunbeam Islam, Mohd Hafiz Baharuddin, Norbahiah Misran, and Mohammad Tariqul Islam**

**Abstract** This article highlights the development of a simple and cost-effective metamaterial absorber (MMA) to solve manufacturing limitations and unleash the tremendous potential of the visible and infrared spectrum in many applications. The MMA design employs low-cost gold metal and silicon dioxide as dielectric spacers, allowing operation from 661 to 1082 nm. This absorber achieved 95.29% average absorption throughout the whole wavelength, including two resonant points where maximum absorption attained about 99.98% and 99.4% for 724.93 nm and 1002 nm, respectively. The design is distinguished by its simplicity and permits straightforward fabrication, distinguishing it from earlier studies. Furthermore, the absorber's small size (200 nm × 200 nm × 220 nm) helps to its adaptability. Comprehensive performance testing in various modes illustrates the absorber's resistance to TE, TM, and TEM modes. Because of their unique qualities and capabilities, the suggested metamaterial absorbers have extensive use in the defense, energy harvesting, imaging, sensing, and communication industries.

M. Raihan · S. E. A. Himu · A. C. Raihan · S. S. Islam
Department of Electrical and Electronic Engineering, International Islamic University Chittagong (IIUC), Chattogram 4318, Bangladesh

S. Hannan
Department of Electronic and Telecommunication Engineering, International Islamic University Chittagong (IIUC), Chattogram 4318, Bangladesh

M. H. Baharuddin · N. Misran · M. T. Islam (✉)
Department of Electrical, Electronic and Systems Engineering, Faculty of Engineering and Built Environment, Universiti Kebangsaan Malaysia (UKM), 43600 Bangi, Selangor, Malaysia
e-mail: tariqul@ukm.edu.my



M. T. Islam et al. (eds.), *Proceedings of the 8th International Conference on Space Science and Communication*, Springer Proceedings in Physics 303,
https://doi.org/10.1007/978-981-97-0142-1_17

## 17.1 Introduction

In recent, a new term, metamaterials, came to light using which researchers can manipulate and make use of EM waves. Unlike natural materials, this type of material shows some unusual behavior that can't be seen in those present in nature. Negative permittivity, negative permeability, and negative refractive index are some extraordinary characteristics. Veselago's landmark study in 2000 outlined the theoretical feasibility of such materials, which would display counterintuitive effects including negative refraction. This sparked an upsurge of scientific interest in fulfilling Veselago's goal. He first conceptualized this in 1968 [1]. Later, in 2006, Pendry, Schurig, and Smith conducted a crucial experiment in which they successfully showed negative index metamaterials [2]. Because of their wide range of applications such as fault detection [3], sensing [4], satellite application [5], and solar harvesting [6], optical wavelength absorbers are attracting a lot of attention in the new research field. Solar EM waves are classified into NIR, Visible, and UV. Several works has been published regarding microwave [5, 7], NIR, Visible [8], and in UV [9] range. Researchers are trying to push the limitation of previous work to make MMA more effective. However for this wideband [10], ultrawideband [11], mechanically tunable [12], thermally tunable [13], and bendable [14] features were studied.

Various types of designed (MMA) in the infrared frequency range were studied. In this paper [15] they use square-shaped copper patches of different sizes fabricated on GaAs substrate that works in the 220–360 THz range. Yang et al. [16] introduces an infrared metamaterial absorber with impressive optical transparency, utilizing an indium tin oxide (ITO) substrate. ITO's unique attributes, characterized by both high infrared reflectivity and optical transparency, render it an apt choice for Surface Plasmon Polaritons applications. Within this design, three distinct variants of infrared metamaterial absorbers were conceived.

Another author demonstrates [17] a large-scale, broadband, polarization-insensitive, and tunable optical metamaterial absorber, realized with a low-cost, time-saving, and lithography-free fabrication method centimeter-scale total size.

To overcome fabrication limitations, cost and better absorption in this paper, we suggest an absorber in the infrared region using gold and silicon dioxide where the absorption level is above 90% for the wavelength spanning 661–1082 nm, which is in the infrared region [18] along with average absorption 95.29%. Moreover, we have got two maximum absorption peaks at 724.93 nm and 1002 nm with 99.98% and 99.4% values, respectively. In addition, the structure is insensitive in both TE and TM mode.

## 17.2 Design of Unit Cell

Like the other conventional MMA our proposed design is also three-layer sandwich-type structure that is made with metal-dielectric-metal. We use gold (conductivity = 4.561e + 007 [S/m]) as the top metal layer and the bottom blocking plate which are separated by spacer $SiO_2$ (silicon dioxide) having permeability 1. Instead of utilizing a Drude model, it is discovered that metals with a dielectric function from experimentally observed data should be used to appropriately predict the behavior of optical metamaterials [19] (Fig. 17.1).

For excellent absorption proper material size and geometric design carries crucial part. Our structure combination comes with thickness of top, mid dielectric, and bottom are 15 nm, 55 nm, and 150 nm respectively. Bottom plate thickness were chosen in a manner that it can properly block the transmission as the absorption depends on transmission and reflection of wave expressed as an Eq. (17.1)

$$A(w) = 1 - [S11]^2 - [S21]^2. \tag{17.1}$$

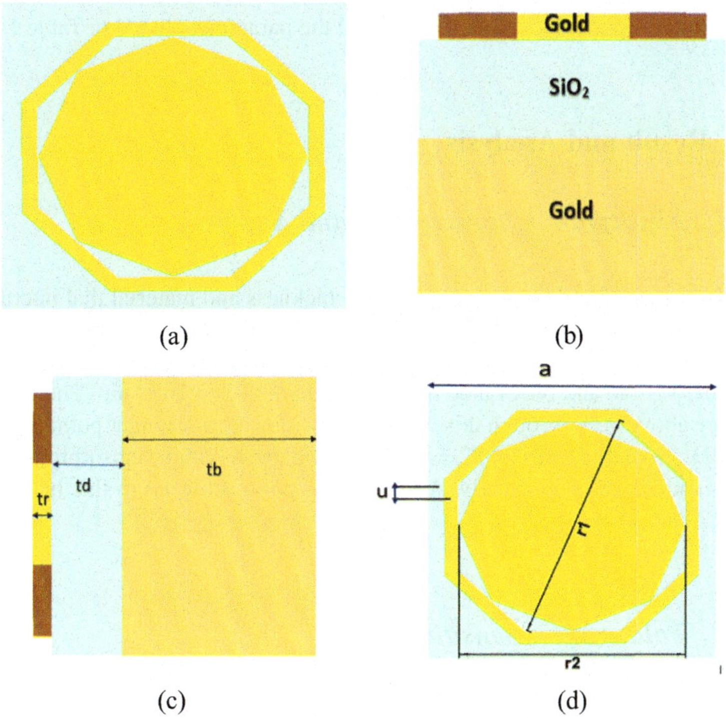


**Fig. 17.1** Demonstration of proposed structure's **a** front view, **b** side view with material layer, **c** thickness of different layer and **d** parameter of the structure

**Table 17.1** Parameter list of the structured

| Parameter | $a$ | tb | tr | ts | $r1$ | $r2$ | $p$ |
|---|---|---|---|---|---|---|---|
| Value (nm) | 200 | 150 | 15 | 55 | 176 | 156 | 10 |

$S11$ is reflection coefficient and $S21$ is transmission coefficient. If we block transmission wave propagation then the equation becomes

$$A(w) = 1 - [S11]^2. \tag{17.2}$$

$S11$ can be manipulated by choosing proper geometric shape and thickness. Back metal gold layer thickness must be greater than skin depth, $\delta = \sqrt{\frac{2\rho}{2\pi f \mu R \mu 0}}$ to perfectly stop the EM wave. TEM mode is implemented by using periodic boundary conditions encompassing perfect electric conductors on the $x$-axis and perfect magnetic conductors on the $y$-axis. A Floquet port is engaged along the $Z$-axis when operating in transverse electric (TE) or transverse magnetic (TM) modes. In addition, the $x$-axis and $y$-axis are assigned as the slave and master axes, respectively. The other structural values are $u = 10$ nm width of the outer circle, $r1 = 170$ nm and inner patch resonator parameter $r2$ is 156 nm. All the parameters listed in Table 17.1.

## 17.3 Result and Analysis

### *17.3.1 Absorption Characteristic and Analysis*

As early discussed proper geometric shape thickness and material that interact with EM wave plays important role for better absorption. Keeping these in mind we chose these values carefully that it matches. Our designed structure shows excellent absorption in the Visible and NIR range from wavelength 661 to 1082 nm. The absorption levels are above 90% through this whole bandwidth. Two resonant point found at $f1 = 724.93$ nm and another is $f2 = 1002$ nm where maximum absorption achieved 99.98% and 99.1% respectively. Average absorption achieved in this bandwidth is about 95.29%.

### *17.3.2 Polarization Sensitivity*

In all the types of mode TE, TM, and TEM, absorber should be insensitive in all of these wave modes to make it more effective in every wave that light incident it in Fig. 17.2b, we have depicted our proposed structure's absorption which is insensitive to polarization wave both in TE, TM, and TEM mode. To perform simulation in these

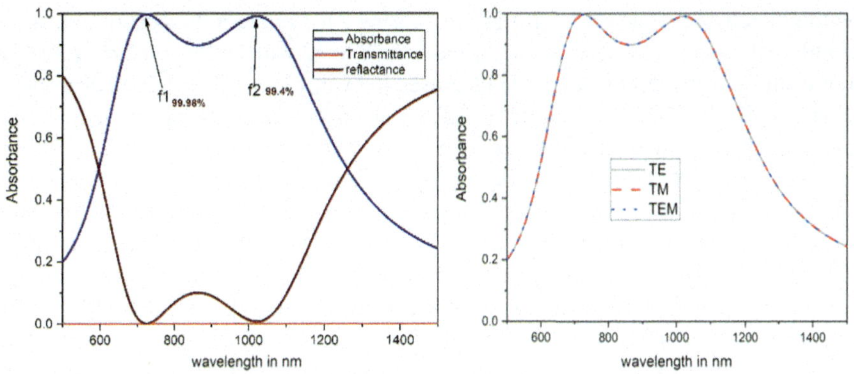


(a) (b)

**Fig. 17.2** Representation of the **a** characteristic curve of absorption, transmission, and reflection and **b** absorption in different modes

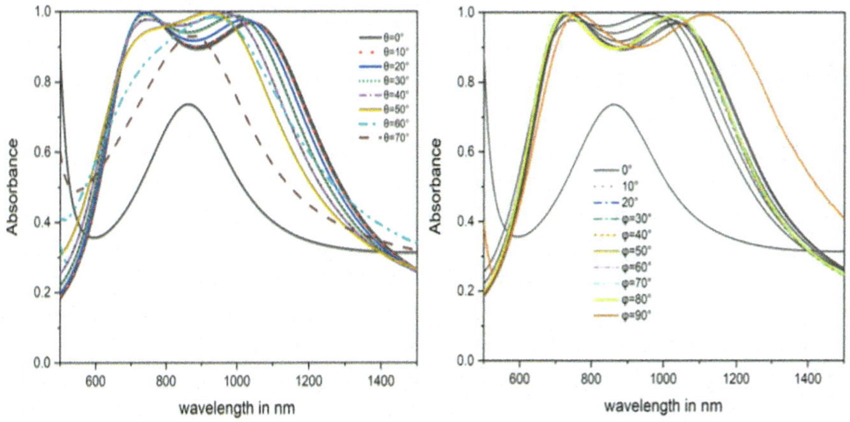


(a) (b)

**Fig. 17.3** Absorbance in different **a** polarization angle and **b** incident angle

modes proper boundary condition is required. In all of these modes absorber's average absorption remains same along with peak values (Fig. 17.3).

### *17.3.3 Parameter Sweep*

We have simulated the design varying the parameter that we used to analyze the effect of these geometric parameters in absorption characteristic. Parameter *r*1 changes 85–105 nm to see its effect. Best absorption found with 95 nm. Absorption increases while increasing value 85–95 further increasing $r1$ it reduces the absorption. Also, width of the outer octagonal shape varied from 8 to 12 with increment 2 nm resonant point shift 630.9–656.83 nm while increased the value $u = 8$–9 nm. Peak absorption changes 99.88%, 99.72%, and 99.98% for 8, 9, and 10 nm respectively. In addition, average absorption is also greater with $u = 10$ nm. It is because at $u = 10$ nm impedance matches perfectly. For changing resonator thickness tr = 5 nm peak absorption point 99.9% at 1002 nm, resonant point 99.85% found at 1002 nm and 98% at 700 nm for 10 nm. Resonant point shift to 745 nm where maximum peak found 99.75%, however other one down to level 98% at 1045 nm. 94.88% average absorption achieved for thickness 20 nm and for 25 nm average found 95%. Best absorption found using thickness 15 nm which is almost 95.29% and two maximum resonant point showed at 724.93 nm and 1002 nm including values 98.98% and 99.4% respectively. Varying the thickness of substrate absorption curve for different value of ts represent in Fig. 17.4e where we can see the best absorption achieved at 55 nm. $r2$ also varied from 148 to 164 nm, however impedance matched at the value $r2 = 156$ nm.

### 17.3.4 Absorption Using Different Substrate

Absorption can be changed or manipulate using different dielectric material. We have proposed silicon dioxide substrate as best absorption achieved and impedance matched perfectly with this material. In this part we analyze the effect of different substrate in absorption that is depicted in Fig. 17.5. 99.65% maximum absorption found at 1000.5 nm using Rogers RO3006. From 1341 absorption above 90%, however absorption is very low to rest of the wavelength. Other substrate does not exhibit good absorption aluminum nitride has a single peak at 522 nm with a value lower than 82%. This component along with Si (amorphous) and GaAs has a wide-band but very low absorption rate thus not suitable for better absorption. On the other hand, $SiO_2$ meets necessary requirement for an absorber to be best in this frequency band and geometric shape.

Average absorption with proposed dielectric spacer is about 95.29% along with two resonant point at 724.93 and 1002 nm.

### 17.3.5 Comparative Study

From the comparative study given in Table 17.2 it can be seen that our proposed absorber is superior than other previous work. The proposed one has excellent average absorption and wide incident angle stability. Average absorption is important factor for an absorber that operates in visible or IR region specially that was designed for solar harvesting, sensing, thermo-photovoltaics, and other applications. Moreover, incident angle is also a distinct feature that makes an absorber more efficient and

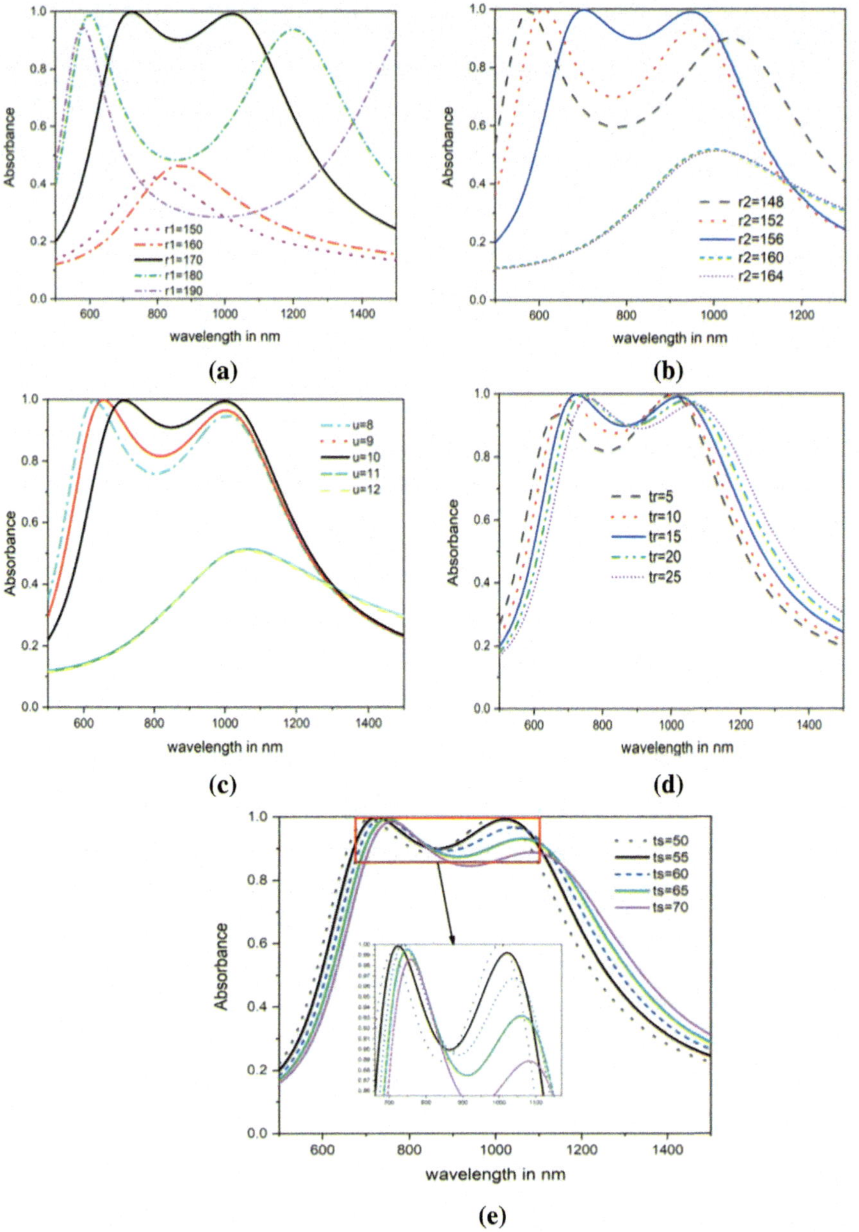


**Fig. 17.4** Graphical representation of **a** the effect of changing the parameter $r1$, **b** parameter $r2$ change, **c** change of absorption with circle width $u$, **d** change of the resonator thickness 'tr' and **e** varying substrate thickness 'td'

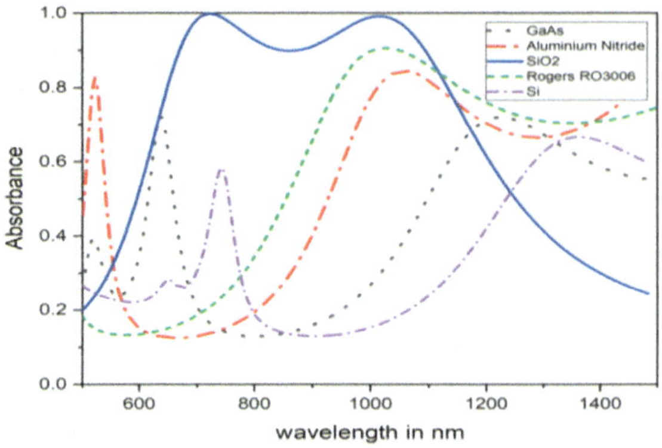


**Fig. 17.5** Absorption with different substances

**Table 17.2** Comparative study between previous work

| Reference | Frequency range (THz) | Material used | Layer | Dimension | Polarization and angular insensitivity | Average absorption |
|---|---|---|---|---|---|---|
| [15] | 220–360 | Cu, GaAs | 3 | 1010 × 1010 × 141.2 | Yes, ≤ 35° | 82.65% |
| [20] | 250–750 | W, $SiO_2$, Si | 4 | 200 × 200 × 60 | No, ≤ 45° | 82.65% |
| [21] | 100–400 | Si, Al | 3 | 980 × 980 × 130 | Yes, ≤ 45° | N/A |
| [22] | 6000–20,000 | Al, $Al_2O_3$, ZnSe | 6 | 6000 × 6000 × 900 | Not mentioned | Not mentioned |
| Proposed | 270–460 | $SiO_2$, Au | 3 | 200 × 200 × 220 | Yes, ≤ 60° | 95.29% |

suitable. So, from the view of these features our proposed one is superior than the other suggested absorber.

## 17.4 Conclusion

In summary, overcoming fabrication obstacles requires the development of a simple and cost-effective metamaterial absorber (MMA). Infrared has significant potential for a wide range of application where MMA can be employed effectively. In this study, we designed an MMA that uses low-cost gold metal and performs in the 661–1082 nm wavelength range. In comparison with previous research included in the literature study, our design stands out since it is simple, making it easier to manufacture. In addition, the absorber is modest, measuring the size as 200 × 200 × 220 nm. We examined the absorber's performance in a variety of modes, including

TE (Transverse Electric), TM (Transverse Magnetic), and TEM (Transverse Electromagnetic), that makes it suitable for absorption of every polarization wave. The findings indicate that the architecture is still relatively immune to summarize, overcoming fabrication obstacles requires the development of a simple and cost-effective metamaterial absorber (MMA). Infrared region has significant potential for a diverse array of uses, where MMA can be employed effectively such as solar harvesting, thermal imaging, sensing, military defense, and in communication sector.

**Acknowledgements** This work was supported by the Universiti Kebangsaan Malaysia research grant DIP-2021-007.

## References

1. A. Alu, N. Engheta, Plasmonic and metamaterial cloaking: physical mechanisms and potentials. J. Opt. A Pure Appl. Opt. **10**(9), 093002 (2008)
2. D. Schurig et al., Metamaterial electromagnetic cloak at microwave frequencies. Science **314**(5801), 977–980 (2006)
3. S. Fang, S. Zhou, D. Yurchenko, T. Yang, W.-H. Liao, Multistability phenomenon in signal processing, energy harvesting, composite structures, and metamaterials: a review. Mech. Syst. Signal Process. **166**, 108419 (2022)
4. A. Tamer et al., Metamaterial based sensor integrating transmission line for detection of branded and unbranded diesel fuel. Chem. Phys. Lett. **742**, 137169 (2020)
5. S. Hannan, M.T. Islam, N.M. Sahar, K. Mat, M.E. Chowdhury, H. Rmili, Modified-segmented split-ring based polarization and angle-insensitive multi-band metamaterial absorber for X, Ku and K band applications. IEEE Access **8**, 144051–144063 (2020)
6. M. Bagmanci, M. Karaaslan, E. Unal, O. Akgol, M. Bakır, C. Sabah, Solar energy harvesting with ultra-broadband metamaterial absorber. Int. J. Mod. Phys. B **33**(08), 1950056 (2019)
7. S. Hannan, M.T. Islam, A. Hoque, M.J. Singh, A.F. Almutairi, Design of a novel double negative metamaterial absorber atom for Ku and K band applications. Electronics **8**(8), 853 (2019)
8. M.M.K. Shuvo, M.I. Hossain, S. Rahman, S. Mahmud, S.S. Islam, M.T. Islam, A wide-angle, enhanced oblique incidence, bend-able metamaterial absorber employed in visible region with a sun shape resonator. IEEE Access **9**, 126466–126480 (2021)
9. M.K. Hedayati, A. Zillohu, T. Strunskus, F. Faupel, M. Elbahri, Plasmonic tunable metamaterial absorber as ultraviolet protection film. Appl. Phys. Lett. **104**(4), 041103 (2014)
10. R. Sekar, S. Inabathini, An ultra-thin compact wideband metamaterial absorber. Radioengineering **27**(2), 365 (2018)
11. J. Liu et al., Numerical analysis of an ultra-wideband metamaterial absorber with high absorptivity from visible light to near-infrared. Opt. Express **28**(16), 23748–23760 (2020)
12. F. Zhang et al., Mechanically stretchable and tunable metamaterial absorber. Appl. Phys. Lett. **106**(9), 091907 (2015)
13. Y. Pang et al., Thermally tunable water-substrate broadband metamaterial absorbers. Appl. Phys. Lett. **110**(10), 104103 (2017)
14. M.M.K. Shuvo et al., Polarization and angular insensitive bendable metamaterial absorber for UV to NIR range. Sci. Rep. **12**(1), 4857 (2022)
15. A. Alam, S.S. Islam, M.H. Islam, A.F. Almutairi, M.T. Islam, Polarization-independent ultra-wideband metamaterial absorber for solar harvesting at infrared regime. Materials **13**(11), 2560 (2020)
16. J. Yang, C. Xu, S. Qu, H. Ma, J. Wang, Y. Pang, Optical transparent infrared high absorption metamaterial absorbers. J. Adv. Dielectr. **8**(01), 1850007 (2018)

17. N. Mou et al., Large-scale, low-cost, broadband and tunable perfect optical absorber based on phase-change material. Nanoscale **12**(9), 5374–5379 (2020)
18. S. Mahmud, S.S. Islam, A.F. Almutairi, M.T. Islam, A wide incident angle, ultrathin, polarization-insensitive metamaterial absorber for optical wavelength applications. IEEE Access **8**, 129525–129541 (2020)
19. N.-H. Shen, T. Koschny, M. Kafesaki, C.M. Soukoulis, Optical metamaterials with different metals. Phys. Rev. B **85**(7), 075120 (2012)
20. M.S. Mia, M.M. Alam, R. Azim, M.J. Singh, T. Alam, M.T. Islam, Triangular-shaped metamaterial-based EM absorber for visible and infrared regime, in *2022 12th International Conference on Electrical and Computer Engineering (ICECE)* (IEEE, 2022), pp. 397–400
21. B. Mulla, C. Sabah, Multi-band metamaterial absorber topology for infrared frequency regime. Phys. E **86**, 44–51 (2017)
22. N. Zhang, P. Zhou, D. Cheng, X. Weng, J. Xie, L. Deng, Dual-band absorption of mid-infrared metamaterial absorber based on distinct dielectric spacing layers. Opt. Lett. **38**(7), 1125–1127 (2013)

# Chapter 18
# An Analysis of the Optimal Metamaterial Absorber for Satellite Applications

**Najeeb Ullah, Md. Shabiul Islam, Ahasanul Hoque, Wong Hin Yong, and Mohammad Tariqul Islam**

**Abstract** This study offers a concise overview of various structures that utilize metamaterial absorbers for satellite applications. Metamaterials possess various unique properties that conventional materials lack. Metamaterial devices have found applications in various fields, such as wireless communication systems, satellite communication systems, and radar systems. A novel compact metamaterials unit cell has been designed and compared to previous works in the literature. The results indicate that the proposed unit cell demonstrates high absorptivity, making it suitable for satellite communication.

## 18.1 Introduction

The need for materials with unusual properties not found in nature led to the development of a new discipline of electromagnetics known as metamaterials [1, 2]. It has been shown that the constituent elements' shapes and configurations affect the metamaterials' optical properties. The optical resonance of metamaterials has been studied

N. Ullah · Md. S. Islam (✉) · W. H. Yong
Faculty of Engineering (F.O.E.), Multimedia University (MMU), 63100 Cyberjaya, Selangor, Malaysia
e-mail: shabiul.islam@mmu.edu.my

N. Ullah
e-mail: 1221400623@student.mmu.edu.my

W. H. Yong
e-mail: hywong@mmu.edu.my

A. Hoque (✉)
Institute of Climate Change, Universiti Kebangsaan Malaysia, 43600 Bangi, Malaysia
e-mail: ahasanul@ukm.edu.my

M. T. Islam (✉)
Department of Electrical, Electronic and Systems Engineering, Faculty of Engineering and Built Environment, Universiti Kebangsaan Malaysia, 43600 Bangi, Malaysia
e-mail: tariqul@ukm.edu.my



M. T. Islam et al. (eds.), *Proceedings of the 8th International Conference on Space Science and Communication*, Springer Proceedings in Physics 303,
https://doi.org/10.1007/978-981-97-0142-1_18

for a variety of applications, including perfect absorption, clocking, sensing, superlenses, and satellite applications [3]. A negative refractive index, which necessitates a negative effective dielectric permittivity and a negative magnetic permeability, is one such attribute that can improve the use of metamaterial [4]. Negative index metamaterial (NIMs) applications include "perfect" lenses, nanometer-sized antennas, subwavelength transmission lines, and resonators. Several components comprised of conventional materials, such as plastic or glass, are combined to form metamaterials. These components are organized in a small, repetitive pattern. Metamaterials' direction and organization give them their properties. Negative index metamaterials are substances with a negative index of refraction at a specific wavelength.

## 18.2 Literature Survey

In [5], the authors have proposed a polarization-independent quasi-TEM metamaterial absorber for sensing applications in the X and Ku bands. The proposed metamaterial comprises two square ring resonators and a mirror-reflected C-shape. It exhibits two distinct absorption bands across the electromagnetic spectrum. Altering the thickness of the metallic ring in the frequency domain allows for controllable absorption. The CST microwave studio simulation program was used for the design and study of the proposed meta-absorber, which is based on the finite-integration technique (FIT), as shown in Fig. 18.1. Two distinct absorption peaks were observed, with values of 99.6% and 99.14% attained at frequencies of 13.78 GHz and 15.3 GHz, respectively.

A metamaterial absorber for microwave application is proposed in [6]. The designed metamaterial demonstrates a high absorption level, specifically 99.99%, at resonance frequencies of 27.5 and 31 GHz. The frequency range of 27.325–27.705 and 29.505–31.264 GHz exhibits a 90% absorption bandwidth, encompassing two absorption bandwidths of 0.38 and 1.76 GHz. The physical mechanism underlying the dual-band absorption behaviour has been elucidated through the presentation of the absorption, electric field, magnetic field, and surface current distribution. In addition, the metamaterial absorber (MMA) has been examined using various polarisation angles, and the impact of its size has been investigated to enhance comprehension of the absorption properties of the metamaterial. Figure 18.2 shows the simulated environment of the proposed design.

Reference [7] introduces a portable, wideband metamaterial absorber optimized for use in the (Ku band). The proposed structure is small, measuring only $10 \times 10\,\text{mm}^2$. The (FR4) substrate is used in the fabrication process, while (ANSYS HFSS) is used for simulation. With an absorption range of (11.39–20.15 GHz), the suggested wideband microwave absorber has an overall bandwidth of 8.76 GHz. This absorber effectively covers the (Ku band), as its absorptivity is 88%. The resulting FWHM is 11.32 GHz, covering a frequency spectrum of 10.57–21.89 GHz. Figure 18.3 shows the unit cell structure of the proposed design.

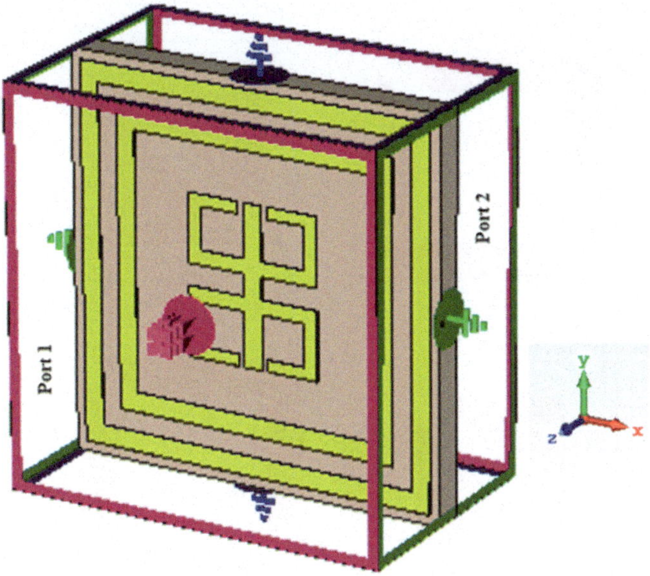


**Fig. 18.1** Boundary condition of the design simulated in CST [5]

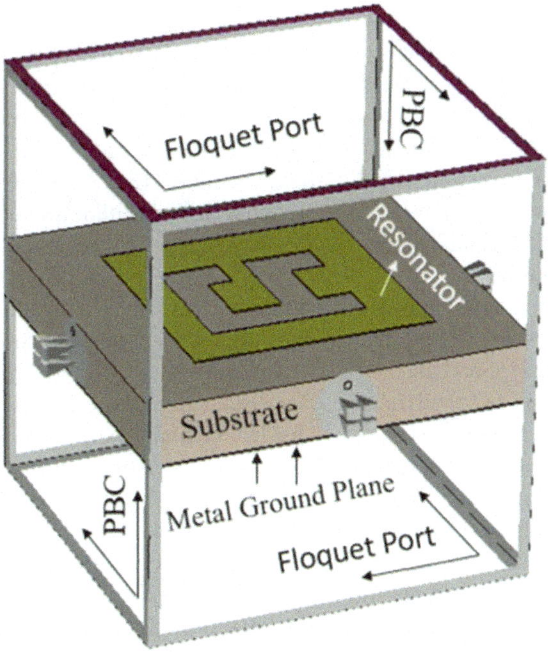


**Fig. 18.2** Proposed simulated MMA [6]

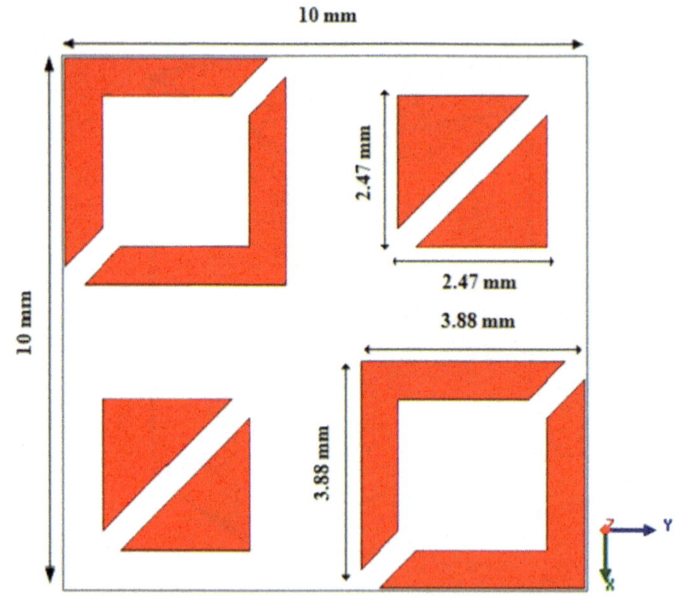


**Fig. 18.3** Proposed unit cell design structure [7]

A small and simple double square-shaped metamaterial absorber (MA) design is presented in [8]. The square measures $1.41 \times 1.41\ mm^2$, while the unit cell structure has an overall dimension of $5 \times 5\ mm^2$. The absorption covers a wide range, from 14.44 to 27.87 GHz, with a bandwidth of 13.43 GHz at a level of 10 dB. This range effectively covers the Ku (12–18 GHz) and K (18–27 GHz) frequency bands. The proposed structure demonstrates potential utility within the radar and satellite communications domains. Figure 18.4 shows the top-view structure of the proposed metamaterial unit cell.

Wang [9] has assembled four rectangular metallic rings, a metallic ground plane, and a polarization-insensitive terahertz metamaterial absorber, as shown in Fig. 18.5. Four distinct absorption bands with average peak intensities of approximately 97% are found to be part of the structure. The distribution of the electric field is used to examine the absorption process. It is considered that the mechanism of the four-band absorber involves the overlapping of four resonance frequencies. Flexible tuning of each absorption peak's frequency is made possible by the size of the associated metallic ring. The design may be scaled up to operate in the microwave frequency range and is adaptable enough to use alternative absorber architectures. The proposed structure uses technologies for imaging, detection, and stealth.

A paradigm for creating an ultra-wideband ultrathin metamaterial absorber was presented by Ghosh and Srivastava [10], as shown in Fig. 18.6. Two concentric circular split rings are carved on a metal-backed dielectric substrate to make up the suggested construction. A typical incidence computer simulation shows that the entire X-band is covered by a 10 dB absorption bandwidth between 7.85 and 12.25 GHz.

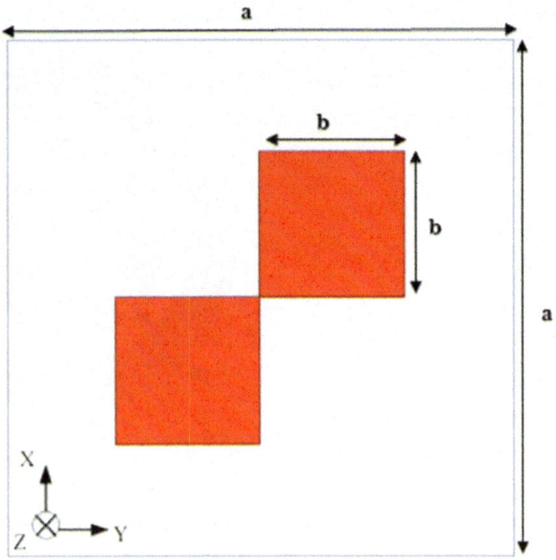


**Fig. 18.4** Top-view structure of the proposed MMA [8]

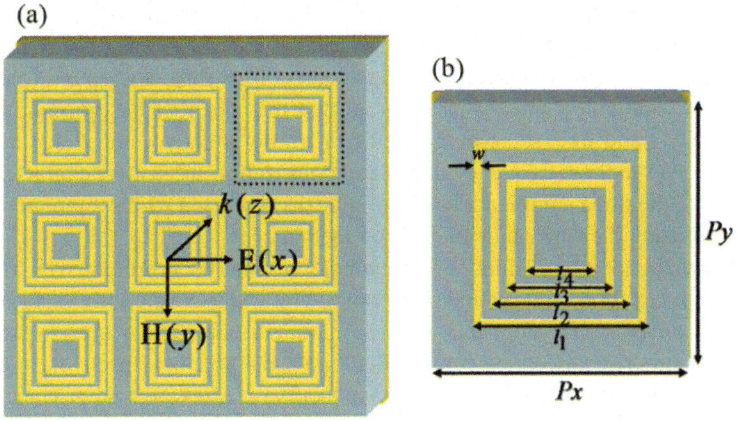


**Fig. 18.5** **a** Proposed metamaterial absorber structure. **b** Unit call of a structure [9]

Furthermore, different polarization angles and oblique incidence have been used to test the suggested structure's absorptivity. In addition, the proposed structure's absorption process has been investigated using surface current plots and electromagnetic field distribution. Finally, the suggested absorber was built, and experimental testing was done to evaluate how well it performed at various electromagnetic wave incident polarization and incidence angles. The resulting absorber is ultra-thin and

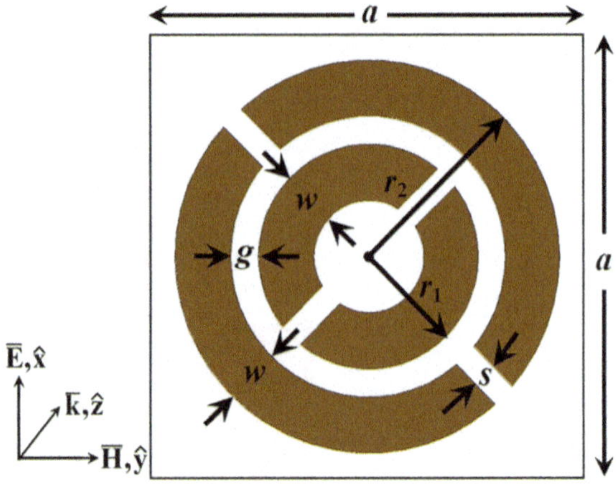


**Fig. 18.6** Ultra-wideband ultrathin unit call structure [10]

extremely small, representing a new approach to building broadband absorbers for various applications.

A wide-angle and polarization-stable triple-band efficient ultrathin metamaterial is given in Fig. 18.7. A metallic background plane, four groups of dipole arranged around a metallic ring, four pins connecting them, and a thin, inexpensive dielectric layer measuring only 1 mm make up the small single-unit cell [11]. The critical design parameters for the absorber are reviewed and optimized to show that they can be successfully separated and altered at each of the three absorption frequency points. The simulation results demonstrate that this absorber has good transverse electric (T.E.) and transverse magnetic (T.M.) absorption rates at wide angles of incident waves in three frequency bands and polarization-insensitive characteristics. In addition, the three significant absorption peaks predicted by the simulation were confirmed using the waveguide measuring method.

A design for an infrared dual-band metamaterial absorber is presented by Ma [12], as shown in Fig. 18.8. The ground plane of the unit cell is covered in a layer of $SiO_2$ dielectric, surrounded by four split-ring resonators that are spaced apart from it. The absorber shows two absorption peaks at 4.17 um and 4.86 um, respectively, of 90.3% and 88.4%. The multireflection interference theory explains the gold ground plane's absorption mechanism in the Drude model. The developed absorber's wide range of incidence angles and polarization direction insensitivity are helpful in real-world applications. For example, the power reflection of a dual-band metamaterial absorber was studied using a microscope and FTIR spectrometer. The incoming light's angle with regard to the sample's surface was around 30′. When the incidence angle rises from 0 to 60° for both T.E. and T.M. polarizations, the two absorption maxima stay higher than 90% with barely any positional displacement.

As shown in Fig. 18.9, an alloy ground plane, a patch formed of a layer of insulation, and a low-conductivity alloy are joined to form a metamaterial absorber. Wang and Wang [13] suggest a straightforward method for increasing the bandwidth of this absorber. Using this method, a polarization-insensitive, ultra-broadband absorber can

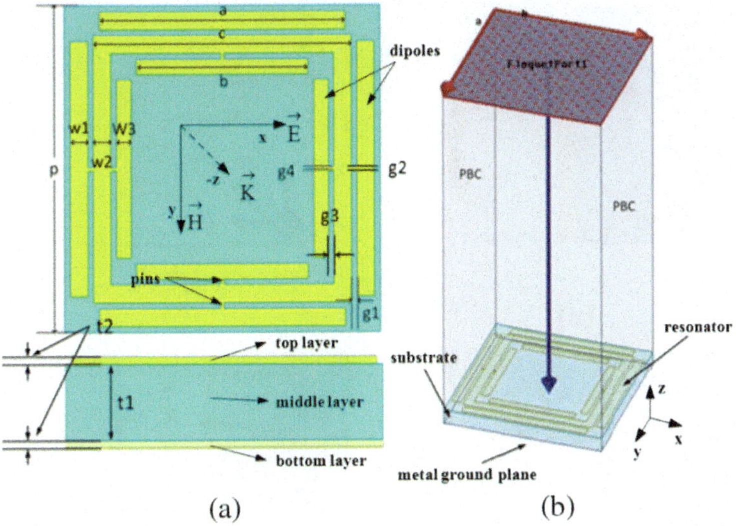


**Fig. 18.7** **a** Physical properties of the periodic absorber cell. **b** Model of the HFSS simulation cell made by Floquet port [11]

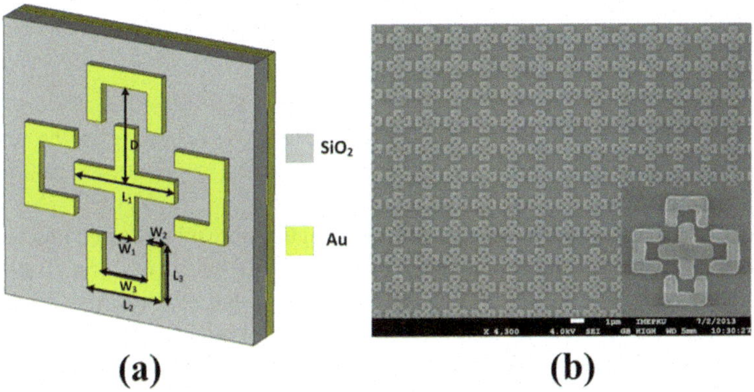


**Fig. 18.8** **a** Dual-band absorber unit cell. **b** S.E.M. picture of the dual-band absorber [12]

be created by stacking three patches of an alloy of various sizes. The frequency range of 1.34 THz achieves more than 90% coverage, with the centre frequency approximately at 1.90 THz. The device's relative absorption bandwidth has dramatically grown to 70.4%. Four distinct but closely spaced resonance frequencies are thought to overlap to produce ultra-wideband absorption. The results of the alloy metamaterial absorber idea seem to be highly promising for solar cells, sensing, and imaging applications.

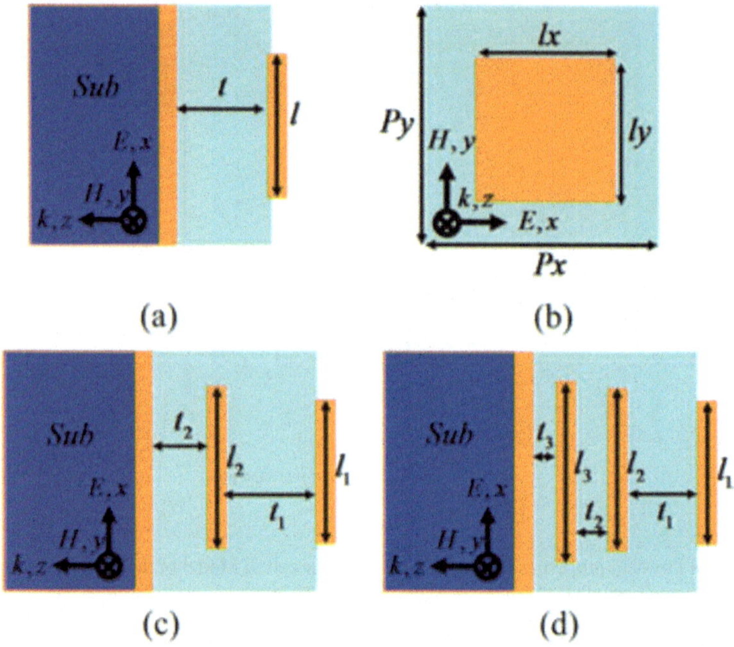


**Fig. 18.9** Cross-section view **a** top, **b** single band alloy, **c** and **d** broadband and ultra-broadband [13]

In Ref. [14], a novel metamaterial absorber (MMA) operating in the terahertz range has been introduced for potential use in sensing applications, as shown in Fig. 18.10.

The configuration comprises a pair of indistinguishable metallic patches, a dielectric medium that separates them, and a metallic plane that serves as a ground. The finite element method has been employed to model and evaluate the design. The study reveals that the MMA exhibited a significant resonant peak, displaying an absorbance rate of nearly 100% at a frequency of 4.5 THz. This phenomenon can be attributed to the coupling effect between the two identical patches. Furthermore, the distribution of surface currents, absorption mechanisms, and structural parametric analysis has been examined.

Later, efforts were undertaken to broaden M.M.A.'s absorption bandwidth. The construction of multiband, broadband, and frequency-tunable metamaterial absorbers received much attention from researchers [15–18]. A co-planar structure combined several resonant unit cells to create the first M.M.A.'s with multiband absorption properties [19–21].

A flexible, optically transparent M.M.A. was proposed by Min et al. [22]. A topology optimization method based on a genetic algorithm (G.A.) was used in the suggested design to demonstrate broadband absorption properties. The design,

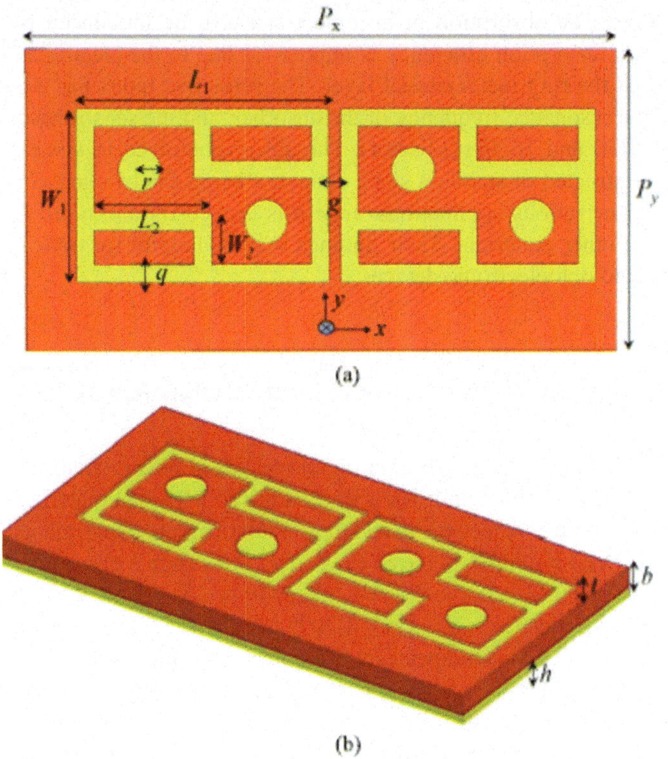


**Fig. 18.10** Proposed absorber design **a** top view and **b** side view [14]

fabrication, and testing of a metamaterial absorber with optical transparency with broadband absorption spanning 5.3–15 GHz illustrate the technique's effectiveness.

Further evidence of their successful theoretical research and planning comes from the excellent concordance between numerical simulation and experimental data. The M.M.A. is a favoured choice for practical applications because both studies show remarkable absorption capabilities.

In order to research various metamaterial absorbers relevant to IoT, Amiri et al. [23] conducted their research. Furthermore, the authors of this work conducted extensive research on practical M.M.A. applications such as sensing, optical switching, and energy harvesting. Therefore, the practical advice for using M.M.A. in industrial applications is the most valuable data in this review.

A stainless steel absorber for terahertz radiation was demonstrated by Yu et al. [24]. This absorber consists of a substrate and layers of all-metal metamaterial, with a layer of metamaterial shaped like a ring geometrically. The work not only demonstrates its theoretical validity but also enhances sensing capabilities. According to the model, the absorber exhibits a narrow peak absorption of 99.95% at a frequency of 1.563 THz. The absorption of an object can also be influenced by its geometry. The high absorption efficiency is impressive due to the electric and magnetic resonances of the ring metamaterial layer. The refractive index has a considerable impact on the resonance frequency of the absorber, and the sensor has a sensitivity of 74.18 m/RIU and 36.35 ($Q$-Factor). The absorber utilized in this experiment is similarly constructed of stainless steel, albeit not being sandwiched. Stainless steel may be used in various applications since it is less costly and more accessible to produce than other materials. Their findings may be helpful for optical switching, biosensors, and optoelectronic devices.

## 18.3 Analysis of the Proposed Metamaterial Absorber

The proposed study entailed developing and testing a novel, small, and perfectly absorbing Complementary Split Ring Resonator (CSRR) metamaterial. This novel metamaterial's performance was assessed in numerical and experimental setups. In addition, a C.S.T. simulator was used to conduct the design's electromagnetic analysis. The total size of the unit cell is calculated to be $15 \times 15 \times 1.575$ mm$^3$. Three unique resonant frequencies, at 3.7275, 4.7075, and 5.9325 GHz, have been observed in the produced unit cell. The spectrum of frequencies includes both the S-band and the C-band. The proposed device has a unique design allows for the manifestation of double negative characteristics despite its small dimensions and simple resonator configuration. The proposed design is classified as a Double Negative Material (D.N.G.) due to the negative values of both permittivity ($\varepsilon$) and permeability ($\mu$). The investigation then focuses on the capacity of the structure to absorb electromagnetic energy. The experimental findings indicate that the suggested configuration demonstrates 97, 99, and 98% absorption efficiencies at the resonance frequencies. This research proved that the proposed design is superior to those found in the literature. The small size of the proposed unit cell provides the necessary resonance frequency and the near-unity absorption at these frequencies. As a result, the proposed unit cell is the best candidate to be used in satellite applications. Figure 18.11 shows the unit cell structure, while Fig. 18.12 depicts the proposed structure's absorption and reflection co-efficient responses. Table 18.1 compares the proposed structure with the previous literature designs.

## 18.4 Conclusion

This article provides a brief overview of metamaterial absorbers in the context of satellite applications. It covers various structures that make use of metamaterials and the principles associated with them. Furthermore, the research efforts regarding metamaterial absorbers and their applications have been briefly mentioned. The proposed structures have potential applications in various fields, including satellite

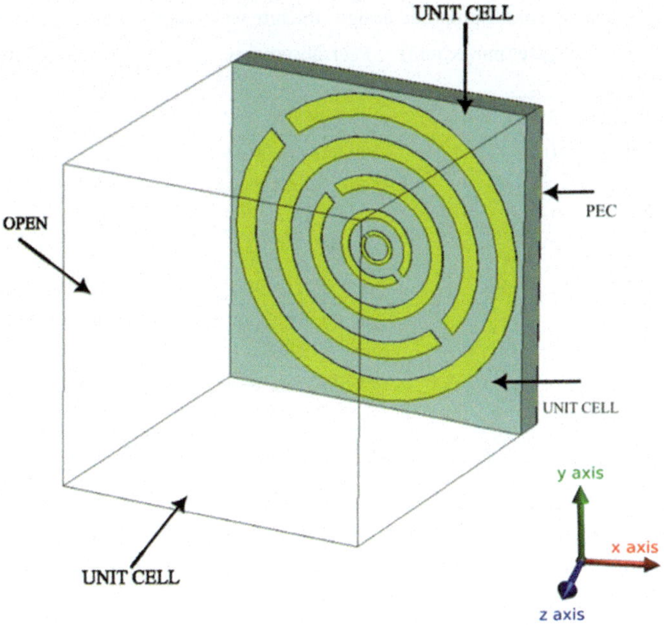


**Fig. 18.11** Proposed structure of the unit cell with its boundary conditions

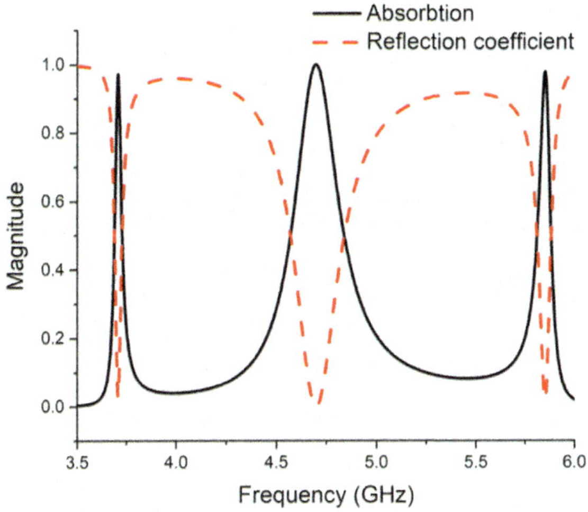


**Fig. 18.12** Magnitude of absorption and reflection versus frequency

**Table 18.1** Comparison of the proposed design structure with other designs in the literature

| Reference | Unit cell size (mm × mm) | Peak absorption | Resonance frequencies (GHz) |
|---|---|---|---|
| [6] | 10.69 × 10.69 | 99.99% | 27.5, and 31.00 |
| [25] | 10.5 × 10 | 99.95, and 99.61% | 8.006, and 11.525 |
| [9] | 8.5 | 99.41, 97.53, 99.06, and 95.59% | 0.777, 1.13, 1.53, and 2.06 |
| [10] | 7.1 × 7.1 | > 90% | 10.05 |
| [11] | 11 | > 90% | 3.25, 9.45, and 10.90 |
| [15] | 10 × 10 | 99.59, 95.49, 97.5, 99.25 and 93.89% | 3.35, 5.1, 9.5, 11.55, and 16.9 |
| This work | 15 × 15 | 97, 99, and 98% | 3.7275, 4.7075, and 5.9325 |

technology, phase imaging, electromagnetic compatibility (E.M.C.), electromagnetic interference (E.M.I.), and stealth technologies. The networks used in this study were designed using Design Pro software, which incorporates the computer simulation technique (C.S.T.).

**Acknowledgements** This work is supported and funded by a Telekom Malaysia Research & Development grant, RDTC 221039, MMUE/220015, TM, Malaysia.

## References

1. M.A.W. Nordin, M.T. Islam, N. Misran, Design of a compact ultrawideband metamaterial antenna based on the modified split-ring resonator and capacitively loaded strips unit cell. Prog. Electromagn. Res. **136**, 157–173 (2013)
2. S.H. Blumenthal, Medium earth orbit Ka band satellite communications system, in *MILCOM 2013—2013 IEEE Military Communications Conference* (2013), pp. 273–277
3. M. Amiri, F. Tofigh, N. Shariati, J. Lipman, M. Abolhasan, Ultra wideband dual polarization metamaterial absorber for 5G frequency spectrum, in *2020 14th European Conference on Antennas and Propagation (EuCAP)*, Copenhagen, Denmark (2020), pp. 1–5
4. W.J. Padilla, D.N. Basov, D.R. Smith, Negative refractive index metamaterials. Mater. Today **9**(7), 28–35 (2006)
5. A. Hoque, M.T. Islam, A.F. Almutairi, T. Alam, A polarization independent quasi-TEM metamaterial absorber for X and Ku band sensing applications (2018)
6. A. Musa, Dual-band metamaterial absorber for Ka-band satellite application, in *2021 7th International Conference on Space Science and Communication* (2021), pp. 151–155
7. C. Barde, A. Choubey, R. Sinha, S. Kumar, M. Prakash, A compact wideband metamaterial absorber for Ku band applications. J. Mater. Sci. Mater. Electron. (2020)
8. C. Barde, A. Choubey, R. Sinha, Wide band metamaterial absorber for Ku and K band applications. J. Appl. Phys. **175104** (2019)
9. X. Zhai, G.Z. Wang, Design of a four-band and polarization-insensitive terahertz metamaterial absorber. IEEE Photonics J. **7**(1), 1–8 (2015)

10. S. Ghosh, S. Bhattacharyya, D. Chaurasiya, K.V. Srivastava, An ultrawideband ultrathin metamaterial absorber based on circular split rings. IEEE Antennas Wirel. Propag. Lett. **14**, 1172–1175 (2015)
11. H. Zhai, C. Zhan, Z. Li, C. Liang, A triple-band ultrathin metamaterial absorber with wide-angle and polarization stability. IEEE Antennas Wirel. Propag. Lett. **14**, 241–244 (2015)
12. W. Ma, Y. Wen, X. Yu, Theoretical and experimental demonstrations of a dual-band metamaterial absorber at mid-infrared. IEEE Photon. Technol. Lett. **26**(19), 1940–1943 (2014)
13. B. Wang, L. Wang, G. Wang, W. Huang, X. Li, X. Zhai, Metamaterial-based low-conductivity alloy perfect absorber. J. Lightwave Technol. **32**(12), 2293–2298 (2014)
14. A. Mohanty, O.P. Acharya, B. Appasani, S.K. Mohapatra, M.S. Khan, Design of a novel terahertz metamaterial absorber for sensing applications. IEEE Sens. J. **21**(20), 22688–22694 (2021)
15. P. Garg, P. Jain, Novel ultrathin penta-band metamaterial absorber. AEUE Int. J. Electron. Commun. **116**, 153063 (2020)
16. X. Dai, S. Lin, S. Mao, G. Qing, Broadband metamaterial absorber with double resonant modes using single-layer substrate. AEUE Int. J. Electron. Commun. **119**, 153179 (2020)
17. S.U. Rahman, H. Ullah, M. Sajjad, M.I. Khan, Q. Cao, Comment on 'Design and characterization of an ultrabroadband metamaterial microwave absorber'. IEEE Photonics J. 1 (2020)
18. M. Bakır, M. Karaaslan, E. Unal, O. Akgol, C. Sabah, Microwave metamaterial absorber for sensing applications. Opto-Electron. Rev. **25**(4), 318–325 (2017)
19. H.Y. Zheng, X.R. Jin, J.W. Park, Y.H. Lu, J.Y. Rhee, W.H. Jang, Tunable dual-band perfect absorbers based on extraordinary optical transmission and Fabry-Perot cavity resonance. Opt. Express (2012)
20. W. Kim, J.W. Park, P. Van Tuong, J.Y. Rhee, K.W. Kim, W.H. Jang, Multi-band metamaterial absorber based on the arrangement of donut-type resonators. Opt. Express (2013)
21. N. Wang, M. Absorber, Novel quadruple-band microwave metamaterial absorber. IEEE Photonics J. **7**(1), 1–6 (2015)
22. P. Min, Z. Song, L. Yang, V.G. Ralchenko, J. Zhu, Optically transparent flexible broadband metamaterial absorber based on topology optimization design. Micromachines (2021)
23. M. Amiri, F. Tofigh, N. Shariati, J. Lipman, M. Abolhasan, Review on metamaterial perfect absorbers and their applications to IoT. IEEE Internet Things J. **8**(6), 4105–4131 (2021)
24. J. Yu, T. Lang, H. Chen, All-metal terahertz metamaterial absorber and refractive index sensing performance. Photonics (2021)
25. J.D. Harrison, Design and analysis of compact perfect metamaterial absorber for X-band applications (2019)

# Chapter 19
# Preliminary Analysis on the Effect of Temperature on LSTM-Based PWV Forecasting

**Christian David C. Aquino and Ernest P. Macalalad**

**Abstract** Precipitable water vapor (PWV) can be applied in meteorological and climatological studies as it is a reliable indicator of the state of the atmosphere. However, the availability of PWV is unreliable. Having a more complete repository of PWV data allows for climatological studies as well as provides another avenue for rainfall prediction. Hence, there is a need to forecast and impute data to improve its robustness. To do this, the use of machine learning (ML) techniques such as Long Short-Term Memory (LSTM) can be used as it specializes in forecasting time-series values such as PWV. The effect of adding other meteorological parameters was analyzed and compared to a baseline model. One-step forecasting performance was similar throughout all of the models tested, having a spread within 0.3 mm with three outliers for the PWV model given by 1.9560, 1.9565, 1.9684; four outliers for the PWV-Temperature model given by 1.9573, 1.9521, 1.9523, 1.9569; and one outlier each for the PWV-Time and PWV-Time-Temperature models given by 1.9765 and 1.9815, respectively. It can be also noted that for data imputation, and the addition of other meteorological parameters had influenced the behavior of the imputation to capture the intricacies of PWV. The median and mean values for the PWV model, PWV-Time model, PWV-Temperature model, and PWV-Time-Temperature model are as follows: 10.869, 14.553; 15.661, 14.400; 11.724, 16.685; and 9.9262, 14.176, respectively. No outliers were observed for the data imputation models. It is recommended that further studies look into inputting other meteorological variables into the ML model. Moreover, it is also recommended that optimizations regarding the inputs and outputs (hyperparameters, lead and lag times) be done in order to improve the forecasting results.

---

C. D. C. Aquino · E. P. Macalalad (✉)
Department of Physics, Mapúa University, Manila, Philippines
e-mail: epmacalalad@mapua.edu.ph

 
M. T. Islam et al. (eds.), *Proceedings of the 8th International Conference on Space Science and Communication*, Springer Proceedings in Physics 303,
https://doi.org/10.1007/978-981-97-0142-1_19

## 19.1 Introduction

Precipitable water vapor (PWV) is widely described as the amount of water vapor found in a vertical column of atmosphere that has the potential to condense into rain or snow. Sherwood et al. [1] define PWV as the main transport mechanism of the atmosphere. Moreover, Stocker et al. [2] report that PWV is the most abundant naturally occurring greenhouse gas. The easiest method of measuring PWV is by radiosondes, an in situ measurement system on a weather balloon, twice a day. Radiosonde data, comprised of temperature, pressure, and relative humidity, along the radiosonde's rising path results in a computed PWV reading. While easy to do, this method proves to be expensive and unsustainable as it leaves non-recoverable apparatus into the atmosphere [3]. In addition, it has a limited spatial resolution due to the nature of the measurement process.

With the use of the Global Navigation Satellite System (GNSS), however, PWV data can be obtained more frequently and allows for a significant improvement in its temporal and spatial resolution. GNSS-based data gathering works by extracting PWV data from signal delays caused by atmospheric lensing. As a result, PWV can be obtained more frequently in contrast to radiosonde-based data collection [4, 5].

To further improve the availability of PWV data, machine learning (ML) models can be used to forecast PWV since it is a time series of PWV values. Traditionally, forecasting PWV can be achieved by using statistical methods such as the Naïve-Bayes, averaging, and exponential smoothing [6]. The main downside for using classical statistical methods is that it does not capture the long-term nuances of time-series data. On the other hand, the Long Short-Term Memory (LSTM) algorithm can handle complex and nonlinear relationships between weather variables over long periods of time, making it highly suitable for PWV forecasting. It can also function despite the presence of incomplete and noisy data, meaning that the resulting model can also be applied toward the imputation of missing data on GNSS-PWV archives [7].

At its core, LSTM is a recurrent neural network that uses its output as an input for its next iteration. By implementing 'memory' gates that control the flow of information, the LSTM network allows for a state-based prediction. These gates can be thought of as filters that decide which information is useful or irrelevant to the system [8].

The goal of this paper is to leverage the LSTM network alongside Temperature to forecast PWV. Specifically, this paper would like to train an LSTM Model for forecasting PWV and measure the capability of the model in data imputation. For this paper, the researchers only focused on the forecasting and imputation of PWV out of the variety of physical quantities estimated and/or derived from GNSS-based meteorology.

## 19.2 Methodology

### *19.2.1 Data Preparation*

The data used to train the model were obtained from the University Corporation for Atmospheric Research's (UCAR's) SuomiNet, an international network of GPS receivers intended to provide global atmospheric PWV measurements [9]. This dataset was used due to its relative availability and accessibility to the researchers. The text files taken from UCAR contain columns for time, PWV Value (mm), PWV Error (mm), Zenith Delay (mm), Surface Pressure (Pa), Surface Temperature (°C), and Surface Relative Humidity (%). These parameters will be explored to determine their feasibility as inputs in forecasting PWV. The gathered data set has a semi-hourly temporal resolution of observations from the PIMO station (14° 38′ 8.59″ N, 121° 4′ 39.84″ E) located in Quezon City, Philippines. Data from the years 2008 to 2011 were chosen for their relative completeness.

Additionally, since the magnitude of the meteorological variables vary from tens to thousands, all variables were normalized within the range of − 1 to 1 in order to make sure that all of the variables contribute equally to the forecasting of PWV. For this study, linear normalization (Min–max Scaling) was used. This is defined by

$$x' = \frac{x - \min(x)}{\max(x) - \min(x)}, \tag{19.1}$$

where $x$ is the unnormalized value, min($x$) and max($x$) are the global minimum and maximum values for the dataset, respectively, and $x'$ is the normalized value.

From the study of Manandhar et al. [10], Eqs. 19.2–19.5 were utilized as features:

$$\mathrm{DoY}_x = \cos\left(\frac{2\pi \cdot \mathrm{DoY}}{365}\right), \tag{19.2}$$

$$\mathrm{DoY}_y = \sin\left(\frac{2\pi \cdot \mathrm{DoY}}{365}\right), \tag{19.3}$$

$$\mathrm{HoD}_x = \cos\left(\frac{2\pi \cdot \mathrm{HoD}}{24}\right), \tag{19.4}$$

$$\mathrm{HoD}_y = \sin\left(\frac{2\pi \cdot \mathrm{HoD}}{24}\right), \tag{19.5}$$

where DoY and HoD are the days of year and hours of day, respectively. These mapping functions served as continuous representations of dates and times which helped the model to associate the correct timestamps with their corresponding PWV values. Furthermore, the use of sine and cosine functions allowed for the normalization of timestamps across a range of − 1 to 1.

This was done so that the LSTM network can encapsulate the time-dependent nuances such as seasonality and daily variation into the model [10]. Furthermore, as the study delves into time-series forecasting, taking the past $N$ observations to forecast future values of PWV was necessary in training the LSTM Model. The model was trained using the data from 2008 to 2010 then tested using the data from 2011.

Since the observations had a bihourly temporal resolution, there were 48 recorded PWV values taken each day. From this, an LSTM algorithm was used to predict the 49th value, giving a lead of one entry or a lead time of 30 min. Moreover, the Sliding Window Method [11] was used before model training to group the data into "windows" that contain 48 consecutive observations. These observations amount to a total lag time of 24 h. Each window corresponded to one output prediction.

### *19.2.2 LSTM Model Training*

Moving forward, Fig. 19.1 shows the architecture of the network used in the study. The network is comprised of a layer of LSTM followed by two dense layers. This architecture was chosen as it utilizes the LSTM network alongside one hidden dense layer and one output dense layer. There were 64 nodes each for the LSTM and hidden dense layer. This value came from a rough trial and error that considered training time and prediction accuracy.

To optimize the weights and biases of the model, the Stochastic Gradient Descent (SGD) optimization [12] was used. SGD is a variant of the common Gradient Descent algorithm where a random subset of the data is taken each iteration to save time instead of considering the error for the whole training set. The Gradient Descent algorithm is defined by the following equation:

$$p_{n+1} = p_n - \eta \nabla f(p_n). \tag{19.6}$$

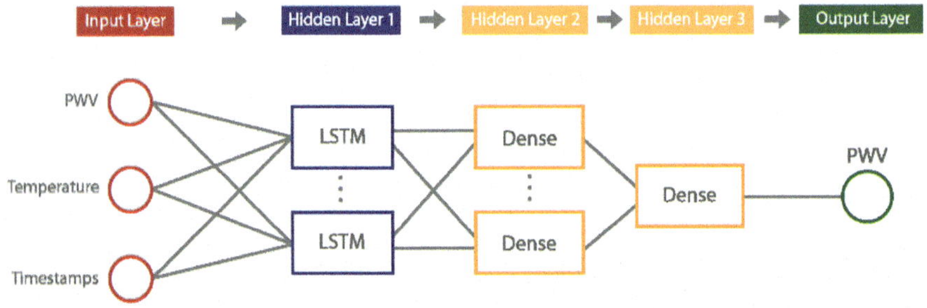


**Fig. 19.1** Design of the neural network model containing the LSTM network alongside two dense layers

Here, the learning rate $\eta$ was used to successfully find the best values for the weights and biases used in the model fitting process. This constant controls the step size of the descent—too big a value might result in the model skipping the minima, while a value that is too small might result in an overly extended training time. In general, the gradient descent algorithm is the numerical equivalent of getting the derivative of a function and equating it to 0 (optimization).

Furthermore, the Huber loss [13] training metric was utilized to achieve robust regression by disregarding outliers in the data. Huber loss was designed to work like the sum of squared error for small values of error and act like the absolute value error for large values of error. This was done in order to create a loss function that was immune to both outliers all while performing well for small values of error. The model was trained using a maximum of 300 epochs.

However, an early stopping mechanism was implemented which halted training whenever the loss of the model did not change significantly after 10 epochs. Moreover, since the LSTM algorithm assumes random weights and biases prior to training, the LSTM network is said to be non-deterministic. Meaning, training two models bearing the same configurations will not necessarily result in the same prediction. To fix this, the researchers hard-coded seed values prior to training as the TensorFlow and Numpy libraries allow for the setting of these seed values. Here, a seed value is an integer that translates to a set of weights and biases for the network to use as initial parameters. In the spirit of being reproducible, the researchers trained the LSTM model 50 times with the same configuration, the only difference per run being the seed values. This resulted in 50 models with different Root Mean Square Error (RMSE) values.

To test the effect of adding meteorological inputs to the LSTM model, a baseline model needed to be defined. Here, the LSTM model used was based on a similar model done by Jain et al. [6] in their 2020 paper. Their model was trained with PWV as its sole input to forecast future PWV values. This paper, on the other hand, will be using a combination of time and temperature alongside PWV as inputs to forecast future PWV values. Lastly, one-step forecasting was recursively applied for data imputation. Here, the outputs from prior predictions were used as inputs to forecast 200 time-steps into the future.

## 19.3 Analysis of Results

As shown in Fig. 19.2, the PWV model with the highest (Fig. 19.2a) and lowest (Fig. 19.2e) RMSE show that the prediction levels out after about 50 points and proceeds to adapt a constant output. This behavior can be attributed to saturation of the model. Since the previous output of the model is being used as the input of the very next timestep, there exists a point wherein the differences of the inputs are not significant enough that it changes the output. The model assumes that because the previous inputs are very close in terms of value, the output should also obey that

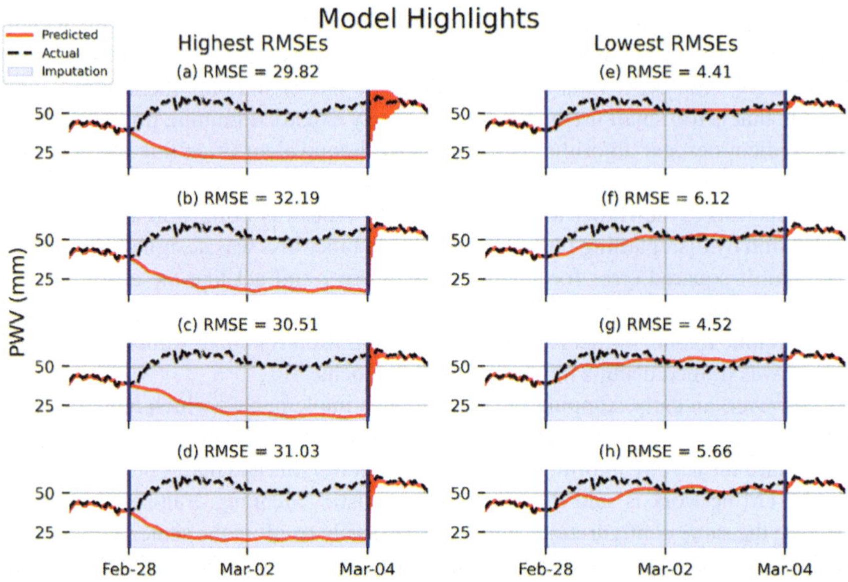


**Fig. 19.2** Data imputation of each model type with the highest RMSEs (left): **a** PWV model, **b** PWV-time model, **c** PWV-temperature model, and **d** PWV-time-temperature model; and the lowest RMSEs (right): **e** PWV model, **f** PWV-time model, **g** PWV-temperature model, and **h** PWV-time-temperature model

trend. This behavior is the result of only using one input, PWV in this case, as a predictor.

The combination of the cyclical interpretation of time with PWV, however, shows a different behavior. Unlike the PWV model, the PWV-Time model (Fig. 19.2b, f) does not stabilize to a constant for either highest or lowest RMSE values. Instead, this model exhibits a periodic behavior that lasts throughout the imputation period. This behavior can be attributed to the cyclical nature of the timestamps. Because the values of the mapped timestamps change with respect to time constantly, the model output also exhibits a periodic behavior.

Meanwhile, the imputation for the PWV-Temperature model starts out with erratic behavior but levels out to a constant value after around 100 points for the model with the highest RMSE (Fig. 19.2c). However, the model with the lowest RMSE did not level out (Fig. 19.2g). This behavior can be attributed to the erratic nature of Temperature. Since the model accepts temperature as another output, it accounts for another complex relationship between the two meteorological variables. This results in a prediction that relies heavily on the PWV input (such as in Fig. 19.2c) or on the temperature (such as in Fig. 19.2g). The difference between the model with the highest and lowest RMSE falls on the weights assigned by the model to the input parameters—associating importance onto the PWV input results to a model that stabilizes, while associating more importance to the temperature input results to a more erratic model.

Lastly, the highest RMSE model (Fig. 19.2d) for the PWV-Time-Temperature model performed similarly. The model exhibited periodicity but leveled out to a constant while the lowest RMSE model (Fig. 19.2h) showcased more pronounced ups and downs which lasted throughout the imputation period. The same interpretation can be made as in the PWV-Temperature model, except this time the cyclical effect of the timestamps can also be observed.

Figure 19.3 shows the results after training the model 50 times for each set of parameters. The boxplot for forecasting (Fig. 19.3a) shows that the median values (in mm) from all of the runs are 1.9351, 1.9400, 1.9289, and 1.9337 for PWV, PWV-Time, PWV-Temperature, and Time-Temperature-PWV, respectively. In addition, the mean values (in mm) for the model runs are 1.9367, 1.9410, 1.9310, and 1.9348, respectively. Moreover, the Interquartile Range (IQR) (in mm) for PWV alone ranges from 1.9305 to 1.9398, PWV-Time ranges from 1.9309 to 1.9489, PWV-Temperature ranges from 1.9248 to 1.9355, and PWV-Time-Temperature ranges from 1.9235 to 1.9446. From this, we can see that the spread of the RMSEs from the different models are within a range of about 0.3 mm across all the model types.

The forecasting resulted in three outliers (in mm) for the PWV model runs. It was recorded as 1.9560, 1.9565, and 1.9684. Meanwhile, the PWV-Time model recorded an outlier RMSE equal to 1.9765 mm and the PWV-Temperature model recorded the following outliers (in mm): 1.9573, 1.9521, 1.9523, and 1.9569. Lastly, the PWV-Time-Temperature model recorded an outlier equal to 1.9815 mm.

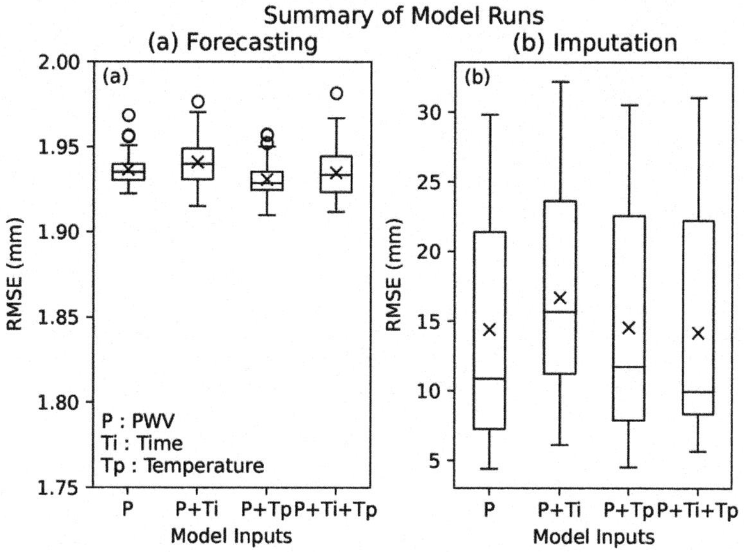


**Fig. 19.3** Boxplots for **a** forecasting and **b** data imputation containing the performance of the 50 model runs for the PWV, PWV-time, PWV-temperature, and PWV-time-temperature models

On the other hand, the performance of the models for data imputation is presented on the right (Fig. 19.3b). The data show that the median value (in mm) generated from all of the runs are 10.869, 15.661, 11.724, and 9.9262 for PWV, PWV-Time, PWV-Temperature, and PWV-Time-Temperature, respectively. Meanwhile, the mean values acquired for the model runs are 14.400, 16.685, 14.553, and 14.176, respectively. The IQR (in mm) for the models are 7.2583–21.392, 11.220–23.624, 7.8676–22.5624, and 8.3298–22.224, respectively. It is also noted that there were no outliers found for all of the model runs. Under the recursive application of a one-step step forecasting for data imputation, larger RMSE values were observed.

## 19.4 Conclusions and Recommendations

To summarize, it can be observed that in terms of RMSE, the LSTM model shows similar results for one-step forecasting for all of the models tested. However, recursively applying the one-step forecast for data imputation reveals that additional input variables can change the behavior of the prediction to mimic the properties of PWV but do not necessarily minimize RMSE.

The researchers recommend that work on the optimization of the hyperparameters (e.g., node count, hidden layers, optimizer parameters) of the model be done in order to further improve the RMSE and generalization of the models. Furthermore, varying the imputation length and the lead and lag periods can also be investigated so as to push the prediction capabilities of the model. Lastly, it is suggested that future researchers consider exploring the effects of other GNSS-based parameters such as pressure and relative humidity alongside their combinations as inputs to the model.

## References

1. S.C. Sherwood, R. Roca, T.M. Weckwerth, N.G. Andronova, Tropospheric water vapor, convection and climate: a critical review. Crit. Rev. **2009**, 1–93 (2009)
2. T. Stocker, D. Qin, G. Plattner, M. Tignor, S. Allen, J. Boschung, A. Nauels, Y. Xia, B. Vincent, P. Midgley, Climate change 2013: the physical science basis. Technical Report (2013)
3. D. Gaffen, W. Elliott, A. Robock, Relationships between tropospheric water vapor and surface temperature as observed by radiosonde. Geophys. Res. Lett. **19**, 1839–1842 (1992)
4. J. Saastamoinen, Atmospheric correction for the troposphere and stratosphere in radio ranging satellites, in *Geophysical Monograph Series*, Mar 1972 (1972), pp. 247–251
5. M. Bevis, S. Businger, T.A. Herring, C. Rocken, R.A. Anthes, R.H. Ware, GPS meteorology: remote sensing of atmospheric water vapor using the global positioning system. Ohio State University, Massachusetts Institute of Technology, Technical Report D14 (1992)
6. M. Jain, S. Manandhar, Y.H. Lee, S. Winkler, S. Dev, Forecasting precipitable water vapor using LSTMs, in *2020 USNC-URSI Radio Science Meeting (Joint with AP-S Symposium), USNC/URSI 2020—Proceedings*, June 2020, pp. 147–148
7. D. Hosahalli, I. Gad, A generic approach of filling missing values in NCDC weather stations data, in *2018 International Conference on Advances in Computing, Communications and Informatics, ICACCI 2018* (2018), pp. 143–149

8. S. Hochreiter, J. Schmidhuber, Long short-term memory. Neural Comput. **9**(8), 1735–1780 (1997)
9. R.H. Ware, D.W. Fulker, S.A. Stein, D.N. Anderson, S.K. Avery, R.D. Clark, K.K. Droegemeier, J.P. Kuettner, J.B. Minster, S. Sorooshian, SuomiNet: a real-time national GPS network for atmospheric research and education. Bull. Am. Meteorol. Soc. **81**(4), 677–694 (2000)
10. S. Manandhar, S. Dev, Y.H. Lee, Y.S. Meng, S. Winkler, A data-driven approach for accurate rainfall prediction. IEEE Trans. Geosci. Remote Sens. **57**(11), 9323–9330 (2019)
11. Y. Yao, L. Shan, Q. Zhao, Establishing a method of short-term rainfall forecasting based on GNSS-derived PWV and its application. Sci. Rep. **7**(1), 1–11 (2017)
12. H. Robbins, S. Monro, A stochastic approximation method. Ann. Math. Stat. **22**(3), 400–407 (1951)
13. P.J. Huber, Robust estimation of a location parameter. Ann. Math. Stat. **35**(1), 73–101 (1964)

# Chapter 20
# Comparison of Precipitable Water Vapor and Sea Surface Temperature Anomalies Over the Tanay and Legazpi, Philippines from 2013 to 2022

**Andrei V. Bibaoco, Princess Tucio, and Ernest P. Macalalad**

**Abstract** Precipitable Water Vapor (PWV) is a meteorological parameter used to characterize the amount of Integrated Water Vapor (IWV) present in vertical column of the atmosphere. The El Nino Southern Oscillation (ENSO) is a meteorological phenomenon that causes weather variability. ENSO is categorized as a warm phase and a cold phase, which are El Nino and La Nina, respectively. The impact of ENSO is not uniform and varies, so predicting and understanding its influence requires taking into account meteorological variables and their connection with it. This paper presents the analysis of the Precipitable Water Vapor anomaly (PWVa) and Sea Surface Temperature anomaly (SSTa) from 2013 to 2022 using radiosondes Precipitable Water Vapor (PWV) over Tanay (14.5797° N, 121.3313° E), and Legazpi (13.1506° N, 123.7283° E) from the Integrated Global Radiosonde Archive (IGRA) and Sea Surface Temperature (SST) from the National Oceanic and Atmospheric Administration (NOAA). The anomaly method was used for the PWV and SST values to calculate the PWVA and SSTa. Through the PWV radiosonde data over Tanay and Legazpi from 2013 to 2022, this study investigates the PWVa and SSTa. Moreover, the PWVa shows variation during the identified ENSO events from 2013 to 2022. A high standard deviation was found in Legazpi compared to Tanay, which is attributed to its location and distance to the sea. The negative correlation indicates the PWVa increases as the SSTa decreases, while the SSTa increases while the PWVa decreases, relative to the La Nina and El Nino events. Additional stations and a long-term analysis of the PWVa and SSTa are proposed for future studies to explore and refine our understanding of these relationships for improved forecasting and climate modeling purposes.

A. V. Bibaoco · P. Tucio
Department of Earth and Space Sciences, Rizal Technological University, Mandaluyong City, Philippines
e-mail: avbibaoco@rtu.edu.ph

E. P. Macalalad (✉)
Department of Physics, MAPUA University, Manila City, Philippines
e-mail: epmacalalad@mapua.edu.ph

 
M. T. Islam et al. (eds.), *Proceedings of the 8th International Conference on Space Science and Communication*, Springer Proceedings in Physics 303,
https://doi.org/10.1007/978-981-97-0142-1_20

## 20.1 Introduction

Water vapor is integral to significant chemical and atmospheric processes that occur in the atmosphere. In terms of the vertical stability of the atmosphere and other factors, the distribution of water vapor is analogous to cloud formation and rain occurrence [1]. Precipitable Water Vapor (PWV) is a meteorological parameter used to characterize the amount of Integrated Water Vapor (IWV) present in a vertical column of the atmosphere [2]. PWV data is quantified and retrieved using a variety of techniques and methods, including GNSS meteorology, reanalysis products, MODIS, and the traditional radiosonde approach. The PWV data from the radiosonde is mainly utilized for data calibration and is frequently used to evaluate the outcomes of other approaches [3]. The radiosonde used meteorological balloons to perform high-level and accurate profiling of the atmosphere, including significant meteorological parameters such as relative humidity, pressure, and temperature [4]. The radiosonde can reach an altitude of about 30 km in the atmosphere and usually operates twice a day (0000UT and 1200UT) [5].

The El Nino Southern Oscillation (ENSO) is a meteorological phenomenon that causes weather variability. ENSO is categorized as a warm phase and a cold phase, which are El Nino and La Nina, respectively. The warm phase, or El Nino, is identified by a Sea Surface Temperature anomaly (SSTa) equal to or greater than 0.5 °C. La Nina is considered the cold phase of ENSO, which is marked by SSTs less than or equal to − 0.5 °C [6]. The ENSO is a period of fluctuation of air pressure and Sea Surface Temperature (SST) in the equatorial Pacific region, making it a significant factor and indicator for climatic and regional prediction of meteorological patterns considering the periodicity and predictability of the event [7]. The eastern Pacific Ocean's surface water heats more than usual during an El Nino event, increasing the rate of evaporation. This cause an increase in PWV, which causes heavy precipitation and the creation of weather disturbances in places that are generally dry [8]. Moreover, the cold phase of ENSO depicts wind movement in the Pacific Ocean rising from the east to the west, which influences the western Pacific Ocean [9].

The Clausius–Clapeyron equation states that a 1 K increase in the temperature ($T$) of the atmosphere will lead to a 7% rise in water vapor [10]. Additionally, a 1 K increase in SST will result in a rise in PWV of 1.5 mm in other regions and 5.2 mm in tropical regions [9]. The Sea Surface Temperature Anomaly (SSTa) has been used in numerous studies to show the ENSO peak and development in the Pacific Region, leading to an extensive investigation of ENSO and successfully demonstrating relationships to other meteorological variables like water vapor, temperature, and precipitation. The impact of ENSO is not uniform and varies. Predicting and determining the influence and effect of ENSO requires taking into account some variables, including its strength, year and time of occurrence, and potential interactions with other climatic patterns [3, 8, 9, 11, 12]. To anticipate and analyze the evolution of ENSO and its effects, it is essential and significant to take into consideration both atmospheric and oceanic factors [10, 13].

This study aims to analyze the Precipitable Water Vapor anomaly (PWVa) and Sea Surface Temperature anomaly (SSTa) using radiosonde over Tanay and Legazpi from 2013 to 2022. These stations were used based on the availability of data supported by previous studies [5, 14]. The PWVa and SSTa values were computed using the anomaly method using the PWV and SST values. The time series of PWVa and SSTa, including the monthly variation of PWV and relationship of PWVa and SSTa will be presented as well.

## 20.2 Methodology

### *20.2.1 Data*

In this study, the Precipitable Water Vapor (PWV) data were acquired from the Integrated Global Radiosonde Archive (IGRA v2) [15] from 2013 to 2022. The selection of the station is based on the availability of the data and location. In this study, two radiosonde stations were utilized, mainly Tanay (14.5797° N, 121.3313° E), and Legazpi (13.1506° N, 123.7283° E), as presented in Fig. 20.1.

The variations in PWVa values were analyzed together with the Sea Surface Temperature anomaly to investigate its response during ENSO events. The Sea Surface Temperature (SST) of the Nino 3.4 region and the Nino 4 region, derived from the National Oceanic and Atmospheric Administration (NOAA), was used to determine the event of an El Nino Southern Oscillation (ENSO) and the variation in the Sea Surface Temperature anomaly (SSTa).

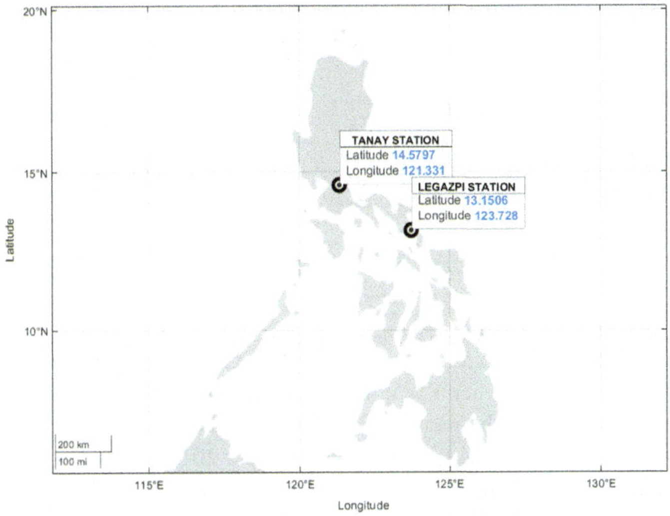


**Fig. 20.1** Map of location of the radiosonde stations used in the study

### 20.2.2 Data Processing

In this paper, the 12-hourly PWV data from the radiosonde stations were used to calculate the anomaly of the PWV. The anomaly of the PWV can be calculated through the anomaly method using the formula:

$$\mathrm{PWV_a} = \underline{x}_{\mathrm{m}} - \underline{x}_{\mathrm{M}}, \tag{20.1}$$

where $\underline{x}_{\mathrm{m}}$ is monthly mean PWV, and $\underline{x}_{\mathrm{M}}$ is the monthly-based period. The monthly mean is the summation of all of the monthly values of the variable, which is acquired by averaging all the daily values of the variables. The monthly-based period is acquired from the average of all the monthly mean values from the covered years in a station [16]. The anomaly is acquired through the difference between the monthly mean and the monthly-based period. Anomalies of the PWV and SST are acquired by observing out of the normal or usual behaviors, values, or trends of the PWV and SST. The data for the Sea Surface Temperature from the Nino 3.4 and Nino 4 regions will also be used by the anomaly method to determine the anomaly specifically for the years 2013–2022. The Nino 3.4 is a region in the Central Pacific (5° N–5° S, 120–170° W) which has a threshold of + 0.5 and − 0.5 °C. As it provides a more accurate representation of SST fluctuation, which affects patterns of deep tropical convection and atmospheric circulation, this region has been utilized to identify the existence of El Nino. Meanwhile, the Nino 4 region (5° N–5° S, 160° E–150° W) captures anomalies of SST in the Western and Central Pacific, which tend to have less variance than other regions [17]. In this study, a linear time series will be used to present the PWVa data. The monthly mean and standard deviation are also calculated to validate the monthly mean average of PWVa over the 10-year period wherein, correlation between PWVa and SSTa is explored.

## 20.3 Results and Discussion

### 20.3.1 Time Series of PWV

Figures 20.2 and 20.3 represent the monthly mean variation of PWV from Tanay and Legazpi from 2013 to 2022 from the standard deviation of each month. The PWV in Tanay and Legazpi is generally high from May to October, with a PWV ranging from 32 to 45 mm and 42 to 57 mm, respectively. Meanwhile, low PWV is observed from April to November, when PWV in Legazpi ranges from 29 to 55 mm and 20 to 40 mm in Tanay. Despite the increase in PWV from the months of May to October, a lower standard deviation was observed from years 2015 to 2021, which indicates less spread data for PWV, which is above 50 mm in Legazpi and 40 mm in Tanay.

Overall, a higher standard deviation was observed in Legazpi compared to Tanay, which is associated with the location of the station closer to the sea and its elevation.

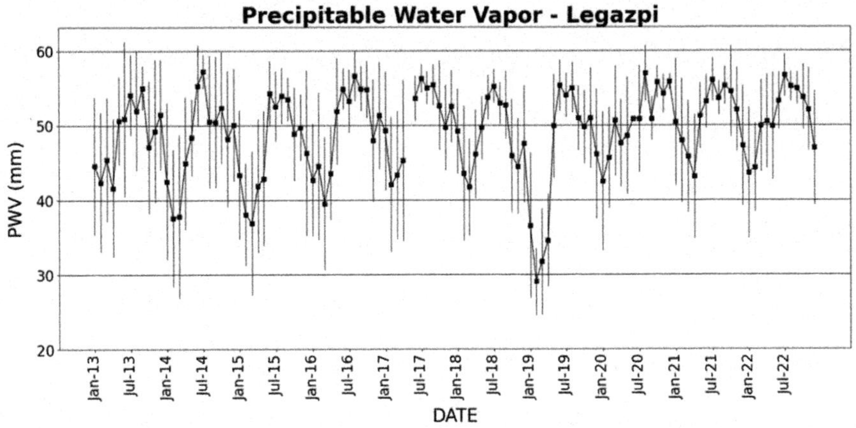


**Fig. 20.2** Monthly mean variation of PWV over Legazpi from 2013 to 2022

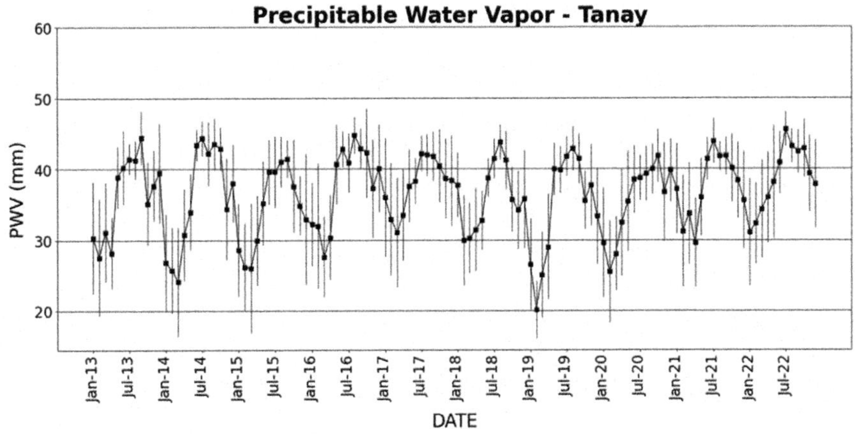


**Fig. 20.3** Monthly mean variation of PWV over Tanay from 2013 to 2022

Similar studies show station heights and distance to the sea contribute to the variation of PWV. Despite these differences, stations on land and near the coast show a good response to PWV [9, 12].

### *20.3.2 PWV and SST Anomalies*

In this study, the anomaly method was used for the PWV from the identified radiosonde station in Philippines. Figure 20.4 shows the monthly time series of radiosonde PWVa in Tanay and Legazpi from 2013 to 2022.

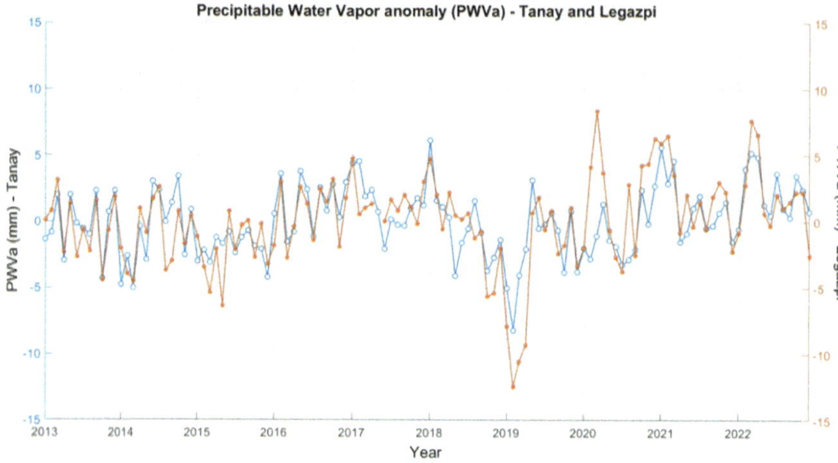


**Fig. 20.4** Monthly mean variation of PWV over Tanay from 2013 to 2022

A positive anomaly in PWV means PWV is higher than the calculated 10-year base period, while a negative anomaly indicates a lower PWV. A consistent positive anomaly in PWV was observed from 2016 to 2018. An increase in PWVa was observed from January to April in the years 2020 to 2022, when it surpassed the 5 mm mark. Meanwhile, a decrease in PWVa was observed in 2015–2016 and in 2018–2019, where it was lower than − 0.5 mm. In 2019, a PWVa of − 12 mm in Legazpi and − 8 mm in Tanay was the lowest observed from 2013 to 2022. The negative feedback from PWVa in 2016–2018 and 2020–2022, respectively, was due to a La Nina event in these years. Moreover, the increase in PWVa in 2015–2016 and 2018–2019 was due to an El Nino event. Similar to the studies of Domingo and Macalalad [14], PWV shows recurring patterns without distinctive behaviors while showing a variation of PWV occurrences during pre-monsoon periods. Moreover, PWV and rainfall are generally low during the dry season (November–April) and high during the wet season (May–October).

To assess the ENSO event in relation to PWVa, the SSTa was utilized from the Nino 3.4 and Nino 4 regions in the Pacific Ocean. Figure 20.5 presents the SSTa from the Nino 3.4 and Nino 4 regions from 2013 to 2022. The NOAA characterizes ENSO events by a five consecutive three-month running mean of SSTa in the Niño 3.4 region. A SSTa higher than or equal to 0.5 °C is classified as El Nino, and a SSTa lower than or equal to − 0.5 °C is considered La Nina.

The figure shows an increase in SSTa from 2014 to 2016 and from 2018 to 2019, which is associated with El Nino events in the identified years. A greater response was observed in the Nino 3.4 region, where there was a significant fluctuation from 2015 to 2016, where it reached 2.74 and 1.39 °C for the Nino 4 region. Meanwhile, a decrease in SSTa was observed in late 2016 to early 2017, and in 2018, where an SSTa of − 0.74 °C and − 0.99 °C was observed in 2016 and 2018, respectively. From 2020 to 2022, a decrease in SSTa was also observed, with − 1.40 °C and − 1.15 °C

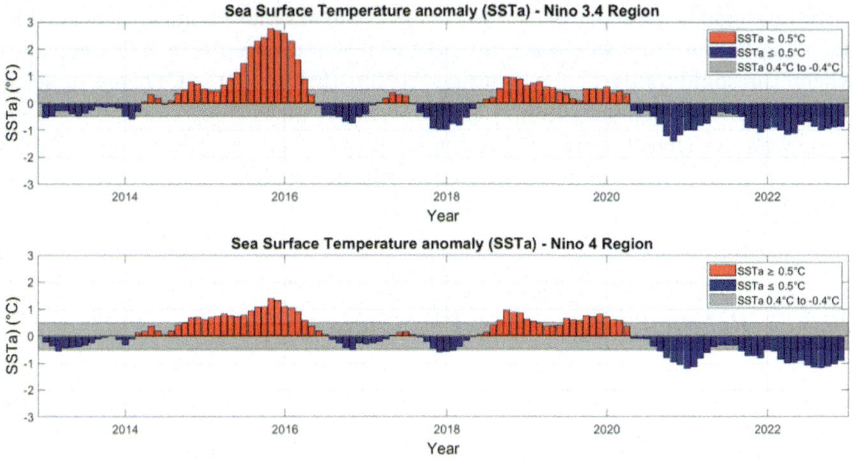


**Fig. 20.5** SSTa from **a** Nino 3.4 and **b** Nino 4 region from 2013 to 2022

in 2020 and 2022, respectively. The increase and decrease in SSTa are responses to temperature variation due to the presence of El Nino or La Nina. The SSTa response is relative to its intensity; the significant fluctuation in 2015–2016 was attributable to a very strong El Nino. Meanwhile, the 2020–2022 moderate La Nina shows lower values from the threshold of − 0.5 °C of SSTa.

The 2014–2016 increase in SSTa in both the Nino 3.4 and Nino 4 regions shows a decrease in PWV in Tanay and Legazpi in the same years. A similar trend was observed in 2018–2019, where a decrease in PWV was observed while the SSTa increased. Meanwhile, in 2018, 2020–2021, and 2021–2022, an increase in PWV was observed with a decrease in SSTa. This trend is mainly associated with the El Nino and La Nina events over the years, where an inverse trend between PWVa and SSTa were identified. In the study of Zhao et al., the decrease in the diurnal anomaly of PWV variation in 2014–2016 at PIMO station was associated with El Nino [18]. A similar result shows a decrease in PWV with an increase in SSTa, which is relative during an El Nino event [11]. There is also an increase in PWV during La Nina, when the air mass in the central Pacific moves to the west, causing an increase in evaporation, which results in an increase in PWVa [19].

The SSTa from the Nino 3.4 region shows a better response during an El Nino event, especially if the event is strong, as can be observed in the 2014–2016 significant increase in SSTa in that region. Meanwhile, SSTa in the Nino 4 region shows a good response to moderate La Nina events during the 2020–2021 and 2021–2022 events, where a noticeable increase in PWVa was seen and a decrease in SSTa was observed compared to SSTa in the Nino 3.4 region. Similar results were observed in the study of Suparta et al., where the SSTa from the Nino 4 region indicated a better response in comparison with the SSTa from the Nino 3.4 region during a La Nina event. This condition is mainly due to its location in the West Pacific [20]. The Nino 4 covers SST anomalies in the western and central sections of equatorial Pacific. This area often responds better to La Nina since its SSTs are at or above the level needed for deep convection. This would lead to a better reaction in terms of SST anomalies, where Threshold for the anomaly during La Nina (below − 0.5 °C) being sufficient to reduce the water temperature and cause a change in the Pacific's convection pattern to the west. As a result, there will be an increase in Precipitable Water Vapor (PWV). The drop in SSTa in the Nino 4 region during the 2020–2022 La Nina periods, when SSTa declined while PWV increased, constitutes proof of this.

### 20.3.3 Relationship Between PWVa and SSTa

The magnitude of an effect is measured by correlation. In order to further understand what the *R*-value means, we can characterize the correlation's strength using the guide of Evans [21].

The SSTa and PWVa relationship is presented in Table 20.1 using the *r* value over 2013–2022 with the ENSO event identified.

The correlation coefficient (*r*) in the 2014–2016 El Nino ranged from 0.02 to 0.12, which is a very weak positive relationship with SSTa and PWVa. This is attributed to a strong El Nino in 2014–2016, which caused variation in climate in Philippines, as observed in the differences in PWV anomalies [18]. A weak positive and a very weak negative correlation were observed in 2018–2019 in Nino 3.4 for Tanay and Legazpi stations, respectively. Meanwhile, the Nino 4 region shows a moderate positive correlation in Tanay and a weak positive correlation in Legazpi. The negative correlation in PWVa, which can be interpreted as a decrease in PWVa while SSTa increases, means a decrease in PWVa means water vapor is minimal in the atmosphere, leading to less rain as the moisture moves to the central Pacific during an El Nino event [19]. The difference between the two relationships is associated with the elevation of the station and its distance from the sea [9, 12]. A negative correlation coefficient (*r*) was observed in 2017–2018 and 2020–2022. The negative relationship can be attributed to the trend of PWVa, which increases during La Nina as the SSTa decreases. A moderately negative to strongly negative correlation was observed in 2017–2018, and a weakly negative to moderately negative correlation was observed in 2020–2022. The weak negative correlation is attributed to an increase in SSTa in mid-2021 during the moderate La Nina event in 2020–2022. A greater response in

**Table 20.1** Range of *R* values for the correlation coefficient

| Years (ENSO event) | Nino 3.4 and Tanay | Nino 3.4 and Legazpi | Nino 4 and Tanay | Nino 4 and Legazpi |
|---|---|---|---|---|
| 2014–2016 | 0.08 | 0.12 | 0.02 | 0.12 |
| 2017–2018 | − 0.53 | − 0.22 | − 0.30 | − 0.67 |
| 2018–2019 | 0.34 | − 0.17 | 0.41 | 0.34 |
| 2020–2022 | − 0.08 | − 0.47 | − 0.46 | − 0.41 |

the Nino 4 region was observed both in Tanay and Legazpi, where *r* values are − 0.30 and − 0.67 in 2017–2018 and − 0.46 and − 0.41 in 2020–2022. This shows a similar result to Suparta et al., where the Nino 4 region shows a stronger correlation with Nino 3.4 due to its location in the West Pacific Ocean [20]. Though the strength of the ENSO event plays a role and contributes to the variation of PWV affecting its response during this event [16].

## 20.4 Conclusion

The variation of the Precipitable Water Vapor and Sea Surface Temperature anomalies using radiosondes from 2013 to 2022 over Tanay and Legazpi has been analyzed using the anomaly method and time series analysis. The identified ENSO events from 2013 to 2022 showed observable effects on the variation of PWVa at the two stations, Tanay and Legazpi. The standard deviation from 2015 to 2021 in Tanay and Legazpi shows a low standard deviation, while the standard deviation is higher during lower PWV. A higher standard deviation was observed in Legazpi compared to Tanay due to its elevation and distance to the sea. The PWVa increases when the SSTa values are low (La Nina) and decreases when the values are high (El Nino). The Nino 3.4 fluctuates more with SSTa during an El Nino, and the Nino 4 shows a better response during a La Nina event. A negative correlation is seen during La Nina when PWV increases and SSTa decreases, which is observable in both Tanay and Legazpi stations. All of these findings suggest further investigation of the relationship between PWVa and SSTa and their possible effects on one another (e.g., lag or delay). Additional stations and a long-term analysis of the PWVa and SSTa are proposed for future studies to explore and refine our understanding of these relationships for improved forecasting and climate modeling purposes.

## References

1. M. Bevis, S. Businger, T.A. Herring, C. Rocken, R.A. Anthes, R.H. Ware, GPS meteorology: remote sensing of atmospheric water vapor using the global positioning system. J. Geophys. Res. Atmos. **97**, 15787–15801 (1992)
2. E.P. Macalalad, R.V. Macalalad, Variation of GNSS-derived precipitable water vapor over Manila in 2017, in *2019 6th International Conference on Space Science and Communication (IconSpace)*, Johor Bahru, Malaysia, July 2019 (2019), pp. 235–238
3. Q. Zhao, Y. Liu, W. Yao, X. Ma, Y. Yao, A novel ENSO monitoring method using precipitable water vapor and temperature in southeast China. Remote Sens. **12**(4), 649 (2020). https://doi.org/10.3390/rs12040649
4. Z. Baldysz, G. Nykiel, B. Latos, D.B. Baranowski, M. Figurski, Interannual variability of the GNSS precipitable water vapor in the global tropics. Atmosphere **12**, 1698 (2021)
5. Y. Rivera, K.C. Capacete, S.K. Rodriguez, A.R. David, E. Macalalad, Variations of precipitable water vapor from 2012 to 2019 over the Philippines using radiosondes. IOP Conf. Ser. Earth Environ. Sci. **880**(1), 012001 (2021). IOP Publishing

6. W. Suparta, A. Iskandar, M.S.J. Singh, M.A.M. Ali, B. Yatim, F. Tangang, A study of El Niño–Southern oscillation impacts to the South China Sea region using ground-based GPS receiver. J. Phys. Conf. Ser. **423**(1), 012043 (2013). IOP Publishing
7. A. Lv, L. Fan, W. Zhang, Impact of ENSO events on droughts in China. Atmosphere **13**, 1764 (2022)
8. W. Suparta, A. Iskandar, F.L. Gaol, GPS PWV and its response to ENSO activities in the western Pacific Region during 2009–2011. Space Sci. Commun. Sustain. 157–172 (2018). https://doi.org/10.1007/978-981-10-6574-3_14
9. X. Wang, K. Zhang, S. Wu, Z. Li, Y. Cheng, L. Li, H. Yuan, The correlation between GNSS-derived precipitable water vapor and sea surface temperature and its responses to El Niño–Southern Oscillation. Remote Sens. Environ. **216**, 1–12 (2018)
10. K.E. Trenberth, J. Fasullo, L. Smith, Trends and variability in column-integrated atmospheric water vapor. Clim. Dyn. **24**, 741–758 (2005)
11. W. Suparta, A. Iskandar, Monitoring of GPS water vapor variability during ENSO events over the Borneo Region. Asian J. Earth Sci. **5**(3), 88 (2012)
12. C. Trakolkul, C. Charoenphon, C. Satirapod, Impact of El Niño–Southern oscillation (ENSO) on the precipitable water vapor in Thailand from long term GPS observation. Int. J. Geoinform. **18**(3), 13–20 (2022)
13. S. Wang, X. Yuan, Y. Li, Does a strong El Niño imply a higher predictability of extreme drought? Sci. Rep. **7**, 40741 (2017)
14. A.L.S. Domingo, E.P. Macalalad, Multi-year precipitable water vapor observations in East Manila, Philippines using radiosonde and global navigation satellite system. E3S Web Conf. **199**, 00002 (2020). EDP Sciences
15. I. Durre, R.S. Vose, D.B. Wuertz, Overview of the integrated global radiosonde archive. J. Clim. **19**, 53–68 (2006)
16. F.A. Rahim, J.S. Mandeep, M.M. Ali, B. Yatim, W. Suparta, Evaluation of GPS PWV anomaly during ENSO, in *Proceeding of the 2011 IEEE International Conference on Space Science and Communication (IconSpace)*, Penang, Malaysia, July 2011 (2011), pp. 107–111
17. K.E. Trenberth, The definition of El Nino. Bull. Am. Meteorol. Soc. **78**(12) (1997)
18. Q. Zhao, Y. Yao, W.Q. Yao, Z. Li, Near-global GPS-derived PWV and its analysis in the El Niño event of 2014–2016. J. Atmos. Solar Terr. Phys. **179**, 69–80 (2018)
19. W. Suparta, A. Iskandar, M.S.J. Singh, A new technique to observe ENSO activity via ground-based GPS receivers, in *Recent Trends in Physics of Material Science and Technology* (Springer Singapore, Singapore, 2014), pp. 173–186
20. W. Suparta, A. Iskandar, M.S.J. Singh, M.A.M. Ali, B. Yatim, A.N.M. Yatim, Analysis of GPS water vapor variability during the 2011 La Niña event over the western Pacific Ocean. Ann. Geophys. **56**(3), 0330 (2013)
21. J.D. Evans, *Straightforward Statistics for the Behavioral Sciences* (Brooks/Col Publishing, Pacific Grove, CA, 1996)

# Chapter 21
# Intercomparison of MODIS- and GNSS-Based Precipitable Water Vapor Over Manila, Philippines in 2015

**Arvee M. De Castro, Princess Tucio, and Ernest P. Macalalad**

**Abstract** Precipitable water vapor (PWV) plays a crucial role in the climate system and its impact on atmospheric activities, weather, and climate changes. MODIS-PWV products are often compared to ground-based retrieved PWV data over different areas around the globe. However, there haven't been as frequent studies similar to it over any areas of Philippine territory. This study assessed the accuracy of MODIS-PWV against GNSS-derived PWV using correlational and linear regression analysis. It provided results that showed a very high positive correlation of $R = 0.91$ and an accuracy of RMSE = 4.83 mm, $R^2 = 0.83$ between MODIS-PWV and GNSS-derived PWV in 2015 over Manila. MODIS-PWV is found to be significantly underestimating the GNSS-derived PWV; which could be due to their disparate spatial positioning, the location of the region, and surface elevation. During the dry and wet season, there is also an underestimation but a very high positive correlation. The PWV peak values at the site in 2015 ranged from 50 to 70 mm for both its daily and monthly mean. In the dry season, there was a greater variability between MODIS-PWV and GNSS-derived PWV, with a range of 5.31–9.55 mm. To the contrary, during the wet season, the variability was lower and ranged from 5.48 to 7.63 mm; which agree well with the climate type of Manila. The study infers that the results are significant in the Philippines, where the understanding of MODIS-PWV accuracy against GNSS-derived PWV and its limitations helps provide outright improved PWV observations.

A. M. De Castro · P. Tucio
Department of Earth and Space Sciences, Rizal Technological University, Mandaluyong City, Philippines

E. P. Macalalad (✉)
Department of Physics, Mapua University, Manila, Philippines
e-mail: epmacalalad@mapua.edu.ph

 
M. T. Islam et al. (eds.), *Proceedings of the 8th International Conference on Space Science and Communication*, Springer Proceedings in Physics 303,
https://doi.org/10.1007/978-981-97-0142-1_21

## 21.1 Introduction

Precipitable water vapor is a parameter well known in meteorology as it is utilized to describe the water vapor content in the atmosphere. Its role in our climate system is essential in studying its impacts on atmospheric activities, such as radiative energy balance, hydrological cycle, and precipitation patterns [6]. Thus, it also plays a role in the almost unpredictable behavior of weather, and even changes in climate. The water vapor content is only 5% of the air and yet possesses a high variability in its distribution; hence the fluctuation that it has regionally along with seasonally [5]. This caveat makes it challenging to conduct the said essential investigation into PWV, so as to predict potential weather trends and even as far as climate changes [7]. As a consequence, various instruments and methods have been developed to do the difficult job of measuring PWV. Two of the effective instruments that are predominantly used to measure PWV are radiosondes and research satellites: Particularly, radiosonde stations, Global Positioning System (GPS) or more generally known as the Global Navigation Satellite System (GNSS), and NASA's Terra and Aqua satellites. Numerous networks of GPS ground receivers have been developed that make quantifying and measuring PWV possible from these networks [3, 14].

The Moderate Resolution Imaging Spectroradiometer (MODIS) is one of the payloads that was built to measure cloud properties—one of these properties is water vapor content. Both Terra and Aqua are equipped with MODIS payload. MODIS has a sensitivity to cloud cover or cloudy skies. As a solution to this, MODIS also provides data for cloud masks: MODIS product for cloud mask (MOD35) is used to identify the observations made during increased cloud cover and clear skies. In order to further analyze the variation of PWV, a combined effort of two or more instruments is necessitated to be investigated together. Hence, the validation of satellite-derived PWV using ground-based derived PWV essentially permits further studies on the discipline. Reference [13] conducted a study on the validation of MODIS-PWV over IAO-Hanle, in the trans-Himalayan region. They found that MODIS-PWV had a good correlation with the GPS-derived data; a bias of − 0.018 cm, with RMSE of 0.137 cm, and $R^2$ of 0.91 were computed. Moreover, [17] and their study on comprehensive validation over Mainland China inferred that the validation of MODIS-PWV over China indicates its capability to make both seasonal and annual observations of its variations.

Precipitable water vapor possesses an interesting behavior, especially throughout the seasons in a year. Because of this PWV variation can be further explored through temporal and seasonal observation. There have been studies that have demonstrated how radiosondes and other tools, such remote sensing, may be used to analyze short-term fluctuations in PWV [16]. This is precisely what [10] did in their study on PWV variations observations over Manila in 2017. While [7] found in their paper on inter-annual variation of PWV in Manila, the same consistency between the monthly mean variations and the climate classification of the site. Observations and findings on PWV's behavior, especially its variability can be improved through the use of other data sources and different methods. Therefore, through the accuracy assessment of MODIS-PWV, a further step in observation can be made to investigate the variation of PWV over Manila, Philippines. The aim of this study is to assess the accuracy, limitations, and variability MODIS-PWV using GNSS-derived PWV over Manila, Philippines for the year 2015. The combined power of MODIS-PWV, a satellite-derived PWV with high spatial resolution, and GNSS-derived PWV with high temporal resolution allows this to be a big potential for further studies in the field. This could help to improve and as well as advance the weather observations that the Philippines has and is currently utilizing. By using NASA's MODIS products and its data for PWV, we are able to have a unique perspective on this study, as we progressively make use of different sources and methods.

## 21.2 Study Area and PWV Datasets

### *21.2.1 Study Area Description*

We have set the study area over Manila, also known as the capital city of the Philippines; located in Southeast Asia, within the West Pacific Ocean. It has a territory of 619.5 $km^2$ and a population of 14.66 million [11]. The climate of the Philippines is tropical, generally characterized by the amount of monthly rainfall received in a year [12]. The two distinct seasons of Type I climate are wet from May to October, and dry from November to April. Modified Corona Classification provides distinct climate types for the country: Climate of Type II have a defined maximum period from December to February with no dry season. Although there is a shorter dry season in Type III, the climate is very similar to Type I. While Type IV climate sees evenly spaced rainfall annually [2, 12]. Hence, the area of this study is found and classified to be a Type I climate.

### *21.2.2 GNSS-Derived PWV Data*

PWV data acquired from new observation methods are deemed coveted for one to attain sufficient observational coverage globally. Numerous networks of GPS ground receivers have been created that make quantifying and measuring PWV possible from these networks [3, 14]. One of the known sources that is trustworthy for estimating PWV found in the atmosphere is GPS [4]. The meteorological community has paid a lot of attention to how GPS is becoming a more practical instrument for monitoring precipitable water vapor. No matter what the weather is over a certain location, GNSS run constantly with a 30 s sampling period and do not require manual operation [7]. The chosen study region and period of one (1) year, which is for the year 2015, and the availability of GNSS observations are used to determine which GNSS station

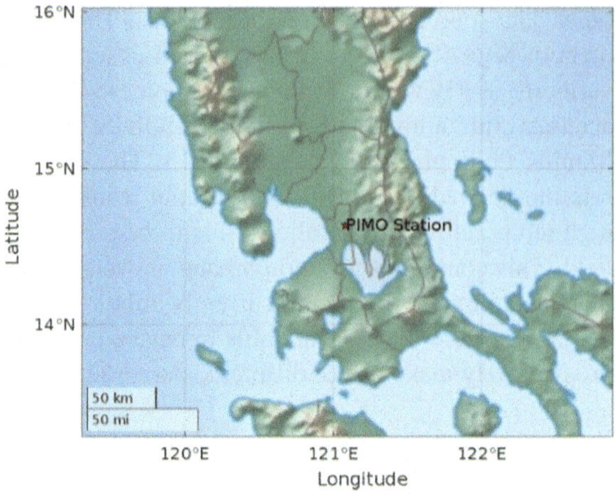


**Fig. 21.1** Map of Manila, Philippines and the location of PIMO station

should be used. SuomiNet is the database where the PWV from GNSS data was obtained. SuomiNet is an academic network that delivers PWV values from IGS GNSS receivers [7]. The data resolution is 30 min; which are to be computed to get the daily and monthly average, in relation to and for alignment with MODIS-PWV's data resolution. Ultimately, the daily mean values are utilized in the assessment. The chosen station is the PIMO station located in Quezon City (Station ID: PIMO, location: 14.635719° N, 121.077731° E, 100 masl). Figure 21.1 shows the location of the chosen GNSS station in Quezon City: PIMO station.

### *21.2.3 MODIS-PWV Data*

In recent years, the amount of atmospheric water vapor has been continually measured with the two MODIS devices from NASA. On 18 December 1999, the Terra Platform was launched, while it was successfully followed by the Aqua Platform on 4 May 2002. Its evaluation has been researched utilizing global ground-based PWV data, such as GNSS and radiosonde [10, 15]. On NASA's sun-synchronous near-polar orbiting satellite known as Earth Observing System Terra platform, a MODIS sensor is currently operational in Earth's orbit. MODIS-retrieved PWV is a parameter that provides the necessary and unique PWV estimation to be trained by PWV data derived from ground-based measurements, such as GNSS. It is one of the Level-2 and Level-3 PWV products of MODIS that are generated at $1 \times 1$ km$^2$ spatial resolution using its near infrared (NIR) and at $1 \times 5$ km$^2$ using its infrared (IR) equipped on both Terra and Aqua satellites [15]. MODIS data products and its primary sources of errors are found to be in the spectral reflectance associated with various surface types, temperature in the atmosphere, moisture profiles, and pixel registration between channels. Thus, the method MODIS measures PWV is by focusing only on clear pixels, which essentially means it fails to capture the moisture present in the cloudy areas [13]. Therefore, the clear-sky measurements with the quality control system are necessary to be considered for MODIS atmospheric and surface parameter retrievals. In the data processing of MODIS-PWV, it uses a cloud-mask method to detect free cloud pixels [1]. The MODIS near-infrared PWV's accuracy ranges from 5 to 10% as it was cited in [8] paper. As a result, the accuracy assessment of MODIS-PWV heavily involves temporal and spatial alignment, and taking into account cloud cover over a specific area. Hence, for proper alignment of the data comparison, the MODIS-PWV data collected are the daily mean. The MODIS-PWV data utilized in this study was obtained from January to December of 2015 for the grids over Manila, Philippines. The data are acquired on Giovanni online data system; developed and maintained by the NASA GES DISC for EARTHDATA repository. The required study area is manually selected, along with the study period on the website before we can generate the data. The data values are then estimated based on the provided spatiotemporal maps; done by getting the average value which will be found as the closest pixel to the chosen GNSS station.

## 21.3 Comparison of GNSS-Derived PWV and MODIS-PWV

The comparison of PWV data derived from space-based observations, such as MODIS-PWV, is an integral component of PWV observations. Conversely, there is a misalignment in time and space of PWV data between MODIS and other techniques that results in a significant issue that impacts the intended comparisons. To address this issue in our study, we take the MODIS PWV products and average them over the grids centered around the location of the chosen GNSS station. The data are estimated similar to how [9] conducted their estimations. MODIS-PWV estimations are done using the average value which will be found as the closest pixel to the chosen GNSS station and its surrounding pixels. Moreover, temporal misalignment is also considered. Daily and Monthly average were computed to accomplish the comparisons with proper temporal accuracy.

The MODIS-PWV data that we have used in this study include specifically the Level-3 MODIS Terra products of Total Column: Mean of Level-3 QA Weighted Mean (MOD08); including lower scale for daily (MOD08_D3) and larger scale for monthly (MOD08_M3) global grid with a spatial resolution of 1°; the MODIS products [13] used in their study. Figure 21.2 shows the correlation between MODIS-PWV and GNSS-derived PWV for the year 2015 over Manila, Philippines. The MODIS-PWV consisting of 328 days' worth of data is compared to the GNSS-derived PWV, which consists of 292 days' worth of data. The correlation coefficient of is found to be $R = 0.91$; it shows a very high positive correlation. The coefficient of determination is at $R^2 = 0.83$; this shows a large positive linear association between the two datasets. The positive correlation and linear association indicate ideal validity and applicability of the MODIS-PWV data against GNSS-derived PWV data. Consequently,

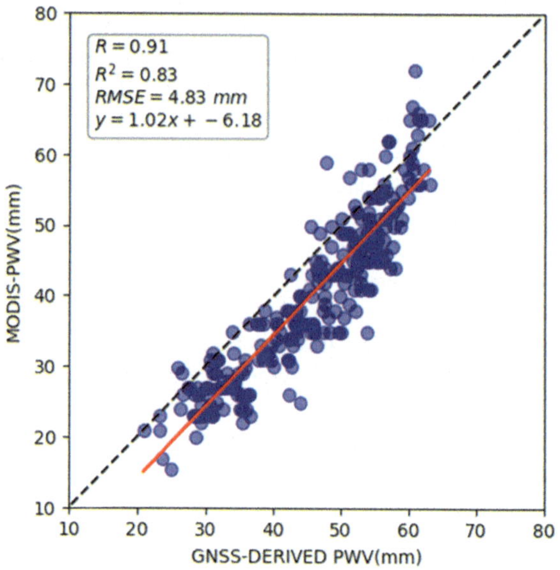


**Fig. 21.2** Comparison between the MODIS-PWV and GNSS-derived PWV in 2015

this implies an ideal observation, given the previous validations and comparisons over different study areas over the continent of South and East Asia. The RMSE is computed to be 4.83 mm; with the bias being − 6.18 at *y*-intercept. Moreover, the linear regression analysis shows that the MODIS-PWV is underestimating the GNSS-derived PWV. This could be due to the different spatial positioning of the instruments and the surface elevation of the region. Thus, it is consistent with the previous study conducted by [13] over trans-Himalayan region in China, which is essentially located in East Asia, as the city of Manila is also located somewhere in East Asia; more specifically in South East Asia.

### 21.3.1 Seasonal Comparison of MODIS-PWV and GNSS-Derived PWV

In [13], a seasonal comparison was employed to further assess the correlation between MODIS-PWV and GNSS-derived PWV, hence enhancing the validation for any seasonal observations that could be done subsequently. Figure 21.3 shows the correlation between MODIS-PWV and GNSS-derived PWV during the dry season of 2015: A correlation coefficient of $R = 0.85$ was computed; along with the coefficient of determination of $R^2 = +0.73$. It still shows a very high positive correlation and a large positive linear association. The RMSE is at 4.11 mm; not significantly different from the computed annual RMSE of 4.83. It has a bias of + 2.00 at *y*-intercept; still indicating an underestimation, the same as of the annual comparison. The correlation between MODIS-PWV and GNSS-derived PWV during the wet season of 2015 is also shown in Fig. 21.3: The correlation coefficient is $R = 0.81$, and the coefficient of determination is $R^2 = +\,0.65$. This tells us that during the wet season, there is still a high positive correlation and a large positive linear association. The RMSE was computed to be at 4.85 mm, with a bias of $-$ 10.44 at $y$-intercept. The underestimation is still prominent for MODIS-PWV during the wet and dry season Fig. 21.4 shows the daily MODIS-PWV and GNSS-derived PWV in mm unit for the year 2015. As it is observed, the peak values of MODIS-PWV and GNSS-derived PWV over the site in Manila, Philippines were recorded to have varied as the daily mean from 50 to 70 mm, which took place from around June to the first quarter of July. This observed trend from both MODIS and GNSS seems to agree well with each other. There are two distinct seasons in the Philippines: These are grouped as the months considered to be the dry season of November–April, and the wet months of May–October.

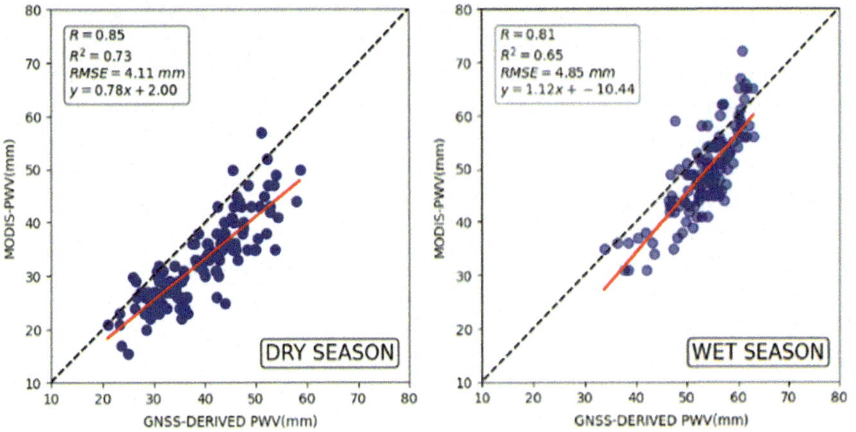


**Fig. 21.3** Comparison between the MODIS-PWV and GNSS-derived PWV during the (left) dry and (right) wet season in 2015

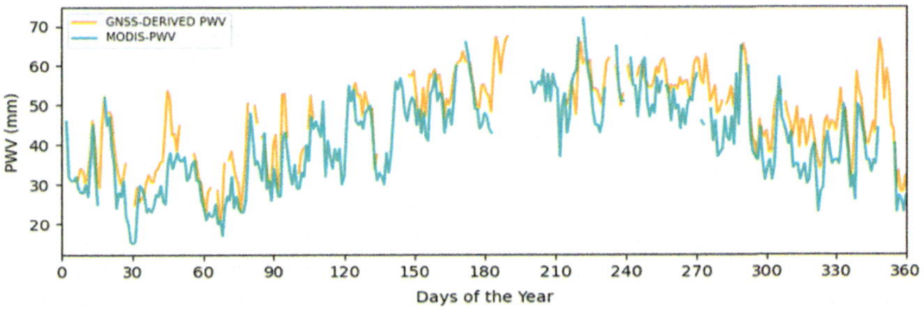


**Fig. 21.4** Daily comparison between the MODIS-PWV and GNSS-derived PWV in 2015

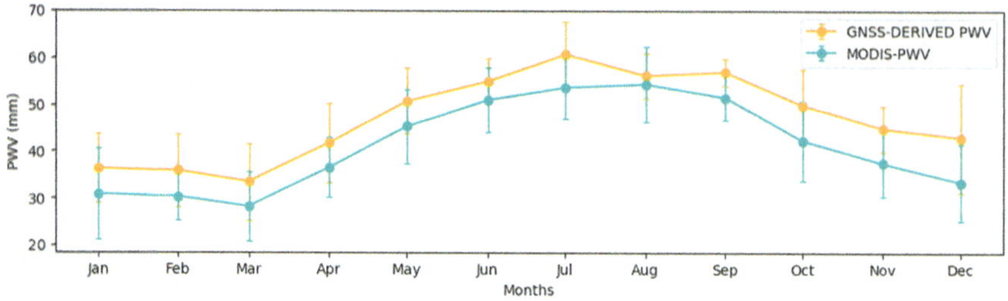


**Fig. 21.5** Seasonal variation of the estimated MODIS-PWV and GNSS-derived PWV in 2015

### *21.3.2 Analysis of Seasonal PWV Variations*

Figure 21.5 shows the seasonal variation of MODIS-PWV and GNSS-derived PWV for 2015. It is shown that there is an upward trend during the wet season, particularly for the months of May–July. This trend can be explained by the Type 1 climate of the area where the site is located, and the onset of the known Southwest Monsoon season in the country that normally starts around that time. As [7, 10] found in their respective studies on PWV observations, the analysis in this study also found that despite the PWV values being low during the dry season, it is when the variability is observed to be relatively high. Vice versa, during wet season, this is when the PWV is high, but the variability is observed to be relatively low. The variability of MODIS-PWV and GNSS-derived PWV during dry season ranged from 5.31 to 9.55 mm. While the variability for the wet season ranged from 5.48 to 7.63 mm. This denotes that during the wet season, particularly at the height of the southwest monsoon and the typhoon season in Manila, there is an uptick in moisture level in the atmosphere; making the PWV values relatively higher. Additionally, the higher variability of PWV levels could be attributed to the occasional uptick in moisture caused by isolated thunderstorms during the dry season; giving more variability to PWV values in the region. The observed variability accentuates the impacts of the abovementioned synoptic climate forcings that occur in the region: The monsoons and the climate type being the main indicators of the variability.

## 21.4 Conclusion

The MODIS-PWV products are often compared to ground-based retrieved PWV data over different areas around the globe. Yet there haven't been as frequent studies similar to it over areas in the territory of the Philippines. This is why this study is important and holds significant relevance in the country. The very high positive correlation of the annual comparison of MODIS-PWV and GNSS-derived PWV for 2015 and its accuracy of RMSE = 4.83 mm, along with the bias of − 6.18, a correlation coefficient of $R = 0.91$ and coefficient of determination of $R^2 = 0.83$

indicate the promising utilization of MODIS-PWV data for further validation and long-term seasonal variation observations of PWV over the Philippines. Contrastingly, MODIS-PWV is found to be significantly underestimating the GNSS-derived PWV over Manila, Philippines in 2015; which could be attributable to their disparate spatial positioning, and the location of the region, and as well as its surface elevation. As expected for the seasonal comparisons, they showed an underestimation but a very high positive correlation for the dry season with the corresponding correlation coefficient of $R = 0.85$, and an accuracy of RMSE = 4.11 mm, $R^2 = 0.73$ along with its bias of + 2.00 at $y$-intercept. The underestimation is again prominent, but there is a consistent high positive correlation for the wet season with the corresponding $R = 0.81$, $R^2 = 0.65$, and an accuracy of RMSE = 4.85 mm, and a bias of − 10.44 mm at $y$-intercept.

The underestimation of MODIS-PWV during the wet and dry season indicates the overall underestimation observed over Manila in 2015. The GNSS-PWV values exhibit a general higher increase in value compared to those of MODIS-PWV measurements, also ostensibly attributable to their disparate spatial positioning. The PWV peak values of MODIS-PWV and GNSS-derived PWV at the site in 2015 ranged from 50 to 70 mm for both its daily and monthly mean; which took place in July. This is due to the occurrence of wet season and the start of Southwest Monsoon during this time period; when rainfall events are more frequent. They agree well with the expected PWV values for the region and the climate type of Manila. During the dry season, the variability was higher between the MODIS-PWV and GNSS-derived PWV; which varied between 5.31 and 9.55 mm. In contrast, during the wet season, the variability was observed to be low and ranged from 5.48 to 7.63 mm. This pattern is consistent with the climate type of Manila and what [7, 10] found in their studies on the variation of PWV over Manila. Given that there haven't been many papers published on MODIS-PWV validation and PWV observations using MODIS-PWV data, we recommend that future and prospective researchers focus on the long-term validation and assessment of MODIS-PWV over Manila and the rest of the Philippine archipelago.

## References


1. S.A. Ackerman, K.I. Strabala, W.P. Menzel, R.A. Frey, C.C. Moeller, L.E. Gumley, Discriminating clear sky from clouds with MODIS. J. Geophys. Res. Atmos. **103**(D24), 32141–32157 (1998)
2. G. Bagtasa, Contribution of tropical cyclones to rainfall in the Philippines. J. Clim. **30**(10), 3621–3633 (2017)
3. M. Bevis, S. Businger, T.A. Herring, C. Rocken, R.A. Anthes, R.H. Ware, GPS meteorology: remote sensing of atmospheric water vapor using the global positioning system. J. Geophys. Res. Atmos. **97**, 15787–15801 (1992)
4. M. Bevis, S. Businger, S. Chiswell, GPS meteorology: mapping zenith wet delays onto precipitable water. AMETSOC (1994)
5. S. Choy, C.S. Wang, T.K. Yeh, J. Dawson, M. Jia, Y. Kuleshov, Precipitable water vapor estimates in the Australian region from ground-based GPS observations. Adv. Meteorol. (2015)

6. J.P.O. De Galisteo, V.E. Cachorro, C. Toledano, B. Torres, N.S. Laulainen, Y. Bennouna, A.M. De Frutos, Diurnal cycle of precipitable water vapor over Spain. Q. J. R. Meteorol. Soc. **137**(657), 948–958 (2011)
7. A.L.S. Domingo, E.P. Macalalad, Multi-year precipitable water vapor observations in East Manila, Philippines using radiosonde and global navigation satellite system. E3S Web Conf. **199** (2020)
8. B.C. Gao, Y.J. Kaufman, Water vapor retrievals using moderate resolution imaging spectroradiometer (MODIS) near-infrared channels. J. Geophys. Res. Atmos. **108**(13) (2003)
9. G. Gurbuz, S. Jin, Long-time variations of precipitable water vapour estimated from GPS, MODIS and radiosonde observations in Turkey. Int. J. Climatol. **37**(15), 5170–5180 (2017)
10. E.P. Macalalad, R.V. Macalalad, Variation of GNSS-derived precipitable water vapor over Manila in 2017, in *IEEE International Conference on Space Science and Communication* (2019)
11. Manila population 2023, https://worldpopulationreview.com/world-cities/manila-population. Accessed 23 Mar 2023
12. J. Matsumoto, L.M.P. Olaguera, D. Nguyen-Le, H. Kubota, M.Q. Villafuerte, Climatological seasonal changes of wind and rainfall in the Philippines. Int. J. Climatol. **40**(11), 4843–4857 (2020)
13. S.S. Ningombam, S. Jade, T.S. Shrungeshwara, H.J. Song, Validation of water vapor retrieval from moderate resolution imaging spectro-radiometer (MODIS) in near infrared channels using GPS data over IAO-Hanle, in the trans-Himalayan region. J. Atmos. Solar Terr. Phys. **137**, 76–85 (2015)
14. J.P. Ortiz de Galisteo, V. Cachorro, C. Toledano, B. Torres, N. Laulainen, Y. Bennouna, A. de Frutos, Diurnal cycle of precipitable water vapor over Spain. Q. J. R. Meteorol. Soc. **137**(657), 948–958 (2011)
15. A.K. Prasad, R.P. Singh, Validation of MODIS Terra, AIRS, NCEP/DOE AMIP-II Reanalysis-2, and AERONET Sun photometer derived integrated precipitable water vapor using ground-based GPS receivers over India. J. Geophys. Res. Atmos. **114**(5) (2009)
16. Y.J. Rivera, K.C. Capacete, S.K. Rodriguez, A.R. David, E.P. Macalalad, Variations of precipitable water vapor from 2012 to 2019 over the Philippines using radiosondes. IOP Conf. Ser. **880**(1), 012001 (2021)
17. Y. Zhao, H. Zhao, J. Li, G. Xiao, Comprehensive validation and calibration of MODIS PWV over Mainland China. Atmosphere **13**(11) (2022)

# Chapter 22
# Vertical Total Electron Content Variation Due to Lightning Activities Near Malaysia

**Sulaiman Ali Mohammmad, Mardina Abdullah, Siti Aminah Bahari, and Mohd Riduan Ahmad**

**Abstract** Lightning events could be one of the causes of disturbance during a transmission of radio signals that could lead to errors. Additionally, the changes of time, seasons and geomagnetic fields could affect the ionospheric layer. By determining the total electron content (TEC), it could be a useful indicator for the disturbance in the ionospheric layer. In this study, TEC was obtain from the Global Positioning System (GPS) receiver in Receiver Independent Exchange Format (RINEX) format. The vertical TEC (VTEC) was observed together with the lightning occurrences. The lightning strike distribution was obtain from the World-Wide Lightning Location Network (WWLLN). Quiet day based on the planetary index number ($K_{\text{p}}$) which needs to be less than or equal to four determines the day of observation. Four days were chosen, comprise of two days for each day with and without lightning strike occurrences, respectively. Due to higher VTEC value during equinox, it could be observed the sample date closer to the equinox season have higher VTEC value even on a quiet day without lightning activities.

## 22.1 Introduction

Ionospheric disturbance can be caused by multiple reasons; evidently, the solar events [1–3], geomagnetic storms [4, 5], and lightning activities [6–8]. The total electron content (TEC) can be recognized as the number of electron along the path of signal that is transmitted by the satellite in the Global Positioning System (GPS)

S. A. Mohammmad · M. Abdullah (✉) · S. A. Bahari
Space Science Centre (ANGKASA), Institute of Climate Change (IPI), Universiti Kebangsaan Malaysia (UKM), 43600 Bangi, Selangor, Malaysia
e-mail: mardina@ukm.edu.my

M. R. Ahmad
Atmospheric and Lightning Research Laboratory, Fakulti Kejuruteraan Elektronik dan Kejuruteraan Komputer, Centre for Telecommunication Research and Innovation (CeTRI), Universiti Teknikal Malaysia Melaka, 76100, Durian Tunggal, Melaka, Malaysia

 
M. T. Islam et al. (eds.), *Proceedings of the 8th International Conference on Space Science and Communication*, Springer Proceedings in Physics 303,
https://doi.org/10.1007/978-981-97-0142-1_22

to the receiver on Earth. By determining the TEC, it could be a useful indicator to find the disturbance in ionospheric layers. Error effected by ionosphere toward the transmission of radio signal could also be observed by determining the TEC [9].

The ionospheric layer could change based on time, season, solar cycle, and geomagnetic field [10]. Due to the absence of solar radiation during night time, the density of free electron increases as the sun rise while slowly decrease when the sun set. Such effects could also be seen when the season change during equinox and solstice. During equinox, the sun traverse along the equator with similar period of day and night. Meanwhile, during solstice, the sun position furthest to the north or south of the equator. The more sunlight the earth received, the denser the electron accumulated [11].

Considerable amount of attention could be observed over the past several decades relating the effects of lightning and thunderstorms to the middle or upper atmosphere. Among the studies includes [12] that found correlation of intensify in sporadic E layer with lightning activities. Through the modeling of atmospheric gravity waves (AGWs) originating from thunderstorms, the variation of TEC is predicted [13]. Days with high lightning activities in South Africa area show high TEC variation when compared to low lightning activity day [14]. The study of TEC is important in the modern technology world where human rely heavily on communication, navigation, and surveillance using satellite. Since Malaysia is one of the most active regions of thunderstorms [15], it is motivated in this study to observe the relationship between thunderstorms and variations of ionosphere over Malaysia.

## 22.2 Data Sources and Methodology

Total electron content (TEC) is obtained from one Global Positioning System (GPS) receiver station; station ANMG situated in Malaysia at geographical latitude and longitude of $2^\circ\,47'\,7.07''$ N, $101^\circ\,30'\,26.81''$ E, respectively. The GPS data is obtained from the Crustal Dynamics Data Information System (CDDIS) in Receiver Independent Exchange Format (RINEX) format at (https://cddis.nasa.gov/) for daily 30 s data.

Selected day of observation was determined by low geomagnetic activity known as quiet day; guided by the planetary index number ($K_\mathrm{p}$) that is less than or equal to 4 [7]. The index of $K_\mathrm{p}$ could be obtain from OMNIWeb Data Documentation (https://omniweb.gsfc.nasa.gov/) where specific day of year (DOY) and the desired indices could be obtained. The TEC processing was conducted using software developed at Boston College [16], which differences the phase and code values at L1 and L2 frequencies. The slant TEC (STEC) measurements is converted to vertical TEC (VTEC) values in the software by dividing single layer mapping function, $M(\varepsilon, h) = \sec\left[\sin^{-1}\left(\frac{R_\mathrm{e}\cos E}{R_\mathrm{e}+h}\right)\right]$ where $R_\mathrm{e}$ is the Earth radius, $\varepsilon$ is the elevation of the satellite, and $h$ is the ionospheric height, assumed at 350 km. The World-Wide Lightning Location Network (WWLLN) data is observed for its lightning strikes around 50 km from the GPS station to ensure the observed storm is near the GPS receiver station.

## 22.3 Results and Discussion

Figure 22.1a shows the $K_p$ index values throughout the year 2016 with red line indicating quiet days. Meanwhile, Fig. 22.1b show the distribution of WWLLN lightning strikes on days with $K_p$ value less or equal to 4 throughout the year of 2016. A total of 20 quiet days observed; four with quiet day with no lightning strikes while the rest of the quiet day have lightning strikes. Considered day of observation includes 3 June and 9 April 2016 for quiet day with lightning strikes; while 28 March and 3 May 2016 for quiet day without lightning strikes.

Figure 22.2 shows the day of observation for VTEC value and the occurrences of lightning indicated with the present of WWLLN lightning strikes. Figure 22.2a, b presents the two quiet days with lightning strikes on 9 April and 3 June 2016, while Fig. 22.2c, d present two quiet days without lightning strikes on 28 March and 3 May 2016.

From Fig. 22.2a, the peak of lightning strikes was observed between 10:00 and 11:00 UTC with 169 lightning strikes. During the intense lightning strikes, the VTEC value of PRN 7 was average to 25.52 TECU at around 10:00 UTC. For Fig. 22.2b, there were two peak to be considered. The first peak was between 10:00 and 11:00 UTC with 126 lightning strikes. Meanwhile, between 11:00 and 12:00 UTC, a total of 185 lightning strikes were recorded. Between 5:53 and 9:56 UTC, PRN 8 recorded VTEC between 31.4 and 36.08 TECU. On the other hand, between 9:47 and 13:47

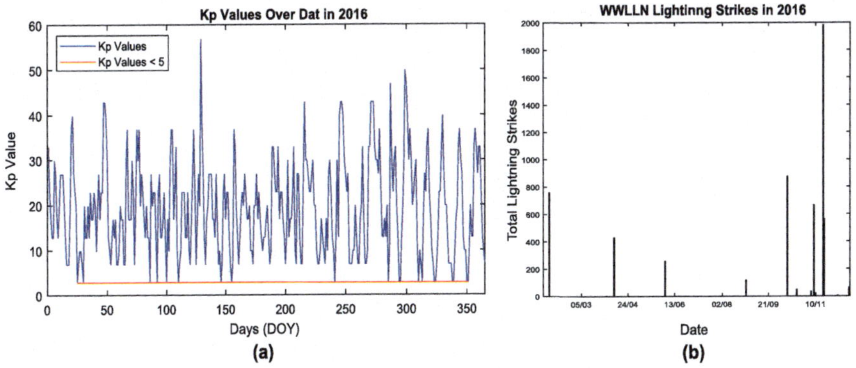


**Fig. 22.1** **a** Show day with $K_p$ index throughout the year of 2016 (red line indicated $K_p \leq 4$) while **b** show day with WWLLN lightning strikes

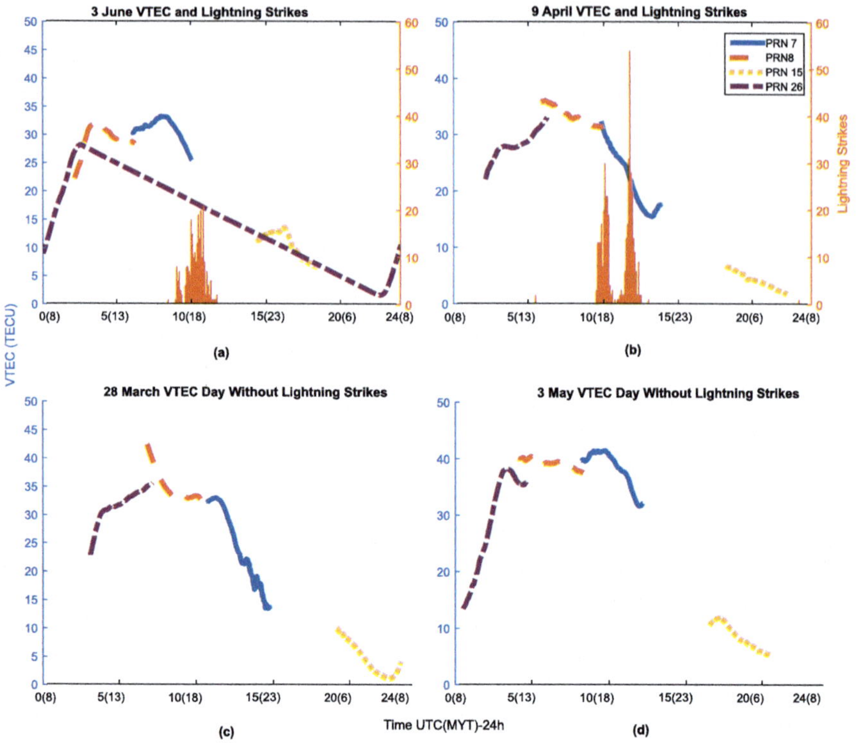


**Fig. 22.2** Show the time against the VTEC (TECU) on the left side and total number of lightning strikes from WWLLN on the right side. VTEC values are differentiated by their PRN number. **a**, **b** Compared the VTEC and lightning strikes. **c**, **d** Show VTEC on quiet day without lightning strike occurrences

UTC, PRN 7 recorded VTEC between 15.43 and 32.38 TECU. At 10:00 UTC, PRN 7 has VTEC at 30.41 TECU which gradually decreases over time. Meanwhile, at 11:00 UTC PRN 7 has VTEC at 25.9 TECU which also decrease gradually over time.

Figure 22.2c shows day with no lightning strike on 28 March 2016. Between 10:00 and 11:00 UTC, PRN 8 show VTEC value between 31 and 33.39 TECU that peak at 10:00 UTC with 33.39 TECU. Meanwhile, for PRN 7, between 10:45 and 11:00 UTC, the VTEC was between 32.34 and 32.79 TECU. By continuing the observation of VTEC at PRN 7, between 11:00 and 12:00 UTC, the VTEC value was between 30.25 and 33.07 TECU, peak at 11:00 UTC at 33.07 TECU that decreases over time. Figure 22.2d also shows day with no lightning strikes on 3 May 2016. At similar timing of observation when lightning strikes occurred, between 10:00 and 11:00 UTC, PRN 7 show VTEC value between 37.59 and 40.55 TECU that peaks at 10:00 UTC at 40.55 TECU. Meanwhile, between 11:00 and 12:00 UTC, the VTEC value was between 31.61 and 37.56 TECU, peak at 11:00 UTC with 37.56 TECU.

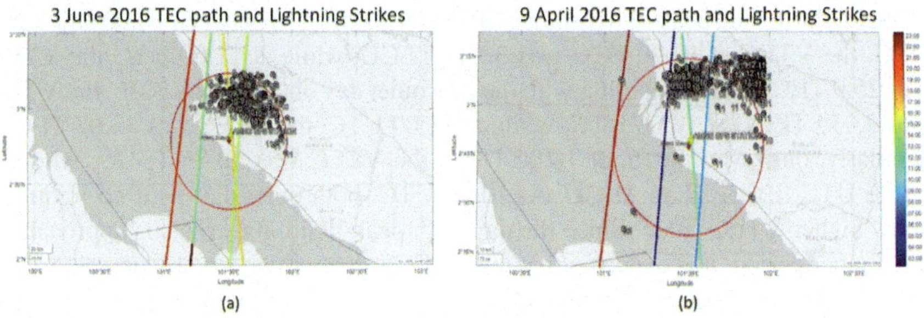


**Fig. 22.3** Color points show the latitude and longitude of PRN path based on color bar contour to differentiate the hour (UTC). Red circle represent 50 km from ANMG GPS station. White points represents WWLLN lightning strikes indicated with their hour (UTC). **a** Present lightning occurrences on 3 June 2016 while **b** on 9 April 2016

Figure 22.3 presents TEC measurement path (at 350 km) where the contour color represent the hour in UTC. A 50 km red circle that act as a guide, centered at the ANMG GPS station represent with diamond shape. Additionally, the WWLLN lightning strikes near the GPS station is also observed based on its latitude and longitude. In the previous observation in Fig. 22.2a, the lightning strike peaks between 10:00 and 11:00 UTC. The peak of the lightning strike is approximately represented by the dashed black circle. During that period, PRN 7 was present between 9:00 and 10:00 UTC (turquoise to light green). The second observation has two peaks of lightning strikes, seen in Fig. 22.2b between 10:00 and 12:00 UTC which approximately within the dashed black circle in Fig. 22.3b. PRN 7 and PRN 8 were present during the peak lightning strikes presented by the turquoise to light green color in the map.

Based on [17] during the minimum solar activity in Malaysia, the variation of total electron content differentiates between seasons. The four seasons (TEC value) (day) includes; equinox (36 TECU) (7 March 2007), summer (24 TECU) (8 June 2007), fall (26 TECU) (23 September 2007), and winter (22 TECU) (3 December 2007). During the equinox, it is the highest TEC recorded compared to other seasons. Additionally, the TEC value increases throughout the day and gradually decrease late afternoon.

During the peak of lightning strikes between 10:00 and 11:00 UTC on 3 June 2016, the only visible VTEC was from PRN 7 with peak VTEC at 25.52 TECU at 10:00 UTC. By comparing with days without lightning strikes on 28 March 2016, during similar time, the VTEC value was 33.39 TECU. Meanwhile on 3 May 2016, VTEC at 10:00 UTC is at 40.55 TECU. Both quiet day without lightning strike show higher VTEC value at 10:00 UTC. Since both quiet day is closer to the equinox as mention by [13] it is possible that the VTEC during that time is higher when compared to the day with lightning strike. By comparing the minimum solar during summer period [13] at 24 TECU, there is an increase of 1.52 TECU during the occurrence of lightning strikes.

Next, during the peak of lightning strike on 9 April 2016 between 10:00 and 12:00 UTC, the VTEC at 10:00 UTC was 30.41 TECU, while at 11:00 UTC the VTEC was 25.9 TECU. By comparing with the first quiet day on 28 March 2016, the VTEC was 33.39 TECU at 10:00 UTC, while at 11:00 UTC, the VTEC was 33.07 TECU. Similarly, on 3 May 2016, at 10:00 UTC, the VTEC was 40.55 TECU, while at 11:00 UTC it was 37.56 TECU. At 10:00 UTC, both quiet day with no lightning show higher VTEC value. Since the day with peak lightning strikes is approaching to summer period, it is expected that the VTEC value will be lower when compared to period closer to equinox. Nonetheless, at 11:00 UTC, during both quiet day, it has a higher value of VTEC.

## 22.4 Conclusion

In this study, the quiet day with lightning strike occurrences were on 9 April and 3 June 2016. Meanwhile, the quiet day without lightning occurrences were on 28 March and 3 May 2016. By comparing at similar time for both days with and without lightning occurrences, it shows that the days without lightning has higher VTEC value. Such observation could be contribute by the different day and seasons; since the closer the day to equinox, the higher the value of VTEC during solar minimum. Additional indices could be observed to further filter the quiet day with and without lightning occurrences. More sample day could also give a better understanding for the variation of vertical total electron content and lightning activities.

**Acknowledgements** This research was funded by YPASM Grant (ZF-2019-001). The authors would like to thank Institut Perubahan Iklim (IPI), Universiti Kebangsaan Malaysia (UKM) and Universiti Teknikal Malaysia Melaka (UTeM) for their supports in this project. The authors would also like to thank the World-Wide Lightning Location Network (WWLLN) (http://wwlln.net) accessed on 17 March 2021, a collaboration among over 50 univer-sities and institutions, for providing the lightning location data used in this paper.

## References

1. Y. Jiao, Y.T. Morton, Comparison of the effect of high-latitude and equatorial ionospheric scintillation on GPS signals during the maximum of solar cycle 24. Radio Sci. **50**(9), 886–903 (2015)
2. M. Mendillo, C. Narvaez, A.G. Marusiak, Are ionospheric storms the same during different solar cycles? J. Geophys. Res. Space Phys. **118**(10), 6795–6805 (2013)
3. L. Qian, A.G. Burns, S.C. Solomon, P.C. Chamberlin, Solar flare impacts on ionospheric electrodynamics. Geophys. Res. Lett. **39**(6), 1–6 (2012)
4. C.M. Ngwira, L.A. McKinnell, P.J. Cilliers, A.J. Coster, Ionospheric observations during the geomagnetic storm events on 24–27 July 2004: long-duration positive storm effects. J. Geophys. Res. Space Phys. **117**(1), 1–13 (2012)

5. E.G. Thomas, J.B.H. Baker, J.M. Ruohoniemi, A.J. Coster, S. Zhang, The geomagnetic storm time response of GPS total electron content in the North American sector. J. Geophys. Res. Space Phys. **121**, 1744–1759 (2016)
6. S. Kumar, D. Siingh, R.P. Singh, A.K. Singh, A.K. Kamra, Lightning discharges, cosmic rays and climate. Surv. Geophys. **39**(5), 861–899 (2018)
7. E.H. Lay, X.-M. Shao, C.S. Carrano, Variation in total electron content above large thunderstorms. Geophys. Res. Lett. **40**(10), 1945–1949 (2013)
8. E.H. Lay, X.-M. Shao, A.K. Kendrick, C.S. Carrano, Ionospheric acoustic and gravity waves associated with midlatitude thunderstorms. J. Geophys. Res. Space Phys. **120**, 6010–6020 (2015)
9. R.M. Akir, M. Abdullah, K. Chellappan, A.M. Hasbi, S.A. Bahari, Comparative study of TEC for GISTM stations in the Peninsular Malaysia region for the period of January 2011 to December 2012. Adv. Sci. Lett. **23**, 1304–1309 (2017)
10. P.P. Chaitanya, A.K. Patra, N. Balan, S.V.B. Rao, Ionospheric variations over Indian low latitudes close to the equator and comparison with IRI-2012. Ann. Geophys. **33**, 997–1006 (2015)
11. S.A. Bahari, M. Abdullah, A.M. Hasbi, A review of ionospheric studies in Malaysia using GPS, in *Proceeding of the 2015 International Conference on Space Science and Communication (IconSpace)* (2015), pp. 95–100
12. C.J. Davis, K.-H. Lo, Lightning-induced intensification of the ionospheric sporadic-E layer. Nature **435**, 799–801 (2005)
13. S.L. Vadas, H.-L. Liu, Generation of large-scale gravity waves and neutral winds in the thermosphere from the dissipation of convectively generated gravity waves. J. Geophys. Res. Space Phys. **114**, A10 (2009)
14. M.M. Amin, M. Inggs, P.J. Cilliers, Influence of lightning on total electron content in the ionosphere using WWLLN lightning data and GPS data. Proc. SA Inst. Phys. 411–416 (2014)
15. E.H. Lay, A.R. Jacobson, R.H. Holzworth, C.J. Rodger, R.L. Dowden, Local time variation in land/ocean lightning flash density as measured by the world wide lightning location network. J. Geophys. Res. Atmos. **112**, D13 (2007)
16. G.K. Seemala, C.E. Valladares, Statistics of total electron content depletions observed over the South American continent for the year 2008. Radio Sci. **46**, RS5019 (2011)
17. S.A. Bahari, M. Abdullah, B. Yatim, Variasi Jumlah Kandungan Elektron Ionosfera di Malaysia ketika Solar Minimum. Sains Malays. **46**(10), 1987–1995 (2017)

# Chapter 23
# A Case Study on Magnetic Signature of Ionospheric Disturbance Dynamo During an Intense Geomagnetic Storm on 6 August 2011

**Raja Adibah Raja Halim Shah, Nurul Shazana Abdul Hamid, Mardina Abdullah, Adlyka Annuar, Idahwati Sarudin, Zahira Mohd Radzi, and Akimasa Yoshikawa**

**Abstract** Coronal mass ejection (CME) occurs in the Sun's corona that travels and reaches Earth's magnetosphere, and it can trigger a geomagnetic storm. In this work, we presented a case study of a geomagnetic storm triggered by CME and observed the presence of ionospheric disturbance dynamo (Ddyn) at the equatorial station. Our analysis utilized CME data from the SOHO/LASCO CME Catalog, which contains information such as CME speed, solar source location, digital data, and the CME LASCO/C2 movies. The Disturbance storm time (Dst) index calculated from ground-based magnetometer data describes the geomagnetic storm phases. On the other hand, the variation of the Ddyn at the equatorial region was derived from the *H* component of Earth's magnetic field. Four stations at the dip equator cover the longitude sectors of Asia (DAV and TIR) and Africa (AAB and ILR). This study shows that halo

R. A. R. H. Shah · N. S. A. Hamid (✉) · A. Annuar
Department of Applied Physics, Faculty of Science and Technology, Universiti Kebangsaan Malaysia, 43600 Bangi, Malaysia
e-mail: zana@ukm.edu.my

N. S. A. Hamid · M. Abdullah
Space Science Centre (ANGKASA), Institute of Climate Change, Universiti Kebangsaan Malaysia, 43600 Bangi, Malaysia

M. Abdullah
Department of Electrical, Electronic and Systems Engineering, Faculty of Engineering and Built Environment, Universiti Kebangsaan Malaysia, 43600 Bangi, Malaysia

I. Sarudin
School of Physics, Universiti Sains Malaysia (USM), 11800 Pulau Pinang, Malaysia

Z. M. Radzi
Malaysian Space Agency (MYSA), 42700 Banting, Malaysia

A. Yoshikawa
International Research Center for Space and Planetary Environmental Science (i-SPES), Kyushu University, Fukuoka 819-0395, Japan

 
M. T. Islam et al. (eds.), *Proceedings of the 8th International Conference on Space Science and Communication*, Springer Proceedings in Physics 303,
https://doi.org/10.1007/978-981-97-0142-1_23

CME on 4 August 2011 triggered intense geomagnetic storms (Dst = − 115 nT) on 6 August 2011. During this geomagnetic storm, after a few hours of Dst peak, the signature of Ddyn can be clearly observed in two stations of the African sector (ILR and AAB), while no presence of Ddyn in two other stations (TIR and DAV) in the Asian sector.

## 23.1 Introduction

Solar activity such as solar flares, coronal mass ejection (CME), and high-speed solar wind streams can lead to a space weather phenomenon near the Earth's environment called geomagnetic storm events [1]. CME is a sudden release of large amounts of plasma and magnetic field from the Sun's corona. The change in corona structure can be observed through the coronagraphy image as its outward motion shows bright and white light features [2]. When a CME travel directed toward Earth through space, its magnetic field can interact with our planet's magnetosphere, causing disturbances that can contribute to geomagnetic storms. CME can take about 1–5 days to reach the Earth, which depends on its speed. Halo CME had a full 360° width and appeared as a circular 'halo' around the Sun in the coronagraphy image [3]. On the other hand, a geomagnetic storm is a temporary disturbance in Earth's magnetosphere that happens when Earth encounters solar activity from the Sun. There are four types of geomagnetic storms based on Disturbance storm time (Dst) index; moderate (Dst $\leq -$ 50 nT), intense (Dst $\leq -$ 100 nT), severe (Dst $\leq -$ 200 nT), and great (Dst $\leq -$ 350 nT) [4, 5]. Previous studies found that CMEs trigger the majority of intense storms, while moderate storms can be attributed to other activities, including corotating interaction regions [6].

The ionospheric disturbance dynamo (Ddyn) represents the variation of electric currents in the Earth's ionosphere during disturbances, particularly during the presence of geomagnetic storms. During the stormy period, strong ionospheric electric current was enhanced in the auroral zone. The joule heating at the auroral area produces thermospheric storm winds. Consequently, it spreads toward mid and low latitudes and changes the movement of ionospheric electric currents from a regular eastward direction into a westward flowing current [7]. The signature of the Ddyn was highlighted when a southward *H* component of the Earth's magnetic field was observed [8]. In this study, we select one case study of halo CME that has been the driver of an intense geomagnetic storm with Dst $\leq -$ 100 nT and observed its effect on Ddyn at equator stations covering the Asian and African sectors.

**Table 23.1** Coordinates of the observational *H* data station used in this study

| Sector | Station name (code) | Geographic coordinate (°) | | Geomagnetic coordinate (°) | | Local time (h) |
|---|---|---|---|---|---|---|
| | | Latitude | Longitude | Latitude | Longitude | |
| Asia | Davao (DAV) | 7.00 | 125.40 | – 2.22 | 197.90 | UT + 8 |
| | Manado (TIR) | 1.44 | 124.84 | – 7.80 | 197.63 | UT + 5.30 |
| Africa | Addis Ababa (AAB) | 9.04 | 38.77 | 5.41 | 112.54 | UT + 3 |
| | Ilorin (ILR) | 8.50 | 4.68 | 10.50 | 78.90 | UT + 1 |

## 23.2 Methodology

The data of planetary K (Kp) index, interplanetary magnetic field IMF, Bz and By representing the vertical and horizontal direction, respectively, and solar wind speed were acquired from the OMNIWeb Data [9]. The intensity of geomagnetic storms is identified using the Dst index, calculated based on recorded hourly data of horizontal magnetic variations by the observatory network near the equator [10]. This calculated Dst index data was provided by the World Data Center for Geomagnetism, Kyoto, Japan. Subsequently, the identification of CME is based on data provided by SOHO LASCO CME Catalog [3]. A geomagnetic storm with an associate CME candidate has a time window of 30–120 h backward, as the transit time is from 1 to 5 days. During the recovery phase, a day after the storm when the auroral activity is weak (< 500 nT), the Ddyn happen when the ionospheric disturbance dynamo process acts in the ionosphere. The expression of the ionospheric disturbance dynamo (Ddyn) is represented as [8, 11]:

$$\mathrm{Ddyn} = H - \mathrm{SR} - \mathrm{Dst},$$

where *H* is the variation of observation northward component of geomagnetic field and SR is *H* component during nearest quiet day. Table 23.1 provides the coordinates of four stations located in Asian and African sectors, from the observational *H* data was obtained for this study.

## 23.3 Results and Discussion

The analysis found that solar source CME, which occurred on 4 August 2011, 04:12 UT, was the corresponding source of a geomagnetic storm on 6 August 2011, 03:00 UT, with peak Dst at – 115 nT. Figure 23.1 shows that the CME, a frontside halo CME, can be observed using an EUV difference image [3] as pointed with a black arrow. The diffuse structure can interpret the shock sheath; the main body is the bright part. This CME has asymmetrical brightness as the main body moves toward

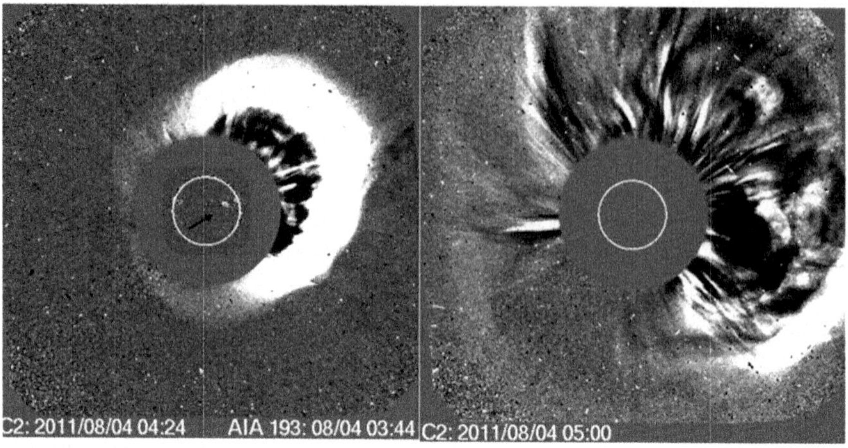


**Fig. 23.1** The coronagraphy image of the CME on 4 August 2011 that observed by the SOHO/ LASCO C2 FOV

Earth, mainly to the west. This halo CME, with a linear speed of 1315 km/s directed toward the Earth, produced an intense geomagnetic storm two days later.

We further analyze solar wind, interplanetary magnetic field (IMF), and global indices data to elucidate how the halo CME led to the occurrence of the intense geomagnetic storm. Figure 23.2 shows the variation of solar wind speed parameter (Vsw), IMF components of Bz and By, and the global magnetic indices of AU, AL, AE, Dst, and Kp. On 4 August 2011, when the CME was currently produced at the Sun's corona, we observed a steady solar wind speed as it was near-Earth solar wind data. However, solar wind speed showed a sudden increase from ~ 400 km/s (05:00 UT) to ~ 600 km/s (20:00 UT) on 5 August 2011, representing a fast solar wind starting to interact with the Earth's magnetosphere. The IMF Bz variation turns southward from ~ 0 nT (16:00 UT) to ~ − 19 nT (21:00 UT), and IMF By turns westward from ~ 17 nT (19:00 UT) to ~ − 14 nT (10:00 UT), which this interaction of IMF components and the Earth's magnetic field leads to the beginning of magnetic reconnection. The variation of AU, AL, and AE indices show a disturbance starting from 5 August 2011 (14:00 UT) and steadily stable around 6 August 2011 (10:00 UT). Consequently, on 5 August 2011 at 23:00 UT, the Dst reached − 96 nT for the first step (first green arrow) and dropped to the Dst minimum value of − 115 nT on 6 August 2011 at 03:00 UT as the second step development of the geomagnetic storm (second green arrow). At the same time, the Kp index increased up to 8 nT. Based on the Dst index, this storm interval started with the sudden storm commencement (SSC) at 19:00 UT, indicating the arrival of the CME. The main phase lasts approximately ~ 8 h, with the main phase onset on 5 August 2011 at 19:00 UT and ending on 6 August 2011 at 03:00 UT. The long recovery lasts ~ 7 days, ending on 13 August 2011 at 22:00 UT with Dst is 0 nT.

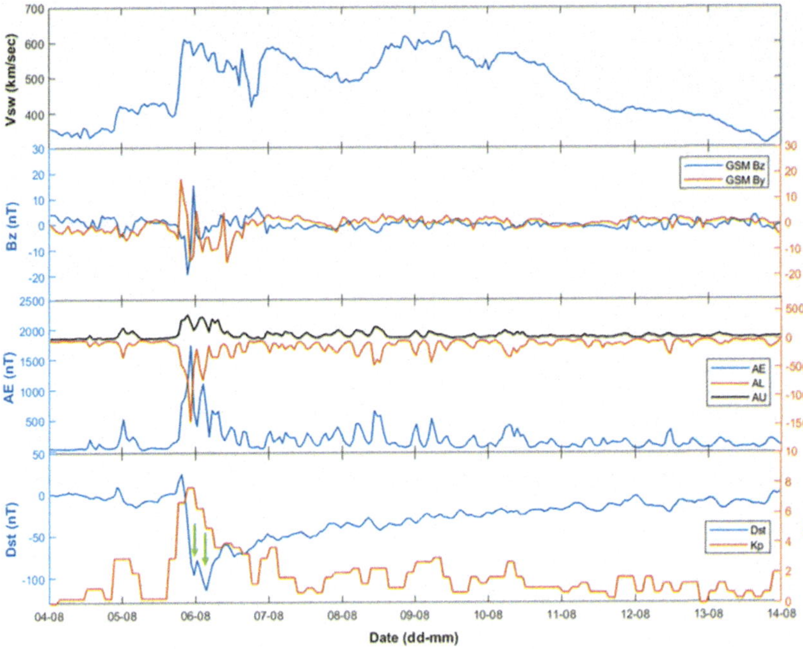


**Fig. 23.2** Magnetic indices and interplanetary parameters from 5 to 14 August 2011. From top to bottom there are the components of solar wind speed (km/s), component *y* and *z* of the interplanetary magnetic field (IMF), AE, AU, AL, Kp, and Dst indices. The green arrow represents the peak of geomagnetic storms

In the following part, studies continue to observe the present ionospheric disturbance dynamo (Ddyn) associated with the geomagnetic storm. Figure 23.3 depicts the variation of the Dst index and Ddyn in the Asian sector, represented by data from (a) TIR (India) and (b) DAV (Philippines) as well as the African sector, encompassing (c) ILR (Nigeria) and (d) AAB (Ethiopia). The shaded area in Ddyn panel plots represents the nightside, while the non-shaded area is the dayside of the local time stations. During the Dst peak with a value of − 115 nT (03:00 UT), these two stations, TIR and DAV (Asian sector), are on the dayside (08:30 LT and 11:00 LT). Meanwhile, stations in the African sector AAB on dayside (06:00 LT) and ILR at nightside (04:00 LT). The location of all stations during this Dst peak is illustrated in Fig. 23.3f, where the shaded area represents nightside.

On the first dayside part of the local time for each station immediately after the Dst peak of the geomagnetic storm (dashed black line), the auroral activity started to weak around 08:00 UT with an AE index below 500 nT. There was no presence of the Ddyn at the TIR and DAV stations in the Asian sector when the auroral started to weaken around 13:30 LT and 16:00 LT, respectively (first-day side, D1). Meanwhile, the southward disturbance of Ddyn can be observed in the two other stations in the African sector (D1). In the ILR station, the presence of Ddyn just started to begin

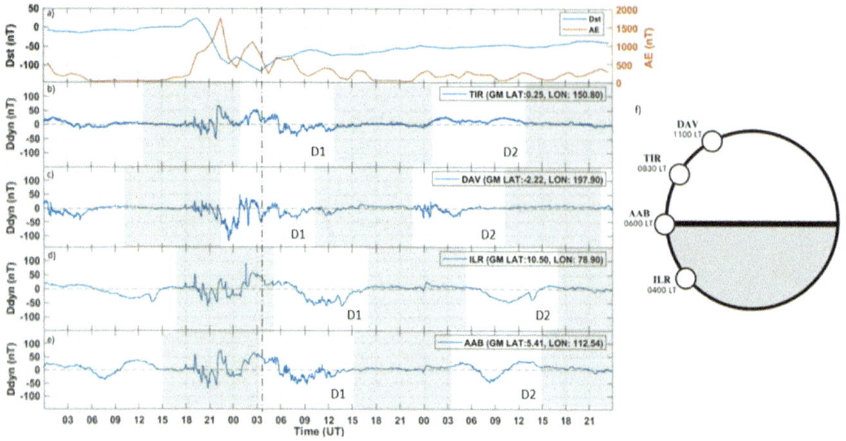

**Fig. 23.3** **a** The Dst and AE index for geomagnetic storm from 5 to 7 August 2011, the variation of *H* component of Ddyn in two sectors where Asian sector is **b** TIR station, **c** DAV station, and African sector is **d** ILR station, **e** AAB station. **f** The location of the equatorial stations in local time during the peak of Dst is − 115 nT on 6 August 2011 (03:00 UT). The shaded area represents the nightside of the local time station. The dashed line represents the Dst peak

around 09:00 LT (08:00 UT) to 16:00 LT (15:00 UT). However, the presence of the Ddyn at AAB stations began to end around 13:00 LT (10:00 UT) when the auroral started weakening. The value of Ddyn in this sector is around − 50 to − 100 nT and lasts for several hours. On the second dayside part (D2), in the Asian sector, no presence of Ddyn can be spotted at TIR station, as in DAV station, no clear signature of the Ddyn is observed. On the other hand, in the African sector (D2), the southward deviation of the Ddyn can be clearly perceived at ILR station around 08:00 LT (07:00 UT) to 16:00 LT (15:00 UT) and AAB station from 09:00 LT (06:00 UT) to 13:00 LT (10:00 UT) which the southward correspond was due to the present of westward ionospheric electric current in the dynamo layer. The duration of the Ddyn at the ILR station was more prolonged than the AAB station, which was around ~ 8 h and ~ 4 h, respectively. This southward variation represents the signature of the ionospheric disturbance dynamo (Ddyn). This shows that the attenuation of the Ddyn varies for different longitude stations [8].

## 23.4 Conclusion

This study examined an intense geomagnetic storm on 6 August 2011 due to the halo CME on 4 August 2011. The geomagnetic storm had sudden storm commencement (SSC) Dst is ~ 25 nT, two-step development of geomagnetic storm with the first peak Dst is ~ − 96 nT (23:00 UT) and second peak Dst is ~ − 115 nT (03:00 UT), and normal interval with the long recovery phase which last ~ 7 days. Corresponding to this geomagnetic storm, during weak auroral activity (AE < 500 nT), the $H$ component disturbance of Ddyn on both dayside parts (D1 and D2) shows an attenuation only on the observatory station in African sector and no clear signature of Ddyn in Asian sector can be observed during dayside of local time.

**Acknowledgements** This work is funded by Fundamental Research Grant Scheme (FRGS) under Ministry of Higher Education (KPT) Malaysia with the grant code number FRGS/1/2023/STG07/UKM/02/1. The authors would like to express our gratitude to the data provider from the GSFC/SPDF OMNIWeb interface at https://omniweb.gsfc.nasa.gov/ and World Data Center for Geomagnetism, Kyoto, Japan (WDC Kyoto) at https://wdc.kugi.kyoto-u.ac.jp/ for providing the magnetic indices and interplanetary parameter data. An appreciation was also given to NASA and The Catholic University of America, in collaboration with the Naval Research Laboratory, for developing and managing the CME catalogue at the CDAW Data Centre. SOHO is an international collaboration project between ESA and NASA (https://cdaw.gsfc.nasa.gov/CME_list/). We gratefully acknowledge the MAGDAS observation data that was provided by the International Research Center for Space and Planetary Environmental Science.

## References

1. P.I. Reyes, V.A. Pinto, P.S. Moya, Geomagnetic storm occurrence and their relation with solar cycle phases. Space Weather **19**(9) (2021)
2. A.P. Rouillard, C. Mostl, J.A. Davies, D. Bewsher, C.J. Eyles, M.A. Hapgood et al., On the shapes and widths of halo coronal mass ejections: results from the STEREO mission. Astrophys. J. **876**(1), 14 (2019)
3. N. Gopalswamy, S. Yashiro, G. Michalek, H. Xie, P. Mäkelä, A. Vourlidas, R.A. Howard, A catalog of halo coronal mass ejections from SOHO. Sun Geosph. **5**(1), 7–16 (2010)
4. C.A. Loewe, Classification and mean behavior of magnetic storms. J. Geophys. Res. Space Phys. **102**(A7), 14209–14213 (1997)
5. P.I. Reyes, V.A. Pinto, P.S. Moya, Geomagnetic storm occurrence and their relation with solar cycle phases. Space Weather **19**(9) (2021)
6. N. Gopalswamy, Halo coronal mass ejections and geomagnetic storms. Earth Planets Space **61**(5), 595–597 (2009)
7. C. Amory-Mazaudier, O.S. Bolaji, V. Doumbia, On the historical origins of the CEJ, DP2, and Ddyn current systems and their roles in the predictions of ionospheric responses to geomagnetic storms at equatorial latitudes. J. Geophys. Res. Space Phys. **122**(7), 7827–7833 (2017)
8. M. Le Huy, C. Amory-Mazaudier, Magnetic signature of the ionospheric disturbance dynamo at equatorial latitudes: "Ddyn." J. Geophys. Res. Space Phys. **110**(A10) (2005)
9. J.H. King, N.E. Papitashvili, Solar wind spatial scales in and comparisons of hourly wind and ACE plasma and magnetic field data. J. Geophys. Res. Space Phys. **110**(A2) (2005)
10. T.L. Gulyaeva, A.J. Mannucci, Echo of ring current storms in the ionosphere. J. Atmos. Sol.-Terr. Phys. **205** (2020)
11. J. Bulusu, R.K. Archana, K. Arora, Spatial variations in disturbance ionospheric currents—inference from 12 geomagnetic storms. J. Geophys. Res. Space Phys. **125**(8) (2020)

# Chapter 24
# Cumulative Distribution Function Analysis on Radio Frequency Spectrum Survey for Solar Monitoring Applications

**Radial Anwar, Dwi Andi Nurmantris, Hasanah Putri, Norbahiah Misran, Mohammad Tariqul Islam, and Peb Ruswono Aryan**

**Abstract** Solar activity has been observed since ancient times to anticipate its direct impact on life on earth. One of the instruments developed to monitor solar activity is the Callisto spectrometer developed by ETH Zurich, Switzerland. This instrument is spread in many countries, collaborating in the e-Callisto network. To build this system, it is necessary to conduct a radio frequency spectrum survey to see whether the spectrum conditions in a certain place allow Callisto to be installed or not. In this study, a radio frequency spectrum survey was conducted at Telkom University, Bandung, Indonesia. The RF signal strength in several frequency bands protected for radio astronomy observations is recorded. The obtained data was analyzed by using Cumulative Distribution Function (CDF) analysis to determine whether solar activity monitoring can be carried out at this location, within the Callisto frequency range. Observational results show that the signal level in all investigated frequency bands is relatively low, well below − 85 dBm for all bands, and with 80% samples are lower than − 100 dBm. Therefore, the observational sites are suitable for solar activity monitoring.

R. Anwar (✉) · D. A. Nurmantris · H. Putri
School of Applied Science, Telkom University, Bandung, Indonesia
e-mail: radialanwar@tass.telkomuniversity.ac.id

N. Misran · M. T. Islam
Department of Electrical, Electronic and Systems Engineering, Universiti Kebangsaan Malaysia, Bangi, Malaysia

P. R. Aryan
Vienna University of Technology, Vienna, Austria

 
M. T. Islam et al. (eds.), *Proceedings of the 8th International Conference on Space Science and Communication*, Springer Proceedings in Physics 303,
https://doi.org/10.1007/978-981-97-0142-1_24

## 24.1 Introduction

Solar activity has been observed since 1610 with the invention of the telescope [1]. Since then, a lot of information has been collected, which has improved researchers' understanding of the sun. One of the most researched aspects of solar observation is the effect of solar activity on the earth. This study includes the relationship between solar activity and earthquakes [2], the effect of solar activity on the variability of the tropospheric circulation mode [3], the effect of the solar wind on the earth's geomagnetic activity [4], the relationship between ozone and the number of sunspots [5], and even for research in biology that provides solutions to the problem of how solar activity affects biological objects on earth [6].

One of the solar activities that is always observed and studied is solar flare, which is a phenomenon where the sun shoots energy with high intensity in certain directions. It is usually associated with the presence of sunspots [7]. When the intensity of a solar flare is very high, the effect can disrupt wireless telecommunications systems on earth, damage satellites in outer space, and cause damage to the installation of the electricity distribution network (power grid). Therefore, the sun is always monitored, and its intensity measured to collect data that can be analyzed to produce a prediction system (early warning system) with high accuracy.

There are many instruments that have been designed and operated to monitor the sun. One of them is the Compound Astronomical Low-Cost Low-Frequency Instrument for Spectroscopy and Transportable Observatory (CALLISTO), a spectrometer developed by the Swiss Federal Institute of Technology (ETH) Zurich, Switzerland, to observe the sun [8]. Callisto can be operated on a frequency of 45–870 MHz. ETH together with other institutions around the world form a global network that observes and monitors solar activity, called the e-Callisto network. The purpose of this networking is so that the sun can be monitored 24 h a day, and thus, all activities that occur on the sun can be recorded. The instrument itself has even been used for geomagnetic disturbance prediction [9].

One of the challenges in conducting solar monitoring is the interference of telecommunication signals in the frequency band used for the monitoring. This phenomenon is known in the world of radio astronomy as radio frequency interferences (RFI). Therefore, solar observation stations or other radio astronomical objects are usually located in areas far from residential areas. However, locations with minimal RFI (RFI Quiet Zone) are already very difficult to find. Even if the place is available, other problems will arise, such as access and accommodation. Therefore, another alternative in conducting radio astronomy observations, including monitoring the sun, is to record data on the frequencies protected by the International Telecommunication Union (ITU) for radio astronomy purposes.

However, these frequency bands often experience interference which may be caused by unauthorized signal transmission, or telecommunication signals that have leaks in their filters so that the signal bandwidth is widened. Therefore, before building a radio astronomy observation station or solar monitoring system, it is necessary to conduct a radio frequency spectrum survey. This survey is to ensure that in the area, the frequency bands allocated for radio astronomy observations are free from interference. To see if a frequency band is polluted by RFI, ambient observations can be made by measuring the average power received in that band [10–16]. If the received power is high enough, it can be concluded that the band has been polluted by RFI, making it less than ideal for astronomical observations.

## 24.2 Methodology

Data acquisitions were conducted in two different places within the Telkom University (Bandung, Indonesia) area. The first location is outside the Selaru Building (FIT). Coordinate of this site is − 6.97303, 107.63295. The second measurement was conducted at the parking lot of the Tokong Nanas Building (GKU), with coordinate − 6.97245, 107.62984. RF Explorer spectrometer was employed to measure the signal level, while the data was recorded in a laptop, as shown in Fig. 24.1. On each signal recording, the antenna was oriented to all directions in both horizontal and vertical polarization, with more than 1000 recorded samples on each recording session. Since the spectrometer is equipped with a monopole antenna, it inherits a more equal coverage on a certain axis and hence can be used to get better coverage for terrestrial interferences when the antenna is pointed upward.

**Fig. 24.1** Spectrometer (RF Explorer) and laptop used for signal measurement

**Table 24.1** List of protected frequencies for astronomy at HF–UHF bands

| No | Band | Frequency (MHz) |
|---|---|---|
| 1 | I | 19.1–21.1 |
| 2 | II | 25.55–25.67 |
| 3 | III | 37.5–38.25 |
| 4 | IV | 73.0–74.6 |
| 5 | V | 150.05–153.0 |
| 6 | VI | 322.0–328.6 |
| 7 | VII | 406.1–410.0 |
| 8 | VIII | 608.0–614.0 |

The observed frequency bands are listed in Table 24.1. Band I–III are not within the Callisto spectrometer frequency range. However, they are also the protected bands for radio astronomy application, and can be observed by utilizing Callisto or other receiver (i.e., Radio JOVE) with some modification on the systems.

After the data were collected, they are processed and some statistical analyses were derived, including the overall spectrum display, average signal levels and the Cumulative Distribution Function (CDF) on the particular bands. The CDF of $FX(x)$ describes the probability that a random variable $X$ with a given probability distribution will be found at a value less than or equal to $x$ [17]. This statistical analysis provides a more flexible point of view in interpreting a distribution. Excel Add-in, KuTools, was employed to obtain these statistical data.

## 24.3 Results and Discussion

Due to the frequency span limitation of the spectrometer, the wideband Callisto spectrum should be taken in a split observation. The first recording covers the 15–615 MHz (Fig. 24.2), while the second recording obtain the 615–870 MHz data, as shown in Fig. 24.3. However, there is no protected frequency in the 615–870 MHz band and hence data from this second recording is not analyzed any further. The spectrum of averages and maximum peaks from the two different locations at 15–615 MHz band were then combined and compared, as depicted in Fig. 24.4.

Data obtained from both locations are relatively similar. The signal is strong in the range 87–108 MHz originating from FM radio signals, and several peak at the frequencies of 480 and 510 MHz which may originate from terrestrial communications. However, due to the very wide spectrum span compared to the bandwidth in each observed frequency band, the spectrum data in these bands has low resolution and seems quite high in signal level ($\approx -90$ to $-80$ dBm). In order to obtain more accurate data, further observations are carried out with narrower spans and larger samples for each observed frequency band.

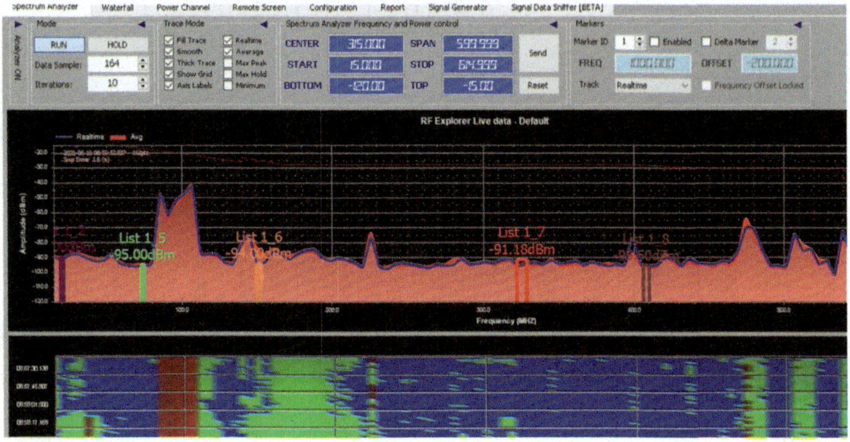


**Fig. 24.2** Recorded spectrum at 15–615 MHz

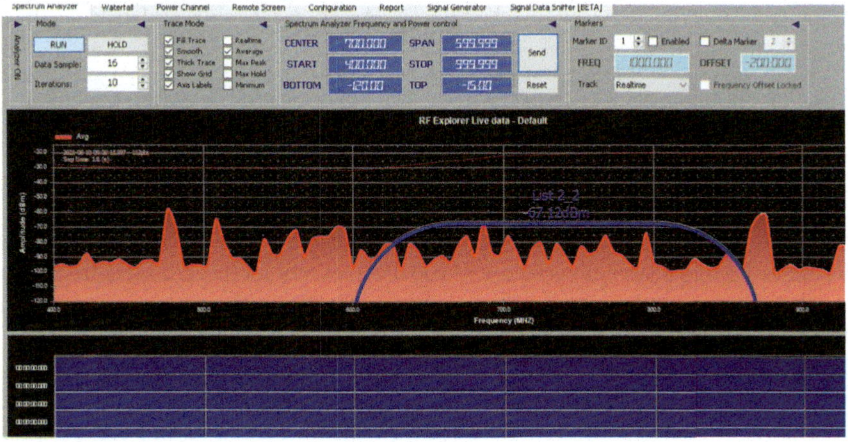


**Fig. 24.3** Recorded spectrum at 615–870 MHz (mark in blue line boundary)

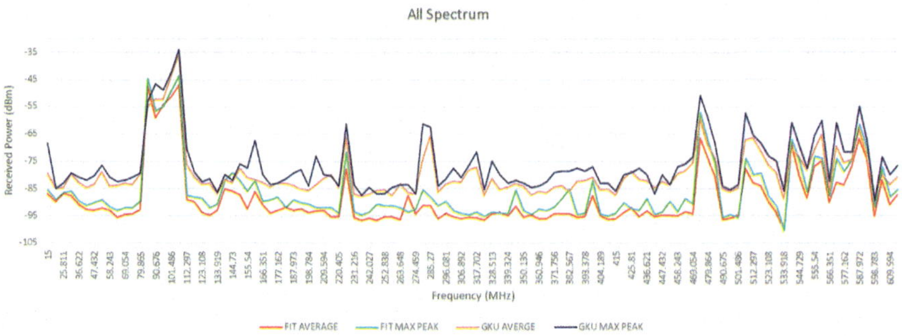


**Fig. 24.4** All recorded spectrum at FIT and GKU, shown in average and maximum peak level

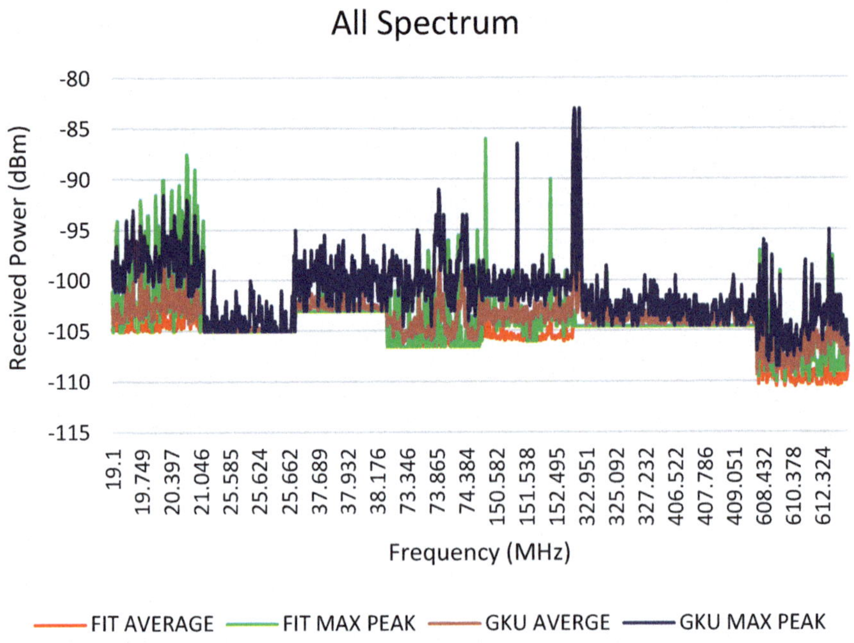


**Fig. 24.5** The combined high-resolution spectrum for the observed frequency bands

Figure 24.5 shows the combined spectrum of bands I–VIII. It can be seen that on average, the signal strength is below − 100 dBm, except for the 322 MHz frequency. At that frequency, the signal peak reaches − 90 dBm. This value is not too far from the average value across the spectrum range, so it is considered suitable as a frequency band for astronomical observations or solar monitoring. To get a more valid conclusion, the numerical data obtained is then processed statistically.

Figure 24.6 shows the Cumulative Distribution Function (CDF) curve for the signal strength obtained in band I to band VIII, at the two observation sites. It can be seen that in all observed bands, 80% of the data samples obtained were below − 100 dBm. Even for bands II, III, and VII, all samples show signal strength below − 100 dBm. This shows that the Selaru Building (FIT) and the Tokong Nanas Building (GKU), or the Telkom University environment in general, are suitable for being a place for observing radio astronomy or monitoring the sun.

From Fig. 24.6, none of the spectrum has a signal level more than − 80 dBm. Band II, III, and VII detect the lowest signal level, well below − 100 dBm. Band VI receives the highest signal level, up to − 83 dBm. These recorded signal levels are much lower compared to data obtained in [18]. If compared to [19], the values are uniformly higher. However, signal levels in [19] are presented in dB units, which might refer to another signal strength measurement units differ to dBm. Thus, Telkom University can be concluded as a suitable location to conduct radio astronomy observation, including solar monitoring.

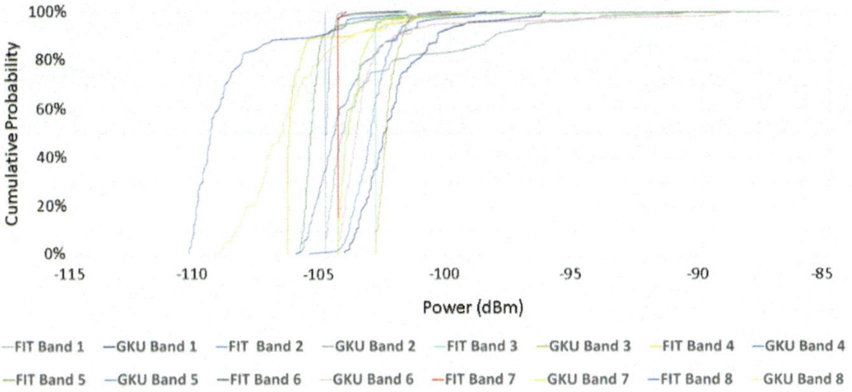


**Fig. 24.6** CDF for all bands at both observation sites

## 24.4 Conclusion

From the results of the analysis of the data obtained, it can be concluded that the environment at Telkom University is suitable to be used as a place for radio astronomy observations or solar monitoring. The frequency bands that are protected for the benefit of radio astronomy observations and intersect with the working frequency range of the Callisto spectrometer, are not polluted by radio frequency interferences (RFI), which is indicated by a low and relatively equal signal strength. The presence of a peak at a frequency of 322 MHz needs to be observed further, whether it is a constant signal or just a spike that might be caused by the instrument used.

In astronomical observations, the beam antenna is directed up into the sky. However, if there is a strong enough terrestrial RFI, it can contaminate the astronomical signal you want to observe. This study shows that even though the spectrometer antenna is pointed in all directions, the signal obtained is quite low. So that the use of directional antennas for radio astronomy observations or monitoring solar activity can provide data that is free from RFI.

## References


1. F. Steinhilber, J.A. Abreu, J. Beer, Solar modulation during the Holocene. Astrophys. Space Sci. Trans. **4**, 1–6 (2008)
2. M.N. Gousheva, K.Y. Georgieva, B.B. Kirov, D. Atanasov, On the relation between solar activity and seismicity, in *Proceedings of International Conference on Recent Advances in Space Technologies, 2003. RAST'03* (2003), pp. 236–240
3. I. Dorotovic, I. Trigo, Influence of solar activity on modes of tropospheric circulation variability, in *Proceedings of EGU General Assembly 2010*, Vienna, Austria, 2–7 May 2010, p. 4509
4. D. Jankovicova, Z. Voros, J. Simkanin, The influence of solar wind turbulence on geomagnetic activity. J. Nonlinear Process. Geophys. **15**, 53–59 (2008)

5. R.S. Selvaraj, S.T. Selvi, S.P.V. Priya, Association between surface ozone and solar activity. Indian J. Sci. Technol. **3**(3), 332–334 (2010)
6. V.K. Evstafyev, How solar activity influences earth's molecular processes. Open Biol. J. **2**, 38–41 (2009)
7. NASA, *Sunspots and Solar Flares* (2012). Available: https://www.nasa.gov/multimedia/imagegallery/image_feature_2201.html
8. A.O. Benz, C. Monstein, H. Meyer, CALLISTO—a new concept for solar radio spectrometers. Sol. Phys. **226**, 143–151 (2005)
9. N.M. Ansor, Z.S. Hamidi, N.N.Md. Shariff, The effectiveness of E-CALLISTO system in predicting geomagnetic disturbance. Int. J. Recent Technol. Eng. (IJRTE) **9**(2) (2020)
10. A. Zavvari, M.T. Islam, R. Anwar, A.M. Hasbi, M.F. Asillam, C. Monstein, CALLISTO radio spectrometer construction at Universiti Kebangsaan Malaysia. IEEE Antennas Propag. Mag. **56**(2) (2014)
11. A. Zavvari, M.T. Islam, R. Anwar, Z.Z. Abidin, M.F. Asillam, C. Monstein, Analysis of radio astronomy bands using CALLISTO spectrometer at Malaysia-UKM station. Exp. Astron. (2015). https://doi.org/10.1007/s10686-015-9480-z
12. P.P. Sitompul, T. Manik, M. Batubara, B. Suhandi, Radio frequency interference measurements for a radio astronomy observatory site in Indonesia. Aerospace **8**, 51 (2021)
13. J.A. Rengifo, V. Loaiza-Tacuri, J. Bazo, W.R. Guevara Day, Commissioning of CALLISTO spectrometers in Peru and observations of type III solar radio bursts. Res. Astron. Astrophys. **21**(6), 145 (2021)
14. T. Dyson, H.C. Chiang, E. Egan, N. Ghazi, T. Menard, R.A. Monsalve, T. Moso, Radio-frequency interference at the McGill Arctic Research Station. J. Astron. Instrum. **10**(2) (2021)
15. M.M. Shafie, R. Umar, N.H. Sabri, Statistical analysis of radio frequency interference (RFI) caused by solar radiation during wet and dry. Malays. J. Fundam. Appl. Sci. **17**, 430–436 (2021)
16. N. Prasert, A. Phakam, K. Asanok, P. Jaroenjittichai, A. Chumthong, T. Thongmeearkom, CALLISTO spectrometer for solar radio bursts monitoring in Chiangmai. J. Phys. Conf. Ser. **1380**, 012037 (2019)
17. J.S. Arora, *Introduction to Optimum Design*, 3rd edn. (Academic Press, 2012). Available: https://www.sciencedirect.com/book/9780123813756/introduction-to-optimum-design
18. R. Umar, N.H. Sabri, Z.A. Ibrahim, Z.Z. Abidin, A. Muhamad, Measurement technique in radio frequency interference (RFI) study for radio astronomy purposes. Malays. J. Anal. Sci. **19**(5), 960–965 (2015)
19. Z.N. Zainuddin, N.H. Sabri, R. Umar, C. Monstein, The study on the effect of population density on radio frequency interference (RFI) analysis on dynamic spectrum at selected CALLISTO stations. Phys. Sci. Technol. **8**(1–2), 53–61 (2021)

# Chapter 25
# A Novel Compact Ultrawideband Microstrip Patch Antenna for Satellite Communications System

**Md. Afzalur Rahman, Samir Salem Al-Bawri, Mohammad Tariqul Islam, Mandeep Jit Singh, and Md. Asifur Rahman**

**Abstract** The present study outlines the development and execution of a unique, condensed, ultrawideband microstrip patch antenna through the utilization of CST Microwave Studio software. The aforementioned antenna, with dimensions of 15 × 18.30 × 1.6 $mm^3$, has been designed with a particular focus on its utility in the context of satellite communication applications. The substrate utilized in its construction is Rogers RT5880 material. The antenna exhibits a noteworthy characteristic of achieving a wide range of bandwidth coverage, estimated to be around 140.39%. The system exhibits versatility in fulfilling diverse communication needs as it functions through three distinct frequency bands, specifically the C, X, and Ku bands. The antenna demonstrates exceptional transmission capabilities, achieving maximum efficiency of 93%. Remarkably, the antenna demonstrates a noteworthy gain of 4.8 dBi, indicating its exceptional ability to amplify signals, despite its diminutive size. The device's expansive frequency range, condensed physical dimensions, and noteworthy amplification render it appropriate for a diverse array of wireless implementations. Given these characteristics, the aforementioned antenna presents itself as a highly appropriate option for applications in satellite communication.

Md. A. Rahman · S. S. Al-Bawri (✉)
Space Science Centre, Climate Change Institute, Universiti Kebangsaan Malaysia (UKM), 43600 Bangi, Malaysia
e-mail: s.albawri@gmail.com; samir@ukm.edu.my

M. T. Islam · M. J. Singh
Department of Electrical, Electronic and Systems Engineering, Faculty of Engineering and Built Environment, Universiti Kebangsaan Malaysia, UKM, 43600 Bangi, Selangor, Malaysia

Md. A. Rahman
Department of Electrical and Electronic Engineering, School of Science and Engineering, United International University, UIU, United City, Madani Ave, Dhaka 1212, Bangladesh

 
M. T. Islam et al. (eds.), *Proceedings of the 8th International Conference on Space Science and Communication*, Springer Proceedings in Physics 303,
https://doi.org/10.1007/978-981-97-0142-1_25

## 25.1 Introduction

The advancement of wireless communication has generated significant interest among scholars in compact patch antennas that possess wideband capabilities. The increasing demand for higher data rates, improved spectral efficiency, and wider bandwidth has propelled the development of antennas with the objective of meeting the requirements of forthcoming wireless communication systems [1]. Microstrip antennas are extensively employed in diverse high-performance domains, including satellite systems, industrial applications, and communication via wireless systems. This can be attributed to their possession of crucial attributes such as significant bandwidth, diminutive dimensions, lightweight composition, and economic viability. Microstrip antennas exhibit certain limitations, such as low gain [2]. To cater to these requirements, UWB patch antennas have gained significant preference among wireless communication researchers. These antennas offer attractive attributes such as ultrawide bandwidth, simple design, cost-effectiveness, easy fabrication, and consistent radiation patterns [3]. The frequency band spanning from 3.1 to 10.6 GHz has been designated for employment in ultrawideband applications (UWB) [4]. The utilization of UWB technology [5, 6] is prevalent in wireless communication, and it is concomitant with the increasing significance of the C, X, and Ku bands in diverse communication systems. The electromagnetic spectrum is divided into specific frequency ranges, known as bands, as designated by the International Telecommunication Union (ITU). The frequency spectrum utilized by communication systems can be divided into various bands. The C band is characterized by frequencies ranging from 4 to 8 GHz, while the X band involves frequencies about 8–12.5 GHz. The Ku band, on the contrary, is defined by frequencies ranging from 12.5 to 18 GHz [7]. The C band is utilized in a multitude of services, including Fixed Satellite Service (FSS) and terrestrial microwave for fixed services. The X band finds application in a wide range of domains, including but not limited to satellite communication, radar systems, and earth observation satellites. The Ku band is utilized in satellite communication contexts, specifically in Fixed Satellite Service (FSS) and Direct Broadcast Satellite (DBS) applications [7]. The scarcity of antennas exhibiting ultrawideband characteristics is a noteworthy challenge in the field of wireless communication research.

Researchers have introduced a variety of patch antenna designs that exhibit ultrawideband (UWB) characteristics. The article cited as Ref. [8] presents a Compact Ultrawideband Circularly Polarized Microstrip Patch Antenna, which exhibits a noteworthy bandwidth of 85%. The authors conducted a research study in [9] to investigate an innovative printed microstrip antenna that demonstrated broadband capabilities. The aforementioned antenna was produced on a substrate possessing a distinct dielectric constant and was able to attain a bandwidth of 34.9%. The authors of a certain research paper [10] presented a circular low-profile monopole antenna that was specifically developed for ultrawideband applications. The antenna design was enhanced to achieve wideband performance through the incorporation of a slotted circular patch and a slotted ground structure. Furthermore, in a separate scholarly article [11], the authors introduced an innovative configuration for a monopole microstrip antenna that boasts a remarkable bandwidth of 107%, functioning within the 4.5–15 GHz frequency spectrum. In [12], authors have presented an Ultrawideband Monopole Antenna, which has a bandwidth of 122% and a pick gain of only 2.55 dBi. In another paper, authors have presented another antenna that has covered only 3.08–14.1 GHz bandwidth with an average gain of 4.54 dBi [13].

This study presents a novel layout for a microstrip patch antenna with a tree-like shape and slots featuring a partial ground with a stub. The antenna is intended for use in ultrawideband (UWB) satellite communication. The antenna being proposed appears to have a high level of versatility and could be well-suited for a range of applications, including Fixed Satellite Service (FSS), Direct Broadcast Satellite (DBS), radar systems, earth observation satellites, and other UWB applications.

## 25.2 Antenna Design and Configuration

The paper introduces a new antenna design that utilizes a substrate made of Rogers RT5880 material, which is known for its high level of loss and has a thickness of 1.6 mm. The dimensions of the antenna are $15 \times 18.30 \times 1.6$ mm$^3$, indicating a compact form factor. The design features a patch in the shape of a tree with a slot structure, antenna performance can be increased by slot structure [14], as well as a partial ground with a rectangular stub. The antenna's ability to function within the ultrawideband frequency range is made possible by utilizing the partial ground structure. The ground structure of the antenna incorporates a rectangular stub to improve its bandwidth and optimize impedance matching. The process of designing and simulating the antenna involves the utilization of the CST Microwave Studio software. The antenna was simulated in the frequency range of 3–18 GHz. The user employs standard microstrip antenna formulas to identify the physical size of the antenna, specifically its length ($L$) and width ($W$). Other parameters are normalized in relation to the envisioned operating frequency. The statement highlights the importance of optimization in achieving optimal performance of all antenna parameters. Figure 25.1 presents a graphical illustration of both the top and bottom components of the antenna that was created. Where Wf = 2.48, Lf = 5.5, L1 = 4.05, L2 = L3 = 3.35, L4 = 1.12, L5 = 1.6, L6 = 3.69, L7 = 0.8, L9 = 2.06, L10 = 3.2, Ws = Wg = 15, Lg = 3.60, Ls = 18.30, Lstub = 1.9 (mm).

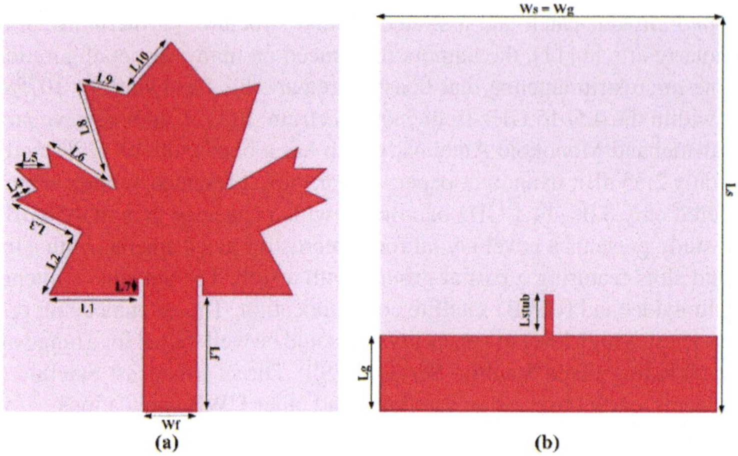


**Fig. 25.1** Dimensional front (**a**) and back (**b**) view of the proposed antenna

## 25.3 Result Analysis and Discussion

A microstrip antenna for broadband applications was designed and simulated utilizing CST Microwave Studio. The dimensions of the antenna in its initial state were derived from the traditional microstrip antenna formulas and subsequently employed in the simulator. This section presents and discusses all simulated outcomes gathered, encompassing a parametric evaluation, reflection coefficient, gain, radiation efficiency pattern, and current distribution.

The antenna design incorporates two symmetrically positioned slots adjacent to the feed line on the patch. The length of these slots plays a crucial role in enhancing the antenna's bandwidth. A parametric study is done to realize the effect of a slot [15]. The impact of slot length on bandwidth improvement is illustrated in Fig. 25.2. Optimal results are gathered when the slot length is set at 0.8 mm. Increasing the slot length results in a narrower bandwidth compared to the proposed design, while decreasing the slot length from the proposed value leads to inferior performance.

Figure 25.3 depicts the input coefficient of reflection (S11) of the antenna under consideration, which has been optimized for optimal performance. Return loss is a key parameter that is used to measure how electromagnetic waves bounce at the end of an antenna or transmission line [16]. The findings suggest that the antenna exhibits a broad frequency spectrum spanning from 3.5 to 17.7 GHz, with a center frequency of 10.1 GHz. The antenna exhibits four resonant frequencies, namely 4.3, 8.6, 13.4, and 16.2 GHz. The antenna's reflection coefficient maintains a value lower than − 10 dB throughout its entire range of functioning frequencies. The measured return loss values at frequencies of 4.3 GHz, 8.6 GHz, 13.4 GHz, and 16.2 GHz are − 42.3 dB, − 33.7 dB, 34.8 dB, and 23.3 dB, respectively. The antenna's percentage bandwidth was determined to be approximately 140.59% by using the following Eq. (25.1).

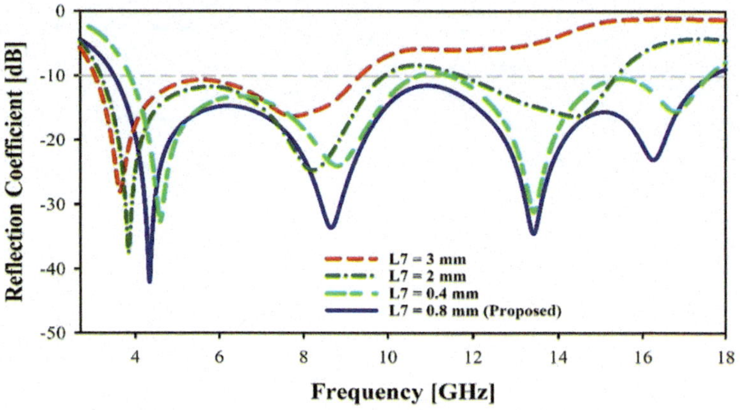


**Fig. 25.2** Reflection coefficient of parametric study of slots

$$\text{Percentage Bandwidth} = \left(\frac{\text{UF} - \text{LF}}{\text{Center Frequency}}\right) \times 100\% \tag{25.1}$$

Here, UF = Upper Frequency and LF = Lower Frequency.

Figure 25.4 illustrates the antenna's gain and efficiency across the entire frequency spectrum. Upon examination of the operational frequency spectrum, it has been determined that the apex gain is 4.8 dB isotropic at the median frequency. The variability of radiation efficiency ranges from 54 to 93%. The gain and efficiency

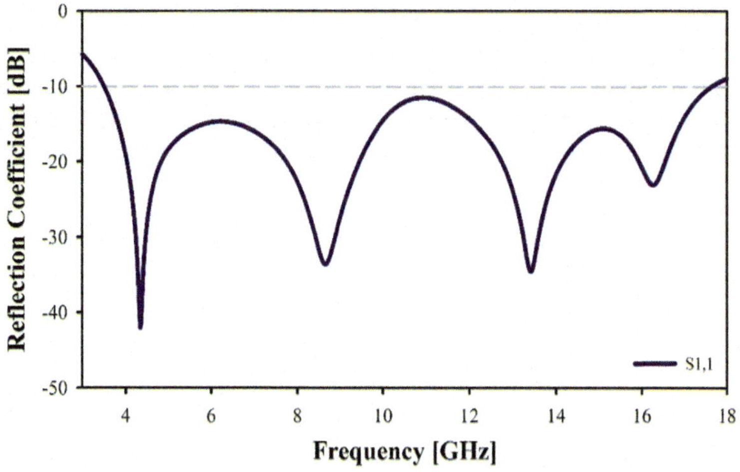


**Fig. 25.3** Reflection coefficient and bandwidth of proposed antenna

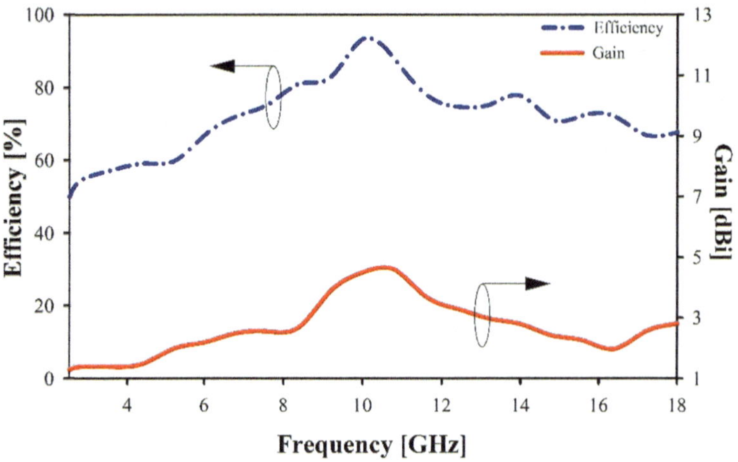


**Fig. 25.4** Gain and efficiency curve of proposed antenna

obtained from the cheap and compact structure make it especially suited for wireless communication uses. Gain and efficiency will be improved by using metamaterial techniques in the future [17, 18].

The E-field radiation patterns at different frequencies are depicted in Fig. 25.5. At lower frequencies of 4.3 and 8.6 GHz, the radiation pattern is observed to be omnidirectional when the azimuth angle (Phi) is set to 0. However, at 13.4 GHz, there is a slight change in the radiation pattern. When the azimuth angle is set to 90°, the radiation pattern becomes bidirectional. At 4.3 GHz, the radiation pattern is slightly omnidirectional, whereas at 8.6 and 13.4 GHz, it is fully bidirectional.

The distribution of electrical current moving through the structure of an antenna is commonly referred to as its current pattern. The distribution of current over the patch of the antenna is depicted in Fig. 25.6 at frequencies of 4.3, 8.6, and 13.4 GHz. The examination of current density indicates that the current attains its maximum value

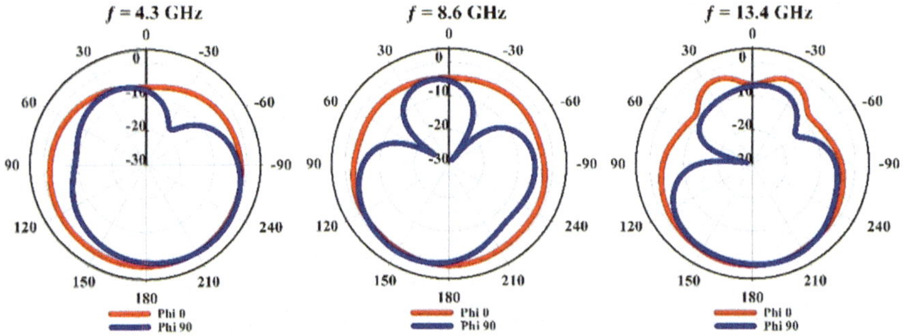


**Fig. 25.5** 2D radiation pattern for $\varphi = 0$ and $\varphi = 90$ in E-field

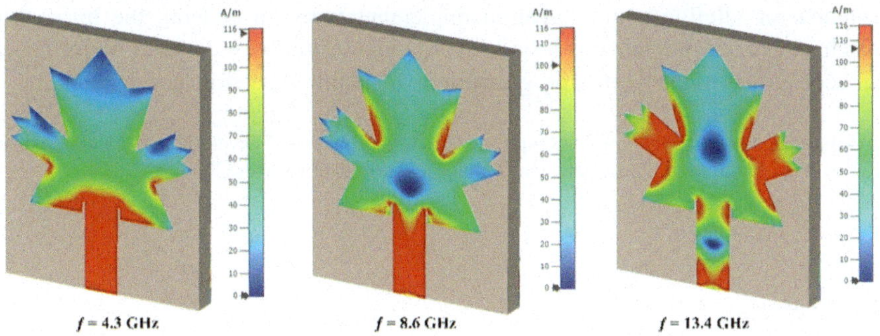


**Fig. 25.6** Current distribution of proposed antenna at 4.3, 8.6, and 13.4 GHz

**Table 25.1** Performance comparison of the proposed antenna

| References | [3] | [9] | [11] | Proposed |
|---|---|---|---|---|
| Reflection coefficient | − 30 | − 45 | − 31 | − 42.3 |
| Bandwidth (%) | 114 | 34.9 | 107 | 140.59 |
| Realized gain (dBi) | 1–4 | 8 | 5.3 | 4.8 |
| Antenna size (mm$^2$) | 24 × 12 | 42 × 55 | 21 × 15 | 15 × 18.30 |
| Substrate | RT/Duroid 5880 | FR-4 | RT/Duroid 5880 | RT/Duroid 5880 |

at the feed line when operating at a frequency of 4.3 GHz. Subsequently, the current gradually diminishes across the entire patch. The highest recorded value of current is 114.35 A/m. A current of 99.79 A/m is observed at both the feed line and the root of the tree-shaped patch when the frequency is 8.6 GHz. At a frequency of 13.4 GHz, the branch of the tree-shaped patch exhibits a maximum current of 105.83 A/m. Performance comparison with current literature is presented in Table 25.1.

## 25.4 Conclusion

This research paper presents a novel design for a microstrip patch antenna, which has been developed using the CST Microwave Studio Software. The antenna exhibits a condensed configuration characterized by a diminutive physical stature and maximal spatial efficiency, thereby facilitating ultrawideband transmission across three distinct frequency ranges for satellite communication. By enabling high data rates and providing a suitable gain, a significant bandwidth of 140.59% is attained. The antenna exhibits noteworthy directional characteristics. The proposed design encompasses the C, X, and Ku frequency bands, rendering it a highly appropriate antenna for satellite communication implementations. In general, the microstrip patch antenna design that has been put forth demonstrates exceptional performance and constitutes a noteworthy progression in the realm of satellite communication.

**Acknowledgements** This work was supported by Geran Universiti Penyelidikan (GUP), Universiti Kebangsaan Malaysia (UKM), Selangor, Malaysia under grant number GUP-2022-075.

## References

1. K. Ramahatla, M. Mosalaosi, A. Yahya, B. Basutli, Multi-band reconfigurable antennas for 5G wireless and CubeSat applications: a review. IEEE Access (2022)
2. T. Tewary, S. Maity, S. Mukherjee, A. Roy, P.P. Sarkar, S. Bhunia, High gain miniaturized super-wideband microstrip patch antenna. Int. J. Commun. Syst. **35**, e5181 (2022)
3. H. Yang, X. Xi, Y. Zhao, L. Wang, X. Shi, Design of compact ultrawideband slot antenna with improved band-edge selectivity. IEEE Antennas Wirel. Propag. Lett. **17**, 946–950 (2018)
4. T. Shabbir, R. Saleem, S.S. Al-Bawri, M.F. Shafique, M.T. Islam, Eight-port metamaterial loaded UWB-MIMO antenna system for 3D system-in-package applications. IEEE Access **8**, 106982–106992 (2020)
5. A.N. Biswas, S. Ballav, A. Chatterjee, S.K. Parui, Evolution of low-profile ultra-wideband frequency selective surface with a stable response and sharp roll-off at lower band for C, X and Ku band applications. Radioengineering **29**, 494–503 (2020)
6. K. Hossain, T. Sabapathy, M. Jusoh, P.J. Soh, M.N. Osman, S.S. Al-Bawri, A compact wideband CSRR near zero refractive index and epsilon negative metamaterial for wearable microwave applications. J. Phys. Conf. Ser. 012019 (2021). IOP Publishing
7. B.R. Elbert, *The Satellite Communication Applications Handbook* (Artech House, 2004)
8. C. Sun, A design of compact ultrawideband circularly polarized microstrip patch antenna. IEEE Trans. Antennas Propag. **67**, 6170–6175 (2019)
9. V.P. Sarin, M.S. Nishamol, D. Tony, C.K. Aanandan, P. Mohanan, K. Vasudevan, A wideband stacked offset microstrip antenna with improved gain and low cross polarization. IEEE Trans. Antennas Propag. **59**, 1376–1379 (2011)
10. T.K. Saha, C. Goodbody, T. Karacolak, P.K. Sekhar, A compact monopole antenna for ultra-wideband applications. Microw. Opt. Technol. Lett. **61**, 182–186 (2019)
11. H.N. Awl, B.A. Karim, Y.I. Abdulkarim, L. Deng, D.A. Hassan, M. Karaaslan, Broadband microstrip antenna for C-band, X-band, and KU-band applications, in *2019 International Conference on Information and Telecommunication Technologies and Radio Electronics (UkrMiCo)* (IEEE, 2019), pp. 1–5
12. A.J.A. Al-Gburi, Z. Zakaria, M. Palandoken, I.M. Ibrahim, A.A. Althuwayb, S. Ahmad, S.S. Al-Bawri, Super compact UWB monopole antenna for small IoT devices. Comput. Mater. Contin. **73**, 2785–2799 (2022)
13. S.S. Al-Bawri, H. Hwang Goh, M.S. Islam, H.Y. Wong, M.F. Jamlos, A. Narbudowicz, M. Jusoh, T. Sabapathy, R. Khan, M.T. Islam, Compact ultra-wideband monopole antenna loaded with metamaterial. Sensors **20**, 796 (2020)
14. S.S. Al-Bawri, M.T. Islam, M.J. Singh, E. Alyan, M. Jusoh, T. Sabapathy, S. Padmanathan, K. Hossain, Broadband sub-6 GHz slot-based MIMO antenna for 5G NR bands mobile applications. J. Phys. Conf. Ser. 012038 (2021). IOP Publishing
15. M.A. Haque, M.A. Zakariya, N.S.S. Singh, M.A. Rahman, L.C. Paul, Parametric study of a dual-band quasi-Yagi antenna for LTE application. Bull. Electr. Eng. Inform. **12**, 1513–1522 (2023)

16. M.A.K. Jilani, M.A. Hossain, F.Y. Lamia, M.A. Rahman, A.H. Sharker, M.R. Ahsan, Design of 2*1 patch array antenna for 5G communications systems using mm-wave frequency band, in *2022 IEEE 12th Annual Computing and Communication Workshop and Conference (CCWC)* (IEEE, 2022), pp. 0857–0862
17. T. Shabbir, M.T. Islam, N. Misran, S.S. Al-Bawri, S. Singh, Broadband single-layer reflectarray antenna loaded with meander-delay-lines for X-band applications. Alex. Eng. J. **60**, 1105–1112 (2021)
18. M.A. Haque, N. Sarker, N.S. Sawaran Singh, M.A. Rahman, M.N. Hasan, M. Islam, M.A. Zakariya, L.C. Paul, A.H. Sharker, G.E.M. Abro, Dual band antenna design and prediction of resonance frequency using machine learning approaches. Appl. Sci. **12**, 10505 (2022)

# Chapter 26
# Nickel-Based Ultra-Thin Nano Architect Metamaterial Absorber for Optical and NIR Regime

**Abu Hanif, Touhidul Alam, Tanveer Ahsan, and Mohd Fais Mansor**

**Abstract** This article proposed ultra-thin square ring-shaped metamaterial absorber nano-architectures for optical and near-infrared regime (NIR). The absorber structure comprises three layers of nickel-polyimide-nickel (Ni–Pi–Ni) material and accomplishes better absorption performance. It shows an absorption rate of 99% across a large bandwidth ranging from 990 to 1210 nm wavelength. Additionally, it shows high average absorptivity of 93.97% in the optical regions (380–700 nm) and 96.95% in the NIR regime (700–2100 nm) at 0° incidence angle of electromagnetic waves. The absorber exhibits polarization insensitivity and remains stable at oblique incidence angles up to 70° for both transverse electric (TE) and transverse magnetic (TM) modes. The study also investigates metamaterial properties, electric and magnetic characteristics, as well as surface current distribution to understand the absorption behavior. The main notable features of this metamaterial absorber are its wideband near-perfect absorptivity in the NIR regime and its structurally simple configuration.

## 26.1 Introduction

Metamaterials have captivated researchers and engineers due to their remarkable properties, offering unique control over electromagnetic waves through their precisely engineered structures [1, 2]. These materials, carefully designed on a subwavelength scale, exhibit extraordinary electromagnetic responses not found in naturally occurring substances [3]. By exploiting the principles of metamaterials, researchers have unlocked the potential to manipulate light and tailor its interaction with matter. Among the vast array of metamaterial applications, absorbers

A. Hanif · T. Alam (✉)
Pusat Sains Angkasa (ANGKASA), Institut Perubahan Iklim, Universiti Kebangsaan Malaysia (UKM), 43600 UKM Bangi, Selangor, Malaysia
e-mail: touhidul@ukm.edu.my

T. Ahsan · M. F. Mansor
Department of Electrical, Electronic and Systems Engineering, Faculty of Engineering and Built Environment, Universiti Kebangsaan Malaysia, 43600 UKM Bangi, Selangor, Malaysia



M. T. Islam et al. (eds.), *Proceedings of the 8th International Conference on Space Science and Communication*, Springer Proceedings in Physics 303,
https://doi.org/10.1007/978-981-97-0142-1_26

have garnered considerable attention for their ability to efficiently capture and convert incident radiation into heat or electrical signals. They find extensive use in numerous fields, ranging from camouflage technology [4], absorber [5], sensor [6–8], energy harvesting [9], and optical communication [10–12] systems. However, developing absorbers that possess broadband absorption, polarization independence, and stability across different incident angles has remained a significant challenge [13]. In recent times, researchers have conducted numerous studies to achieve wideband absorption characteristics using various configurations of metamaterial structures. One such study [4] proposed a metamaterial composed of multiple ellipse-shaped elements, which demonstrated polarization-sensitive absorption of over 90% across a small bandwidth ranging from 400 to 750 nm wavelength in the visible regime. Additionally, a star-shaped metamaterial [14] was developed to achieve polarization-insensitive absorption with near-perfection from 389.34 to 697.19 nm, encompassing a size of $1000 \times 1000 \times 225$ nm$^3$. Another investigation [15] explored a polarization-insensitive multilayer metamaterial structure, which achieved less than 90% absorption over a wavelength range of 300–680 nm. However, the angular stability was limited to 50° only. In the pursuit of angular stability, different shapes such as cross, rectangular, and cylindrical were utilized in multilayer metamaterial structures [16–18], achieving stability up to 50° and 70°, respectively.

However, these methods either suffered from a narrower absorption bandwidth or complexity in design. Therefore, this article proposed MMA absorber breaks free from the limitations of traditional approaches by incorporating a square ring shape and leveraging the unique properties of metamaterial structures. By employing square geometries, the absorber achieves polarization independence, ensuring efficient absorption regardless of the incident light's polarization state [9]. Furthermore, its incident angle stability guarantees consistent performance even incident light varies, broadening its potential applications and enhancing its versatility [13]. The nano-architectural nature of this absorber is a key distinguishing feature. By harnessing the power of nanoscale design principles, the absorber exhibits unique control over light-matter interactions [19]. The precise arrangement of nanostructures allows for exceptional light trapping and absorption, enabling efficient utilization of incident radiation across a wide range of frequencies within the optical and mid-infrared regime [20]. Compactness and size reduction are vital aspects in many practical scenarios where space constraints pose a challenge. The proposed MMA boasts an impressively compact size, making it an attractive solution for applications where miniaturization is a priority. By minimizing the absorber's footprint without compromising performance, this advancement opens new possibilities across visible to mid IR region applications.

## 26.2 Structural Analogy

The suggested square shape split ring metamaterial absorber consists of a three-layer configuration, namely metal-dielectric-metal. Figure 26.1a depict perspective views of the $(3 \times 3)$ unit cell boundary condition and design parameters of the MMA. Nickel (Ni) is used as a top and bottom layer metal because it has a high electrical conductivity and is a good heat reflector, which makes it useful for a number of thermal applications [9]. And, polyimide is preferred as dielectric for its excellent thermal stability, mechanical flexibility, and electrical insulation properties, endure temperatures of up to 350 °C without losing structural integrity [13]. Additionally, the bottom layer, substrate layer, and resonator layer of the proposed structure have thicknesses of 10 nm, 24 nm, and 8 nm, respectively. To design and simulate the structure, the CST Studio suite software was utilized [21]. The unit cell boundary conditions were implemented along the $x$ and $y$ axes, while open add-space boundary conditions were defined along the negative $z$ axis. The total absorption of the square resonator structure was determined by employing Eq. (26.1) [17]

$$\begin{aligned} A(\omega) &= 1 - R(\omega) - T(\omega) \\ &= 1 - R(\omega) \end{aligned} \tag{26.1}$$

$A(\omega)$, $T(\omega)$, and $R(\omega)$ represent absorbance, transmittance and reflection, respectively. In Fig. 26.2, $T(\omega)$ equal to zero due to nickel metal use as a bottom layer, where no wave passes through the proposed MMA structure. Also, the average absorptivity is 95.97% for the entire operational wavelength. And, the average absorption is 93.97% for optical and 96.95% for near-infrared regime for this MMA (Table 26.1).

## 26.3 Polarization Angle and Oblique Incident Angle

The polarization insensitivity and incident angle stability of the designed MMA illustrated in Fig. 26.2. The polarization insensitivity is depict in Fig. 26.2a, b for TE and TM mode, where the polarization angle ($\varphi$) is varied from 0 to 180° in 30° increments. The designed structural symmetry of the MMA results in uniform absorption characteristics being observed for different polarization angles [22]. Figure 26.2c, d represent the investigation of the incident angle ($\theta$) ranging from 0° to 70° for Transverse Electric and Transverse Magnetic modes, respectively. The absorptivity gradually decreases from 95 to 93% as the angle of incidence increases from 0 to 30°. However, for oblique incident angles below 60° and 70° for transverse electric and transverse magnetic modes respectively, the average absorptivity can reach up to 80%.

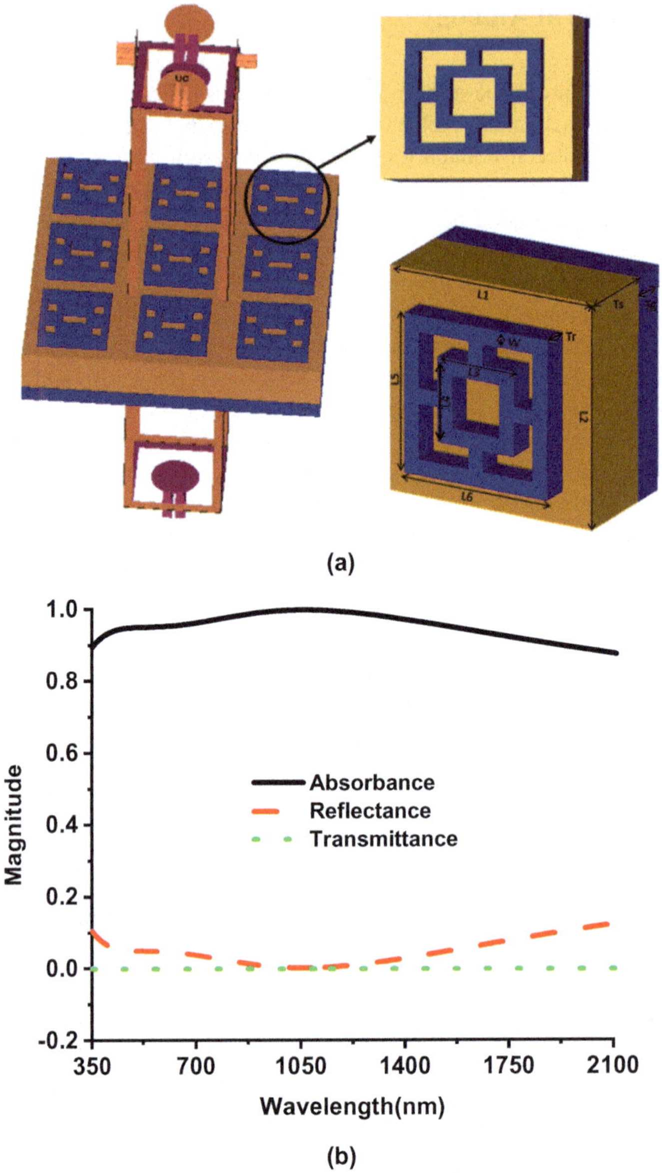


**Fig. 26.1** **a** Schematic view and **b** absorption, reflection, and transmittance of the proposed MMA structure

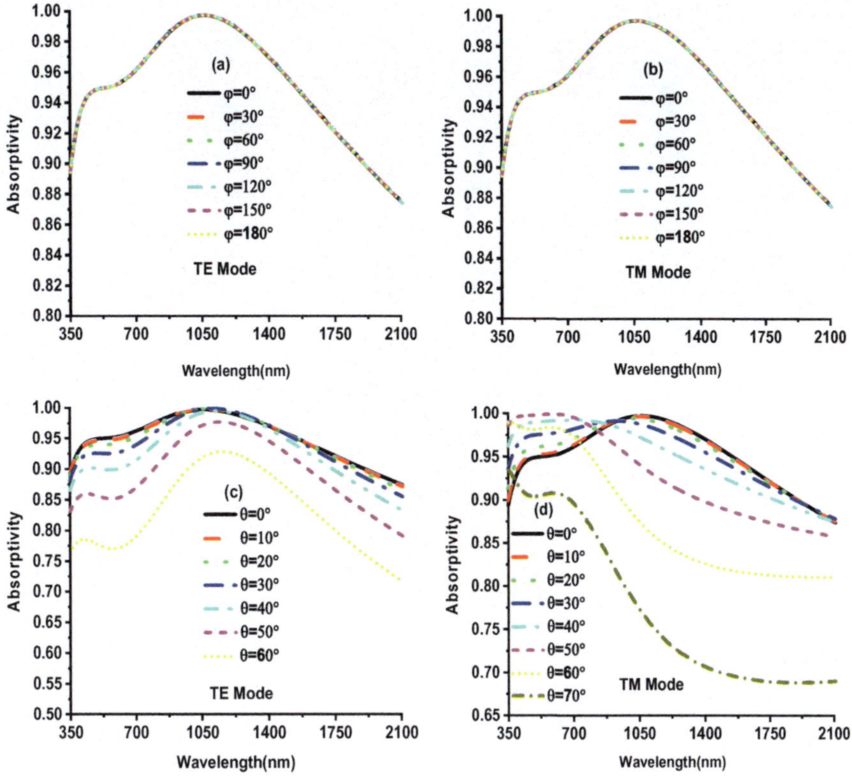


**Fig. 26.2** Absorption spectra for various polarization angles **a** TE, **b** TM mode, and incident angle **c** TE and **d** TM mode of the proposed MMA

**Table 26.1** Proposed MMA parameters

| Name | *L*1 | *L*2 | *L*3 | *L*4 | *L*5 | *L*6 | $T_r$ | $T_s$ | $T_g$ | *w* |
|---|---|---|---|---|---|---|---|---|---|---|
| Value (nm) | 64 | 60 | 22 | 22 | 42 | 42 | 8 | 24 | 10 | 6 |

## 26.4 Electric Field, Magnetic Field, and Surface Current

Figure 26.3 illustrate how electric field (E-field), magnetic fields (H-field), and surface current behave at wavelengths during absorption. The analysis revealed that the electromagnetic field becomes highly concentrated and amplified within the absorber at specific points. When the electric and magnetic responses align perfectly with the incoming wave, it leads to complete absorption [13, 23]. The electric field at resonance wavelength 1150 nm mostly confined square ring resonator patch comparing other portion of the MMA. Also, the magnetic field of the proposed MMA behave like E-field distribution.

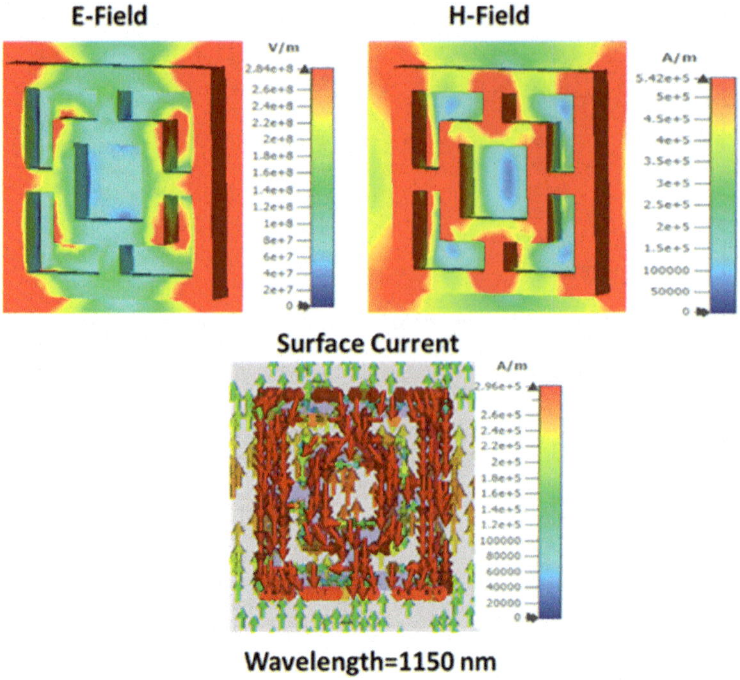


**Fig. 26.3** Electric field, magnetic field, and surface current diagram of the proposed MMA at perfect absorption wavelength

Moreover, the nickel square resonator exhibits a strong presence of circulating current. The current circulates in parallel and anti-parallel directions along the nano ring, leading to electrical resonance within the absorber. Consequently, this circulating current produces magnetic resonance, which activates surface plasmons in the substrate layer. This interplay of magnetic resonance and surface plasmons contributes to broad absorption.

## 26.5 Material Properties

The impedance matching condition is essential for determining how well a metamaterial absorber can absorb EM waves. When an impedance match with free space impedance, the transmission is completely suppressed. For proposed square ring-based MMA, the nickel layer at the bottom is thicker than the skin depth of the incident wave, resulting in minimal loss of transmission [23].

As a result, the WMMA demonstrates nearly perfect impedance matching and negligible transmission across the entire operational wavelength range, as shown in Fig. 26.4. In addition, the connection among absorption, impedance, and the

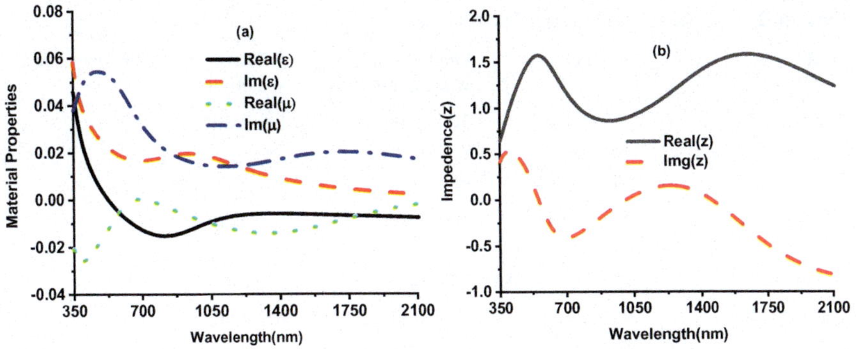


**Fig. 26.4** Material properties of the proposed MMA including **a** permittivity, permeability, and **b** impedance

characteristics of metamaterials such as permittivity ($\varepsilon$) and permeability ($\mu$) can be described using Eq. (26.2) [9].

$$\begin{aligned} A(\omega) &= 1 - R(\omega) \\ &= 1 - |Z_e/Z_o|^2 \\ &= 1 - \left|\sqrt{\mu} - \sqrt{\varepsilon}/\sqrt{\mu} + \sqrt{\varepsilon}\right| \end{aligned} \tag{26.2}$$

Figure 26.4a display the magnitude of the MMA. The magnitudes of both parameters are close to each other throughout the operational wavelength range, spanning from 350 to 2100 nm. In Fig. 26.4b, the impedance shows that the real value is close to 1 from 800 to 1200 nm and the imaginary value is close to 0. As a result, there is perfect absorption from 990 to 1200 nm wavelength, which is a near-perfect match with the impedance of free space.

Table 26.2 distinguished previously reported metamaterial absorber with the proposed MMA. Where the designed structure suffered due to large and complex design resulting lower absorptivity and degradation of incident angle stability. Most of them cover only part of the optical and infrared regime. Comparing other the proposed ultra-thin simple MMA structure achieves 95.97% absorptivity over a large wavelength from 350 to 2100 nm with up 70° incident angle tolerance.

**Table 26.2** Comparison with recent MMA work

| Shape | Dimension ($nm^3$) | Operational wavelength (nm) | Polarization sensitivity | Incident angle |
|---|---|---|---|---|
| Sandwich [15] | 1555 × 1555 × 560 | 300–680 | Insensitive | 50° |
| Nano disk [24] | 400 × 400 × 130 | 450–600 | Sensitive | N/A |
| Cuboid [25] | 1000 × 1000 × 560 | 500–1800 | Insensitive | 60° |
| Cross [17] | 350 × 350 × 385 | 368–1178 | Insensitive | 50° |
| Frustum [26] | 230 × 230 × 500 | 480–1600 | Insensitive | 60° |
| Square [this work] | 64 × 60 × 42 | 350–2100 | Insensitive | 70° |

## 26.6 Conclusion

This article presents a polarization-insensitive simple structured metamaterial absorber (MMA) achieving near-perfect absorption from visible to NIR wavelengths. The results demonstrate over 90% absorption across the 350–2100 nm regime, with an average absorption of 95.97%. Moreover, the MMA exhibits absorptivity exceeding 99% within the 990–1210 nm range. The MMA's structural rotational symmetry enables uniform absorption for both transverse electric and transverse magnetic incident waves. Additionally, it shows a high tolerance for incident angles, $\theta \leq 70°$, enhancing its efficiency. The absorption characteristics were analyzed using metamaterial properties, electric field, magnetic field, and surface current distributions. The MMA design exhibits substantial promise for various applications, including light harvesting, thermal imaging, photodetection, and others, owing to its broad absorption range in the optical and NIR regions.

**Acknowledgements** The authors are grateful to Universiti Kebangsaan Malaysia.

**Author Contribution** *Abu Hanif*: Conceptualization, formal analysis, Methodology, Design, Investigation, Writing—original draft. *Touhidul Alam*: Conceptualization, Supervision, Project administration, Review and editing. *Tanveer Ahsan*: Investigation, Review and editing. *Mohd Fais Mansor*: Project administration and Review and editing.

**Funding** This work was funded by the university research grant of Universiti Kebangsaan Malaysia, Grant code: GUP-2022-074.

**Data Availability Statement** The data presented in this study are presented in this article.

**Conflicts of Interest** The authors declare no conflict of interest.

## References


1. Q. Huang, G. Wang, M. Zhou, J. Zheng, S. Tang, G. Ji et al., Metamaterial electromagnetic wave absorbers and devices: design and 3D microarchitecture. J. Mater. Sci. Technol. **108**, 90–101(2022)
2. J. Zhou, Z. Liu, X. Liu, G. Fu, G. Liu, J. Chen et al., Metamaterial and nanomaterial electromagnetic wave absorbers: structures, properties and applications. J. Mater. Chem. C **8**(37), 12768–12794 (2020)
3. N. Misran, S.H. Yusop, M.T. Islam, M.Y. Ismail, Analysis of parameterization substrate thickness and permittivity for concentric split ring square reflectarray element. J. Kejuruter. **23**, 11–16 (2012)
4. Y. Qu, Q. Li, L. Cai, M. Pan, P. Ghosh, K. Du et al., Thermal camouflage based on the phase-changing material GST. Light Sci. Appl. **7**(1), 26 (2018)
5. M.L. Hakim, T. Alam, A.F. Almutairi, M.F. Mansor, M.T. Islam, Polarization insensitivity characterization of dual-band perfect metamaterial absorber for K band sensing applications. Sci. Rep. **11**(1), 17829 (2021)
6. M.Z.B. Chowdhury, M.T. Islam, I. Hossain, H. Alsaif, A.S. Alshammari, A. Alzamil et al., A bendable wide oblique incident angle stable polarization insensitive metamaterials absorber for visible optical wavelength applications. Optik 171016 (2023)
7. M.L. Hakim, T. Alam, M.T. Islam, M.H. Baharuddin, A. Alzamil, M.S. Islam, Quad-band polarization-insensitive square split-ring resonator (SSRR) with an inner Jerusalem cross metamaterial absorber for Ku- and K-band sensing applications. Sensors **22**(12), 4489 (2022)
8. M.L. Hakim, M.T. Islam, T. Alam, S.K. Abdul Rahim, B. Bais, M.S. Islam et al., Triple-band square split-ring resonator metamaterial absorber design with high effective medium ratio for 5G sub-6 GHz applications. Nanomaterials **13**(2), 222 (2023)
9. M.L. Hakim, A. Hanif, T. Alam, M.T. Islam, H. Arshad, M.S. Soliman et al., Ultrawideband polarization-independent nanoarchitectonics: a perfect metamaterial absorber for visible and infrared optical window applications. Nanomaterials **12**(16), 2849 (2022)
10. M.L. Hakim, T. Alam, M.S. Islam, M.M. Salaheldeen, S.H. Almalki, M.H. Baharuddin et al., Wide-oblique-incident-angle stable polarization-insensitive ultra-wideband metamaterial perfect absorber for visible optical wavelength applications. Materials **15**(6), 2201 (2022)
11. M.L. Hakim, M. Faisal (eds.), Design and simulation of a multiband millimeter wave microstrip patch antenna array for 5G wireless communication, in *2019 22nd International Conference on Computer and Information Technology (ICCIT)* (IEEE, 2019)
12. X. Yang, M. Shi, Y. Yu, Y. Xie, R. Liang, Q. Ou et al., Enhancing communication bandwidths of organic color converters using nanopatterned hyperbolic metamaterials. J. Lightwave Technol. **36**(10), 1862–1867 (2018)
13. A. Hanif, M.L. Hakim, T. Alam, M.T. Islam, H. Alsaif, M.S. Soliman et al., Polarization insensitive oblique incident angle stable ultra-thin nano ring resonator-based metamaterial absorber for visible and near-infrared region applications. Opt. Laser Technol. **164**, 109494 (2023)
14. S. Mahmud, S.S. Islam, K. Mat, M.E. Chowdhury, H. Rmili, M.T. Islam, Design and parametric analysis of a wide-angle polarization-insensitive metamaterial absorber with a star shape resonator for optical wavelength applications. Results Phys. **18**, 103259 (2020)
15. J. Fang, B. Wang, Optical capture capability enhancement by right-angled triangular visible absorber. Phys. Lett. A **404**, 127404 (2021)
16. Z. Liu, G. Liu, Z. Huang, X. Liu, G. Fu, Ultra-broadband perfect solar absorber by an ultra-thin refractory titanium nitride meta-surface. Sol. Energy Mater. Sol. Cells **179**, 346–352 (2018)
17. P.T. Dang, J. Kim, T.K. Nguyen, K.Q. Le, J.-H. Lee, Ultra-broadband metamaterial absorber for high solar thermal energy conversion efficiency. Phys. B Condens. Matter **620**, 413261 (2021)
18. J. Liu, W.-Z. Ma, W. Chen, G.-X. Yu, Y.-S. Chen, X.-C. Deng et al., Numerical analysis of an ultra-wideband metamaterial absorber with high absorptivity from visible light to near-infrared. Opt. Express **28**(16), 23748–23760 (2020)

19. R. Yu, Q. Lin, S.-F. Leung, Z. Fan, Nanomaterials and nanostructures for efficient light absorption and photovoltaics. Nano Energy **1**(1), 57–72 (2012)
20. A. Tooghi, D. Fathi, M. Eskandari, Numerical study of a highly efficient light trapping nanostructure of perovskite solar cell on a textured silicon substrate. Sci. Rep. **10**(1), 1–13 (2020)
21. CST Studio Suite 2023 (2023)
22. P.T. Dang, T.V. Vu, J. Kim, J. Park, V.-C. Nguyen, D.D. Vo et al., Efficient broadband truncated-pyramid-based metamaterial absorber in the visible and near-infrared regions. Crystals **10**(9), 784 (2020)
23. A. Musa, M.L. Hakim, T. Alam, M.H. Baharuddin, M.S.J. Singh (eds.), Dual-band metamaterial absorber for Ka-band satellite application, in *2021 7th International Conference on Space Science and Communication (IconSpace)* (IEEE, 2021)
24. Y.-C. Lai, C.-Y. Chen, Y.-T. Hung, C.-Y. Chen, Extending absorption edge through the hybrid resonator-based absorber with wideband and near-perfect absorption in visible region. Materials **13**(6), 1470 (2020)
25. S. Jiao, Y. Li, H. Yang, S.J. Xu, Numerical study of ultra-broadband wide-angle absorber. Results Phys. **24**, 104146 (2021)
26. N.T.Q. Hoa, P.H. Lam, P.D. Tung, T.S. Tuan, H. Nguyen, Numerical study of a wide-angle and polarization-insensitive ultrabroadband metamaterial absorber in visible and near-infrared region. IEEE Photonics J. **11**(1), 1–8 (2019)

# Chapter 27
# Metamaterial-Based L/S/C Band Antenna for Satellite Communication

**Ahmad Musa, Touhidul Alam, Mohd Hafiz Baharuddin, Yousuf Ibn Habib Ullah, and Norbahiah Misran**

**Abstract** This paper presents an Metamaterial (MTM)-based antenna designed for satellite communication. The proposed dual-band antenna exhibits two resonance bandwidths, covering the frequency ranges of 1.7–1.86 and 3.47–5 GHz. Notably, the antenna demonstrates three lowest signal transmission losses occurring at 1.8 GHz, 4 GHz, and 4.6 GHz, with corresponding values of − 29.15 dB, − 29 dB, and − 68 dB, respectively. The compact size of the MTM-based antenna is only 30 mm × 30 mm × 1.6 mm. Additionally, the antenna exhibits directive radiation properties and remarkably low Voltage Standing Wave Ratio (VSWR), averaging at 1.1. The combination of these unique electromagnetic properties offered by the metamaterial-based antenna, alongside the demand for efficient dual-band operation, compact form factor, and desirable radiation characteristics, presents new possibilities for achieving reliable and high-performance communication in satellite applications.

A. Musa · T. Alam (✉)
Institut Perubahan Iklim (IPI), Universiti Kebangsaan Malaysia, 43600 UKM Bangi, Selangor, Malaysia
e-mail: touhidul@ukm.edu.my

A. Musa
e-mail: p117238@siswa.ukm.edu.my

M. H. Baharuddin · N. Misran
Department of Electrical, Electronic and Systems Engineering, Faculty of Engineering and Built Environment, Universiti Kebangsaan Malaysia, 43600 UKM Bangi, Selangor, Malaysia
e-mail: hafizb@ukm.edu.my

N. Misran
e-mail: bahiah@ukm.edu.my

Y. I. H. Ullah
Electronic and Telecommunication Engineering (ETE), International Islamic University Chittagong, Kumira, Chattogram 4318, Bangladesh

 
M. T. Islam et al. (eds.), *Proceedings of the 8th International Conference on Space Science and Communication*, Springer Proceedings in Physics 303,
https://doi.org/10.1007/978-981-97-0142-1_27

## 27.1 Introduction

Satellite communication has become an integral part of our global communication infrastructure, enabling long-distance, and reliable data transmission across the world. Antenna technology plays a critical role in satellite communication systems, as it determines the efficiency and performance of signal transmission and reception [1]. In recent years, the incorporation of MTM in antenna design has opened up new possibilities for enhancing the capabilities of satellite communication systems [2]. The concept of MTMs emerged in the early 2000s when researchers began exploring artificially engineered materials with unique electromagnetic properties [3]. These materials are composed of subwavelength structures that exhibit extraordinary properties not typically found in natural materials [4]. The manipulation of electromagnetic waves through precise control of the MTM's structure and composition has revolutionized antenna design and performance. MTM-based antennas have shown great potential in enhancing the efficiency, bandwidth, and radiation characteristics of satellite communication systems.

Extensive research has been conducted on MTM-based antennas for satellite communication applications. The literature review highlights key studies and advancements in this field, focusing on the design techniques, materials, and performance characteristics of these antennas. Researchers have explored various MTM structures, including resonators [5, 6], periodic arrays [1, 7], and frequency selective surfaces (FSS), to achieve desired electromagnetic properties. These structures enable precise control over wave propagation [8], polarization [9, 10], EMI Shield [11], and radiation pattern, leading to improved antenna performance. Moreover, the integration of MTM elements with traditional antenna designs has demonstrated enhanced characteristics such as increased gain [12], reduced sidelobes, improved radiation efficiency, and extended bandwidth. Furthermore, studies have investigated the use of MTMs in different frequency bands commonly employed in satellite communication systems [13], such as L-band, S-band [14, 15], C-band [16], and Ku-band [17, 18]. The utilization of MTM-based antennas in these frequency ranges has shown promising results in terms of improved signal reception, enhanced data rates, and increased coverage range. Various research efforts have also focused on optimizing the antenna parameters and structure to meet the specific requirements of satellite communication applications. The incorporation of MTM-based antennas in satellite communication systems brings several benefits and offers significant usefulness in this domain. Firstly, MTM-based antennas provide enhanced performance characteristics, such as increased gain, improved radiation efficiency, and reduced sidelobes. These improvements lead to improved signal quality, higher data transmission rates, and better overall system performance. Secondly, MTMs offer the potential for miniaturization, enabling the design of compact and lightweight antennas. This is particularly valuable in satellite communication, where space and weight constraints are critical considerations. Compact antennas with efficient performance can be deployed in various satellite platforms, including large-scale communication satellites and small CubeSat missions [19]. Thirdly, MTM-based antennas can be tailored to operate in specific frequency bands used in satellite communication systems [20, 21]. By optimizing the antenna design for these frequency ranges, efficient signal reception and transmission can be achieved, ensuring reliable and high-quality communication links.

In this paper, an MTM-based antenna for satellite communication has been presented. The proposed dual-band antenna has two resonance bandwidths from 1.7 to 1.86 GHz and 3.47 to 5 GHz, with three lowest signal transmission loss at 1.8, 4, and 4.6 GHz. The signal transmission loss in those pick resonance points are − 29.15 dB, − 29 dB, and − 68 dB, respectively. The size of the MTM-based antenna is 30 mm × 30 mm × 1.6 mm only. The antenna has directive radiation and very low VSWR, 1.1 on average. Those combinations of unique electromagnetic properties offered by MTM-based antenna, along with the demand for efficient dual-band operation, compact size, and desirable radiation characteristics, opens up new possibilities for reliable and high-performance communication in satellite applications.

## 27.2 Methodology and Design

For designing a microstrip patch antenna for satellite communication applications involves several key steps, shown in (Fig. 27.1). Firstly, the design specifications need to be determined, including the operating frequency bands, bandwidth, and radiation pattern requirements. This information serves as a guideline for the antenna design process. Common frequency bands for satellite communication include L-Band (1–2 GHz) and S-Band (2–4 GHz), and C-Band (4–8 GHz). Next, an appropriate microstrip patch antenna configuration is selected, considering factors such as the low-profile nature of microstrip patch antennas and their ease of fabrication and integration. The substrate material is then chosen based on its dielectric constant, thickness, and loss tangent, which directly affect the antenna's performance at the desired operating frequencies. In this experiment, FR-4 substrate material with dielectric constant 2.2 mm and thickness of 1.6 mm has been used as it is commercially available and cheaper. The FR-4 substrate material had 0.035 mm thick copper layer on both side, which has been used as resonating layer. The antenna design is then simulated using CST simulation software to evaluate its performance in terms of radiation pattern, impedance matching, and bandwidth. Finally, the antenna's performance has been improved and optimized by introducing a MTM array at the back of the antenna. Figure 27.2 shows the final structure of the proposed MTM-based antenna. The design parameters are given in the Table 27.1.

### *27.2.1 Antenna Design*

The proposed antenna has a "T" shaped resonating patch layer in the front and a rectangular ground layer at the back. The patch has a cutting slot at its top middle

**Table 27.1** Parameters and their values of the MTM-based antenna

| Parameter | Value (mm) | Parameter | Value |
|---|---|---|---|
| $L$ | 30 | $P_6$ | 0.5 |
| $W$ | 30 | $G_1$ | 20 |
| $P_1$ | 20 | $G_2$ | 7 |
| $P_2$ | 20 | $M_1$ | 4 |
| $P_3$ | 7 | $M_2$ | 2.8 |
| $P_4$ | 9.75 | $M_3$ | 1.2 |
| $P_5$ | 5 | $M_4$ | 0.8 |

and two rectangular parasiting elements beside its lag. This antenna design has two resonance bandwidths from 1.7 to 1.86 GHz and 3.47 to 5 GHz, with the three lowest signal transmission losses at 1.8, 4, and 4.6 GHz. The signal transmission loss in those pick resonance points is – 29.15 dB, – 29 dB, and – 68 dB, respectively. Figure 27.3 shows the initial structure of the proposed MTM-based antenna and its $S_{11}$ pattern.

### 27.2.2 *Metamaterial-Based Antenna Design*

To improve and optimize the antenna's performance, an MTM array has been introduced at the back of the antenna. The MTM array has four-unit cells. The MTM array has been placed at the back of the antenna parallel to the bottom end of the arms of the T-shaped resonating patch. The MTM has decried the signal transmission loss for the second resonance band at 4.6 GHz. Figure 27.4 shows the final back structure of the proposed MTM-based antenna and its $S_{11}$ pattern.

## 27.3 Result Analysis

An antenna designed for satellite communication should exhibit certain characteristics to ensure effective communication including directive radiation pattern and VSWR near to 1:1 [22]. The antenna should have a directional radiation pattern, focusing the majority of the transmitted or received energy in the desired direction. This helps to establish a concentrated communication link with the satellite, maximizing signal strength, and reducing interference from other directions. The ideal value for VSWR is 1:1, which indicates a perfect impedance match between the antenna and the transmission line. In this case, there would be no signal reflection, and all of the power would be efficiently transferred to or from the antenna. In practical scenarios, achieving a VSWR of exactly 1:1 is challenging due to various factors

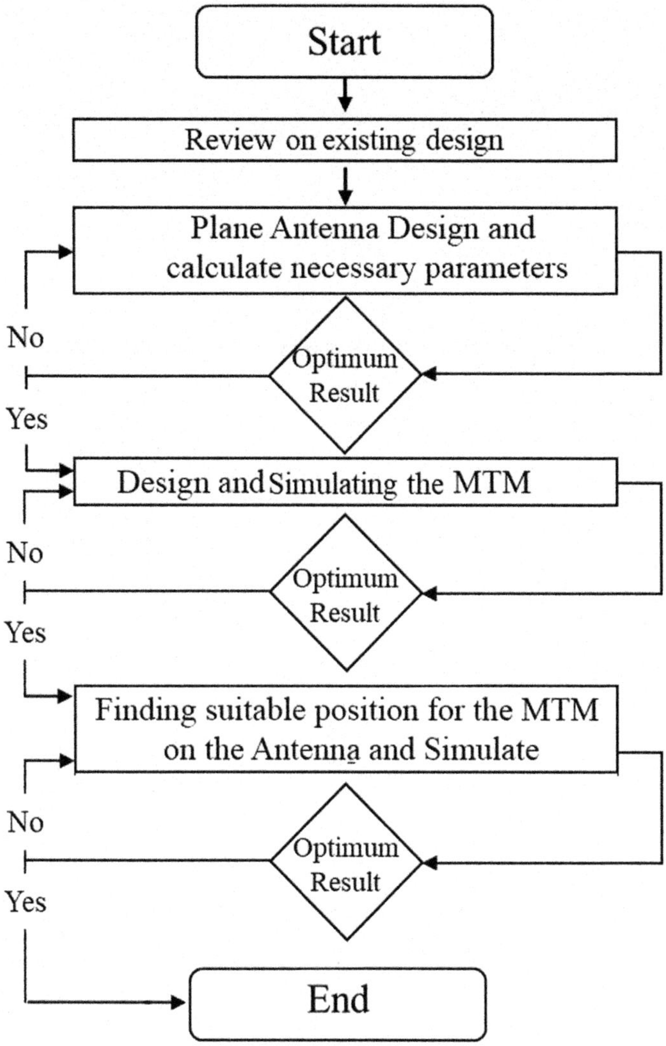


**Fig. 27.1** The flow chart of the MTM-based antenna design

such as impedance mismatches, transmission line losses, and environmental conditions. However, a VSWR below a certain threshold is typically considered acceptable for reliable communication.

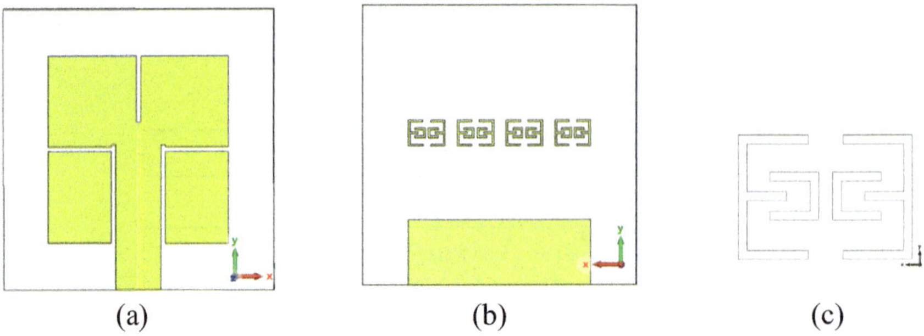


**Fig. 27.2** **a** Front view of the antenna, **b** back view of the antenna and **c** MTM unit cell

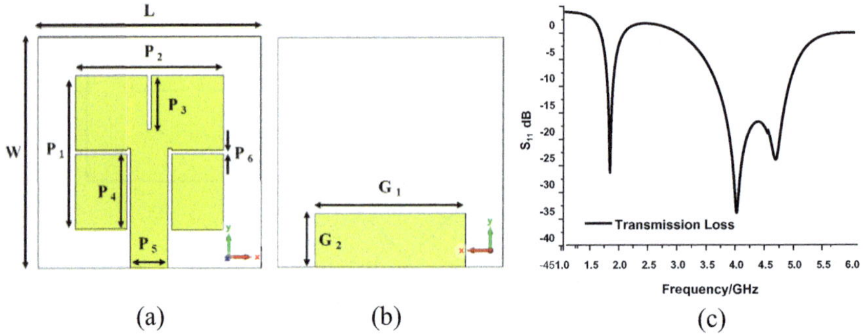


**Fig. 27.3** **a** Front view, **b** back view and **c** $S_{11}$ response of the MTM less antenna

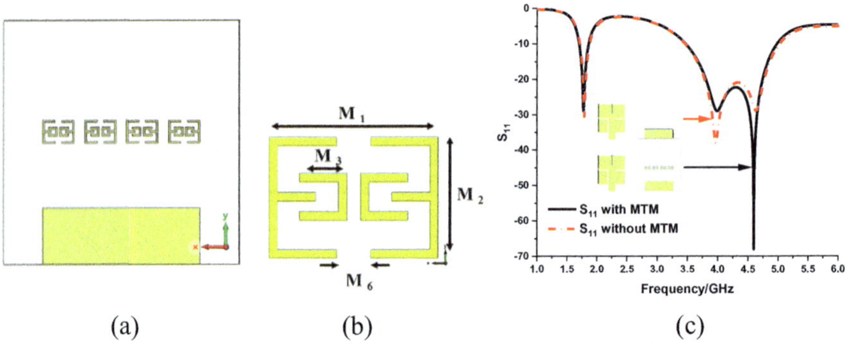


**Fig. 27.4** **a** Back view, **b** MTM view and **c** $S_{11}$ response of the MTM-based antenna

## *27.3.1 Antenna Radiation Pattern*

Figure 27.5 has been illustrated the 2D polar radiation pattern plot for (a) 1.8 GHz, (b) 4 GHz, and (c) 4.6 GHz frequency. Subsequently, it is remarkable that the antenna

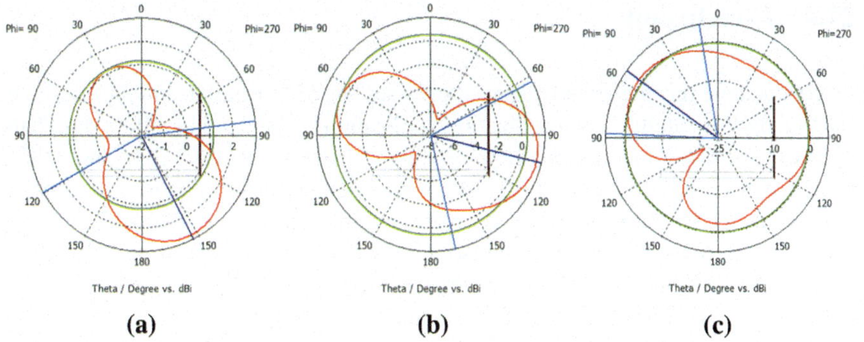


**Fig. 27.5** Radiation pattern of the proposed antenna at **a** 1.8 GHz, **b** 4 GHz, and **c** 4.6 GHz

achieved an almost omnidirectional radiation pattern at frequencies 1.8 and 4 GHz, while a distorted and unstable 2D polar radiation pattern was achieved at frequency band 4.6 GHz.

## *27.3.2 Antenna VSWR*

Figure 27.6 has been illustrated the VSWR plot for (a) 1.8 GHz, (b) 4 GHz, and (c) 4.6 GHz frequency and their corresponding $S_{11}$ response. Subsequently, it is remarkable that the antenna achieved an almost VSWR 1:1 at frequencies 1.8, 4, and 4.6 GHz. while a distorted and unstable 2D polar radiation pattern was achieved at frequency band 4.6 GHz.

**Fig. 27.6** VSWR of the proposed antenna (graph on the right side) $S_{11}$ response of the MTM-based antenna (graph on the left side)

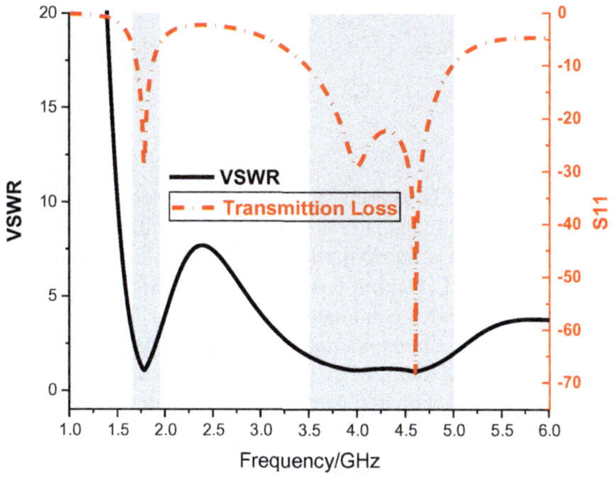

## 27.4 Conclusion

In conclusion, the presented paper showcases an MTM-based antenna designed specifically for satellite communication applications. The antenna offers dual-band operation with resonance bandwidths spanning from 1.7 to 1.86 GHz and 3.47 to 5 GHz. Notably, it exhibits three distinct lowest signal transmission losses occurring at 1.8 GHz, 4 GHz, and 4.6 GHz, with corresponding values of − 29.15 dB, − 29 dB, and − 68 dB, respectively. The compact size of the antenna, measuring only 30 mm × 30 mm × 1.6 mm, further enhances its suitability for satellite communication systems. The MTM-based antenna demonstrates directive radiation characteristics and exceptionally low VSWR, with an average value of 1.1. These unique electromagnetic properties, combined with the demand for efficient dual-band operation, compact form factor, and desirable radiation characteristics, open up new possibilities for reliable and high-performance communication in satellite applications. By harnessing the advantages offered by metamaterial-based antennas, satellite communication systems can achieve enhanced signal transmission, improved data rates, and increased system reliability. The compact size of the antenna allows for efficient integration into satellite platforms, addressing space, and weight constraints. Moreover, the antenna's dual-band capability enables versatile communication across different frequency ranges, catering to the diverse needs of satellite applications.

In summary, the MTM-based antenna for satellite communication presented in this paper exhibits promising performance characteristics, offering a compact size, directive radiation, low VSWR, and dual-band operation. These features pave the way for advanced and efficient satellite communication systems, facilitating reliable, and high-performance communication in various satellite applications.

**Funding Statement** This work is supported by the Universiti Kebangsaan Malaysia, Research Grant code: DIP-2021-007.

## References

1. A. Musa, M.L. Hakim, T. Alam, M.T. Islam, A.S. Alshammari, K. Mat et al., Polarization independent metamaterial absorber with anti-reflection coating nanoarchitectonics for visible and infrared window applications. Materials **15**(10), 3733 (2022)
2. A. Musa, M.L. Hakim, T. Alam, M.H. Baharuddin, M.S.J. Singh (eds.), Dual-band metamaterial absorber for Ka-band satellite application, in *2021 7th International Conference on Space Science and Communication (IconSpace)* (IEEE, 2021)
3. A. Musa, T. Alam, M.T. Islam, M.L. Hakim, H. Rmili, A.S. Alshammari et al., Broadband plasmonic metamaterial optical absorber for the visible to near-infrared region. Nanomaterials **13**(4), 626 (2023)
4. M.N.C. Saddam, M.H. Baharuddin, M.L. Hakim, T. Alam, A. Musa (eds.), Single negative metamaterial absorber of K-band application, in *2021 7th International Conference on Space Science and Communication (IconSpace)* (IEEE, 2021)

5. N. Misran, S.H. Yusop, M.T. Islam, M.Y. Ismail, Analysis of parameterization substrate thickness and permittivity for concentric split ring square reflectarray element. J. Kejuruter. (J. Eng.) **23**, 11–16 (2012)
6. M.L. Hakim, T. Alam, M.T. Islam, N.B.M. Sahar, M.S.J. Singh, H. Alsaif et al., Metamaterial physical property utilized antenna radiation pattern deflection for angular coverage and isolation enhancement of mm-wave 5G MIMO antenna system. Radiat. Phys. Chem. **209**, 110998 (2023)
7. M.A. Rahman, A. Noman, M.J. Uddin, S. Hannan, A. Musa, M.A. Rahman (eds.), Square split enclosed labyrinth maze shaped metamaterial absorber for hydrocarbon sensing application, in *2023 International Conference on Next-Generation Computing, IoT and Machine Learning (NCIM)* (IEEE, 2023)
8. M.L. Hakim, M.T. Islam, T. Alam, S.K. Abdul Rahim, B. Bais, M.S. Islam et al., Triple-band square split-ring resonator metamaterial absorber design with high effective medium ratio for 5G sub-6 GHz applications. Nanomaterials **13**(2), 222 (2023)
9. A. Hanif, M.L. Hakim, T. Alam, M.T. Islam, H. Alsaif, M.S. Soliman, Polarization insensitive oblique incident angle stable ultra-thin nano ring resonator-based metamaterial absorber for visible and near-infrared region applications. Opt. Laser Technol. **164**, 109494 (2023)
10. M.B. Billa, M.L. Hakim, T. Alam, M.T. Islam, N. Misran, Near-ideal absorption high oblique incident angle stable metamaterial structure for visible to infrared optical spectrum applications. Opt. Quant. Electron. (2023)
11. M.L. Hakim, T. Alam, M.T. Islam, Polarization insensitive and oblique incident angle stable miniaturized conformal FSS for 28/38 GHz mm-wave band 5G EMI shielding applications. IEEE Antennas Wirel. Propag. Lett. (2023)
12. M.M. Hasan, M.T. Islam, S.K.A. Rahim, T. Alam, H. Rmili, A. Alzamil et al., A compact mu-near-zero metamaterial integrated wideband high-gain MIMO antenna for 5G new radio applications. Materials **16**(4), 1751 (2023)
13. T. Alam, M.T. Islam, M.L. Hakim, K.H. Alharbi, M.S.J. Singh, M.M. Sheikh et al., Metamaterial based Ku-band antenna for low earth orbit nanosatellite payload system. Nanomaterials **13**(2), 228 (2023)
14. S. Sakib, A. Hoque, S.K.B.A. Rahim, M. Singh, N.M. Sahar, M.S. Islam et al., A central spiral split rectangular-shaped metamaterial absorber surrounded by polarization-insensitive ring resonator for S-band applications. Materials **16**(3), 1172 (2023)
15. M.M. Sheikh, F.A. Shompa, A. Aziz, T. Ahsan, M.S. Alam, A.K.M. Masum et al., Design and analysis of a circular polarized S-band antenna for CubeSat communication system. Optoelectron. Adv. Mater. Rapid Commun. **17**(1–2), 73–77 (2023)
16. M.M. Soliman, M. Alkaeed, M.J.A. Pervez, I.A. Rafi, M.H. Mahfuz, A. Musa (eds.), A comb shape slot UWB antenna with controllable triple band rejection features for WiMAX/WLAN/5G/satellite applications, in *2020 IEEE Student Conference on Research and Development (SCOReD)* (IEEE, 2020)
17. M.L. Hakim, T. Alam, M.T. Islam, H. Alsaif, M.S. Soliman, Polarization-independent fractal square splits ring resonator (FSSRR) multiband metamaterial absorber/artificial magnetic conductor/sensor for Ku/K/Ka/5G (mm-wave) band applications. Measurement **210**, 112545 (2023)
18. M.L. Hakim, T. Alam, M.T. Islam, M.H. Baharuddin, A. Alzamil, M.S. Islam, Quad-band polarization-insensitive square split-ring resonator (SSRR) with an inner Jerusalem cross metamaterial absorber for Ku- and K-band sensing applications. Sensors **22**(12), 4489 (2022)
19. T. Alam, A.F. Almutairi, M. Samsuzzaman, M. Cho, M.T. Islam, Metamaterial array based meander line planar antenna for cube satellite communication. Sci. Rep. **11**(1), 14087 (2021)
20. R. Azim, A.M.H. Meaze, A. Affandi, M.M. Alam, R. Aktar, M.S. Mia et al., A multi-slotted antenna for LTE/5G sub-6 GHz wireless communication applications. Int. J. Microw. Wirel. Technol. **13**(5), 486–496 (2021)
21. T. Alam, M.T. Islam, M. Cho, Near-zero metamaterial inspired UHF antenna for nanosatellite communication system. Sci. Rep. **9**(1), 3441 (2019)
22. M.Z.B. Chowdhury, M.T. Islam, H. Rmili, I. Hossain, M.Z. Mahmud, M. Samsuzzaman, A low-profile rectangular slot antenna for sub-6 GHz 5G wireless applications. Int. J. Commun. Syst. **35**(17), e5321 (2022)

# Chapter 28
# A Four-Element MIMO Antenna System for UWB, C-Band, and X-Band Space Communication Applications

**Nishan Das, Rezaul Azim, Md. Sharif Mia, Pranab Dewan, and Touhidul Alam**

**Abstract** This paper suggested a 4-port Multiple Input Multiple Output (MIMO) antenna arrangement for UWB, cellular (C-band), and interplanetary (X-band) communications. At first, a single component is created and then reconfigured into a $4 \times 4$ MIMO layout. The recommended MIMO antenna is built on a Rogers RT5880 material of dimension $42 \times 42 \times 1.575$ mm$^3$ for a working band of 3.16–13.44 GHz. It provides a peak gain of 5.44 dBi and isolation of $-$ 25 to $-$ 30 dB. The anticipated MIMO array delivers remarkable diversity performance all over the entire operating spectrum, with an Envelop Correlation Coefficient $\leq$ 0.0001 and a Diversity Gain $\geq$ 9.999 dB. Maximum radiation efficiency exceeds 85%, and the antenna's radiation patterns remain constant across the working spectrum. The anticipated MIMO is well-suited to wireless communication due to its wonderful characteristics.

N. Das · R. Azim (✉) · Md. S. Mia · P. Dewan
Research Laboratory for Computational Electromagnetics, Department of Physics, University of Chittagong, Chattogram 4331, Bangladesh
e-mail: rezaulazim@cu.ac.bd

T. Alam
Space Science Centre, Institut Perubahan Iklim, Universiti Kebangsaan Malaysia, 43600 Bangi, Malaysia
e-mail: touhidul@ukm.edu.my

 
M. T. Islam et al. (eds.), *Proceedings of the 8th International Conference on Space Science and Communication*, Springer Proceedings in Physics 303,
https://doi.org/10.1007/978-981-97-0142-1_28

## 28.1 Introduction

In recent decades, the area of technology concerned with wireless communication has seen the most rapid evolution. This has made it possible to develop high-speed mobile networks with greater levels of security [1]. In recent days, ultra-wideband (UWB) technology has gotten much recognition in the wireless sector. The FCC permitted unrestricted frequencies between 3.1 and 10.6 GHz for commercial uses to prevent interference with other bands. These frequencies are used for many kinds of applications, including imaging, surveillance, medical, radio frequency identification (RFID), RADAR, communications, etc. These spectrums can also be employed for X-band (8–12 GHz) RADAR applications when the frequency extends up to 12 GHz. Also, the C-band (4–8 GHz) is widely used in many fields, such as satellite communications, weather RADAR, telecommunications, wireless local area networks (WLANs), military applications, radio astronomy, maritime and aviation communications, etc.

There are a lot of advantages attached to using the UWB band, such as high data speeds with minimal energy usage, simple architecture, low cost, and high accuracy. Multipath fading and intersystem interference induce signal dissipation among the antenna elements in an RF communication system, which is a problem for UWB radio technology [2]. Utilizing antenna arrays with MIMO (multiple input multiple output) technology is one way in which this problem might be resolved. MIMO systems use many antennas on both the transmitting and receiving ends of the communication link to achieve a high data rate, a high SNR, and low latency throughout the entire networking system. The MIMO system's effectiveness may suffer due to the immediacy of the antenna elements and the high degree of mutual interaction between the components. The best value of isolation among two antenna elements is $\leq -20$ dB. This can be done by maintaining the spacing among two elements as $\geq \lambda/2$. But, the antennas for MIMO will need to be larger if this technique is taken. This makes it quite challenging to design a small MIMO antenna system that nevertheless provides adequate isolation. Furthermore, it is more challenging than for tiny bands to maintain effective isolation across the UWB, X-band, and C-band frequency ranges.

Several MIMO antenna configurations for UWB use have been reported. Different UWB-MIMO antennas have been constructed and tested, ranging in size, isolation, and geometry. However, existing narrowband communication technologies, such as WLAN and WiMAX, cause substantial electromagnetic interference for these designs. Designing UWB antennas that may reject specific frequency ranges is one method of suppressing these conflicting signals. Several approaches have been proposed to lessen mutual coupling without significantly altering other required antenna, which includes U-shaped intercoupled isolation wall, defected ground structure (DGS) based monopole with filter and slit/stub in the ground, metamaterial-based super substrate, metamaterial-based EM band gap, metamaterial-based isolator, insertion of slots in the array, stub loaded slot with SRR structure, curved stub with pattern diversity, Sierpinski Knopp fractal UWB antenna with CSRR structure,

Minkowski Fractal with DGS structure [3–9], etc. Some MIMO antenna systems for X-band and C-band communication systems were also reported recently [10–15]. These antennas have shown good performance over the working band, but the problem with them is either their diversity performance is not good (low isolation, large ECC, and small DG) or they have a complex design and 3D profile. Therefore, making it hard to apply in practical devices.

Recent research on the design of a multi-element MIMO antenna has shown that using orthogonal positioning of antenna elements and different isolation strategies, critical compactness can be achieved, such as L-shaped stub, U- and H-shaped slots, polarization diversity, and slit on the ground, coplanar waveguide (CPW) fed MIMO uses a gap-optimized antenna element and ground plane, etc. [16–20]. All the methods discussed above just complicate the design for the production and application in actual devices.

This research intends to create a 4-component UWB-MIMO antenna with good impedance matching, strong isolation, and excellent diversity behavior throughout the UWB, X-band, and C-band (3–12 GHz) communication spectrum.

## 28.2 Design Methodology

### *28.2.1 Single Antenna Element Design*

As in Fig. 28.1a, b, the radiator of the introduced single antenna of the envisioned MIMO is a square patch with a rectangular slot vertically placed on one side and a partial ground on the other side of a Rogers RT5880 substrate of dimension $W_s \times L_s$ mm$^2$ and 1.575 mm thickness. The substrate material has a permittivity of 2.2 and a loss tangent of 0.0009. This patch form's simplicity and diminutive size made it a prime choice. A microstrip feedline of dimension 9 mm × 2.6 mm provides power to the rectangular radiating patch. The feedline width is optimized to have the 50 Ω characteristic impedance by CST microwave studio's in-built computation tool. The patch of the single element was initially chosen as rectangular. The patch has one rectangular slot cut into it, and the ground plane's dimensions have been optimized to provide the appropriate UWB characteristic frequency (3.3–13.6 GHz) with two resonating frequencies of 6.5 and 12.4 GHz, as displayed in Fig. 28.2. In Fig. 28.2, $S_{11}$ of the antenna without slot (dashed line) and with slot (solid line) is portrayed. Clearly, after cutting that slot, the antenna achieved a wider bandwidth and better reflection coefficient. This working bandwidth shows that the footprint of Fig. 28.1 can works over the UWB, C-band, and X-band communication systems. With the help CST simulator's optimization technique, antenna parameters are optimized and tabulated in Table 28.1.

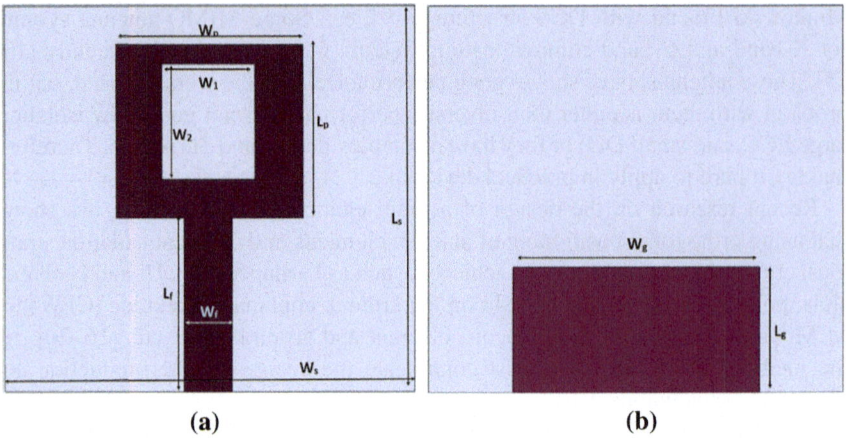


**Fig. 28.1** **a** Top and **b** bottom side of the studied single-element

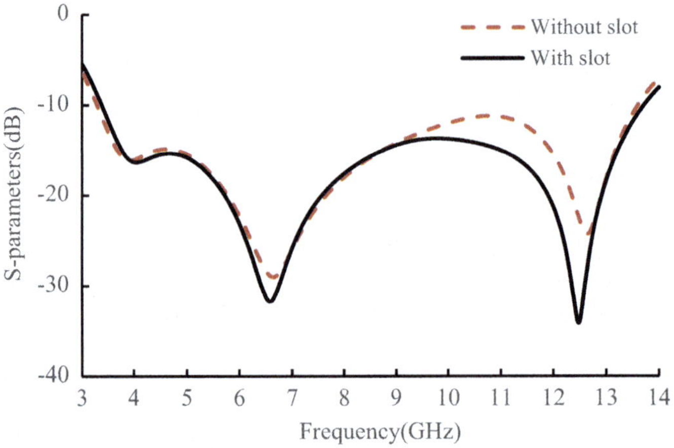


**Fig. 28.2** $S_{11}$ of the studied single-element with and without slot

**Table 28.1** Geometrical parameters of the single antenna element

| Parameters | Value (mm) | Parameters | Value (mm) |
|---|---|---|---|
| $L_s$ | 20 | $W_f$ | 2.6 |
| $W_s$ | 22 | $W_1$ | 5 |
| $L_p$ | 9 | $W_3$ | 6 |
| $W_p$ | 10 | $L_g$ | 6.5 |
| $L_f$ | 9 | $W_g$ | 13 |

### 28.2.2 MIMO Design

A 4-element MIMO array has also been built and tested, as the antenna in Fig. 28.1 can cover all necessary bands. Two perspectives (the top and bottom) on the final MIMO are portrayed in Fig. 28.3. Figure 28.3 depicts how a small MIMO is made by positioning 4 identical antenna elements perpendicular to one another on an $L \times L$ mm$^2$ Roger's substrate. A few simple steps were taken to design the complete prototype of the MIMO. Firstly, a single antenna element was created as the given parameters and characteristics, then taking 4 identical single elements were placed at right angles to each other to complete the full prototype. The spacing between two radiators and the ground plane was optimized using the CST Microwave Studios optimization tool to achieve its desired characteristics. The optimal distance between two radiating elements is kept at $D_1$ mm to make the MIMO smaller. This distance allows for lower mutual coupling without the need for an additional isolator or de-coupler. The antenna elements are excited through the 50-Ω microstrip input lines. $D_3$ and $D_4$ are the optimal separations between ground plane elements. Table 28.2 lists the geometrical specifications for the anticipated MIMO array.

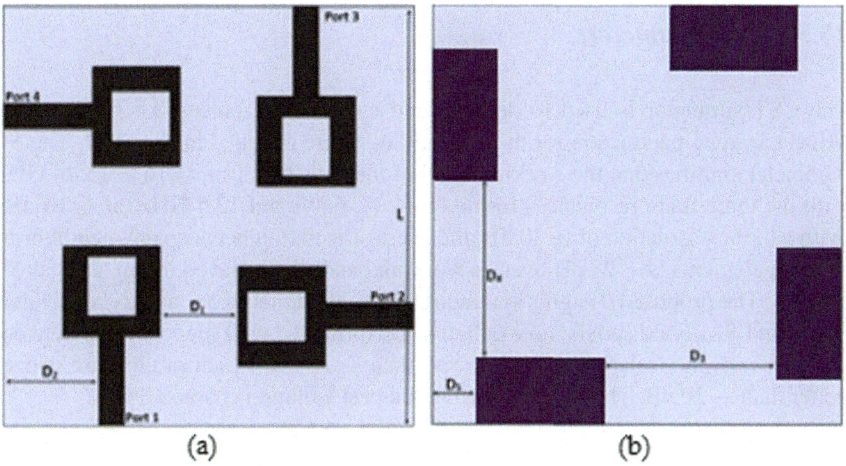


**Fig. 28.3** **a** Top (left) and **b** bottom view (right) of the suggested antenna

**Table 28.2** Parameters of the MIMO

| Parameters | Value (mm) | Parameters | Value (mm) |
|---|---|---|---|
| $L$ | 42 | $D_3$ | 18 |
| $D_1$ | 8 | $D_4$ | 18 |
| $D_1$ | 9.7 | $D_5$ | 4.5 |

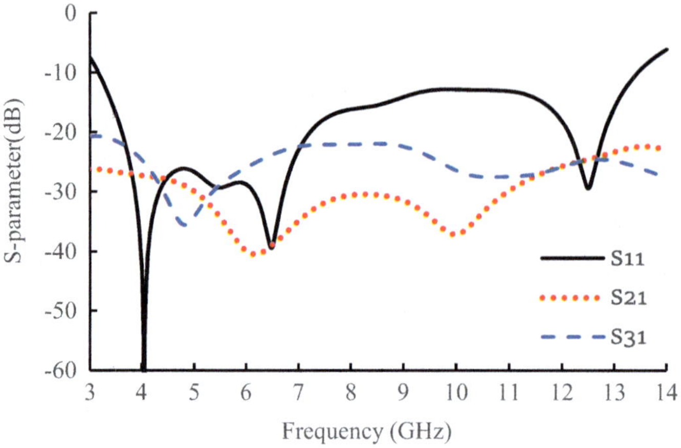


**Fig. 28.4** *S*-parameters of the suggested MIMO array

## 28.3 Results and Discussions

### 28.3.1 S-Parameters

The CST simulator is used to develop and enhance the outcome of the planned MIMO array. *S*-parameters for the studied layout are revealed in Fig. 28.4. The $S_{11}$ parameter signifies that the working band of the MIMO is from 3.16 to 13.44 GHz, with the three main resonances located at 4.05, 6.49, and 12.5 GHz (at − 10 dB). With a highest isolation of − 40 dB, the average isolation between any neighboring antenna elements is − 28 dB over its working band, as displayed in Fig. 28.4 as $S_{12}$ and $S_{13}$. The proposed design's symmetrical structure makes $S_{11}$ and $S_{22}$ equivalent to $S_{33}$ and $S_{44}$. For a satisfactory radiation performance with low coupling between the elements, the isolation should be less than − 20 dB, which in this case is even better than − 20 dB. Therefore, ensuring the best isolation characteristics.

### 28.3.2 Efficiency and Gain

The intended MIMO's peak gain and efficiency over the working frequency are displayed in Fig. 28.5. Maximum gain on the working band is shown to be 5.44 dBi, and maximum efficiency on the operating band is shown to be 85%.

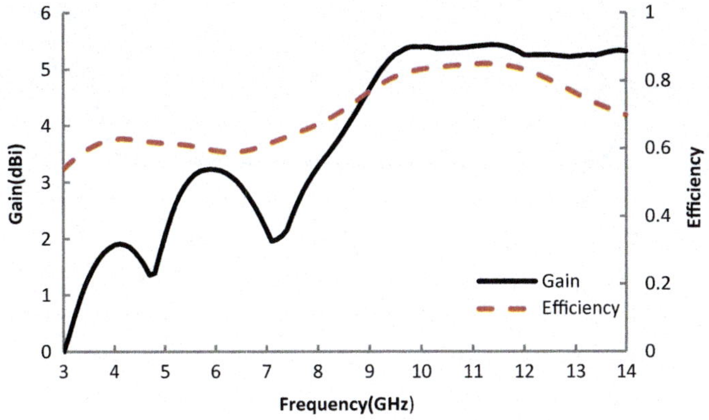


**Fig. 28.5** Peak gain and efficiency of the designed MIMO antenna

## *28.3.3 Diversity Performance*

The diversity behavior of the designed MIMO array is assessed by computing and analyzing several crucial variables, such as the ECC and the DG.

**Envelop Correlation Coefficient**. To work effectively, an array of MIMO antenna systems requires that the radiation characteristics of the array components not correlate with others. Radiation characteristics are correlated in a quantitative sense using a metric called the ECC and can be estimated from either radiation characteristics as well as *S*-parameters. Using the *S*-parameters, we can determine the ECC by Eq. (28.1).

$$\mathrm{ECC} = \frac{\left|S_{ii}^{*}S_{ij} + S_{ji}^{*}S_{jj}\right|^{2}}{\left[1-\left(\left|S_{ii}\right|^{2}+\left|S_{ji}\right|^{2}\right)\right]+\left[1-\left(\left|S_{jj}\right|^{2}+\left|S_{ij}\right|^{2}\right)\right]} \tag{28.1}$$

where (*) is the Hermitian operator.

The ECC value calculated using *S*-parameters is portrayed in Fig. 28.6. As apparent from Fig. 28.6, the ECC value is less than 0.0015 over the 3.16–13.44 GHz band. A good diversity behavior requires an ECC of < 0.5. The ECC is practically zero across the whole shown frequency range. Therefore, it may be reasonable to infer that the MIMO's radiation is tolerable and that interferences between the antenna elements are negligible.

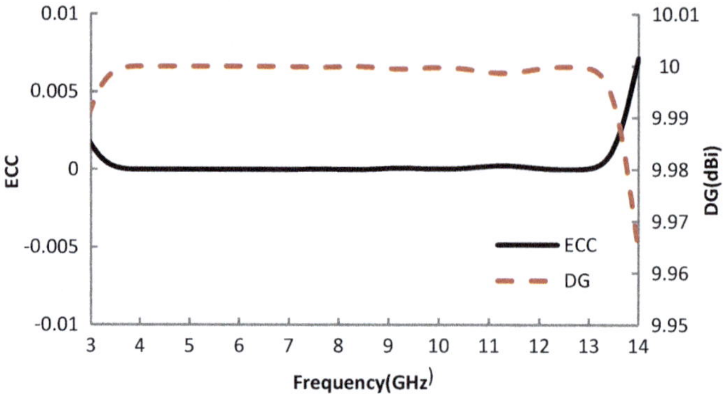


**Fig. 28.6** ECC and DG of the designed MIMO antenna

**Diversity Gain**. The DG is a crucial indicator of MIMO diversity performance. As a small correlation between two antennas results in larger DG, an innate relationship exists between the ECC and DG. The DG quantifies the reliability and efficiency of a communication channel. A DG value of 10 dB or more is recommended for secure communication. Using Eq. (28.2), we can determine the DG given the value of ECC.

$$\mathrm{DG} = 10\sqrt{1 - (\mathrm{ECC})^2} \tag{28.2}$$

Using the simulated ECC values, Fig. 28.6 displays the DG values from where it is revealed that the recommended configuration achieved a DG $\geq$ 9.99 dB and is very close to 10 dB in the working spectrum of 3.16–13.44 GHz.

**Radiation Patterns**. Figure 28.7 depicts the radiation plot of the planned MIMO at 4, 6.5, and 12.5 GHz. Figure 28.7a depicts the pattern at 4 GHz from which it is revealed that the patterns is nearly omnidirectional, but the pattern becomes slightly directed at 6.5 GHz, as in Fig. 28.7b. As seen in Fig. 28.7c, the pattern becomes directional at 12.5 GHz, with primary beam pointing towards the broadside direction. It is possible that activation of higher-order current modes is to blame for the skewing of radiation behavior at higher frequencies.

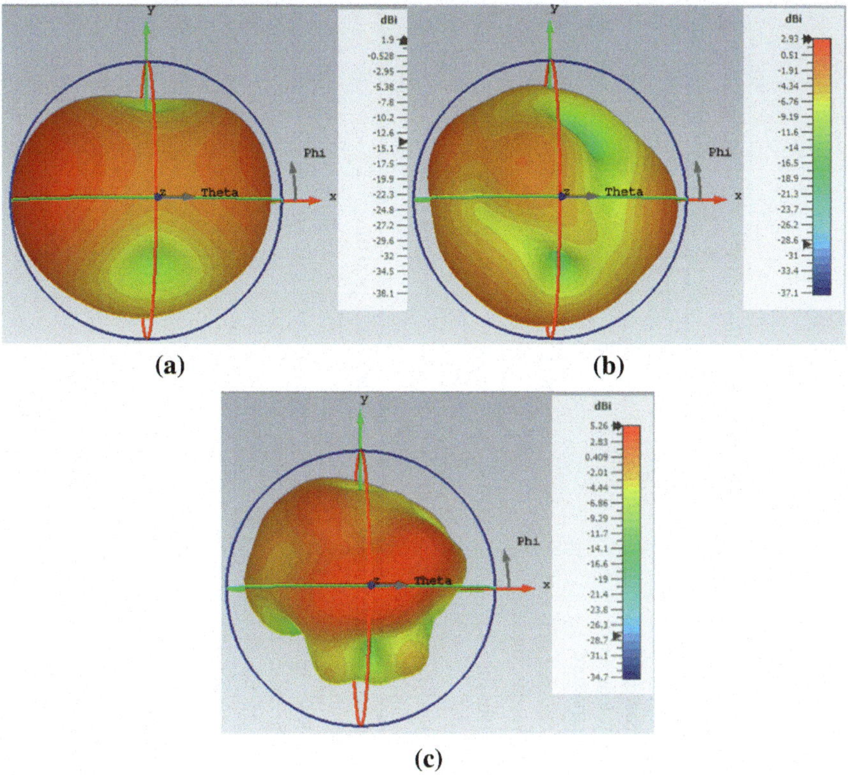


**Fig. 28.7** 3D radiation patterns of the MIMO at **a** 4, **b** 6.5, and **c** 12.5 GHz

Table 28.3 provides a comparison among the proposed MIMO and the previously reported works. Characteristics such as size, working band, design complexity, decoupling method, isolation, DG, ECC, and numbers of antenna elements are considered for the comparison. The designed MIMO has several advantages over many reported systems, including small size, greater bandwidth, better isolation, lower ECC, and greater value of DG, and all of which are achieved without the use of any isolation mechanism. Moreover, the suggested design also covers the entire X-band and C-band which other MIMO's have failed.

**Table 28.3** Comparison of the antenna performances

| References | Size (mm × mm) | Operating band (GHz) | Design complexity | Decoupling method | Isolation (dB) | DG (dB) | ECC | Antenna elements |
|---|---|---|---|---|---|---|---|---|
| [7] | 40 × 40 | 3.18–11.5 | Moderate | U-shaped stubs | < −15 | > 9.99 | < 0.015 | 4 |
| [8] | 40 × 40 | 2.6–10.6 | Complex | Sierpinski–Knopp fractal curve structure | < −20 | > 9.9 | < 0.07 | 4 |
| [16] | 46.7 × 46.7 | 9.75–10.41 | Moderate | Crossed slots and rectangular slots | < −20 | N/A | < 0.003 | 4 |
| [18] | 25.3 × 34. 8 | 5.15–5.85, 6.3–7.27, 7.9–8.4 | Moderate | Parasitic slits | < −16 | > 9.9 | < 0.01 | 2 |
| [20] | 93 × 93 | 3–13 | Complex | Resistance-loaded stub | < −20 | > 9.9 | < 0.003 | 4 |
| [11] | 40 × 25 | 2.4–14 | Moderate | L-shape ground stub | < −15 | N/A | < 0.01 | 2 |
| [12] | 45 × 45 | 2–10.6 | Complex | Koch fractal geometry and grounded stubs | < −17 | N/A | < 0.05 | 4 |
| [13] | 80 × 80 | 3.3–4.1 and 8.2–8.6 | Simple | None | < −25 | ~ 10 | < 0.02 | 4 |
| [14] | 38.3 × 38.3 | 3–13.2 | Moderate | Narrow slits | < −17 | N/A | < 0.02 | 4 |
| [15] | 38 × 38 | 3–20 | Moderate | Matching stubs | < −17 | N/A | < 0.08 | 4 |
| This work | 42 × 42 | 3.16–13.44 | Simple | None | < −28 | ~ 10 | < 0.0001 | 4 |

## 28.4 Conclusion

This work introduces a new MIMO antenna for UWB, C-, and X-band space communications. It is composed of four planar, basic antenna elements placed orthogonally on a Rogers material. The findings demonstrate that the designed antenna is functional between the 3.16 and 13.44 GHz frequency range. The antenna effectively segregated the two elements by more than – 28 dB without using any kind of isolating elements. The estimated MIMO also demonstrates a DG of around 10 dB and an ECC of less than 0.0001. Therefore, the anticipated array is suitable for numerous space applications owing to its low profile, huge bandwidth, good efficiency, good gain, etc.

**Acknowledgements** This work is sponsored by Universiti Kebangsaan Malaysia under the grant number GUP-2022-074. This work is also partly funded by the Research and Publication Cell, University of Chittagong, Bangladesh.

## References

1. A. Goldsmith, *Wireless Communications* (Cambridge University Press, UK, 2005)
2. T. Kaiser, F. Zheng, E. Dimitrov, An overview of ultra-wide-band systems with MIMO. Proc. IEEE **97**, 285–312 (2009)
3. N. Das et al., A dual-band MIMO antenna for 5G sub-6 GHz/WiFi/WiMAX/WLAN/Bluetooth/C-band applications, in *ECCE 2023* (2023), pp. 1–4
4. N. Das et al., A four-element MIMO antenna for WiFi, WiMAX, WLAN, 4G, and 5G sub-6 GHz applications, in *12th ICECE 2022* (2022), pp. 1–4
5. S. Bukkawar, V. Ahmed, Compact slot-loaded ultra-wideband multiple-input multiple-output antenna with fractal-inspired isolator. Int. J. RF Microw. Comput. Aided Eng. **30**, e22036 (2020)
6. S.P. Biswal, S. Das, A compact printed ultra-wideband multiple-input multiple-output prototype with band-notch ability for WiMAX, LTE band43, and WLAN systems. Int. J. RF Microw. Comput. Aided Eng. **29**, e21673 (2019)
7. M.N. Hasan, S. Chu, S. Bashir, A DGS monopole antenna loaded with U-shape stub for UWB MIMO applications. Microw. Opt. Technol. Lett. **61**, 2141–2149 (2019)
8. S. Rajkumar, K.T. Selvan, P.H. Rao, Compact 4 element Sierpinski Knopp fractal UWB MIMO antenna with dual band notch. Microw. Opt. Technol. Lett. **60**, 1023–1030 (2018)
9. A.A. Althuwayb, Low-interacted multiple antenna systems based on metasurface-inspired isolation approach for MIMO applications. Arab. J. Sci. Eng. **47**, 2629–2638 (2022)
10. M. Alsharari et al., A novel design of complementary split ring resonator metamaterial-based low-profile MIMO antenna with defected ground structure for S/C/X/Ka band applications. Micromachines **14**, 1232 (2023)
11. S. Tripathi, A. Mohan, S. Yadav, A compact octagonal fractal UWB MIMO antenna with WLAN band-rejection. Microw. Opt. Technol. Lett. **57**(8), 1919–1925 (2015)
12. S. Tripathi, A. Mohan, S. Yadav, A compact Koch fractal UWB MIMO antenna with WLAN band-rejection. IEEE Antennas Wirel. Propag. Lett. **14**, 1565–1568 (2015)
13. V.S.D. Rekha et al., Dual band notched orthogonal 4-element MIMO antenna with isolation for UWB applications. IEEE Access **8**, 145871–145880 (2020)
14. R.G. Villanueva, H.J. Aguilar, Compact UWB uniplanar four-port MIMO antenna array with rejecting band. IEEE Antennas Wirel. Propag. Lett. **18**(12), 2543–2547 (2019)

15. W. Yin et al., CPW fed compact UWB 4-element MIMO antenna with high isolation. Sensors **21**(8), 2688 (2021)
16. A. Eslami et al., Four-element MIMO antenna for X-band applications. Int. J. Microw. Wirel. Technol. **13**, 859–866 (2021)
17. A.G. Alharbi et al., A multi-slot two-antenna MIMO with high isolation for sub-6 GHz 5G/IEEE802.11ac/ax/C-band/X-band wireless and satellite applications. Electronics **11**(3), 473 (2022)
18. J. Banerjee et al., An orthogonally oriented multiband MIMO antenna for WLAN, C-band and X-band wireless applications, in *IEEE Calcutta Conference 2020* (2020), pp. 328–332
19. B. Feng et al., A dual-wideband and high gain magneto-electric dipole antenna and its 3D MIMO system with metasurface for 5G/WiMAX/WLAN/X-band applications. IEEE Access **6**, 33387–33398 (2018)
20. S.K. Gupta et al., Four element UWB MIMO antenna with improved isolation using resistance loaded stub for S, C, and X band applications. Prog. Electromagn. Res. C **131**, 73–87 (2023)

# Chapter 29
# Improving the Efficiency of Minimum Determinant Computation in Space Time Trellis Code with Optimal Subtree Pruning

**Ungku Azlin Ungku Chulan, Mohammad Tariqul Islam, Norbahiah Misran, and Ungku Azmi Ungku Chulan**

**Abstract** The calculation of minimum determinant plays a crucial role in fulfilling the determinant criterion of a certain code design in space time trellis code. In the heuristic optimization of code construction, the minimum determinant is derived via a variant of the branch and bound algorithm. Although the algorithm is relatively efficient, it is not optimized in terms of the pruning strategy. Search space is pruned when the upper bound is exceeded. No attempt is made on discerning the potential of different structures within the search space. This paper proposes a new pruning approach to improve the computational efficiency of finding the minimum determinant for a particular generator matrix *G*. It builds upon the idea of minimal complete cycles. They are the smallest paths that begin and end with zero. By capitalizing on the minimum complete cycle of the search tree, the structure with the highest potential in the search space can be identified. Consequently, it helps the search process to differentiate subtrees in their capacity of yielding a solution. Search can be focused on a certain subtree while others are pruned altogether. This enables approximately 45% reduction of the overall spatial and temporal cost. Despite its potential, the pruning method is inherently probabilistic. There is a 0.0357 risk that it could provide an erroneous minimum determinant.

## 29.1 Introduction

Data rate and transmission reliability are two of the most important factors in a communication system. To achieve a more reliable high data rate and low transmission error over the wireless channel, efficient modulation and coding schemes must

---

U. A. U. Chulan (✉) · M. T. Islam · N. Misran · U. A. U. Chulan
Department of Electrical, Electronic and Systems Engineering, Faculty of Engineering and Built Environment, Universiti Kebangsaan Malaysia UKM, 43600 Bangi, Selangor, Malaysia
e-mail: proactive.azlin@gmail.com



M. T. Islam et al. (eds.), *Proceedings of the 8th International Conference on Space Science and Communication*, Springer Proceedings in Physics 303,
https://doi.org/10.1007/978-981-97-0142-1_29

be developed. However, this scheme has limitations due to issues such as multi-path fading, interference, and noise. To address this issue, the Space Time Trellis Code technique has been proposed.

In Multi-Input Multi-Output (MIMO) Rayleigh fading channels, Space Time Trellis Code (STTC) achieves both diversity and coding gains. The MIMO antennas provide spatial diversity, while the trellis encoder provides coding gain. The design criteria for STTC's Rayleigh fading channels were developed by [1]. The distance matrices, which are created using pairs of unique codewords, dictate the design criteria. The minimum rank of the distance matrices determines the diversity gain, while the minimum distance of the distance matrices determines the coding gain.

The optimization of code design [2, 3] is one of the most heavily researched topics in Space Time Trellis Code. There are many ways to do this, but it often revolves around the usage of the design criteria. The design criteria are practically a set of governing factors that determine the manner of which the performance of codes can be maximized under a particular situation. For instance, in the case of two transmitters and one receiver (2T-1R), the coding gain is optimal when the generator matrix $G$ has full rank. Apart from the rank [4] and distance criteria [5], the determinant criterion is one of the core criteria of code design. It requires the minimum determinant of the generator matrix $G$ to be found.

Finding the minimum determinant of a particular generator matrix $G$ via the tree-based approach is a practical way of complying with the determinant criterion in code design. The earliest tree-based approach [6] relies on branch and bound [7, 8] with the breadth first search as its traversal strategy. Here, pruning is done when the minimum determinant found at a node in the search tree exceeds the upper bound or largest value that is discovered previously during the search process. This is further improved with depth first search [9] where lower search cost is made computationally feasible. Latest research which is the minimal cycle method (MCM) [10] improves the tree-based approach by introducing a new way of calculating the initial upper bound (IUB).

### *29.1.1 Dominance*

In heuristic search, the employment of dominance or dominant rules [11–13] is a compelling strategy in pruning. The strategy basically postulates the existence of a path $p(i)$, of which superior or at least similar solutions can be found when compared to another path $p(j)$. In this case, it is said that $p(i)$ is dominant over $p(j)$. To illustrate dominance, imagine the following example (Fig. 29.1). Here, two paths are competing for dominance: $p(i) = 0 \rightarrow 1$ and $p(j) = 0 \rightarrow 2$ where the solutions for both are (a) and (b), respectively. Now, the solutions or minimum determinants for (a) are 4, 4, 8 while (b) are 8, 16, 8. Given that smaller minimum determinants are better than larger ones, it is quite easy to see that the series of solutions in (a) from $p(i)$ are superior or at least similar to the solutions in (b) from $p(j)$. If this is so, then it can be surmised that $p(i)$ is dominant over $p(j)$.

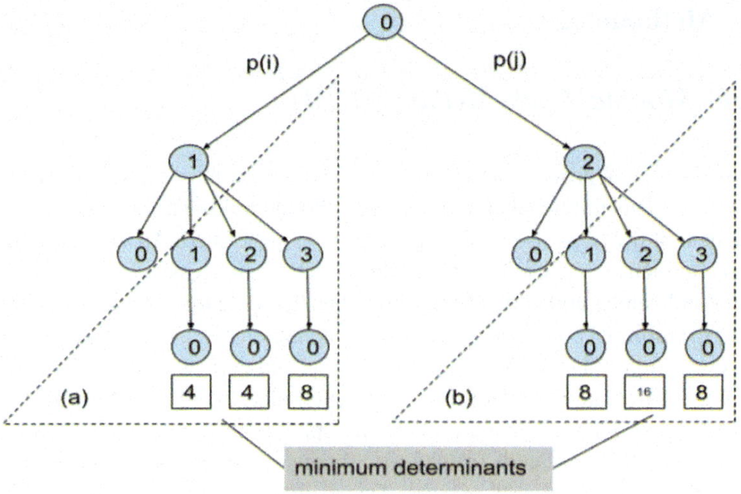


**Fig. 29.1** Dominance between path $p(i)$ and $p(j)$

## *29.1.2 Problem Statement*

Although MCM enables a tighter initial upper bound for the search process, the approach itself is not fully optimized. Further introspection would indicate that MCM does not discriminate between the different minimum complete cycles within the subtrees in considering their subsequent impact. That is, the method is not cognizant upon the potential of dominance in pruning the search space. Contemplating upon these minimum complete cycles, it is not implausible to assume that for a tree with different subtrees, there exists one that is dominant over the rest. Nevertheless, this possibility is ignored altogether by MCM. The pruning mechanism is unable to discriminate between varying subtrees within the search space that offer different potentials in yielding a solution for the minimum determinant (Fig. 29.2). In light of this, the paper proposes a better pruning strategy.

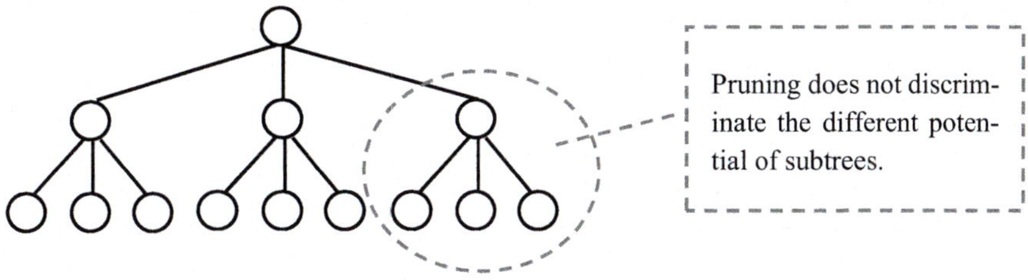


**Fig. 29.2** Weakness of pruning mechanism

## 29.2 Methodology

### 29.2.1 *Minimal Cycle Method (MCM)*

In the tree-based approach, the upper bound (UB) is initially set to infinity when the search starts. This implies that until the upper bound holds a finite value, the pruning mechanism of the tree-based approach cannot be performed. As such, pruning takes time to commence because it requires the search to progress sufficiently before the upper bound can be updated with a valid value. To solve this problem, MCM would traverse the minimal complete cycles within the search tree prior to the actual search. These cycles mark the shortest path that begins and ends with zero. The minimum determinants are calculated at these cycles such that one of them can be chosen as the initial upper bound. This way, the initial upper bound is no longer set at infinity when the search is initiated. Instead, it contains a value that can be immediately used for pruning in the search process. This significantly reduces the computational complexity of the search.

### 29.2.2 *Optimal Subtree Pruning (OSP)*

The optimal subtree pruning (OSP) builds upon the idea in the minimal cycle method (MCM). In MCM, the minimum complete cycles are used to determine the initial upper bound of the search. OSP goes a step further by utilizing them to determine which subtree is optimal within the search space. The algorithm is shown in Fig. 29.3.

The algorithm accepts the generator matrix $G$ and total state $N$ as inputs. Once these are received, the algorithm constructs the corresponding minimal complete cycles in accordance with the number of states $N$. If there are $N$ states, then $N - 1$ minimal complete cycles are generated. Similar to MCM, the minimum determinant at each of the cycle is calculated and then analyzed with one another. The minimal complete cycle with the smallest minimum determinant that is larger than zero is chosen as the representative. If none of the value fits this requirement, then the optimal subtree pruning cannot be enforced. On the other hand, if there are more than one minimal complete cycle with the smallest minimum determinant, then the initial one is given priority and chosen.

As for the pruning strategy [14, 15], the algorithm nominates the subtree with the 'chosen' minimal complete cycle as the optimal subtree [16]. Thus, search would proceed only on this optimal subtree. For the other subtrees that are related to the non-chosen minimum determinant, they are pruned completely. In other words, the search mechanism would not even consider them.

To understand the approach better, observe the given illustration (Fig. 29.4). Three minimal complete cycles are generated for the search tree of the 4 state $4 \times 2$ generator matrix $G$. The minimum determinant for each minimal complete cycle is 0.0, 8.0, and 4.0, respectively. Among them, the last complete cycle with the value of 4.00

```
OSP(generator matrix G, total_state N)
    mcm         = minimal complete cycle
    md                = minimum determinant
    ub                = upper bound

    for i=1 to N-1
        mcm(i) = 0     i      0
    end
    ub = ∞

    for each mcm(i) in mcm(1) .. mcm(N-1)
        md(i) = minimum_determinant(mcm(i))
        if ( md(i) > 0 && md(i) < ub )
            ub = md(i)
        end
    end

    if ( ub != ∞ )
      for each mcm(i) in mcm(1) .. mcm(N-1)
          if(mcm(i) > iub)
            prune(subtree(mcm(i)))
          else
            ub = search(subtree(mcm(i)))
          end
      end
    else
      ub = search(tree(G,N))
    end
    return ub
end
```

**Fig. 29.3** Algorithm of optimal subtree pruning

is the one with the lowest non-zero value. It is therefore chosen as the minimal complete cycle. Not just that, the portion related to it becomes the optimal subtree. This implies that the other subtrees are deemed non-optimal. In effect, they are pruned completely. The search would not even consider other branches that are related to these non-optimal subtrees and would only continue at the optimal subtree.

To sum it up, OSP comprises of three main stages (Fig. 29.5). The first stage is the generation stage which involves the generation of all minimal complete cycles within the search tree. This is followed by the selection stage that determines the optimal

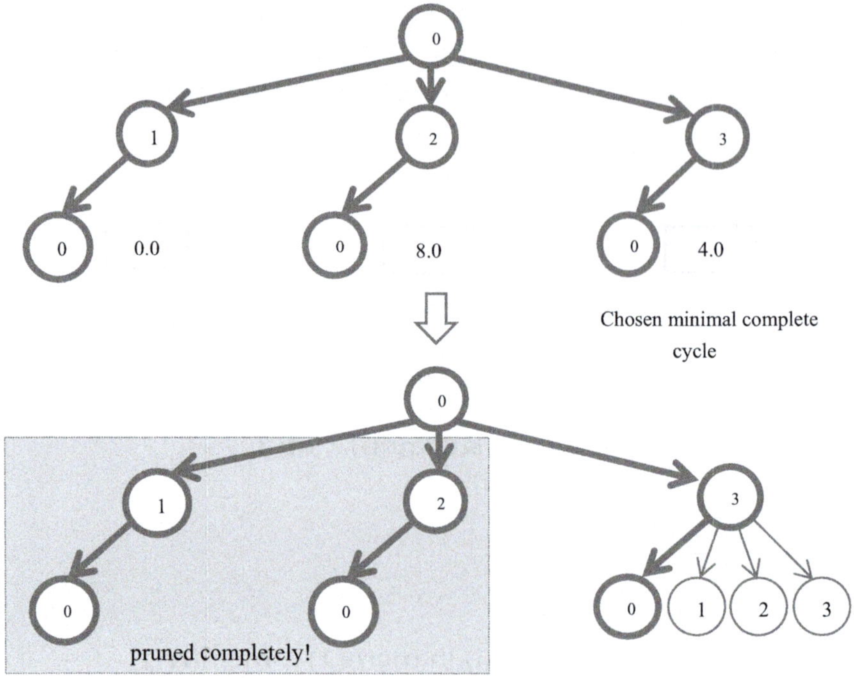


**Fig. 29.4** Conceptualization of optimal subtree pruning

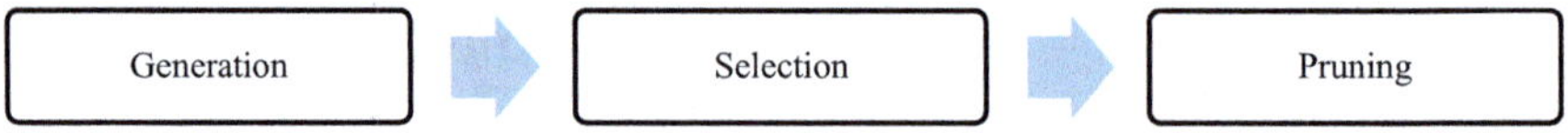


**Fig. 29.5** Main stages of optimal subtree pruning

subtree by analyzing the minimal complete cycle. Finally, the process concludes with the pruning stage that ceases the search at all the non-optimal subtrees and proceeds only at the optimal subtree.

The pruning outcome is highly efficient. It avoids the need to traverse a considerable portion of the search space. However, this benefit often comes with a price. Theoretically speaking, there is a risk that the aggressive pruning method might interfere with the outcome for the search.

### *29.2.3 Experimentation*

The experimentation is done on an Intel i3 processor of 4 GB RAM with Lubuntu as the OS. To conduct the experimentation, the entire 65,536 variations of a 4 state $4 \times 2$ generator matrix are generated via the JAVA programming language. JAVA is

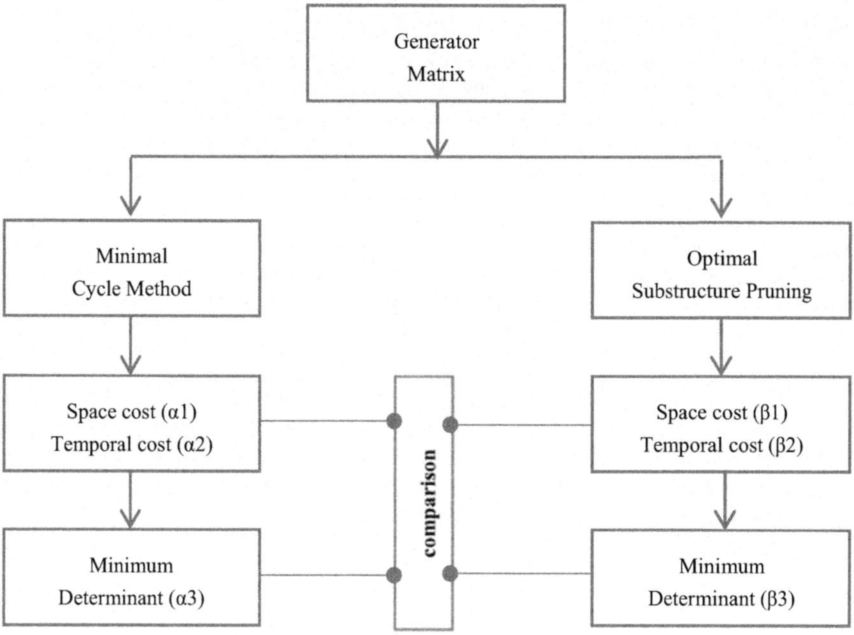


**Fig. 29.6** Algorithm of optimal subtree pruning comparison between minimal cycle method and optimal subtree pruning

used instead of the conventional MATLAB due to its flexibility in forming the graph data structure. MATLAB mainly relies on the array data structure and does not offer the same versatility. The total search space and temporal cost of searching for the minimum determinant between MCM and OSP are then compared (Fig. 29.6). Not just that, the minimum determinants between these two approaches are inspected as well to ascertain whether any discrepancies exist in terms of the search outcome.

The experimentation is also performed to analyze the risk of using OSP in calculating the minimum determinant. It is hypothesized that the approach could lack accuracy in certain cases.

## 29.3 Results and Discussion

### *29.3.1 Analysis of Spatial and Temporal Reduction*

The comparison between the minimal cycle method and optimal subtree pruning in terms of the spatial and temporal cost is shown in Table 29.1. From the spatial perspective, OSP reduces the spatial complexity of MCM by 45.61%. This is consistent with the temporal enhancement where OSP reduces the processing time of MCM

**Table 29.1** MCM versus OSP: spatial and temporal analysis

| Dimension | Total generator matrix | Minimal cycle method (MCM) | Optimal subtree pruning (OSP) | Reduction (%) |
|---|---|---|---|---|
| Spatial | 65,536 | 1,662,040 nodes | 904,000 nodes | 45.61 |
| Temporal | 65,536 | 4633 ms | 2427 ms | 47.61 |

**Table 29.2** Breakdown of enhancement, neutrality, and regression

| | Enhancement | Neutrality | Regression |
|---|---|---|---|
| Probability | 0.9776 | 0.0224 | 0.0000 |

by 47.61%. Apart from the degree of enhancement afforded by OSP, it is also crucial to analyze the breakdown between enhancement, neutrality, and regression. This analysis practically explains how frequent OSP would improve or worsen MCM. As shown in Table 29.2, OSP improves MCM by 0.9776 of the time and does not impact the search process or remains neutral at 0.0224 of the time. It must be noted that OSP does not worsen MCM in any of the instances of the experimentation. There is 0.0000 regression.

### *29.3.2 Analysis of Enhancement*

In addition to the former analysis, the extent of enhancement that is succeeded by OSP is further investigated. This is done by analyzing the average, maximum, and minimum enhancement (Table 29.3). On average, OSP enhances MCM by 45.77%. The maximum enhancement is 64.00% while the lowest one is 21.05%. Besides that, the probability distribution of the intensity for each of the enhancement that is exhibited by OSP when compared with MCM is studied as well (Table 29.4). In simple terms, this form of analysis attempts to figure out how often the proposed approach would be able to improve the current approach within a certain range. As shown from the result, OSP improves MCM most frequently by 50.00–60.00% (Fig. 29.7). This happens at 0.4312 of the time.

**Table 29.3** Average, maximum, and minimum enhancement

| Enhancement | Average | Maximum | Minimum |
|---|---|---|---|
| | 45.77% | 64.00% | 21.05% |

**Table 29.4** Intensity of enhancement

| | Enhancement range | Total search enhanced | Probability |
|---|---|---|---|
| 1 | [00.00, 10.00) | 0 | 0.0000 |
| 2 | [10.00, 20.00) | 0 | 0.0000 |
| 3 | [20.00, 30.00) | 6300 | 0.0983 |
| 4 | [30.00, 40.00) | 4728 | 0.0738 |
| 5 | [40.00, 50.00) | 22,044 | 0.3441 |
| 6 | [50.00, 60.00) | 27,624 | 0.4312 |
| 7 | [60.00, 70.00) | 3372 | 0.0526 |
| 8 | [70.00, 80.00) | 0 | 0.0000 |
| 9 | [80.00, 90.00) | 0 | 0.0000 |
| 10 | [90.00, 100.00) | 0 | 0.0000 |

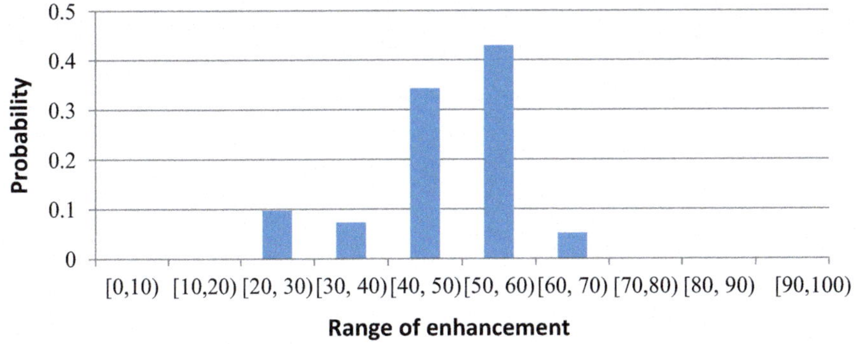


**Fig. 29.7** Intensity of enhancement

### *29.3.3 Analysis of Failure*

Failure is defined as the erroneous minimum determinant that is given by the optimal subtree pruning. In other words, it transpires when pruning interferes with the ability of the search algorithm to find the minimum determinant accurately. From the 65,536 generator matrices that are processed with the optimal subtree pruning, 2340 failures are detected. Thus, the risk of failure is quite marginal, only 0.0357. Failure is seen in the case of the generator matrix $G = [3\ 3\ 0\ 1;\ 3\ 2\ 2\ 0]^{\mathrm{T}}$ where the optimal subtree pruning mistakenly states that the minimum determinant is 8.0 when in actuality, it is 4.0, as found by the minimal cycle method. This happens when the minimum determinant resides within the subtree that is perceived as non-optimal to the optimal subtree pruning. Since non-optimal subtrees are pruned entirely, the search process would never come across the right solution. To understand how it could happen, consider a hypothetical scenario (Fig. 29.8). The shaded nodes are those that are part of the minimal complete cycles. For the sake of argument, assume that OSP

**Fig. 29.8** Failure of OSP

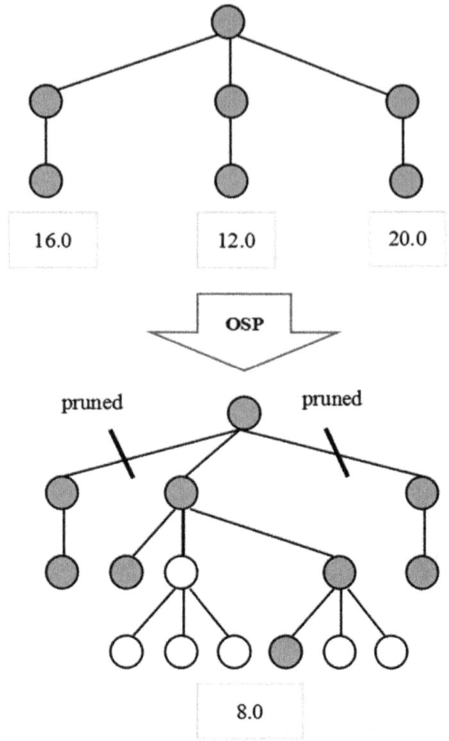


generates three minimal complete cycles and finds their minimum determinants to be 16.0, 12.0, and 20.0, respectively. Here, the minimal complete cycle with the least value lies in the middle part of the tree. As such, search only continues within this subtree while the others are pruned. In doing so, the minimum determinant of 8.0 is found.

Now, consider the same scenario but this time, it is handled by MCM instead (Fig. 29.9). MCM does not enforce any subtree pruning. Therefore, all subtrees would be searched. This implies that the generation stage for MCM is similar to OSP where three minimal complete cycles are also generated. Again, the node with the minimum determinant of 8.0 is found by MCM. Here, however, the search does not simply end. MCM finds a better value of 4.0 for the minimum determinant on a node that belongs to the subtree pruned by OSP.

Fig. 29.9 Success of MCM

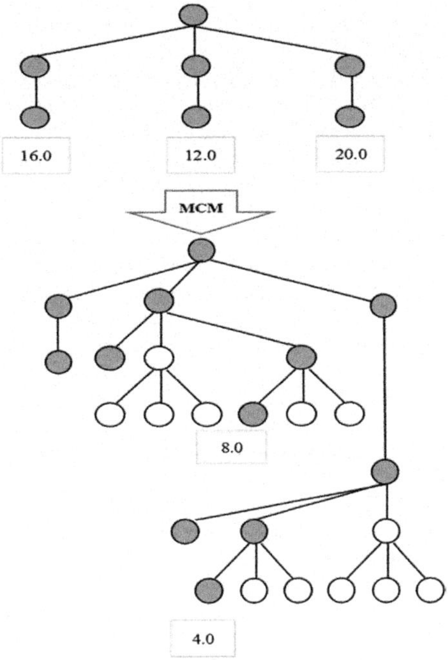


## 29.4 Conclusion

Optimal substructure pruning improves the search efficiency of the current approach from two perspectives, namely the spatial and temporal aspects. Specifically, the spatial efficiency and temporal efficiency are improved by 45.61% and 47.61%, respectively. On average, the spatial enhancement is 45.77% where the enhancement range of 50–60% being the most frequent one with the probability of 0.4312. In spite of these enhancements, the approach suffers a 0.0357 risk of failure. In retrospect, the overall benefits outweigh the risk.

**Acknowledgements** The authors would like to thank Universiti Kebangsaan Malaysia under Grant Number GUP-2022-017 for financial support.

## References

1. V. Tarokh, N. Seshadri, A. Calderbank, Space-time codes for high data rate wireless communication: performance criterion and code construction. IEEE Trans. Inf. Theory **44**(2), 744–765 (1998)
2. S.Ö. Ata, I. Altunbas, STTC design for vehicular communication systems employing fixed-gain AF PLNC over cascaded fading channels. IET Commun. **12**(11), 1283–1289 (2018)

3. H. Harun, K. Dimyati, U.A.I. Ungku Chulan, Optimal generator matrix G. Aerosp. Sci. Technol. **24**(1), 136–140 (2013)
4. S. Banerjee, M. Agrawal, A simple analytical design approach to space time trellis code. Wireless Pers. Commun. **75**, 1141–1154 (2014)
5. P. Viland, G. Zaharia, J.-F. Hélard, Euclidean distance decomposition to generate new 16-QAM and 64-QAM space-time trellis codes, in *21st Annual IEEE International Symposium on Personal, Indoor and Mobile Radio Communications*, Istanbul, Turkey (IEEE, Istanbul, Turkey, 2010), pp. 385–390
6. T. Fukuda, S. Otsu, Y. Tokunaga, H. Zhao, A realization of determinant criterion for STTC design, in *ITC-CSCC 2008: The 23rd International Technical Conference on Circuits/Systems, Computers and Communications* (2008), pp. 61–64
7. W. Jiang, J. Wu, H. Qi, W. Feng, X. Duan, An optimization method for evaluation and decision-making of distribution network planned projects based on branch and bound algorithm, in *2018 2nd IEEE Conference on Energy Internet and Energy System Integration (EI2)* (IEEE, Beijing, China, 2018), pp. 1–9
8. J.-Y. Wang, A branch-and-bound algorithm for minimizing the total tardiness of a three-agent scheduling problem considering the overlap effect and environmental protection. IEEE Access **7**, 5106–5123 (2018)
9. H. Harun, An improved algorithm for fast evaluation of space time trellis code (STTC) generator matrix, Ph.D. dissertation, Faculty of Engineering, University Malaya, Kuala Lumpur, Malaysia, 2010
10. H. Harun, U.A.N. Ungku-Chulan, U.A.I. Ungku-Chulan, K. Khazani, Improving the evaluation of generator matrix G by initial upper bound estimation, in *2013 Computing, Communications and IT Applications Conference (ComComAp)* (IEEE, Hong Kong, China, 2013), pp. 85–89
11. A. Khoudi, A. Berrichi, Branch and bound algorithm for identical parallel machine scheduling problem to maximise system availability. Int. J. Manuf. Res. **15**(3) (2020)
12. I.E. Büyüktahtakın, Stage-t scenario dominance for risk-averse multi-stage stochastic mixed-integer programs. Ann. Oper. Res. **309**, 1–35 (2022)
13. I.E. Büyüktahtakın, Scenario-dominance to multi-stage stochastic lot-sizing and knapsack problems. Comput. Oper. Res. **153**(C) (2023)
14. F. Prasser, F. Kohlmayer, K.A. Kuhn, Efficient and effective pruning strategies for health data de-identification. BMC Med. Inform. Decis. Mak. **16**, 49 (2016)
15. S. Krishnamoorthy, Pruning strategies for mining high utility itemsets. Expert Syst. Appl. **42**(5), 2371–2381 (2015)
16. T. Hirakawa, T. Tamaki, T. Kurita, B. Raytchev, K. Kaneda, C. Wang, L. Najman, Tree-wise discriminative subtree selection for texture image labeling. IEEE Access **5**, 13617–13634 (2017)

# Chapter 30
# Flower-Shaped Resonator-Based Triple-Band Ideal Metamaterial Absorber with Incident and Polarization Angle Independent for Sensing Application

**Mst Ishrat Jahan and Mohammad Rashed Iqbal Faruque**

**Abstract** The triple-band ideal metamaterial absorber with incidence and polarization angle independence for sensing applications in the form of a flower has been proposed. The ideal metamaterial absorber (MMA) and the Rogers RO4003C dielectric substrate with a diameter of 0.58 mm are put between the top metal patch, the middle sensor layer, the bottom copper ground, and the four layers that comprise this concept. The three perfect absorption peaks occurred at frequencies of 5.4281 GHz, 8.5963 GHz, and 14.32 GHz, respectively, with absorbance values of about 99.867%, 99.985%, and 99.96%. The design's guiding principles and maximum absorption have been determined by observing the electromagnetic properties involving the flowing surface current. The ideal MMA can reach maximum absorbance at incidence angles up to 80° and for polarization-independent coefficients of 0, 45, 90, 135, and 180°. The ideal MMA-based sensors, such as pressure, moisture, and density, are presented in this study in several noteworthy applications. The sensing capability has been particularly concentrated on resonance frequency, especially in frequency ranges where resonance transitions evenly or is non-sequentially dependent on differences in pressure, moisture, and density.

## 30.1 Introduction

Metamaterial absorbers are more prone to failure and are available in more significant quantities [1, 2]. These artificial structures that can manipulate electromagnetic waves at frequencies represent a potential strategy to meet this demand in various application areas [3, 4]. The optimal MMA has an absorbance value larger than unity,

M. I. Jahan · M. R. I. Faruque (✉)
Space Science Centre, Institute of Climate Change, Universiti Kebangsaan Malaysia UKM, 43600 Bangi, Selangor D.E., Malaysia
e-mail: rashed@ukm.edu.my



M. T. Islam et al. (eds.), *Proceedings of the 8th International Conference on Space Science and Communication*, Springer Proceedings in Physics 303,
https://doi.org/10.1007/978-981-97-0142-1_30

is polarization-insensitive, and is omnidirectional [5]. According to Bakir et al., a metamaterial absorber has an absorption rate of 91.8% at 4.2 GHz [6]. The unit cell is 30 mm by 24 mm in dimension and made of RR-4 material. The authors describe a polarization-independent MA with broadband absorptivity [7]. Nonetheless, the statistics suggest that the highest absorbance occurs between 8.5 and 15.5 GHz, with 70% and 90% absorption rates, respectively, eventually exhibiting insufficient absorbance. A broadband absorption analysis was carried out by Chen et al. performed, with reinforced MA based on a double circular ring; however, it exhibits polarization independence of more than 90% at 7.60 GHz, which is compatible with observations [8]. Ahamed et al. determined an enhanced SRR in the microwave project with a wide band of negative refractive index and a resonance frequency of 8.84 GHz. The magnitude attained for a typical SRR is just 2.47 GHz, although the bandwidth of its negative refractive index is about 3.40 GHz [9]. Zhao and Cheng used an electric split-ring resonator patch to make an ultra-wideband with a reflection coefficient of less than − 10 dB from 4.4 to 18 GHz [10]. Polarization dependency describes how the absorbance performance of MA varies with polarization angle [11] and shows an example of three bands of enhanced MA with absorbance peaks close to unity absorbance at 7.48, 9.70, and 11.27 GHz. It also has a high absorption at such a wide angle of incidence and is polarization angle insensitive. In three-band MA, the maximum absorbance is exhibited at 6.16, 9.22, and 11.6 GHz [12].

The analysis in this work focuses on the ideal MMA for sensing applications with triple-band polarization and incidence angle insensitivity. The design for the flower-shaped resonator base unit cell is made up of four layers: a metal grounding layer with a diameter of 0.035 mm, a dielectric substrate made of Rogers RO4003C with a height of 0.58 mm, a sensor layer with a density of 10 mm, and a copper patch. The ideal MMA had a unit cell dimension of 7 mm by 7 mm and exhibited exceptional absorption of 99.867, 99.985, and 99.96% over the resonance frequencies of 5.4281, 8.5963, and 14.32 GHz. Furthermore, it delivers higher incidence angle stability up to 80° in normal incident and polarization angle independence at 00, 450, 900, 1350, and 1800. In fact, surface current, electric, and magnetic field effects have all been examined to understand these phenomena better. The ideal MMA provided here can have many sensor implementations, including pressure, moisture, and density. The sensing ability has been concentrated on the resonance frequency, specifically, frequency ranges in which resonance changes equally or non-sequentially depending on pressure, moisture, and density variations.

## 30.2 Design Technique

Figure 30.1 shows the proposed ideal MMA unit cell, which is a square copper patch with a flower-shaped resonator that has been cut into a Rogers (lossy) substrate to give it insensitive polarization characteristics and three different resonances. The proposed MMA in this paper comprises four layers: a substrate between two dielectric layers as the top-level unit cell layer, a sensor layer between the substrate and the

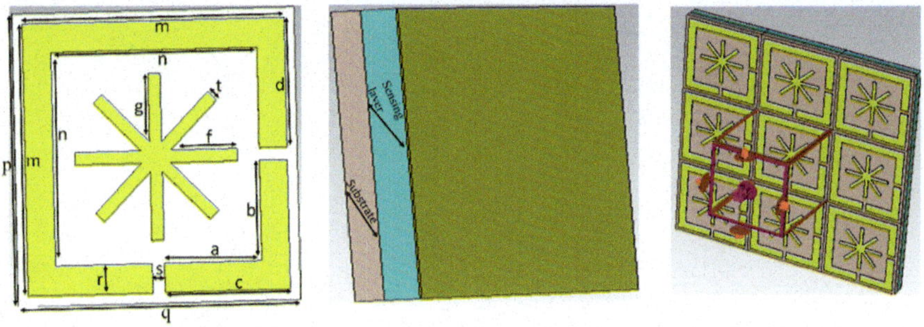


(a) (b) (c)

**Fig. 30.1** Ideal MMA unit cell situations **a** top view, **b** backside view and **c** boundary conditions of the simulation environment

metal ground plate on the bottom layer, and a metal ground plate. The groundsheet and metallic resonator are modeled using copper sheets with a diameter of 0.035 mm and electrical conductivities of 5.8 10 7 S/m. Figure 30.1b displays the view from the bottom. The Rogers RO4003C has a 0.58 mm diameter, a loss tangent of 0.0027, a permeability of 1, and a permittivity of 3.55. The substrate layer and the outer ring are divided by 0.5 mm; the substrate layer has a length and width of 7 mm. The parameters for the created ideal MMA are shown in Fig. 30.1a as $p = q = 7$ mm, $a = b = 2.4$ mm, $c = d = 3.1$ mm, $m = 6.5$ mm, $n = 5.1$ mm, $r = 0.7$ mm, $s = 0.3$ mm, $g = f = 1.64$ mm, and $t = 0.3$ mm (Fig. 30.1). CST Studio Suite® 2019 analyses have been performed to show the attributes of the suggested design. This simulates design using a frequency-domain solution and is based on the single integration approach. The proposed unit cell was implemented utilizing the rules for the edges of the unit cell in both the $x$ and $y$ directions, as shown in Fig. 30.1c, enabling the incident electromagnetic waves to propagate along the $Z$-axis, which is an open direction. The absorber's performance is denoted by the formula $A(\omega) = 1 - R(\omega) - T(\omega)$ [13, 14], where $A(\omega)$ denotes absorptivity, $R(\omega)$ refers to reflection, and $T(\omega)$ denotes transmission that is resonance variable [15, 16].

## 30.3 Results and Discussion

### *30.3.1 Absorption*

Figure 30.2a, b show the ideal MMA triple-band microwave reflection, absorption, and transmission values in TE and TM modes. There appear to be three absorbance peaks at 5.4281, 8.5963, and 14.32 GHz, with maximum absorption of 99.867, 99.985, and 99.96%.

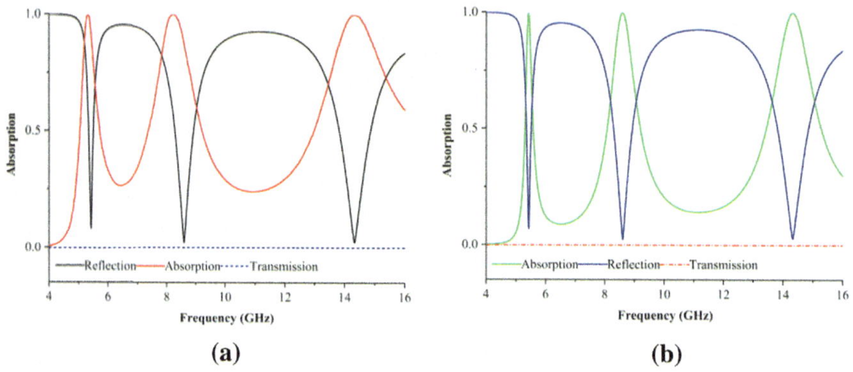


**Fig. 30.2** Results in reflection, absorption, and transmission at **a** TE and **b** TM modes

### *30.3.2 The Ideal MPA Parametric Analysis*

Figure 30.3 shows the unit cell design optimization process. The ideal MPA has three sections. As a result of the ideal MMA unit cell's symmetric capacitance, different copper patch layouts changed the electrical and magnetic resonances. Three-stage absorbance displays used green, red, and blue edges. The initial configuration involving a split ring with two split-gap resonators resulted in 98%, 99%, and 99% absorption at 5.96, 9.56, and 15.58 GHz, respectively. The split-ring and plus-shaped resonator with two split-gap structures have 98% and 99% absorptions at 6.17 and 9.98 GHz, respectively. The unit cell was created by combining the first and second designs. This setup produced perfect triple-band absorption.

### *30.3.3 Angle Stability*

Figure 30.4a, b demonstrate the ideal MMA absorption at various incidence and polarization angles when an incident is normal. As a result, the oblique incidence angle varied from 0 to 80° over the resonant frequency ranges of 5.4281, 8.5963, and 14.32 GHz, with absorbances of 99.867, 99.985, and 99.96%, and the polarization angle was also altered at 0, 45, 90, 135, and 180°. Electromagnetic resonances with triple-band frequencies are created because both polarization modes know their impedance. The ideal MMA resonance effect persisted consistently as the angle value; however, the resonance shifts at a higher frequency of roughly 0.1 GHz when the incidence angle is changed. At each frequency, the proposed absorber has electromagnetic dipoles. The underlying or core aspect source of variances in absorption intensity is a shift in the H-field during the depression of the MMA. As a result, the ideal MMA described here possesses exceptionally strong absorption properties that are incident and polarization angle independent.

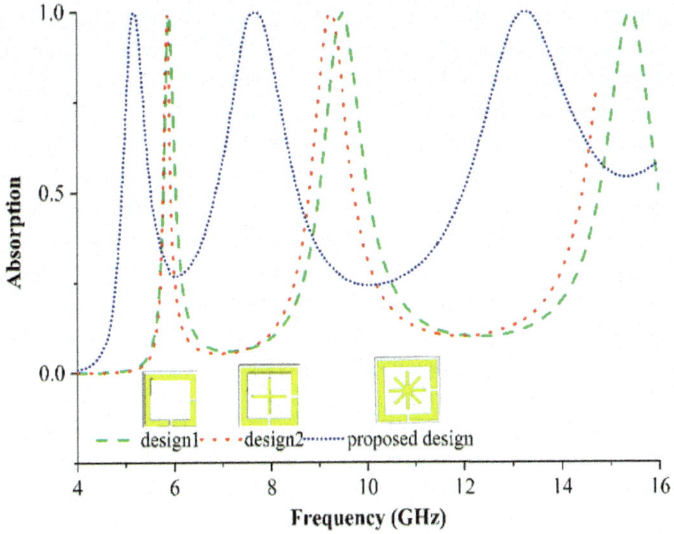


**Fig. 30.3** The scenario of design optimization

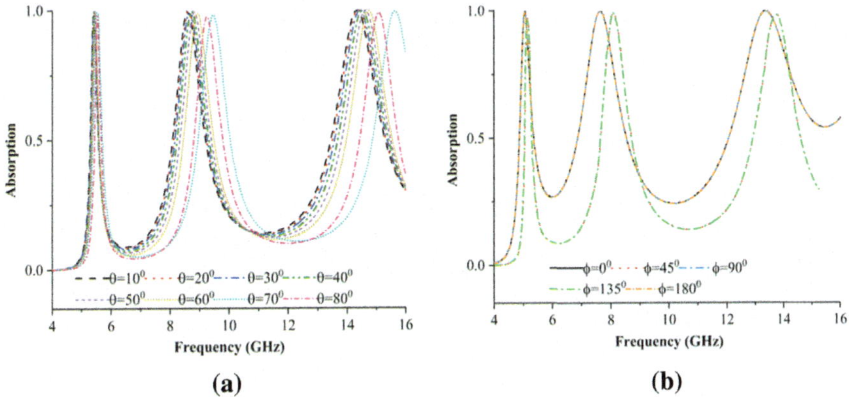


**Fig. 30.4** Absorbance performance in different **a** incident, **b** polarization angles at normal incident

## *30.3.4 Sensing Application*

Multiple sensor devices were examined, utilizing the sensing capabilities of the numerical outputs of the study's principal purpose. This section describes the applications of pressure, moisture, and density sensors. Figure 30.1b shows the proposed ideal MMA construction with a sensor layer. The proposed sensor layer is presumed to be stuffed with air, and its level of thickness appears to vary in response to pressure being applied [17], as shown in Fig. 30.5a. The analysis is carried out statistically by changing sensor layer thickness, resulting in a pressure change. This study examined five different sensor layer thicknesses: 0.1, 0.3, 0.5, 0.7, and 0.9 mm. This study examined five different sensor layer thicknesses: 0.1, 0.3, 0.5, 0.7, and 0.9 mm. The absorption performance based on volumetric moisture is shown in Fig. 30.5b. In this instance, the sensor layer had been intended to be coated in sandy soil, and soil reflects the moisture level where frequencies have been measured [17]. Sensor layer thicknesses of 0.1, 0.3, 0.5, 0.7, and 0.9 mm are used for the relevant volumetric moisture. The resonant frequencies changed to the right level of moisture heightened. The absorption rate is much higher than 99% when sandy soil's loss tangent and permittivity coefficients have been altered. In general, the moisture content of agricultural and textile products is an essential factor influencing their real life. The absorption performance depending on density is shown in Fig. 30.5c. If appropriately designed, the sensor layer determines a material's density. This work presupposed that the sensor layer would be made up of several materials with variable densities [17]. The sensor layer is being kept at a thickness of 0.7 mm. Four distinct materials—Arlon AD260, Arlon AD320, Arlon AD430, and Arlon AD600—make up the sensor layer. When a sensor layer made of material of the AD600 type was utilized, the maximum absorbance rate of almost 99% was attained. AD260, AD320, AD430, and AD600 have dielectric characteristics of 2.6, 3.2, 4.3, and 6. The sensor frequency's absorption coefficient moved to the right as the electrically conducting properties of the substances employed in the sensor's layer exhibited an increase.

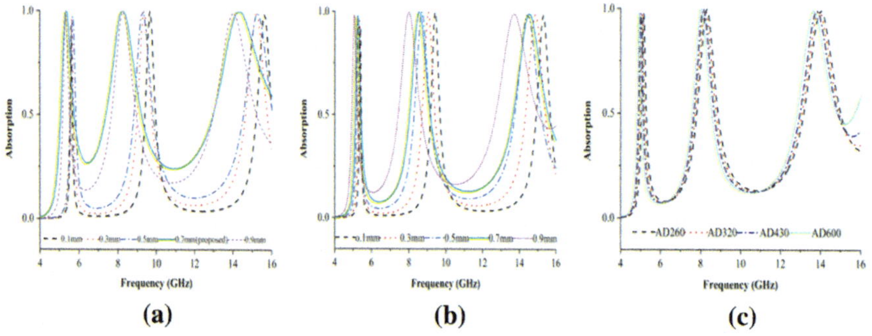


**Fig. 30.5** Measure of absorbance using **a** pressure, **b** moisture, **c** density sensor method at normal incident

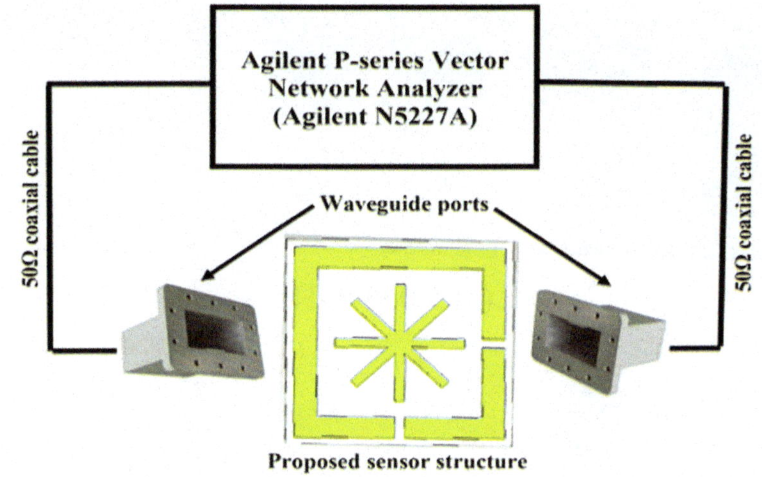


**Fig. 30.6** Proposed experimental arrangement for different sensor

Three structures will be fabricated for pressure, moisture, and density sensor applications for experimental purposes. The measurement setup is displayed in Fig. 30.6, where a fabricated prototype is positioned between the two waveguide ports coupled to the Agilent N5227 PNA Microwave Network Analyzer (to generate an excitation signal) through a 50 Ω coaxial cable. The reflection coefficient ($S11$) will be measured using this arrangement. From this $S11$ result, the absorption behavior is determined. In this investigation, absorption is the primary requirement for all sensor implementations.

### *30.3.5 Analysis of Electric Field, Magnetic Field, and Surface Current*

Figure 30.7i–iii show the electric field, magnetic field, and surface current and illustrate the electric excitation by displaying the electric field at the upper resonating patch for three different absorption bands. The areas at which the most significant electric field is concentrated appear to be (i) the opposite right-side corner of the ring at 5.4281 GHz, (ii) between the substrate and ring gap section at 8.5963 GHz, and (iii) the right-side outside the ring gap at 14.32 GHz. The manufactured magnetization created by anti-parallel current looping is closely connected to transmission magnetism.

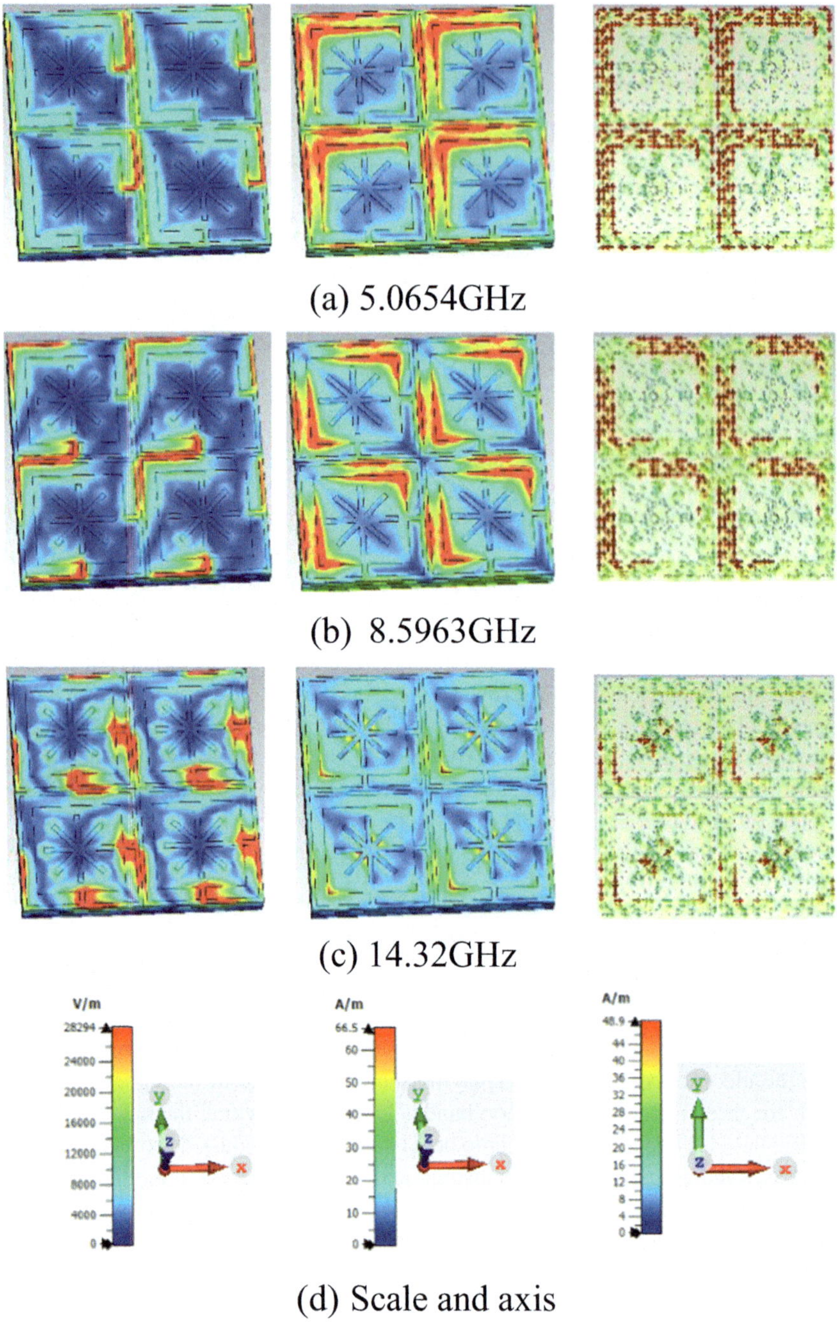


**Fig. 30.7** Electromagnetic field analysis of the proposed structure. **i** Electric field, **ii** magnetic field and **iii** surface current

The simultaneous action of frequencies at 5.4281, 8.5963, and 14.32 GHz is indeed made possible by the ideal MMA design that has been described. Figure 30.7ii, iii show the magnetic field and surface intensity for three absorbances. The magnetic field grows in intensity along with the surface current density. Contrasting field distributions imply that the fields are compressing along the split ring, causing electromagnetic excitation. The ideal MMA configuration being shown exhibits significant absorption because of the control of magnetic and electric stimulation. According to the findings, the highest current is concentrated on the left side of the ring at 5.4281 GHz, the two opposing corners of the ring at 8.5963 GHz, and the flower-shaped resonator at 14.32 GHz.

Table 30.1 compares the proposed optimal results for MMA and the previous research conducted on MA, focusing on various aspects such as absorption, unit cell size, design structure, frequency range, and other relevant factors. In addition to enhancing the absorption level, the suggested absorber structure also reduced both the dimensions and thickness of the substrate. As previously discussed, the triple-band sensing applications of MA are characterized by enhanced absorption levels resulting from the utilization of diverse top resonators and dielectric substrates. The MMA that has been designed is a highly efficient and versatile sensor application, demonstrating exceptional absorption capabilities across three distinct frequency bands. Specifically, 99.867%, 99.985%, and 99.96% absorption rates have been achieved at frequencies of 5.4281 GHz, 8.5963 GHz, and 14.32 GHz, respectively.

## 30.4 Conclusion

A unique design approach for a triple-band, flower-shaped resonator with polarization and incident angle independent ideal MMA has been successfully demonstrated. The absorption rates are 99.867, 99.985, and 99.96% at 5.4281, 8.5963, and 14.32 GHz. The ideal MMA that is being described as polarization independent has angular stability at 0, 45, 90, 135, and 180° and is incident angle insensitive up to 80°. Pressure, moisture, and density sensors, as well as other essential implementations of the ideal MMA-based sensors, are all included in the proposed study. The sensing capability has been mainly concentrated on resonance frequency, particularly frequency ranges where resonance transitions evenly or non-sequentially dependent on variations in pressure, moisture, and density.

**Table 30.1** Comparisons of ideal MMA

| References | Size (mm) | Frequency range (GHz) | Design | Above absorption rate (%) | Polarization independent | Incident angle insensitive | Sensing application |
|---|---|---|---|---|---|---|---|
| [18] | 9 × 9 | 4.37, 7.08, 9.38, 10.65 | X-shaped triple split-ring resonator | 99 | Yes | Yes | No |
| [19] | 9 × 9 | 2.21, 4.1, 5.6 | A double split ring enclosed nested meander-line-shaped | 90 | No | No | Yes |
| [20] | 8 × 8 | 4.87, 7.49, 11.78, 13.09 | Double X-shape ring resonator | 99 | Yes | Yes | Yes |
| [21] | 35 × 35 | 8–12 | Oval-shaped wing resonator | 83 | No | No | Yes |
| Proposed design | 7 × 7 | 5.4281, 8.5963, 14.32 | Flower-shaped resonator | 99 | Yes | Yes | Yes |

**Acknowledgements** The authors acknowledge that the research Universiti Grant, Universiti Kebangsaan Malaysia, Geran Translasi, UKM-TR2022-05, conducted the research work.

## References

1. F. Dincer, M. Karaaslan, E. Unal, O. Akgol, C. Sabah, Design of polarization- and incident angle-independent perfect metamaterial absorber with interference theory. J. Electron. Mater. **43**, 3949–3953 (2014)
2. M.R.I. Faruque, M.T. Islam, Design of miniaturized double-negative material for specific absorption rate reduction in human head. PLoS ONE **9**(10), e109947 (2014)
3. Z. Ozer, A.M. Mamedov, E. Ozbay, Metamaterial absorber based multifunctional sensor application. IOP Conf. Ser. Mater. Sci. Eng. **175**(1), 012059 (2017)
4. M.I. Jahan, M.R.I. Faruque, M.B. Hossain, S. Abdullah, An ultra-thin, triple-band, incident angle-insensitive perfect metamaterial absorber. Materials **16**(4), 1623 (2023)
5. M.R.I. Faruque, M.T. Islam, Novel triangular metamaterial design for electromagnetic absorption reduction in human head. Prog. Electromagn. Res. **141**, 463–478 (2013)
6. M. Bakır, M. Karaaslan, F. Dinçer, K. Delihacioglu, C. Sabah, Tunable perfect metamaterial absorber and sensor applications. J. Mater. Sci. Mater. Electron. **27**, 12091–12099 (2016)
7. M. Bagmancı, O. Akgol, M. Ozakturk, M. Karaaslan, E. Unal, M. Bakır, Polarization-independent broadband metamaterial absorber for microwave applications. Int. J. RF Microw. Comput. Aided Eng. **29**(1), e21630 (2019)
8. J. Chen, X. Huang, G. Zerihun, Z. Hu, S. Wang, G. Wang, M. Liu, Polarization-independent, thin, broadband metamaterial absorber using double-circle rings loaded with lumped resistances. J. Electron. Mater. **44**, 4269–4274 (2015)
9. E. Ahamed, M.R.I. Faruque, M.F.B. Mansor, M.T. Islam, Polarization-dependent tunneled metamaterial structure with enhanced fields properties for X-band application. Results Phys. **15**, 102530 (2019)
10. J. Zhao, Y. Cheng, Ultrabroadband microwave metamaterial absorber based on electric SRR loaded with lumped resistors. J. Electron. Mater. **45**, 5033–5039 (2016)
11. S. Ji, Z. Luo, J. Zhao, H. Dai, C. Jiang, Design and analysis of an ultra-thin polarization-insensitive wide-angle triple-band metamaterial absorber for X-band application. Opt. Quant. Electron. **53**, 1–19 (2021)
12. S. Kalraiya, R.K. Chaudhary, R.K. Gangwar, Polarization independent triple band ultra-thin conformal metamaterial absorber for C- and X-frequency bands. AEU-Int. J. Electron. Commun. **135**, 153752 (2021)
13. M.J. Hossain, M.R.I. Faruque, M.R. Ahmed, M.J. Alam, M.T. Islam, Polarization-insensitive infrared-visible perfect metamaterial absorber and permittivity sensor. Results Phys. **14**, 102429 (2019)
14. S. Ji, C. Jiang, J. Zhao, X. Zhang, Q. He, Design of a polarization-insensitive triple-band metamaterial absorber. Opt. Commun. **432**, 65–70 (2019)
15. M.I. Jahan, M.R.I. Faruque, M.B. Hossain, Flower-shape resonator-based triple-band metamaterial wave absorbers to find the dispersion relation utilizing one dimensional (1-D) periodic waveguide. J. Magn. Magn. Mater. 171202 (2023)
16. Y. Liang, H. Lin, S. Lin, J. Wu, W. Li, F. Meng, Y. Yang, X. Huang, B. Jia, Y. Kivshar, Hybrid anisotropic plasmonic metasurfaces with multiple resonances of focused light beams. Nano Lett. **21**(20), 8917–8923 (2021)
17. M. Bakir, M. Karaaslan, O. Akgol, O. Altintas, E. Unal, C. Sabah, Sensory applications of resonator-based metamaterial absorber. Optik **168**, 741–746 (2018)
18. M.I. Jahan, M.R.I. Faruque, M.B. Hossain, An X-shaped triple split ring resonator-based metamaterial perfect absorber with quad-band incident and polarization angle insensitivity for C, X, and Ku band applications. J. Magn. Magn. Mater. 170940 (2023)

19. A.M. Siddiky, M.R.I. Faruque, M.T. Islam, S. Abdullah, M.U. Khandaker, N. Tamam, A. Sulieman, Body-centered double-square split-ring enclosed nested meander-line-shaped metamaterial-loaded microstrip-based resonator for sensing applications. Materials **15**, 6186 (2022)
20. M.I. Jahan, M.R.I. Faruque, M.B. Hossain, M.U. Khandaker, F. Elsayed, M. Salman, H. Osman, Quad-band metamaterial perfect absorber with high shielding effectiveness using double X-shaped ring resonator. Materials **16**(12), 4405 (2023)
21. Y.I. Abdulkarim, L. Deng, M. Karaaslan, Ş. Dalgaç, R.H. Mahmud, F. Ozkan Alkurt, H. Luo, The detection of chemical materials with a metamaterial-based sensor incorporating oval wing resonators. Electronics **9**, 825 (2020)

# Chapter 31
# Measurement of Cylindrical Lens Antenna for 5G Mobile Base Station

**Farizah Ansarudin, Yoshihide Yamada, Nurul Huda Abd Rahman, and Norbahiah Misran**

**Abstract** In the fifth-generation (5G) mobile system, extensive of new technologies are required such as millimetre wave, massive MIMO technology, small cell size of multibeam base station antenna and advanced multiple access. For 5G multibeam base station application, a dielectric lens antenna is a potential candidate due to more versatile with a simple structure, excellent multibeam shapes, very low network losses and good scanning performance. In this paper, a hollow cylindrical structure composed of a dielectric lens antenna is proposed for practical installation on the base station pole. The lens shaping program was developed by the ray-tracing method and implemented in MATLAB software which helps in designing and optimizing the shape of the dielectric lens. To assess the multibeam radiation characteristics on the cylindrical lens antenna, a full wave analysis is conducted using the FEKO electromagnetic simulator. Experimental measurements are performed on the cylindrical lens antenna at an operating frequency of 28 GHz. The measurements aim to validate the antenna performance in generating multibeam radiation patterns. As a result, the cylindrical lens antenna achieves good multibeam radiation patterns within a scanning angle from 0° to 55°. Good agreement between measured and simulated results with a gain difference of about 1.97 dBi.

F. Ansarudin (✉) · N. Misran
Microwave Technology Research Group, Department of Electrical, Electronic and Systems Engineering, Faculty of Engineering and Built Environment, Universiti Kebangsaan Malaysia, 43650 Bangi, Malaysia
e-mail: farizah.ansarudin@ukm.edu.my

Y. Yamada
Malaysia-Japan International Institute of Technology, Universiti Teknologi Malaysia, 54100 Kuala Lumpur, Malaysia

N. H. A. Rahman
School of Electrical Engineering, College of Engineering, Universiti Teknologi MARA, 40450 Shah Alam, Malaysia

 
M. T. Islam et al. (eds.), *Proceedings of the 8th International Conference on Space Science and Communication*, Springer Proceedings in Physics 303,
https://doi.org/10.1007/978-981-97-0142-1_31

## 31.1 Introduction

Nowadays, the developing of 5G mobile system is driving to support growing demand for broadband services over mobile networks. In 5G mobile system, many new technologies are required such as millimetre wave (mm-wave), massive multiple-in multiple-out (MIMO) technology, small cell size of multibeam base station antenna, low latency and high speed communication. Furthermore, new base stations must support both new and legacy mobile communications standards. To accomplish this performance, a multibeam base station is a promising solution for a 5G mobile communication system. The multibeam base station produces multiple beams in different directions within a defined cell area and each beam is assigned to a particular mobile user [1, 2]. The multibeam antennas are required to increase the capacity of existing cells in cellular networks. In 5G, it is essential to have the capability to provide dedicated beams for individual users or user groups. Each radiation beam serves as a dedicated communication channel as shown in Fig. 31.1, which is crucial for achieving high data rates and efficient connectivity. Therefore, the base station antennas are expected to be compact, with a size of less than 30 cm. This size restriction is particularly relevant when operating in the mm-wave frequency bands, which are used in 5G for the high bandwidth and high gain performance.

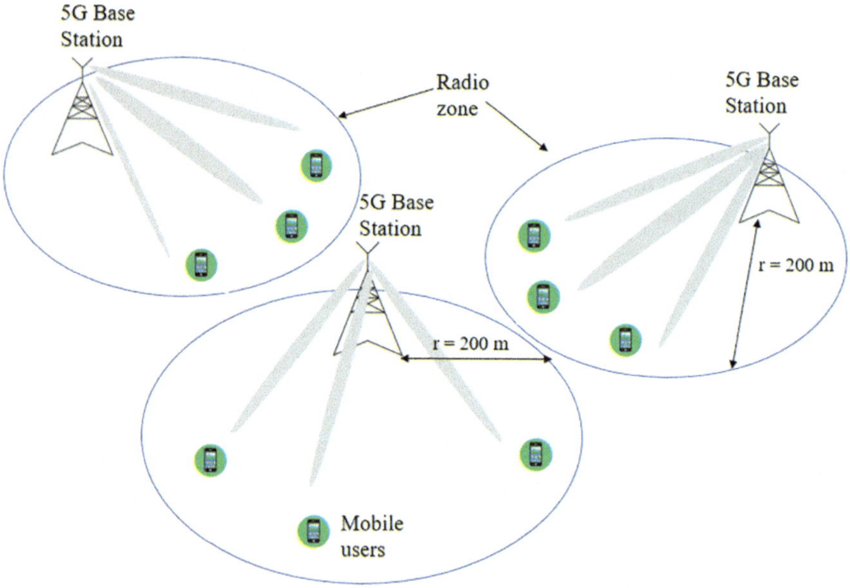


**Fig. 31.1** Multibeam operation in 5G base station application

According to [3–5], the ability to control and manufacture dielectric materials with the required precision and continuity is a significant technical challenge in the development of Luneburg lens antennas. Overcoming these challenges is crucial for realizing the benefits of Luneburg lenses in practical multibeam antenna systems, which are needed for advanced wireless communication technologies like 5G. In addition to that, the reference to bulky spherical-sized Luneburg lenses has traditionally been designed for specific applications where their size and performance characteristic are well suited. However, in the context of 5G and other advanced communications systems, there is a need for more compact and versatile antenna solutions that can provide for multibeam capabilities and high gain performance. A dielectric material with many different types of lens surface curvatures has been studied for various communication applications as well as its feeding networks [6]. However, there is lack of comprehensive discussion of the capabilities of performing wide angular scanning for multibeam applications in those works. The traditional lens shaping methods were primarily designed for the synthesis of single beam radiation patterns. These methods have been used in applications like airborne radar and collision avoidance system, where a single focussed beam is sufficient for their purposes.

In [7], multibeam radiation characteristics are studied sufficiently with the Abbe's sine case. However, for practical base station application in a 5G system, including the ability to generate multiple beams, ease of installation on thin support structure and lightweight design to meet the demand should be satisfied. In this paper, a suitable practical structure is proposed and multibeam radiation pattern characteristics are discussed at the beam direction angular range between 0° and 50°.

## 31.2 Antenna Design and Structure

### *31.2.1 Lens Designing*

The analysis and performance of lens antenna systems comprise three main categories: ray-tracing method algorithm which estimates its radiation characteristic of lens antenna, simulation method for the electrical performance by the design requirements and measurement method. The principle of lens shaping is based on Snell's Law for constructing a surface 1 (S1) and a surface 2 (S2) to shape the lens structure by ray-tracing method. Figure 31.2 shows the lens shaping principle of ray tracing from origin coordinate to the aperture plane. From the figure, the coordinate system is expressed by O as an origin and corresponds to the on-focus of the lens antenna. Initially, the ray emitted from the origin at angle $\theta$ reaches S1 which is indicated by $(r, \theta)$. Then, the ray is refracted at S1 by an angle, $\phi$, and reaches S2 at the point indicated by $(z, x)$. After the ray is refracted at S2, the ray becomes parallel to the $z$-axis at the aperture plane. The maximum angle, $\theta_{\mathrm{m}}$, shown in the figure is important parameter which represents an optimum angle for the power to radiate at the lens

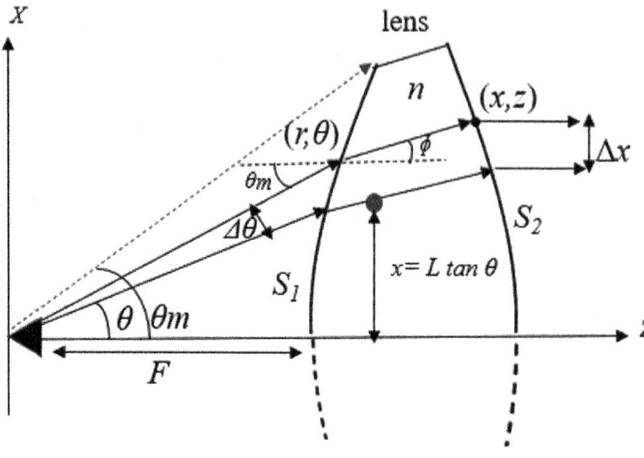


**Fig. 31.2** Design structure and coordinate system

edge with very minimum of spill over. Absorbing material can be used by covering the lens side to suppress the spill over radiations. Therefore, refracted rays only are taken into consideration. In this paper, the Straight-Line Condition (SLC) lens design method is employed to shape a small and thin lens curvature.

The ray-tracing result developed by the Matrix Laboratory (MATLAB) algorithm based on Snell's Law principle is shown in Fig. 31.3. The lens shaping program is developed by using Eqs. (31.1) and (31.2), respectively [8].

$$\frac{\mathrm{d}r}{\mathrm{d}\theta} = \frac{nr\sin(\theta - \phi)}{n\cos(\theta - \phi) - 1} \tag{31.1}$$

$$\frac{\mathrm{d}z}{\mathrm{d}\theta} = \frac{n\sin\phi}{1 - n\cos\phi}\frac{\mathrm{d}x}{\mathrm{d}\theta} \tag{31.2}$$

where refractive index of the lens is indicated as $n$, and $\phi$ refracted angle indicates as $\phi$ for one-dimensional (1-D) shaping in the $xz$-plane. The equations of Eq. (31.1), (31.2) and (31.4) have been solved simultaneously, in which the lens points of the SLC condition are refracted at the straight line. The distance from feed to straight-line curve indicates as $L$.

$$x = L\tan\theta \tag{31.3}$$

$$\frac{dx}{d\theta} = L\sec^2\theta \tag{31.4}$$

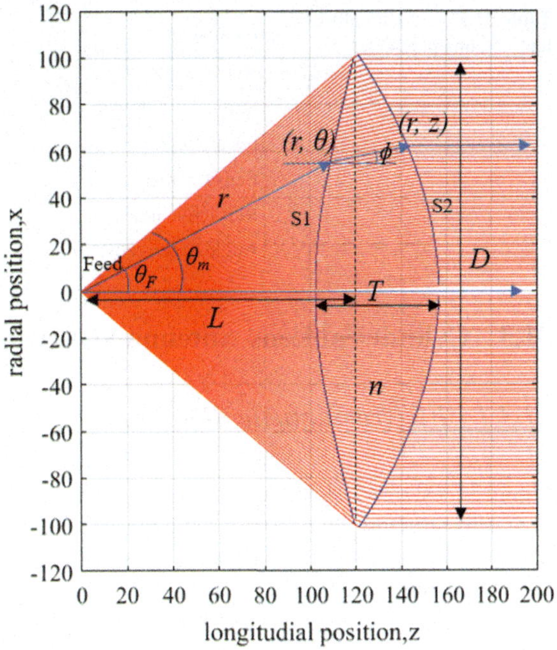


**Fig. 31.3** Ray tracing result for SLC lens shaping

### *31.2.2 Antenna Parameter*

Base station antenna should be provided by considering its practical installation such as simple structure, economical, high performance and reliable. According to [9], manufacturing a cylindrical lens is less problematic than spherical lenses. This configuration also offers a relatively low-profile solution. This paper proposes a cylindrical lens structure with mechanically scanned of multifeed positions to generate multibeam performance for wide scanning angles. An insulator material that is TPS-PC polycarbonate ($n = 1.6$) is selected due to the lowest lens thickness at the lens central obtained for the small radius, $R$, and focal length-to-diameter ($F/D$) ratio. The important parameter of cylindrical SLC lens structure is tabulated in Table 31.1 to observe the electrical performance of multibeam characteristics. These parameters are considered for obtaining the thinnest lens, small curvatures and radius, lightweight, and optimum wide scanning for multibeam directions.

**Table 31.1** Cylindrical SLC lens antenna parameter

| Parameter | Dimension |
|---|---|
| Antenna size, $D$ | 200 mm |
| Radius of cylinder, $R$ | 144.88 mm |
| Focal length, $F$ | 112.89 mm |
| Lens thickness, $T$ | 62.91 mm |
| Refractive index, $n$ | 1.6 |
| Angle to lens edge, $\theta_m$ | 40° |

## 31.3 Cylindrical Lens Antenna Structure

### *31.3.1 Electromagnetic Simulation*

Figure 31.4 shows 284 surface points constructed in FEKO to design a 3D cylindrical SLC lens structure by a polyline method. The 180-degree cylindrical structure is designed vertically along the $x$-axis. The cylindrical lens fed by a pyramidal horn antenna is constructed at the $xz$-plane. The feed horn is located at the origin facing the $z$-axis for the on-focus beam. Electrical characteristics for the ratio of $F/D = 0.6$ are identified as the most practical structure for the multibeam base station application. Table 31.2 shows the simulation parameter in FEKO solved by the Method of Moment (MoM) at the frequency of 28 GHz. Table 31.3 compares the radiation characteristics between theoretical calculation and the electromagnetic simulation at the centre beam direction. Both results are almost in good agreement with a gain difference of about − 2.18 dBi.

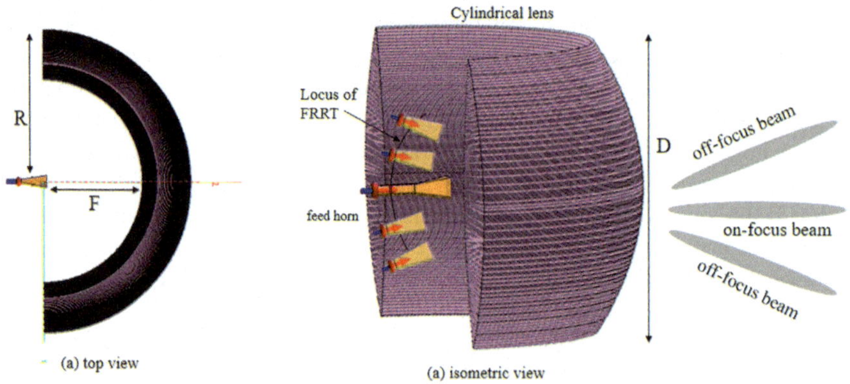


**Fig. 31.4** Cylindrical SLC lens geometry for the ratio of focal length to antenna size, $F/D = 0.6$

**Table 31.2** Simulation SLC lens and feed horn parameter

| Parameter | Description | Specification |
|---|---|---|
| Dielectric lens antenna | Material | Polycarbonate |
| | Loss tangent, δ | 0.0001 |
| | Mesh size | λ/8 |
| | Number of meshes | 78,565 |
| Horn antenna | Material | PEC |
| | Mesh size | λ/12 |
| | Number of meshes | 12,201 |

**Table 31.3** Cylindrical SLC lens antenna performance for on-focus beam

| Antenna parameter | Electromagnetic characteristics |
|---|---|
| Theoretical gain, $G_T$ | 24.86 dBi |
| Theoretical beam width, $\theta_{BT}$ | 3.65° |
| Simulated gain, $G_s$ | 22.98 dBi |
| Simulated beam width, $\theta_{BE}$ | 3.27° |
| Simulated beam width, $\theta_{BH}$ | 45.01° |
| Gain difference | – 2.18 dBi |
| Aperture efficiency, $\eta$ | 64.86% |

Mechanical scanning for feed element is used to scan vertically and produce a beam for a specific mobile user. The base station feed antenna will be placed in the vertical plane to produce beam coverage in the vertical plane. The most critical design parameters for beam shaping in vertical sectorization are the cell radius, the ratio of *F/D*, the lens size, the height of the base station and the beam tilt angle ($\theta_F$) of the base station antenna. In horizontal sectorization, the number of sectors in the cell area was determined by the radiation patterns of the base station antenna in the horizontal plane. The half-power beamwidth (HPBW) of the base station antenna in the horizontal plane is $\theta_{BH} = 45.01°$. The polar radiation patterns for the horizontal (green) and vertical (red) planes are shown in Fig. 31.5 with HPBW value are $\theta_{BE} = 3.27°$ and $\theta_{BH} = 45.01°$, respectively. The gain value for both planes is 22.98 dBi.

### *31.3.2 Multibeam Radiation Pattern*

Figure 31.6 shows the multibeam radiation characteristics of cylindrical SLC simulated for feed angle from $\theta_F = 0°$ to 50°. The dashed line (blue) is the centre beam for the on-focus feed of the base station antenna with a gain value of 22.98 dBi and the simulated half-power beam width (HPBW, $\theta_{BES}$) of 3.27°. The simulated radiation pattern is based on the transmitting mode employed by the MoM solver

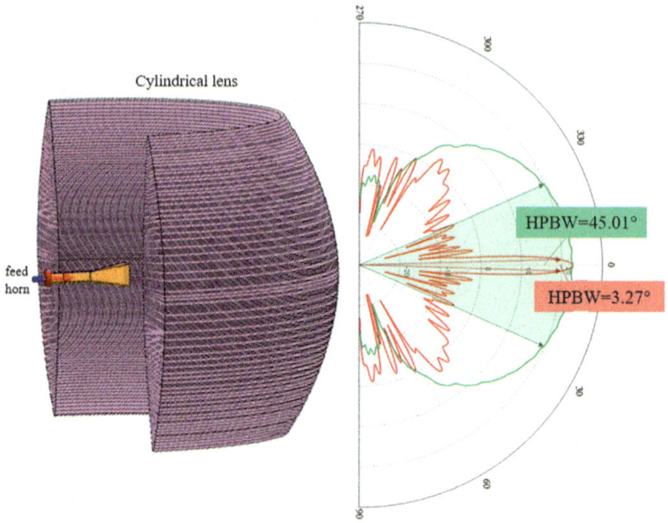


**Fig. 31.5** Polar radiation pattern at on-focus beam

in FEKO at the operating frequency of 28 GHz. From the figure, cylindrical SLC shows good multibeam radiation pattern characteristics for scanning beams from 0° to 30°. In addition, some contribution to the peak gain difference may arise because of the elevation beam tilt between polarizations. The radiation patterns exhibit slightly different gain and large beamwidth can be seen clearly at wider shifted beam direction ($\theta_S \geq 30°$). The transmission rate may be slower because of the increasing values of HPBW and the high possibility of beam sharing among mobile users. There is a practical limit to reduce the elevation beamwidth to maintain high gain value because an antenna with a very small vertical beamwidth is physically large and, because of its size and narrow beamwidth, it requires a very rigid support to avoid unacceptable reflection or refraction rays. For practical application, antennas with large elevation beamwidths (and therefore with low gain) may be desirable if the antenna is mounted on the roofs or nearby high buildings. Table 31.4 tabulates the feed position of multibeam radiation for feed angle, $\theta_F$ from 0° to 50°.

The detailed multibeam characteristics performance is tabulated in Table 31.5. The gain value is dropping gradually from 22.68 to 16.42 dBi for beam direction from $\theta_s = 0°$ to $\theta_s = 47.1°$ with a gain reduction of approximately $-$ 6.56 dBi. This significant result shows that the SLC cylindrical lens is useful for wide-angle scanning beam at the vertical plane. Therefore, the cylindrical SLC lens multibeam base station antenna produces a wider coverage area up to $\theta_s = 94.2°$ at the vertical plane. In this situation, many users are typically located below the horizontal as seen from the BS antenna, and large beam tilts may be used to illuminate correctly the intended users and reduce interference levels in surrounding cells. Therefore, the usefulness of FRRT technique is ensured to determine the optimum feed position for the cylindrical SLC lens multibeam base station antenna performance.

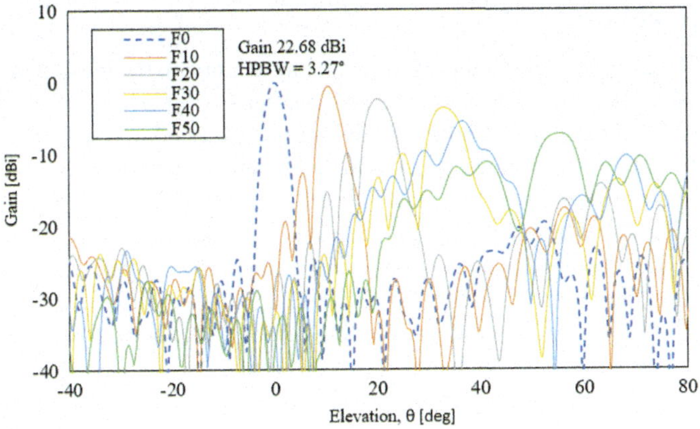


**Fig. 31.6** Multibeam radiation pattern for cylindrical SLC from $\theta_F = 0°$ to $50°$

**Table 31.4** Simulation SLC lens parameter

| Feed position | SLC feed coordinate | |
|---|---|---|
| | $z$ | $x$ |
| $F0$ | 0 | 0 |
| $F10$ | 5 | 22 |
| $F20$ | 22 | 44 |
| $F30$ | 36 | 58 |
| $F40$ | 58 | 68 |
| $F50$ | 78 | 67 |

**Table 31.5** Simulation result for multibeam radiation pattern of SLC lens

| Feed position | Feed angle, $\theta_F$ | Beam direction, $\theta_s$ | HPBW, $\theta_{BES}$ | Gain(dBi) | Gain difference (dBi) |
|---|---|---|---|---|---|
| F0 | 0 | 0 | 3.27 | 22.68 | 0 |
| F10 | 10 | 10.5 | 3.49 | 22.11 | – 0.57 |
| F20 | 20 | 20.1 | 3.96 | 20.24 | – 1.43 |
| F30 | 30 | 33.1 | 5.23 | 19.20 | – 3.16 |
| F40 | 40 | 36.9 | 5.38 | 17.13 | – 5.55 |
| F50 | 50 | 47.1 | 11.03 | 16.42 | – 6.56 |

## Measurement Result

Figure 31.7 shows the fabricated cylindrical lens antenna in the measurement set-up fed by a 28 GHz pyramidal horn antenna. The lens is horizontally clipped with a customized antenna support and the feed horn is positioned in every 10° feed angle for each measurement. The measured and simulated multibeam radiation pattern is compared, as shown in Fig. 31.8, from $\theta_F = 0°$ to $\theta_F = 50°$. The measured results are well agreed with the simulated beam direction from on-focus feed $\theta_F = 0°$ to $\theta_S = 33°$. Table 31.6 tabulates the measured results of the radiation pattern characteristics for beam direction from 0° to 55°.

Table 31.7 tabulates the comparison between SLC cylindrical lens antenna and other works for multibeam application in the frequency range from 28 to 31.5 GHz. It can be seen that the design complexity of those structures was based on beam forming circuit network resulted in mutual coupling between switches. On the other hand, the gain values were quite low as compared to the SLC lens in this work and wider HPBW which one radiating beam is possibly used by the multiple users.

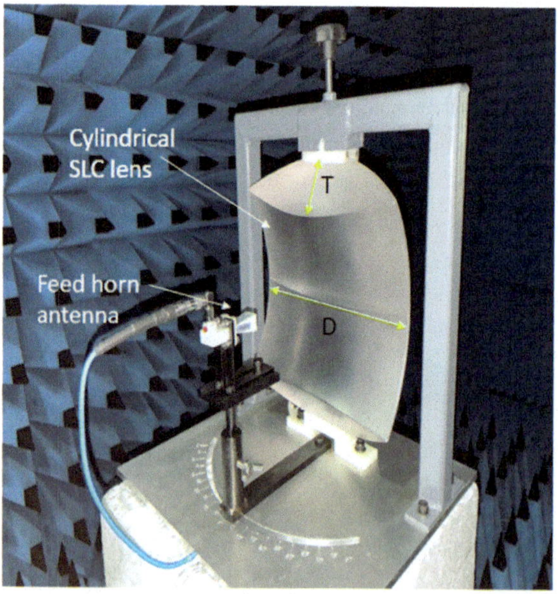


**Fig. 31.7** Fabricated cylindrical SLC lens antenna in anechoic chamber at Antenna Laboratory, MJIIT, UTM

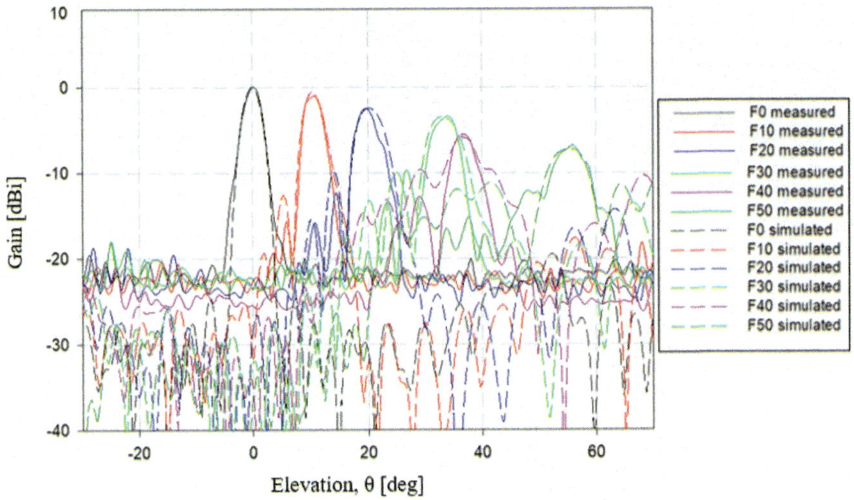


**Fig. 31.8** Comparison between simulated and measured of multibeam radiation pattern characteristic

**Table 31.6** Measured result for multibeam radiation pattern of SLC lens

| Feed position | Feed angle, $\theta_F$ | Beam direction, $\theta_S$ | HPBW, $\theta_{BEM}$ | Gain (dBi) | Gain difference (dBi) |
|---|---|---|---|---|---|
| *F*0 | 0 | 0 | 3.28 | 24.95 | 0 |
| *F*10 | 10 | 10 | 3.49 | 25.00 | – 0.05 |
| *F*20 | 20 | 29 | 3.96 | 23.87 | – 1.08 |
| *F*30 | 30 | 33 | 5.22 | 22.09 | – 2.86 |
| *F*40 | 40 | 36 | 7.48 | 22.15 | – 2.80 |
| *F*50 | 50 | 55 | 10.31 | 18.13 | – 6.82 |

**Table 31.7** Comparison between SLC lens and other multibeam lens antenna

| Authors | Frequency [GHz] | Lens type | Material | Size [$\lambda_o$] | Gain [dBi] | HPBW $\theta 3_{dB}$ | Scan angle $\theta_F$ | Aperture/port efficiency [%] | Design Complexity |
|---|---|---|---|---|---|---|---|---|---|
| This work | 28 | SLC | PC | 10–20 | 22.6–22.7 | 3.27 | ± 50° | 64.8–68.9 | Simple cylindrical |
| [10] | 31.5 | Stacked Luneburg lens | FR4 PCB | 13.8 | 15.4 | 10 | ± 72° | 42 | 11 feeding network elements |
| [11] | 28 | Waveguide-fed lens | Teflon | 10 | 18.5 | 6.42 | 58° | NA | 9 feeding network elements |
| [12] | 28 | Rotman lens (5 port) | RT 5870 | NA | 15.5 | 8.5 | ± 30° | 63–67 | Beam forming network |

## 31.4 Conclusion

This cylindrical SLC lens structure is proposed to simplify the existing multibeam base station with complicated beamforming circuits. SLC is designed for ease and practical installation for the ratio $F/D = 0.6$ with lens material $\varepsilon_r = 2.9$. The measurement multibeam characteristic accuracy is clarified with the simulated multibeam radiation pattern performances with maximum beam direction which is obtained up to $\theta_s = 55°$ at the vertical plane.

**Acknowledgements** This work is supported by Universiti Teknologi MARA and Ministry of Higher Education (MOHE) for funding under the Prototype Development Research Grant Scheme (PRGS) 600-RMC/PRGS 5/3 (001/2022) and Antenna Laboratory, Malaysia-Japan International Institute of Technology, UTM, for the measurement set-up.

## References

1. A. Arun et al., Evolution of mobile communication technology towards 5g network and challenges. Am. J. Electr. Electron. Eng. **7**(2), 34–37 (2019)
2. T.S. Rappaport, et al., Millimeter wave mobile communications for 5G cellular: It Will Work! IEEE Access **1**, 335–349 (2013)
3. D. Angela, H. Yang, Slim Luneburg lens for antenna applications. Opt. Expr. **19**, 19925–19934 (2011)
4. Y. Jay Guo, J. Bevan, *Base Station Antennas: Antenna Engineering Handbook* (McGraw Hil, New York, 2019)
5. P. José-Manuel, P. Francisco, R. Eva, Cost-effective wideband dielectric planar lens antenna for millimeter wave application. Sci. Rep. (2022)
6. M. Liang et al., A 3-D Luneburg lens antenna fabricated by polymer jetting rapid prototyping. IEEE Trans. Antennas Propag. **2014**(62), 1799–1807 (2014)
7. Y. Tajima, Y. Yamada, Design of Shaped Dielectric Lens Antenna for Wide Angle Beam. Steering. Electronics and Communications in Japan. Part 3. 89(2), 286 296(2006).
8. Y.T. Lo, S.W. Lee, Antenna Handbook. Van Nostrand Reinhold, Lens Antenna (2013)
9. J. Thornton, K.-C. Huang, *Modern Lens Antenna for Communications Engineering* (Wiley, New York, 2013)
10. X. Wang, Y. Cheng, Y. Dong: A wideband PCB-stacked air-filled Luneburg lens antenna for 5G millimeter-wave applications. IEEE Antennas Wirel. Propag. Lett. **2**(3), 327–331 (2021)
11. S. Saiedeh, K. Shafaq, M. Hani, Waveguide-fed lens based beam-steering antenna for 5G wireless communications, in *2019 IEEE International Symposium on Antennas and Propagation* (2019), pp. 1149–1150
12. A. Shahi, M. Mohamad-Taheri, I. Aryanian, A 5G multibeam antenna including rotman lens and slot array antenna, in *2018 Fifth International Conference on Millimeter-Wave and Terahertz Technologies (MMWaTT)* (2018), pp. 78–81

# Chapter 32
# Remote Cardiorespiratory Monitoring with an In-Ear PPG Device

**Nurul Izzati Salleh, Khalida Azudin, Kok Beng Gan, Mohd Hasni Ja'afar, and Mohd Shawal Faizal Mohamad**

**Abstract** Remote health monitoring has emerged as a significant area of interest, driven primarily by remarkable advancements in technology that enhanced accessibility and convenience. The rise in remote monitoring has experienced substantial (Pedrana et al. in IEEE Sens. J. 20:14,482–490, 2020), particularly amid the ongoing pandemic, where healthcare services are accessible remotely without the need to physically visit the doctors. The recent technology findings have enabled remote monitoring of patients' heart condition, ensuring that healthcare professionals can assess their cardiac conditions from any location. The article presents a prototype for monitoring heart condition using optical technology where the data is transferred to a mobile device via Bluetooth processing card in real time. The data is stored in a database to be consulted and visualized by the patient and doctor from any computer. The data will be processed, and an alert will be sent to emergency contact if abnormal heart rhythm is detected. The prototype was validated with normal heart condition patients with acceptable results.

N. I. Salleh · K. Azudin · K. B. Gan (✉)
Department of Electrical, Electronic and Systems Engineering, Faculty of Engineering and Built Environment, Universiti Kebangsaan Malaysia, 43600 Bangi, Malaysia
e-mail: kbgan@ukm.edu.my

K. Azudin
e-mail: p111466@siswa.ukm.edu.my

M. H. Ja'afar
Department of Community Health, Faculty of Medicine, Universiti Kebangsaan Malaysia Medical Centre, 56000 Kuala Lumpur, Malaysia

M. S. F. Mohamad
Cardiology, Medical Department, Faculty of Medicine, Universiti Kebangsaan Malaysia Medical Centre, Kuala Lumpur, Malaysia



M. T. Islam et al. (eds.), *Proceedings of the 8th International Conference on Space Science and Communication*, Springer Proceedings in Physics 303,
https://doi.org/10.1007/978-981-97-0142-1_32

## 32.1 Introduction

Recently, there has been a growing interest in remote health monitoring, and the potential of cardiovascular and respiratory monitoring has gained significant attention in research [1–4]. In hospital, there are several devices that are commonly used to assess respiratory functions such as spirometers, pulse oxymeters and polysomnography. However, these devices are not designed to be used for long term and remote monitoring. For cardiovascular system, electrocardiograph (ECG), Holter ECG and implantable cardiac monitors are commonly used devices for long-term monitoring. These devices are invasive, complicated and uncomfortable to use for long term and remote monitoring.

Photoplethysmography (PPG) has emerged as a widely used technique for blood oxygen saturation monitoring in clinical settings. Pulse oximetry is used in multiple situations in hospitals such as in ambulatory care and in surgery for respiratory system monitoring. There has been research using PPG on cardiovascular disease can be estimated. Some studies have also focused on investigating heart rate, heart rate variability, heart functionality and heart rhythm. However, the current position of the PPG sensor on the finger poses challenges due to artifacts caused by hand and finger movement as well as vibrations from various sources. Positioning the sensors at the finger for continuous monitoring may be difficult and inconvenient.

Considering these limitations, there is a need to explore alternative locations for the sensors that offer practicality. The ears are suggested to be in a better location as they are located at a fixed position on the head. By fixing the sensors in the ear, the inconvenience associated with finger PPG can be mitigated [4]. The ear manages to provide a stable signal and reduce artifacts caused by the movements. The placement inside the ear offers additional advantages such as comfort and easy to use. Positioning a device inside the ear is not an uncommon practice; in fact, we have been using the ear to position earphones for communication purposes. The position inside the ear also provides flexibility and freedom for the user to utilize it as part of daily activities (see Fig. 32.1).

Moreover, the position of the sensor isolated from the surrounding environment minimizes exposure to various sources of interference from surrounding. This benefit is essential in prolonged monitoring to avoid distortion in the waveform acquired from the PPG. The physiological signal captured from the ear PPG is more reliable. Previous studies of PPG at the ear canals have been done by Budidha et al. [5–7], Pedrana et al. [1], Venema et al. [2, 8], Vogel et al. [9–11] and Poh et al. [12, 13]. Most of the studies focus on the PPG signal from the ear canal, where they study the relationship between the signal and cardiovascular and respiratory system.

None of these studies have explored the potential of utilizing in-ear PPG to detect atrial fibrillation. This paper, on the other hand, studies the feasibility of capturing wireless in-ear PPG signal through Bluetooth Low Energy (BLE) technology and integrating them into a web application. The PPG signal will be processed to detect atrial fibrillation inside the web application to identify the presence of atrial fibrillation.

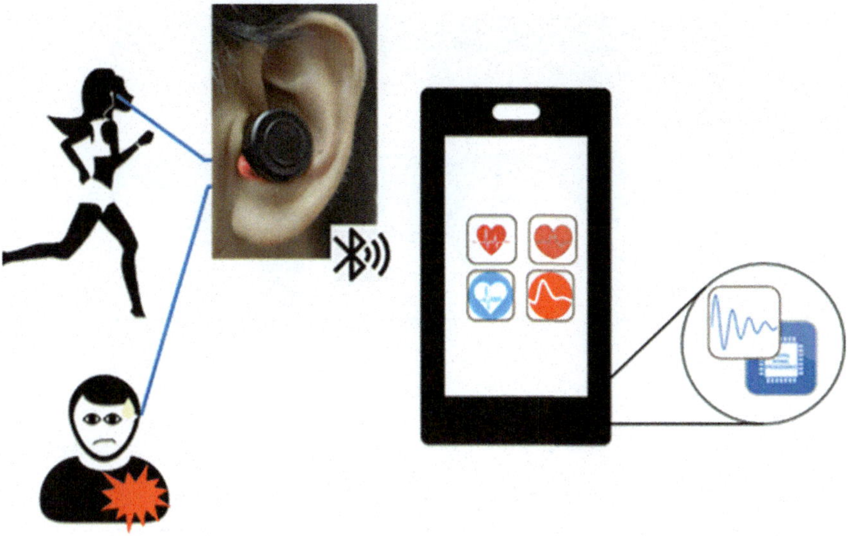

**Fig. 32.1** Example of in-ear PPG system and its connection

## 32.2 Related Works

### *32.2.1 Technology Background*

#### 32.2.1.1 Photoplethysmography (PPG)

PPG is a non-invasive method that uses light to measure blood volume changes by assessing the light absorption by the tissue. The PPG system comprises of a combination of photoemitter and photodetector. The interaction between the tissue and the optical light influenced by multiple factors including tissue penetration depth, isosbestic wavelength and the optical water window [14]. Different wavelengths have different penetration depths enabling PPG to measure signals from different tissue layers (Fig. 32.2).

The PPG waveform signal comprises of two separate current settings: alternating current (AC) and direct current (DC). The AC portion consists of high-frequency component of PPG wave that corresponds with the blood volume fluctuation in the heart. These variations in blood volume cause changes in light intensity detected by the phototransistor when the blood flows through the tissue (see Fig. 32.1). This frequency normally ranges between 0.5 and 25 Hz in healthy individuals. It contains information on blood pressure, and peripheral, vascular tone and artery compliance [14, 15].

The DC component is a baseline combining low-frequency fluctuation mainly due to autonomic nervous system activity such as breathing and thermoregulation [16]. It is a non-pulsatile component that can be impacted by biological attributes

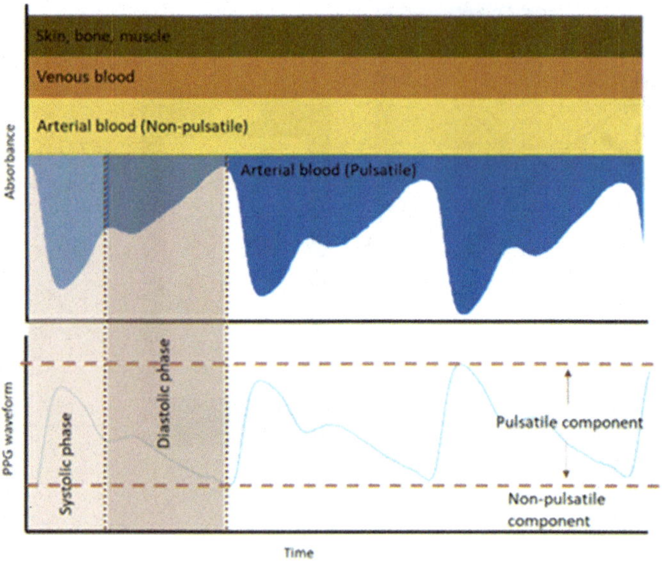


**Fig. 32.2** Principle of photoplethysmography light absorption and waveform generation

such as composition of body tissue as well as external factors like surrounding light and device specification [17]. This low-frequency component contains information from respiration, thermoregulation and vasomotor activities [17–19]. It includes information of the heart at 1–1.67 Hz and respiration rate at 0.25 Hz.

#### 32.2.1.2 Bluetooth

Bluetooth is a wireless telecommunication technology that allows the exchange of data between electronic devices such as laptops, mobile phones and wearables over short distances. The technology is reliable and convenient, and operates on radiofrequency with low power consumption making it efficient in data transmission. Bluetooth occupies the 2.4 GHz spectrum and contains less bandwidth than typical WiFi A-Band. Bluetooth is commonly used for device-to-device communication.

Bluetooth Low Energy (BLE) is a newer version of Bluetooth technology that allows communication between two devices in a broadcast mode. One device permits transmitting to multiple receivers simultaneously with low energy used.

The communication interface of the BLE earbud follows the commonly used Bluetooth LE stack that used all layers of the Open System Interconnection (OSI) reference model (Fig. 32.3). The BLE stack consists of two components which are: host and controller. All BLE devices use the Generic Attribute Profile (GATT) that provides a series of commands for the client to discover information about the BLE server.

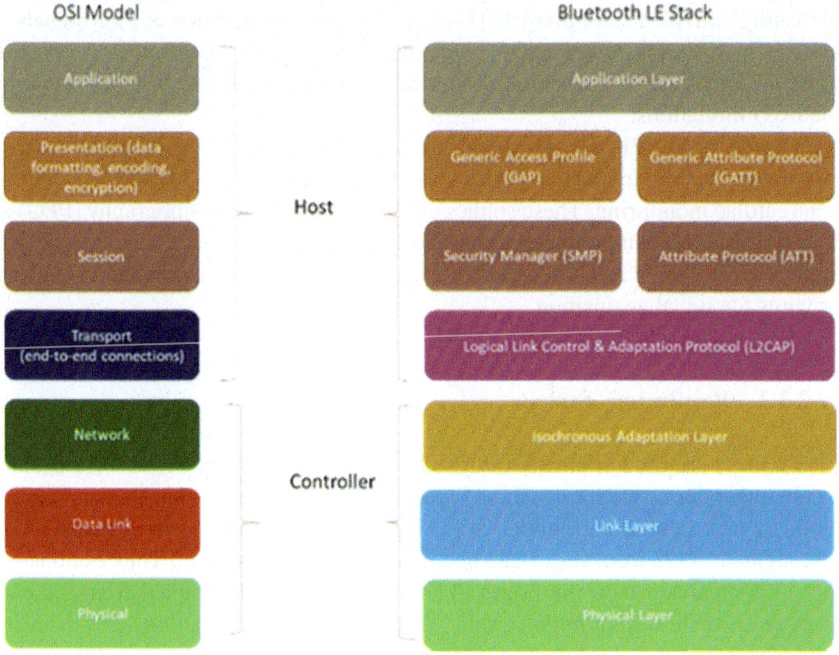


**Fig. 32.3** Bluetooth low energy (BLE) stack and OSI model comparisons

#### 32.2.1.3 Angular Framework

Angular is an open-source framework that is used to develop web applications. The framework was developed by Google and open-source community. Angular is built on Typescript using user-interface components. Each component comprises of CSS styles: HTML templates and Typescript class. This framework enables routing by providing a structured single-page application to navigate multiple views. It has a reactive form that allows synchronous access to data streams. Angular provides easy migration between versions and has a quick component processor.

### *32.2.2 Physiological Monitoring Using PPG*

#### 32.2.2.1 Cardiovascular System

Bolaños et al. [20] and Lu et al. [21] evaluated the variation in the time intervals between successive heartbeats, which can be derived from PPG and ECG signals. They revealed that the PPG signal offers accurate inter-pulse intervals from which heart rate variability (HRV) measures [21–23]. This proves that PPG signal is an exceptional alternative for ECG recordings in extracting HRV signals, particularly in healthy individuals. Mahri et al. [14, 15] utilize the acceleration of PPG signals to classify myocardial infarction by extracting characteristics akin to the QT interval of ECG. Therefore, PPG technology can serve as a substitute solution to overcome the ECG limitations.

Venema et al. [2] have proposed an in-ear PPG system for cardiorespiratory monitoring, and Pedrana et al. [1] have explored a hearable PPG system for continuous cardiac monitoring. These studies established photoplethysmography (PPG) as an alternative heart monitoring technique especially to monitor heart rhythm. The method is affordable and is non-invasive as it does not need skin preparation, patches or wires for signal acquisition.

#### 32.2.2.2 Respiratory System

PPG sensors are used clinically for respiratory monitoring. The sensors are usually put on finger during surgeries. There have been many studies conducted that measure blood oxygen saturation from the ear [2, 3, 10, 23, 24]. Budidha and Kyriacou [5, 7] investigated SPO2 from ear canals in cold temperature condition. They established that in comparison with the finger PPG signal, the signal obtained from the ears is more accurate during hypothermia. Venema et al. [2, 16, 17] engaged with in-ear pulse oximetric sensor under laboratory settings and in operation room during surgery. In-ear SPO2 measurement under laboratory settings met the measurement specificity requirement for pulse oximeter. However, under the clinical conditions, the requirement was not met.

Davies et al. [3, 24] performed a non-invasive respiratory monitoring using in-ear PPG sensors. They extracted respiratory waveform from the in-ear PPG signal where the blood oxygen saturation (SPO2) and breathing rate were calculated while chronic obstructive pulmonary disease (COPD) was estimated [3, 24]. Measuring vital signs from the ear is of interest because of the limitations of the current measurement methods including motion error and reduced peripheral perfusion particularly in life-threatening situations.

## 32.3 System Overview

### *32.3.1 In-Ear PPG System Configuration*

In-ear PPG involves using optical sensors to measure various physiological parameters by detecting blood volume within the blood vessels inside the ear. A general system configuration for in-ear PPG monitoring consists of optical sensors, signal processing unit, microcontroller and analog to digital converter.

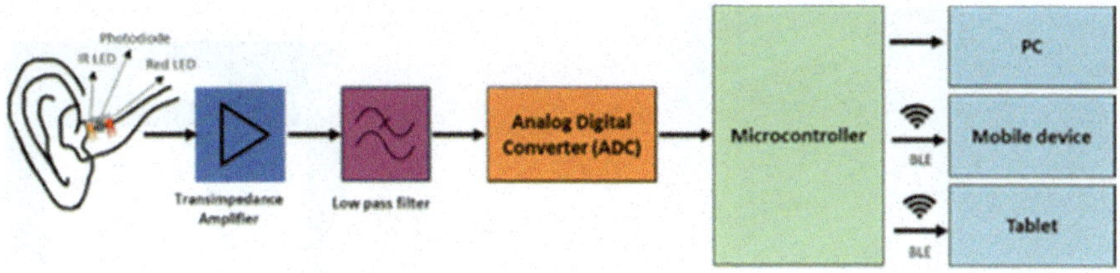


**Fig. 32.4** System block diagram

#### 32.3.1.1 System Design

To acquire signal from the ear, PPG sensors with three different colors are used (green, red and infrared). The sampling frequency is 75 Hz with 16-bit data output. The device also has 3 axis accelerometers; however, in this study the accelerometer is not used. The device is powered by a 60 mAh Li-on rechargeable battery and connected to the host by BLE.

The system consists of multiple components including transimpedance amplifier and low-pass filter to filter the noise and to prevent converting noise into the ADC (Fig. 32.4). The system is using Bluetooth 5.0 for communication with other devices.

- **Transimpedance Amplifier (TIA)**: TIA is one of the essential electronic blocks that converts the current input generated by a photodiode into a voltage output signal.
- **Low-Pass Filter**: The purpose of low-pass filter is to remove unwanted signal components and attenuate high-frequency noise that may present in the amplified output.
- **Analog Digital Converter**: This component converts continuous analog signal to discrete digital values which then can be processed and stored digitally.

### *32.3.2 Web Application Design*

A web application was developed to monitor continuously the data obtained from the in-ear PPG. The acquired PPG was displayed in real time wirelessly using Bluetooth communication. The web application algorithm has the ability to calculate heart rate, respiration rate, storing and receiving data. The software was developed using Java programming. The web application has the ability to transfer the data from the subject to a server and analyze the signal to identify any atrial fibrillation in the signal.

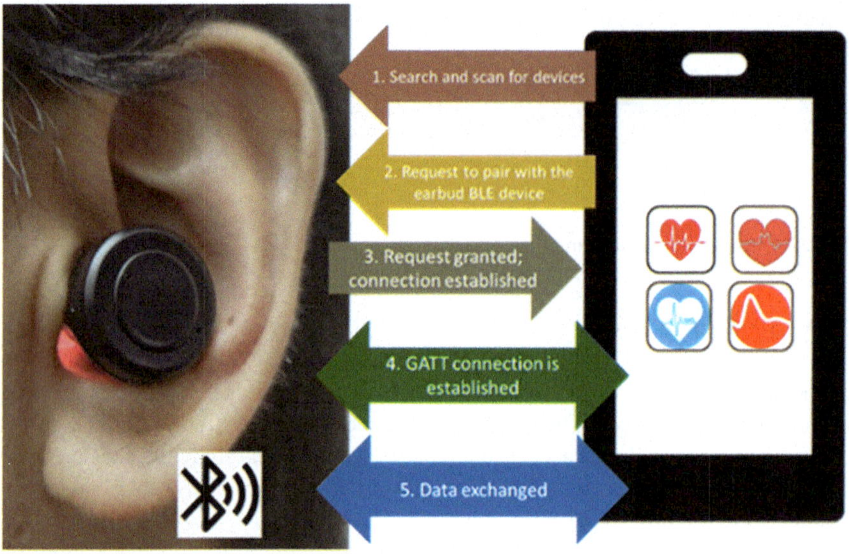


**Fig. 32.5** Web application and earbud BLE device connection

### *32.3.3 Proposed Method*

In the development of this system, the software technology used is Angular. The system is made according to the connection as shown in Fig. 32.5 with the data transferred by BLE.

A connection between the Bluetooth module from the earbuds to the web application is set up as shown in Fig. 32.5. We utilized Node Package Manager (NPM), a package manager that provides an API for BLE communication. This Application Programming Interface (API) on the server serves as a bridge between the web application and the device. This API handles the requests such as establishing connection, sending and receiving data and commands between the earbud BLE device and the web application. This API handles the data formatting to ensure the data exchanged between the web application and earbud is correct.

In the web application, the application initiates a search and scan for available devices in range. The users require Bluetooth permission to access earbud's BLE functionality. A requestDevice() method is used to request to pair with a device, which the user will choose the Bluetooth device from the prompt window. A connection is established when the user selects the device from the listed available devices (see Fig. 32.4) and this method returns a 'Promise' to a 'BluetoothDevice' object representing the selected device.

A Generic Attribute Profile (GATT) connection is established where the specification on how the data exchange and communicate between the web application and the earbud. device.gatt.connect() method is a method used to establish the GATT connection (Fig. 32.6).

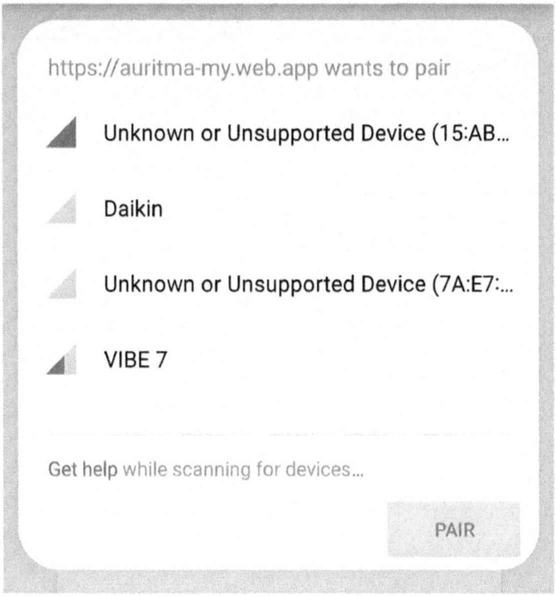


**Fig. 32.6** Example list of devices from the requestDevice() method

Once the GATT connection is established, services and characteristics from the earbud device are available to access. The gatt.PrimaryService() and gatt.getCharacteristic() methods are the methods used to access the specific services and characteristics. We accessed the Universally Unique Identifier (UUID) and heart rate characteristics from the earbud by using the gattCharacteristics.readValue() method to read the data.

The experiment was conducted in a controlled laboratory environment, with the objective of assessing the device communication with the web application. During the experiment, the subject was deliberately instructed to maintain in a resting position, minimizing any unnecessary head and mouth movement.

## 32.4 Results and Discussion

The data from the UUID is captured and saved in a text file. The data is in range of 0–65,535 which refers to the absorption of light reflected by the ear. The raw signal of the data captured from the earbud is shown in Fig. 32.7. The signal will be pre-preprocessed to extract relevant features for atrial fibrillation detection.

As a result of these controlled conditions, we were able to successfully capture a stable and consistent raw PPG signal. However, it is important to note that in real-world scenarios, motion artifacts can significantly impact signal quality. Even minor movements have the potential to introduce interference and compromise signal stability.

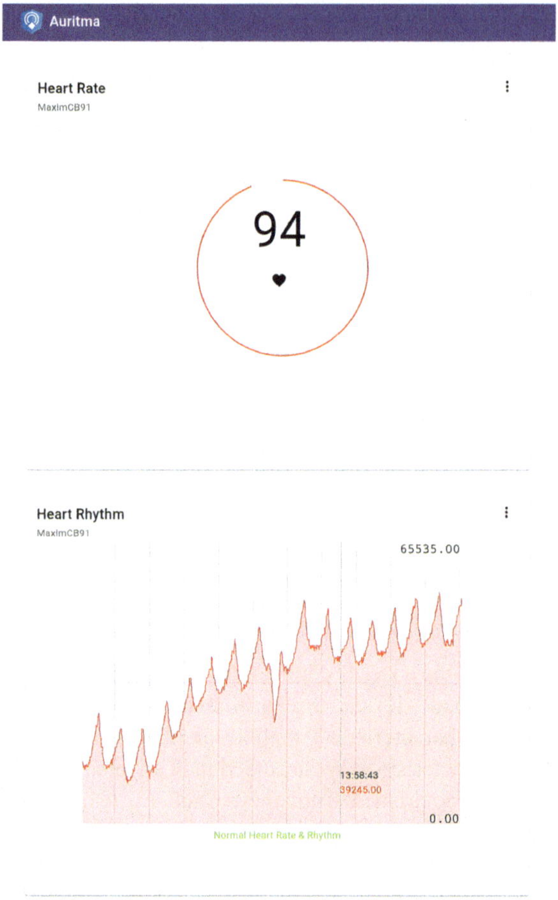


**Fig. 32.7** Raw signal captured in the web application

The PPG waveform obtained from the ear canal has distinct differences in morphology when compared to the PPG pulse derived from the fingertip. The dicrotic notch in the ear PPG signals is nearly imperceptible, and there is a noticeable disparity in the amplitude of the PPG waveform when comparing the in-ear PPG sensor to the finger PPG. The AC amplitude of the in-ear PPG signal is observed smaller than the finger PPG.

Following the acquisition of the raw PPG signal, subsequent processing steps will be carried out to calculate various physiological parameters, including heart rate, breathing rate, heart rate variability (HRV) and the detection of atrial fibrillation.

Noise and artifacts can impact the reliability of data acquired. Some common sources of noise and artifacts in in-ear PPG include motion artifacts, skin contact and baseline noise. In-ear PPG is susceptible to motion artifacts, particularly those related to head and jaw movements. While the in-ear PPG signal is less sensitive to activities like walking or running, jaw movement can introduce inconvenient disturbances into the signal, affecting its accuracy. Proper skin contact between the in-ear PPG sensor and the skin is pivotal for precise measurements. Inadequate contact can result in baseline noise, making it challenging to extract meaningful physiological data.

To mitigate these challenges, the development of an in-ear PPG device that fits securely within the ear canal, acting as both a sensor and a stabilizing support, can enhance skin contact and reduce motion artifacts. This innovation holds promise in improving the accuracy and reliability of in-ear PPG measurements.

Another effective step to mitigate the limitation in noise is by considering Adaptive Noise Cancellation. Adaptive Noise Cancellation has been shown to be effective in reducing the impact of noise on the in-ear PPG signal. This method requires the processing of accelerometer signal which then combined with the LED signal [3, 10].

The well-known limitations in PPG signal that can impact the signal accuracy such as skin pigmentation and ambient light remain challenges for the in-ear PPG. Dark skin tone can act as a filter that attenuates the PPG signal. The attempt to acquire reliable and accurate signal from darker skin pigmentation remains a challenge. Normal PPG signal is susceptible to ambient light sources. The ambient light can interfere with PPG signal acquisition and affect the quality of PPG signal recordings. However, the in-ear PPG might be able to eliminate the ambient light interference. The PPG sensors are located inside the ear canal which is dark and not surrounded by light. This is beneficial for acquiring accurate and quality PPG signal.

## 32.5 Conclusion

In conclusion, this paper explores a novel approach by investigating the practicality of capturing wireless in-ear PPG signals using Bluetooth Low Energy (BLE) technology and seamlessly integrating them into a web application. The primary objective of this study is to explore the potential of these in-ear PPG signals for the detection of atrial fibrillation. By processing the acquired data within the web application, the aim is to identify and confirm the presence of atrial fibrillation, thereby contributing to advancements in non-invasive cardiovascular health monitoring. The use of in-ear PPG with BLE connectivity for atrial fibrillation detection real-time measurement explores the possibility for individuals with heart problems and risk of heart disease as well as physically active people. Additionally, the earbud that is miniature in size and low power consumption enhances its suitability for broad user base.

While this study marks a significant step forward, there remain several limitations. These limitations primarily stem from factors such as poor sensor contact and abrupt amplitude variations caused by movement. Another notable challenge is the presence of baseline noise, which is characterized by low-frequency interference in the in-ear PPG signal. Fortunately, when in a resting state, this low-frequency noise can be effectively mitigated through the application of frequency filters, as its spectral content typically remains below 1 Hz. Understanding and addressing these limitations are crucial for optimizing the accuracy and reliability of PPG ear monitoring systems in various applications. Further refinement of signal processing algorithms, including the incorporation of the Active Noise Cancellation (ANC) algorithm to effectively reduce motion artifacts is required for further improving the accuracy of detection.

This paper can only be considered as the first step for the application of in-ear PPG for atrial fibrillation, and there are a lot more important tasks to obtain the desirable result.

**Acknowledgements** The authors acknowledge the Fundamental Research Grant Scheme (FRGS), Grant Number FRGS/1/2020/TK0/UKM/02/14, funded by the Ministry of Higher Education (MOHE), Malaysia.

## References

1. A. Pedrana, D. Comotti, V. Re, G. Traversi, Development of a wearable In-Ear PPG system for continuous monitoring. IEEE Sens. J. **20**(23), 14482–14490 (2020)
2. B. Venema, V. Blazek, S. Leonhardt, In-ear photoplethysmography for mobile cardiorespiratory monitoring and alarming, in 2015 IEEE 12th *International Workshop on Wearable and Implantable Body Sensor Networks* (BSN 2015)
3. H.J. Davies, P. Bachtiger, I. Williams, P.L. Molyneaux, N.S. Peters, D.P. Mandic, Wearable In-Ear PPG: Detailed Respiratory variations enable classification of COPD. IEEE Trans. Biomed. Eng. **69**(7), 2390–2400 (2022)
4. K. Azudin, K.B. Gan, R. Jaafar, M.H. Ja'afar, The principles of hearable photoplethysmography analysis and applications in physiological monitoring—a review. Sensors **23**(14), 6484 (2023)
5. K. Budidha, P.A. Kyriacou, In vivo investigation of ear canal pulse oximetry during hypothermia. J. Clin. Monit. Comput. **3**(1), 97–107 (2018)
6. K. Budidha, P.A. Kyriacou, Investigation of pulse transit times utilizing multisite reflectance photoplethysmography under conditions of artificially induced peripheral vasoconstriction, in *Annual International Conference on IEEE Engineering in Medicine and Biology* (2014), pp. 1965–1968
7. K. Budidha, P.A. Kyriacou, The human ear canal: investigation of its suitability for monitoring photoplethysmographs and arterial oxygen saturation. Physiol. Meas. **35**(2), 111 (2014)
8. B. Venema, N. Blanik, V. Blazek, H. Gehring, A. Opp, S. Leonhardt, Advances in reflective oxygen saturation monitoring with a novel in-ear sensor system: results of a human hypoxia study. IEEE Trans. Biomed. Eng. **59**(7), 2003–2010 (2012)
9. S. Vogel, M. Hülsbusch, T. Hennig, V. Blazek, S. Leonhardt, In-ear vital signs monitoring using a novel microoptic reflective sensor. IEEE Trans. Inf. Technol. Biomed. **13**(6), 882–889 (2009)
10. S. Vogel, M. Hülsbusch, D. Starke, S. Leonhardt, In-ear heart rate monitoring using a micro-optic reflective sensor, in *Annual International Conference on IEEE Engineering in Medicine and Biology* (2007), pp. 1375–1378
11. T. Wartzek, et al., Analysis of heart rate variability with an in-ear micro-optic sensor in view of motion artifacts, in *6th International Workshop on Wearable and Implantable Body Sensor Networks* (2009), pp. 168–172
12. M.Z. Poh, N.C. Swenson, R.W. Picard, Motion-tolerant magnetic earring sensor and wireless earpiece for wearable photoplethysmography. IEEE Trans. Inf. Technol. Biomed. **14**(3), 786–794 (2010)
13. M.Z. Poh, K. Kim, A. Goessling, N. Swenson, R. Picard, Cardiovascular monitoring using earphones and a mobile device. IEEE Pervasive Comput. **11**(4), 18–26 (2012)
14. N. Mahri, et al., Best pulse selection of photoplethysmography signal through comparative method, in *2012 International Conference on Biomedical Engineering* (2012)

15. N. Mahri, K.B. Gan, M.A.M. Ali, Extracting features similar to QT interval from second derivatives of photoplethysmography: a feasibility study, in *IECBES 2014 Conference on Biomedical Engineering Science* (2014), pp. 470–473
16. B. Venema, J. Schiefer, V. Blazek, N. Blanik, S. Leonhardt, Evaluating Innovative in-ear pulse oximetry for unobtrusive cardiovascular and pulmonary monitoring during sleep. IEEE J. Transl. Eng. Heal. Med. **1**, 2700208–2700208 (2013)
17. B. Venema, H. Gehring, I. Michelsen, N. Blanik, V. Blazek, S. Leonhardt, Robustness, specificity, and reliability of an in-ear pulse oximetric sensor in surgical patients. IEEE J. Biomed. Heal. Inform. **18**(4), 1178–1185 (2014)
18. A.B Hertzman, J.B. Dillon, Applications of photoelectric plethysmography in peripheral vascular disease. Am. Heart J. **20**, 750–761 (1940). https://doi.org/10.1016/S0002-8703(40)90534-8
19. J. Park, H.S. Seok, S.S. Kim, H. Shin, Photoplethysmogram analysis and applications: an integrative review. Front. Physiol. **12**, 2511 (2022). https://doi.org/10.3389/FPHYS.2021.808451/BIBTEX
20. M. Bolanos, H. Nazeran, E. Haltiwanger, Comparison of heart rate variability signal features derived from electrocardiography and photoplethysmography in healthy individuals, in *Annual International Conference on IEEE Engineering in Medicine and Biology*, pp. 4289–4294 (2006)
21. S. Lu et al., Can Photoplethysmography Variability serve as an alternative approach to obtain heart rate variability information? J. Clin. Monit. Comput. **22**(1), 23–29 (2007)
22. A.R. Kiselev, et al., A comprehensive assessment of cardiovascular autonomic control using photoplethysmograms recorded from the earlobe and fingers. Physiol. Meas. **37**(4), 580 (2016)
23. A. Ferlini, A. Montanari, C. Min, H. Li, U. Sassi, F. Kawsar, In-Ear PPG for Vital Signs. IEEE Pervasive Comput. **21**(1), 65–74 (2022)
24. H.J. Davies, I. Williams, N.S. Peters, D.P. Mandic, In-ear SpO2: a tool for wearable, unobtrusive monitoring of core blood oxygen saturation. Sensors **20**(17), 4879 (2020)

# Chapter 33
# A Review on GBAS: Ionospheric Effects at Low-Latitude Regions

**Brelveenraj Kaur Rajwant Singh, Aiffah Mohd Ali, Siti Aminah Bahari, and Mardina Abdullah**

**Abstract** The emergence of the Ground-Based Augmentation System (GBAS) has provided a promising solution for enhancing the accuracy and reliability of GNSS-based positioning, particularly in precision landing and approach procedures. However, the presence of the ionosphere, an upper atmospheric layer surrounding the Earth, introduces significant disruptions towards GNSS signals, specifically in low-latitude regions. Disturbances in the ionosphere, resulting from solar activity and geomagnetic storms, can lead to range errors and ionospheric scintillations, which in turn have adverse effects on the performance of GBAS receivers. Consequently, gaining a comprehensive understanding of the ionospheric influences on GBAS receivers in low-latitude regions holds utmost importance in developing robust and dependable positioning solutions. This review aims to thoroughly examine the impact of ionospheric effects on GBAS receivers in low-latitude areas, primarily focusing on addressing the challenges posed by ionospheric disturbances and exploring potential mitigation techniques. Through an in-depth analysis of existing studies and advancements in this field, this review provides valuable insights and practical recommendations for enhancing GBAS receiver performance in low-latitude regions.

B. K. Rajwant Singh · A. Mohd Ali (✉)
School of Aerospace Engineering, Universiti Sains Malaysia (USM), Engineering Campus, 14300 Nibong Tebal, Pulau Pinang, Malaysia
e-mail: aiffah@usm.my

S. A. Bahari · M. Abdullah
Space Science Centre (ANGKASA), Institute of Climate Change, Universiti Kebangsaan Malaysia (UKM), 43600 UKM Bangi, Selangor, Malaysia

M. Abdullah
Department of Electrical, Electronic and Systems Engineering, Universiti Kebangsaan Malaysia (UKM), 43600 UKM Bangi, Selangor, Malaysia

 
M. T. Islam et al. (eds.), *Proceedings of the 8th International Conference on Space Science and Communication*, Springer Proceedings in Physics 303,
https://doi.org/10.1007/978-981-97-0142-1_33

## 33.1 Introduction

Ground-Based Augmentation System, also referred to as GBAS, plays a crucial role in the aviation industry by significantly enhancing the safety, efficiency, and capacity of aircraft operations. GBAS proves particularly valuable in regions characterized by difficult topography, limited visibility, or congested airspace. By improving the precision and reliability of satellite-based navigation for aircraft, GBAS strengthens safety and operational capabilities in aviation. In essence, GBAS is extensively employed to optimize aircraft operations, ensuring enhanced safety and efficiency in diverse challenging environments within the aviation industry.

Implementing GBAS faces challenges posed by the ionosphere. In typical circumstances, this atmospheric layer introduces signal delays from the Global Navigation Satellite System (GNSS), which vary based on the geographical region. Ionospheric anomalies can have a negative impact on GNSS signals, frequently emerging as quick changes in both amplitude and phase. As the signals travel through ionospheric anomalies, the equatorial plasma bubble (EPB), which regularly forms over low-latitude regions after sunset, causes GNSS scintillations [1]. Besides its scientific significance in ionospheric physics, studying ionospheric irregularities has gained greater importance due to their potential impact on the signal strength of trans-ionospheric links. This is particularly relevant for navigation systems utilizing Global Navigation Satellite Systems (GNSS), including GPS, as these irregularities can lead to signal fluctuations [2].

This paper presents a review of ionospheric effects on GBAS in low-latitude regions. In Sect. 33.2, GBAS operation and the presence of GBAS in different countries are explained. The ionosphere and it's effects at low latitudes are discussed in Sect. 33.3. The effects of ionosphere on GBAS at low latitudes are reviewed in detail in Sect. 33.4. A summary and perspective are given in Sect. 33.5.

## 33.2 GBAS Operation

Ground-Based Augmentation System (GBAS) is designed to enhance the accuracy, integrity, and availability of Global Navigation Satellite Systems (GNSS) for aircraft during the flight's approach, landing, and departure phases.

In GBAS architecture, three or four reference receivers are placed on or close to airport premises, making up the GBAS ground segment [3]. These receivers continuously track the system's performance by constantly monitoring navigation satellite signals and transmitting parameters like pseudorange to a central processing centre which computes corrections for each satellite signal. The system's capabilities encompass computing satellite characteristics, tracking signal integrity, and enabling users to assess signal availability based on specific needs and satellite configurations [4]. This assessment is facilitated through a VHF data broadcast (VDB) transmitter, with the processing unit generating sequential navigation messages intended for

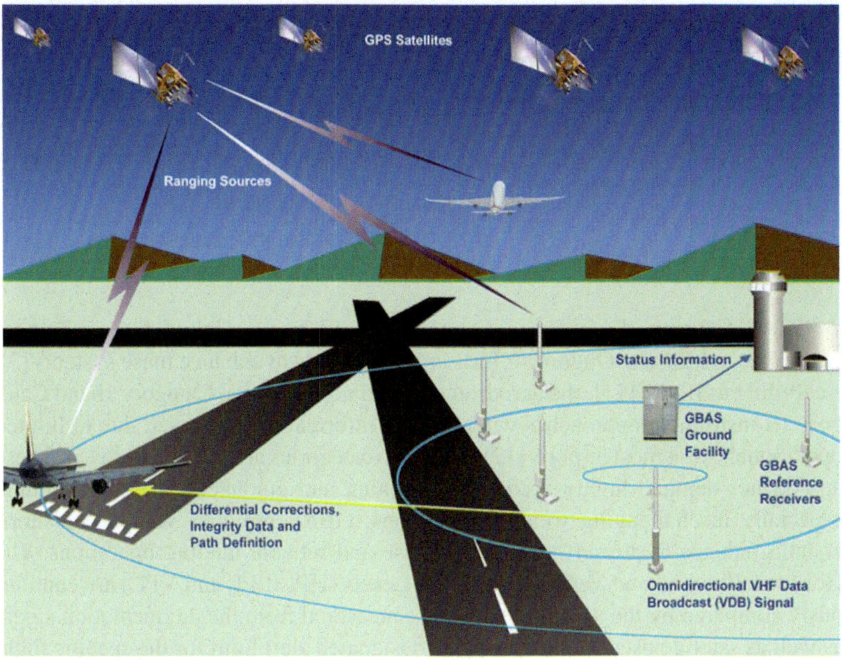


**Fig. 33.1** GBAS architecture [6]

aircraft within a 23 nm radius of the airport [5]. To illustrate the GBAS operation, Fig. 33.1 depicts the GBAS architecture.

The purpose of GBAS is to serve as a precise approach and landing assistance system with the intention of replacing the widely used traditional approach Instrument Landing System (ILS) [7]. GBAS offers a significant advantage compared to ILS due to its capability of supporting multiple runway ends with just one station. To ensure integrity, the current single-frequency GBAS employs position-domain geometry screening as a means to remove potentially hazardous satellite geometries. This is achieved by increasing one or more broadcast GBAS parameters. GBAS also mitigates aircraft disturbances within terminal areas and reduces delays brought on by unfavourable weather. This is helpful for airports that have trouble deploying ILS due to site restrictions [8].

Marini et al. [9] discuss that GBAS offers several benefits over ILS. These include the ability to use a single ground facility to provide information for multiple runways at one aerodrome. There is option to adjust the approach glidepath to accommodate multiple touchdown points on a single runway and there is capability to execute curved approaches. In simpler terms, GBAS enables precise approaches while maintaining the flexibility of GNSS-based navigation procedures.

Another advantage of GBAS is its capability to support Category I, II, and III precision approaches. Table 33.1 lists the different types of precision approach categories [4].

**Table 33.1** GBAS precision approach categories [4, 10]

| Approach Categories | Decision Height (DH) | Runway Visual Range (RVR) |
|---|---|---|
| Category I | > 60 m (200 ft) | > 550 m |
| Category II | 0–60 m (200 ft) | > 350 m |
| Category IIIA | 0–30 m (100 ft) | > 200 m |
| Category IIIB | 0–15 m (50 ft) or not DH | 50–200 m |
| Category IIIC | No DH | No RVR limitation |

GBAS Category I systems are currently used in the United States, Europe, Australia, and Russia. Ongoing efforts are being made to enhance these systems [3]. According to ICAO [11], the aerodrome should not authorize Category II and Category III instrument approaches without RVR information. However, due to limited operational use at most airports [12], these two categories are rarely employed. These approaches demand highly accurate positioning and guidance for a safe landing, especially in challenging weather conditions. GBAS provides vertical and horizontal guidance to aircraft, enabling precise and reliable landing operations. The Horizontal/Lateral and Vertical Protection Levels (HPL/LPL and VPL) are continuously compared by the avionics, which are measured from the augmentation signal as well as satellite pseudorange, with the designated alert limit for the specific flight phase. If either the vertical or horizontal limit is surpassed, the pilot is alerted. When conducting a precision approach, the GBAS receiver on the aircraft solely utilizes satellites that provide corrections [8].

### *33.2.1 Presence of GBAS in Different Countries*

At present, commercially accessible ground stations are operational in multiple global airports, namely Newark and Houston in the United States, Bremen and Frankfurt in Germany, Sydney in Australia, Malaga in Spain, Zurich in Switzerland, and Chennai in India [10]. These ground stations facilitate Category I weather condition approaches, aligning with the GBAS Approach Service Type (GAST) C as per GBAS terminology [13]. Table 33.2 shows the implementation of GBAS around airports globally.

These systems enable precise approaches, landings, and departures, supporting safe navigation in adverse weather conditions and minimizing delays. The widespread adoption of GBAS highlights its significant impact on global aviation, ensuring reliable and accurate positioning for aircraft operations.

**Table 33.2** GBAS Implementation in Airports

| Country | Airports |
|---|---|
| United Stated | Newark Liberty Airport and Houston Intercontinental Airport |
| Germany | Bremen and Frankfurt Airport |
| Australia | Sydney Airport |
| Spain | Malaga Airport/Pablo Ruiz Picasso Airport |
| Switzerland | Zürich Airport |
| India | Chennai International Airport |

## 33.3 The Ionosphere

The ionosphere is situated between 60 and 600 km above the Earth's surface and is a frequently uneven region that has the potential to impact satellite signals passing through its layers. The structure of the ionosphere is primarily shaped by the changing electron density over time and throughout the day. This variability is influenced by solar and geomagnetic activity on Earth [14]. The role of the ionosphere in radio communication is crucial as it allows radio communication to take place over a large distance [15].

These findings demonstrated that ion density increased with altitude, though not uniformly. Throughout the day, there are usually three altitudes where the distribution of ion density reaches zero. These regions were named the E layer, F1 layer, and F2 layer, with their maximum ion density occurring at slightly above 100 km, approximately 200 km, and around 300 km, respectively. The D zone which is located beneath the E layer is considered to be responsible for radio wave absorption [16]. The ionosphere exhibits variability due to various factors such as local time, season, altitude, geographic and geomagnetic location, and solar activity. Only the F2 and E regions are present at night because higher altitudes have a slower rate of ion and electron recombination, which leads to a lower density [17]. This is due to greater ionization during periods of heightened solar activity, which results in a thicker and denser ionosphere. This impact is especially noticeable at the solar maximum portion of the 11-year solar cycle.

### *33.3.1 Ionosphere Effects at Low Latitude*

Solar activity and geomagnetic storms induce ionospheric disturbances, resulting in reduced accuracy of Global Navigation Satellite System (GNSS) positioning. The impact of space weather on GNSS ionospheric models is particularly significant in regions of intense ionospheric activity, predominantly at lower latitudes, leading to substantial perturbations.

One commonly known ionospheric irregularity that occurs frequently in the equatorial region is the Rayleigh–Taylor instability (RTI) [17]. Equatorial Plasma Bubbles (EPB) are frequently observed after sunset in the F region at altitudes ranging from 200 km to over 1000 km. This frequently occurs in the low-latitude ionosphere, especially just before sunset, wherein plasma diffusion is forced downward by plasma cooling and plasma rises upwards by an eastward-oriented electric field. This process is primarily responsible for the equatorial and low-latitude scintillations [2, 18].

Another perturbation is ionospheric scintillation. It is defined as fluctuations in radio signal amplitude and phase as they travel through plasma irregularities within the ionosphere. The small-scale electron density irregularities in the ionosphere give rise to temporary changes [19]. Intense scintillation may clead to loss of lock on radio signals [20]. Research indicates that ionospheric scintillation occurs primarily in equatorial, low, and polar latitudes, characterized by high ionospheric activity.

## 33.4 Effects of Ionosphere on GBAS at Low Latitude

GBAS's users face integrity threats due to severe ionospheric activity that results in extreme ionospheric anomalies. Although the majority of significant spatial gradients in slant ionospheric delay have been detected on high-elevation satellites, there were also instances of extreme gradients examined on satellites with elevations less than 15° [21]. As a consequence, the operation of GBAS can become unsafe since a single ground station is incapable of real-time identification of all the potentially hazardous gradients [22]. Figure 33.2 shows the impacts of the ionosphere on GBAS operations.

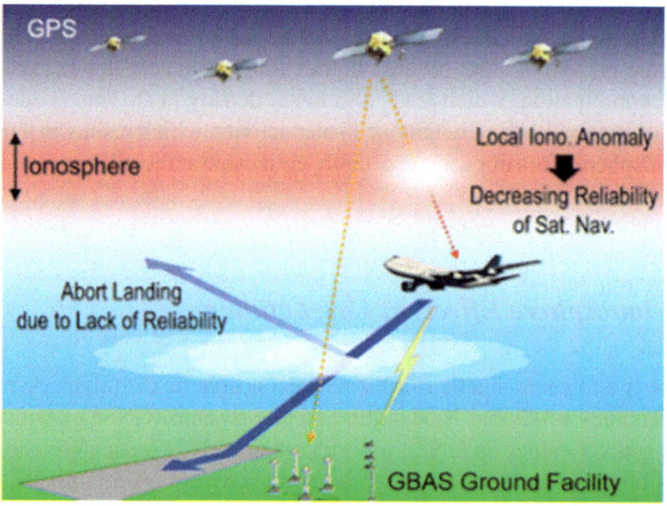


**Fig. 33.2** Effects of the ionosphere on GBAS operations

Ammana et al. [23] conducted a study on the IRNSS S-band signal for GBAS implementations across ionospheric storm events. The investigation took place in low-latitude regions in Hyderabad, India, involved computing a GBAS parameter called sigma-vertical-ionospheric gradient ($\sigma_{vig}$). The study compared the results under calm and stormy ionospheric conditions and found that the $\sigma_{vig}$ with S1 was only 5.08 mm/km, while L5 had 22.80 mm/km. Additionally, the vertical protection levels (VPLs) were lower by 0.65 m compared to L5 and remained within the alert limits. This research is relevant as GBAS implementation is being planned at various low-latitude airports. Meanwhile, in Indonesia, Supriadi et al. [24] proposed a satellite-pair method that is shown to be more appropriate for the Continuous Operating Reference Station (CORS) network over Indonesia located in Java and Sumatra islands and yielded an overall value of $\sigma_{vig}$ (5.21 mm/km).

In another study, Yoon et al. [25] proposed several boundaries to describe the calculation of tropical plasma bubbles (EPBs). The review assessed a greatest ionospheric defer drop of 35 m and a change zone going from 20 to 450 km for EPBs moving towards the east and lined up with the geomagnetic equator at speeds somewhere in the range of 40 and 250 m/s. These boundaries are essential for the improvement of a GBAS ionospheric moderation and well-being case to get functional endorsement in Brazil and other low-scope locales.

Moreover, Rungraengwajiake et al. [26] applied the algorithms from the International Civil Aviation Organization (ICAO) Standards and Recommended Practices (SARPs) to compute GBAS protection levels. They used the GNSS data recorded at the Suvarnabhumi International Airport (AERO) in 2014. The findings from 2014 revealed the occurrences of ionospheric irregularity on various days, along with the corresponding levels of protection. At an altitude of 100 feet, the average availability during the equinox month of March is 99.30%, while during the solstice month of July, it is 99.93%. Notably, the average availability in September during the equinox is even higher at 99.98%.

Furthermore, Yoon et al. [27] proposed an innovative mitigation approach for ionospheric disturbances with Monte Carlo simulation to randomize ionospheric scenarios by generating parameter combinations within a threat model and assessing the collective impacts. By crediting a previous likelihood of an excessive EPB, this technique derives the inflated integrity parameters required to fulfil the safety criterion in the probabilistic definition. The study demonstrated that, by applying a data-driven prior probability of 10–5, the system availability for Category I precision approaches dramatically improved to 89.6% using this method. This work has implications for the development of effective ionospheric mitigation strategies in GNSS applications.

Besides that, Cashin et al. [28] developed a GBAS Ionospheric Threat Model (GITM) by analysing local ionospheric abnormalities identified by two ground stations specifically for Singapore Changi Airport. Data collected from 2011 through early 2017 showed that EPBs were responsible towards the detected anomalous gradients surpassing the threshold of 150 mm/km with a maximum slant ionospheric gradient observed of 526 mm/km on 26 March 2012. The results for gradient, velocity, tilt angle, width, and depth exhibited a close resemblance to findings from previous studies conducted in low-latitude regions.

In another study, Marini-Pereira et al. [29] proposed a firsthand method for estimating the α-μ coefficients constructed on the α-μ autocorrelation function in this work. The suggested technique has been proven using field data and by correlating the empirical autocorrelation function findings with those from the moment-based estimator. The study also demonstrates the method's efficacy by comparing measurements of Level-Crossing Rate (LCR) and Average Fading Duration (AFD) to their theoretical formulations. This work has implications for the development of more accurate and efficient methods for estimating α-μ coefficients in GNSS applications.

Meanwhile all other research showed some significance with other studies, Sanchez et al. [30] found otherwise. They used a specialized automated method for low latitudes to study ionospheric scintillation. Analysing data from dual-frequency receivers across all available days, regardless of geomagnetic indices, they found events not limited by the CONUS threat model. This highlights a need for a broader ionospheric threat model in Ecuador and stresses the insufficiency of geomagnetic indices in identifying relevant anomalies in low-latitude regions. This emphasizes the continuous requirement for vigilant ionospheric monitoring.

Table 33.3 summarizes the discoveries from the findings above.

**Table 33.3** Ionosphere Effects towards GBAS at Low-Latitude Regions

| Authors | Method | Ionospheric Effects | Results |
|---|---|---|---|
| [23] | Utilizing the IRNSS S-band signal for GBAS applications | Ionospheric storm conditions | The VPLs were lower by 0.65 m compared to L5 and remained within the alert limits |
| [24] | Proposed a suitable satellite-pair method for the Continuous Operating Reference Station (CORS) network | Ionospheric gradient | Obtained a total value of $\sigma_{vig}$ (5.21 mm/km) |
| [25] | Monte Carlo simulation was used to randomize ionospheric scenarios by generating parameter combinations within a threat model and assessing the ensemble impacts | Equatorial Plasma Bubbles (EPBs) | The system availability for Category I precision approaches dramatically improved to 89.6% |
| [28] | About 180 anomalous gradient observations were recorded and analysed, encompassing around 50 distinct plasma bubble events | Equatorial Plasma Bubbles (EPBs) | In 2012, the maximum slant ionospheric gradient observed was 526 mm/km |

(continued)

**Table 33.3** (continued)

| Authors | Method | Ionospheric Effects | Results |
|---|---|---|---|
| [29] | Proposed a new method for estimating the α-μ coefficients based on the α-μ autocorrelation function | Ionospheric scintillation, plasma bubbles | The events found above the CONUS threat model took place throughout periods of nominal geomagnetic indices |
| [30] | Used an automated methodology tailored for low latitudes to process data from dual-frequency receiver networks | Equatorial Plasma Bubbles (EPBs) | The insufficiency of geomagnetic indices in indicating significant ionospheric anomalies highlights the importance of continuously monitoring the ionosphere |

## 33.5 Conclusion

This review compared the findings of the ionospheric disturbances in low-latitude regions such as India, Brazil, Singapore, Ecuador, Thailand and Indonesia towards GBAS. The ionospheric effects are due to scintillations, EPBs, ionospheric gradients, and storms. However, further research will be necessary to analyse the behaviour of the ionosphere towards the GBAS at low-latitude regions and to further mitigate the errors for future implementations.

**Acknowledgements** The authors would like to thank everyone who has directly or indirectly provided assistance to support this project. This research was funded by the Ministry of Higher Education Malaysia on Fundamental Research Grant Scheme with Project Code: FRGS/1/2020/TK0/USM/03/8, and the APC was funded by Universiti Sains Malaysia.

## References

1. A. Bumrungkit, P. Supnithi, S. Saito, L.M.M. Myint, A study of equatorial plasma bubble structure using VHF radar and GNSS scintillations over the low-latitude regions. GPS Solut. **26**(4) (2022)
2. P. Abadi, S. Saito, W. Srigutomo, Low-latitude scintillation occurrences around the equatorial anomaly crest over Indonesia. Ann. Geophys. **32**(1), 7–17 (2014)
3. N. Zhu, J. Marais, D. Betaille, M. Berbineau, GNSS position integrity in urban environments: a review of literature. IEEE Trans. Intell. Transport. Syst. **19**(9), 2762–2778 (2018)
4. T. Murphy, T. Imrich, Implementation and operational use of ground-based augmentation systems (GBASs)—a component of the future air traffic management system. Proc. IEEE **96**(12), 1936–1957 (2008)
5. T.J.G. Peter, O. Montenbruck, *Springer Handbook of Global Navigation Satellite Systems* (Teunissen Montenbruck Editors, 2017)
6. ESA, "Navipedia," Jun. 16, 2021. https://gssc.esa.int/navipedia/index.php/File:LAAS_Architecture.png. Last accessed June 01, 2023

7. H. Lee, S. Pullen, J. Lee, B. Park, M. Yoon, J. Seo, Optimal parameter inflation to enhance the availability of single-frequency GBAS for intelligent air transportation. IEEE Trans. Intell. Transport. Syst. **23**(10), 17801–17808 (2022)
8. P. Oliveira, J. Silva, P. Soares, A comparative study between ILS and GBAS approaches: the case of Viseu airfield. J. Airline Airport Manage. **10**(2), 65 (2020)
9. L. Marini-Pereira, S. Pullen, A. de O. Moraes, and J. Sousasantos, "Ground-Based Augmentation Systems Operation in Low Latitudes - Part 1: Challenges, Mitigations, and Future Prospects. Journal of Aerospace Technology and Management 13,(2021).
10. Skybrary," Jun. 13, 2023. https://www.skybrary.aero/articles/gbas-landing-system-gls, last accessed Jun. 15, 2023
11. ICAO, ICAO—Annex 10—Aeronautical Telecommunications, 7th Edition Volume I, (2018)
12. IVAO, IFR Precision Approach and Minima, 2017. https://wiki.ivao.aero/en/home/training/documentation/IFR_non_precision_approach_and_minima. Last accessed June 11, 2023
13. M. Felux, M.-S. Circiu, D. Gerbeth, M. Caamano, Future GBAS Processing-Do we need an ionosphere-free mode?
14. J. Bahru, et al., Total Electron Content (TEC) and Scintillation Measurements
15. R. D. Hunsucker, J.K. Hargreaves, The high-latitude ionosphere and its effects on radio propagation (2003)
16. A.G. Mcnish, Ionospheric Research (1948)
17. Tascione, Introduction to the Space Environment (Orbit, a Foundation Series, 2010)
18. B. Ning, G. Li, K. Liu, Characteristics of GPS ionospheric scintillation and TEC depletion in the Chinese low latitude region, in *2016 URSI Asia-Pacific Radio Science Conference, URSI AP-RASC 2016* (2016), pp. 90–92
19. K. Sahithi et al., Characteristics of ionospheric scintillation climatology over Indian low-latitude region during the 24th solar maximum period. Geod. Geodyn. **10**(2), 110–117 (2019)
20. GBAS safety assessment guidance related to anomalous ionospheric conditions. Adopted by APANPIRG/27 INTERNATIONAL CIVIL AVIATION ORGANIZATION ASIA AND PACIFIC OFFICE (2016)
21. J. Lee, S. Datta-Barua, G. Zhang, S. Pullen, P. Enge, Observations of low-elevation ionospheric anomalies for ground-based augmentation of GNSS. Radio Sci. **46**(6), 1–11 (2011)
22. L. Marini-Pereira, A.O. Moraes, S. Pullen, Advanced warning of threatening equatorial plasma bubbles to support GBAS in low latitudes. IEEE Trans. Aerosp. Electron Syst. 1–14 (2023)
23. S.R. Ammana, M.R. Sujimol, K.K. Songala, A.D. Sarma, Advantage of IRNSS S-band signal for GBAS applications in adverse ionospheric storm conditions. (2022)
24. S. Supriadi, H. Z. Abidin, D.D. Wijaya, P. Abadi, S. Saito, D.U. Prabowo, Construction of nominal ionospheric gradient using satellite pair based on GNSS CORS observation in Indonesia. Earth Planets Space **74**(1) (2022)
25. M. Yoon et al., Equatorial Plasma bubble threat parameterization to support GBAS operations in the Brazilian Region. Navig. J. Inst. Navig. **64**(3), 309–321 (2017)
26. S. Rungraengwajiake, P. Supnithi, S. Saito, N. Siansawasdi, A. Saekow, Ionospheric delay gradient monitoring for GBAS by GPS stations near Suvarnabhumi airport, Thailand. Radio Sci. **50**(10), 1076–1085 (2015)
27. M. Yoon, D. Kim, S. Pullen, J. Lee, Assessment and mitigation of equatorial plasma bubble impacts on category I GBAS operations in the Brazilian region. Navig. J. Inst. Navig. **66**(3), 643–659 (2019)
28. T. Cashin et al., Development and implications of the Singapore GBAS ionospheric threat model (GITM), in *30th International Technical Meeting of the Satellite Division of the Institute of Navigation, ION GNSS 2017* (2017), pp. 1429–1449
29. L. Marini-Pereira, et al., On the field validation of $\alpha$-$\mu$ fading coefficients estimator based on the autocorrelation function for ionospheric amplitude scintillation. Adv. Space Res. **64**(10), 2176–2187 (2019)
30. S. Sánchez-Naranjo, W. Rincón, R. Ramos-Pollán, F.A. González, S. Soley, A comprehensive assessment of ionospheric gradients observed in Ecuador during 2013 and 2014 for ground based augmentation systems. Adv. Space Res. **59**(8), 1992–2006 (2017)

# Chapter 34
# Frontal Plane Gait Assessment Using MediaPipe Pose

**Chang Soon Tony Hii, Kok Beng Gan, Huay Woon You, Nasharuddin Zainal, Norlinah Mohamed Ibrahim, and Shahrul Azmin**

**Abstract** Gait analysis is a vital field in biomechanics, focusing on studying human walking patterns and movement. Traditional methods relied on visual observation and manual measurements, lacking accuracy and objectivity. Instrumented gait analysis improved measurements using specialized equipment like motion capture systems and wearable sensors, but had limitations due to cost, laboratory requirements, and markers or sensors interfering with natural movement. The field underwent a significant transformation with the advent of markerless pose estimation-based gait analysis, which harnessed computer vision and deep learning methodologies to monitor human movement without the requirement for specialized equipment, delivering a cost-effective and easily accessible analysis in real-world environments. However, it still faces challenges, including limitations in the sagittal plane and reliance on computationally demanding models like OpenPose. This study introduced a novel frontal plane gait analysis approach using the lightweight MediaPipe Pose model and a single camera setup. The objective is to evaluate the feasibility and accuracy of MediaPipe Pose by comparing it to the established 3D Vicon motion capture system. The proposed approach tracked body keypoints during gait, detected heel strikes and toe-offs based on ankle depth changes, and calculated gait parameters. The findings demonstrated that MediaPipe Pose-based gait analysis showed promise for accurately analyzing gait parameters in the frontal plane with low mean absolute error (0.04–0.18 s). Combining this method with satellite technology enhances e-health by facilitating teleconsultations, remote diagnostics, and extending healthcare services to disaster-stricken or remote areas.

C. S. T. Hii · K. B. Gan (✉) · N. Zainal
Department of Electrical, Electronic and Systems Engineering, Faculty of Engineering and Built Environment, Universiti Kebangsaan Malaysia, 43600 Bangi, Malaysia
e-mail: kbgan@ukm.edu.my

H. W. You
Pusat GENIUS@Pintar Negara, Universiti Kebangsaan Malaysia, 43600 Bangi, Malaysia

N. M. Ibrahim · S. Azmin
Department of Medicine, Faculty of Medicine, Neurology Unit, Universiti Kebangsaan Malaysia Medical Centre, 56000 Kuala Lumpur, Malaysia

 
M. T. Islam et al. (eds.), *Proceedings of the 8th International Conference on Space Science and Communication*, Springer Proceedings in Physics 303,
https://doi.org/10.1007/978-981-97-0142-1_34

## 34.1 Introduction

Gait analysis is a crucial field in biomechanics that focuses on the study of human walking patterns and movement [1]. It provides valuable insights into various aspects of human locomotion, including balance, coordination, and musculoskeletal function. Traditional gait analysis involved visual observation and subjective assessments by clinicians, which relied heavily on qualitative judgments. This method lacked objective measurements and suffered from inter-observer variability stemming from variations in clinicians' expertise [2]. Furthermore, traditional gait analysis faced constraints in capturing subtle details and providing comprehensive quantitative data related to movement.

Instrumented gait analysis emerged as a more advanced approach, utilizing specialized equipment such as motion capture systems [3, 4], force platforms, wearable sensors [5–7], and cameras. These tools allowed for more precise and unbiased assessments of gait parameters, offering valuable insights into joint angles, ground reaction forces, and muscle activity. Nonetheless, instrumented gait analysis had certain drawbacks, such as the requirement for costly equipment and controlled laboratory environments. Furthermore, affixing markers or sensors to the body might have the potential to impact individuals' natural movement patterns [8].

The advancement in markerless pose estimation-based gait analysis has revolutionized the field of gait analysis. By leveraging computer vision and deep learning techniques, markerless pose estimation algorithms such as OpenPose [9], YOLO-Pose [10], and MediaPipe Pose [11] can accurately track and analyze human movement without the need for markers or specialized equipment. This approach utilizes computer vision algorithms to detect and track key body landmarks, allowing for the estimation of joint angles [12–15], body segment positions, and spatiotemporal gait parameters [14, 16]. Markerless pose estimation-based gait analysis offers several advantages, including cost-effectiveness, accessibility, and the ability to analyze gait in real-world settings without intrusive markers or sensors.

Despite its numerous advantages, markerless pose estimation-based gait analysis still has its limitations. The major constraint is that it confines gait analysis to the sagittal or longitudinal plane, which may be impractical in clinical or home settings due to spatial limitations for video acquisition. Additionally, occlusion risks are higher within the sagittal plane, as keypoints critical for gait analysis may be hidden or obscured by clothing or body parts. Most markerless pose estimation methods heavily rely on the accurate OpenPose model, but this comes at the expense of demanding computational resources.

The present study presents a novel method for analyzing frontal plane gait by leveraging depth information from body keypoints obtained through the lightweight pose estimation model, MediaPipe Pose, and a single camera setup. This study aims to investigate the practicality of utilizing the MediaPipe Pose model for frontal plane gait analysis and to assess its accuracy and effectiveness by comparing it with the widely accepted 3D Vicon motion capture system. The proposed method utilizes the

MediaPipe Pose model to track keypoints on the body during gait analysis. An algorithm is implemented to detect heel strike and toe-off events based on depth changes of the ankle. Gait parameters are then calculated from these events and compared with gait data calculated from the gait analysis software in Vicon system, considered the standard for gait analysis. Integrating this approach with satellite technology enhances e-health capabilities, enabling remote patient monitoring, supporting rehabilitation efforts, facilitating teleconsultations, remote diagnostics, and extending healthcare services to disaster-stricken or remote regions.

## 34.2 Methodology

### *34.2.1 Participants*

A publicly accessible dataset, combining synchronized and calibrated video data from various perspectives with motion capture technology, was employed in this research [17]. The dataset, which comprises 32 individuals, including 10 women and 22 men, can be accessed through the website http://bytom.pja.edu.pl/projekty/hm-gpjatk/ (accessed on 9 September 2022). The dataset comprised healthy young students, with ages ranging from 20 to 25 years, and also encompassed a few academicians whose ages varied from 25 to 62 years, all from the Polish-Japanese Academy of Information Technology (PJATK) at the time of data collection. Individuals with any gait-related injuries or conditions that could potentially impact their participation were excluded from the dataset.

### *34.2.2 Video Data Collection and Processing*

As shown in Fig. 34.1, four calibrated and synchronized digital video cameras: C1 (right view), C2 (front view), C3 (left view), and C4 (back view) were set up to record the walking activity of the healthy participants. The video recordings were acquired at a resolution of 960 × 540 pixels, capturing frames at a rate of 25 frames per second (fps). Participants were instructed to walk in a straight line along a 6.5-m pathway. However, one healthy individual (identified as p16s3) was excluded from the study due to walking diagonally instead of following the prescribed straight path. Consequently, the study included only 31 participants.

For frontal plane gait analysis, videos recorded by camera C4 (back view) and camera C2 (front view) were selected. These choices were made based on their advantages in understanding joint movement, identifying balance and stability issues, and providing a comprehensive assessment of gait mechanics compared to sagittal plane analysis. However, camera C4 (back view) was specifically favored over camera

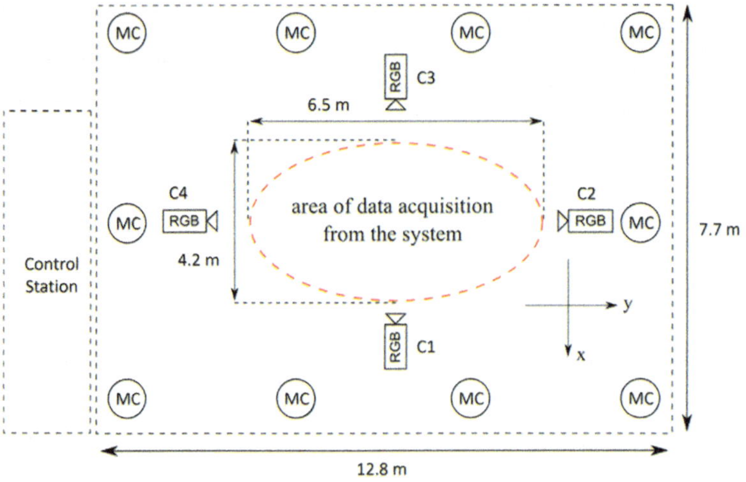


**Fig. 34.1** Video recording setup (where MC stands for the Vicon motion capture camera and RGB denotes the digital camera). The diagram is modified from a source referenced as [17]

C2 (front view) due to the limitations of pose estimation models in accurately tracking body keypoints from the front view.

Simultaneously, the walking sequence was also captured using ten Vicon motion capture cameras. These cameras captured the walking sequence's video frames at a resolution of 2352 × 1728 pixels and a frame rate of 100 fps. The healthy participants were instructed to attach 39 spherical markers on predetermined body keypoints before the walking test (Fig. 34.2). The motion capture system precisely tracked the locations of these anatomical landmarks throughout the walking sessions. The collected location data was subsequently employed in gait analysis tools of the Vicon system to calculate gait parameters, serving as the reference for evaluating the accuracy and reliability of the proposed system.

### *34.2.3 Working Pipeline of Our Proposed System*

Our proposed system was implemented using a Python program running on a laptop featuring a Intel(R) Core(TM) i7-12700H CPU and an NVIDIA GeForce RTX3070Ti Laptop GPU. As depicted in Fig. 34.3, the video from camera C4 (rear view) was initially subjected to processing using the MediaPipe Pose model, precisely tracking the locations of body keypoints. Next, the resulting real-world 3D coordinates of the body keypoints were subjected to both gap-filling, achieved through cubic spline interpolation, and smoothing, accomplished through the application of a Butterworth low-pass filter. By analyzing depth (z-coordinate) variations associated with the ankle joint, the system accurately determined the timings of significant gait events,

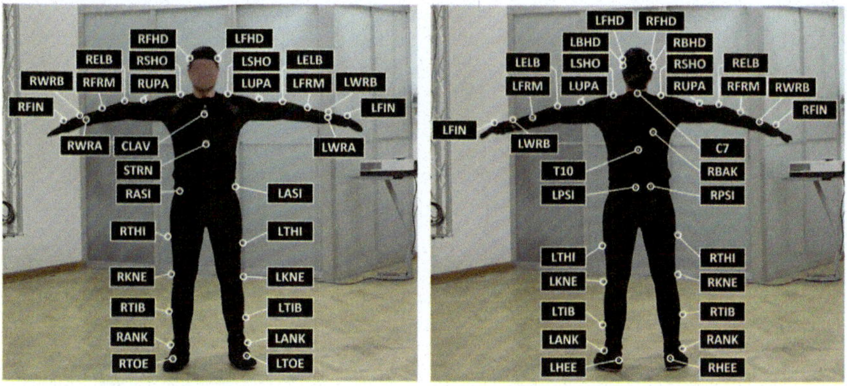


**Fig. 34.2** Positioning of spherical markers on the body keypoints (left: frontal view, right: posterior view). The illustration is adapted from a source cited as [17]

including heel strike and toe-off. Finally, utilizing these identified gait event timings, the system computed various gait parameters, which were then saved in an Excel file for future reference by clinicians.

**Body Pose/Keypoints Tracking using MediaPipe Pose**

MediaPipe Pose [11] is a lightweight machine learning-based solution for real-time detection and tracking of the human body pose. It follows a two-step detector-tracker pipeline. In the detection stage, the model identifies the presence of a human body in an image or video frame and determines the region of interest. The tracking stage utilizes pose keypoints detected in the initial stage as reference points to track and estimate the body's pose over time. By analyzing the temporal coherence of these landmarks, the tracker infers the changes in the body's pose across frames.

The BlazePose GHUM Heavy model was selected for this research due to its superior accuracy in effectively tracking body keypoints. The walking video captured from the back view was initially inputted into the chosen model and executed through a Python program for enabling the tracking of 33 body keypoints' real-world 3D

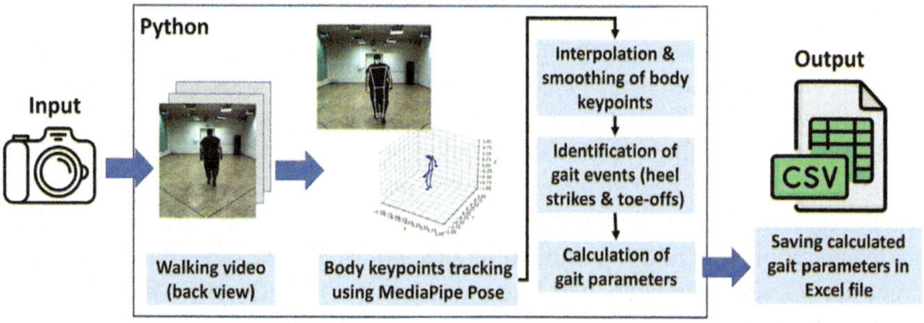


**Fig. 34.3** Working pipeline of our proposed system

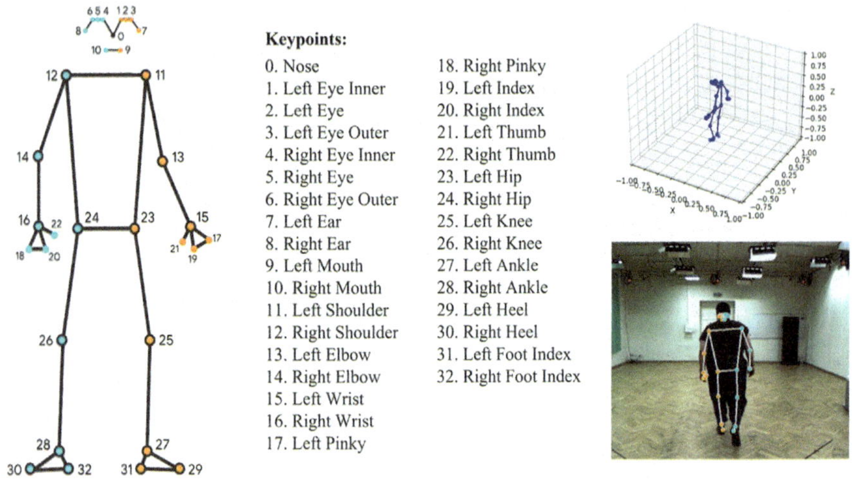


**Fig. 34.4** A 33 body keypoints detected by MediaPipe Pose

coordinates, as shown in Fig. 34.4. Each body keypoint comprises the coordinates data of:

1. x and y: The body keypoints coordinates in meters, relative to the mid-hip origin.
2. z: The depth of the landmark in meters, with the origin set at the mid-hip.

### Interpolation and Smoothing of Body Keypoints

Cubic spline interpolation was employed to estimate values between the known data points of body keypoints extracted using the MediaPipe Pose model, thereby creating a smooth and continuous curve that passed through the given data points. In order to enhance precision and minimize interference, the coordinates of the body keypoints were closely approximated to their actual values through the application of a 10th-order Butterworth low-pass filter, set with a cutoff frequency of 0.8828 Hz.

### Key Gait Events Identification: Heel Strikes and Toe-offs

In Fig. 34.5, the normal gait cycle is visually depicted and comprises two primary phases: the stance phase (consisting of heel strike, loading response, midstance, and terminal stance) and the swing phase (including toe-off, initial swing, mid-swing, and terminal swing). The accurate calculation of gait parameters holds great significance in identifying the timings of heel strike and toe-off events. As depicted in Fig. 34.5, in the gait cycle, the heel strike event takes place when the index toe/ foot index reaches its foremost position forward (indicating the maximum relative depth of the ankle concerning the mid-hip as the reference point). On the other hand, the toe-off event occurs when the index toe/ foot index reaches its furthest position backward (representing the minimum relative depth of the ankle with respect to the mid-hip as the reference point).

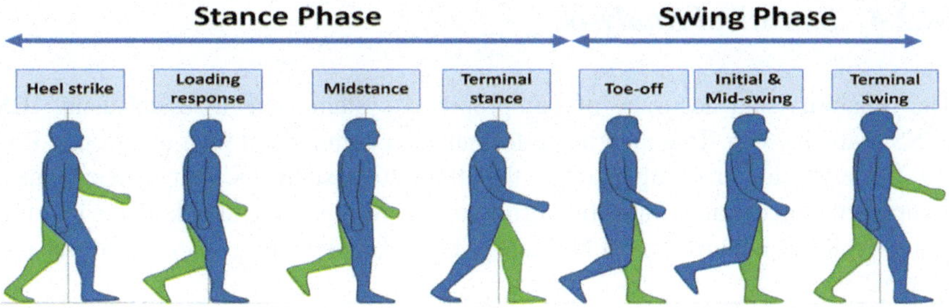


**Fig. 34.5** Normal gait cycle of a healthy individual

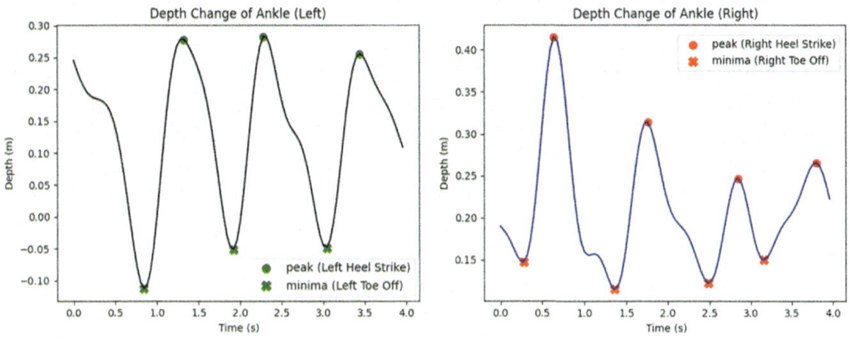


**Fig. 34.6** Graphs of variations in ankle depth over time

To assist in identifying gait events, specifically heel strikes and toe-offs, a graph depicting the changes in ankle depth for both the left and right sides was plotted (Fig. 34.6). The circular markers on the graph indicate the timings of heel strike events, while the cross markers indicate the timings of toe-off events.

**Calculation of Gait Parameters**

The recorded timings of heel strike and toe-off events allowed for the computation of the following gait parameters for each gait cycle:

- Stance time: the duration between heel strike and toe-off of the same leg.
- Swing time: the duration between toe-off and heel strike of the same leg.
- Double support time: the duration between the heel strike of one leg and the toe-off of the opposite leg.
- Step time: the duration between successive heel strikes of both feet.

Following this, the median of the gait parameters was determined and stored in an Excel file for future reference by clinicians.

### 34.2.4 Statistical Analysis

The reliability and accuracy of our proposed system were assessed using IBM SPSS Statistics v26. To gauge its performance in comparison with the ground truth 3D Vicon motion tracking system, the descriptive statistics were carried out to compute the error range, absolute error range, average, and standard deviation of gait parameters acquired from both the Vicon motion capture system and our own system.

## 34.3 Results and Discussions

Based on the results shown in Table 34.1, a comparison between the Vicon motion capture system and our proposed approach (MediaPipe Pose) showed a slight variation in gait parameters for each healthy participant, with an error range of – 0.17 to 0.17 s. The calculated gait parameters, such as stance time, swing time, double support time, and step time, showed average absolute errors spanning from 0.04 to 0.18 s. Notably, the right leg's stance time and swing time displayed the highest error, while the right leg's stance time also exhibited the highest mean absolute error.

**Table 34.1** Summary of descriptive gait parameter statistics for each healthy participant

| Gait parameter (unit) | Leg | mean ± SD (Vicon) | mean ± SD (MediaPipe pose) | mean ± SD (Vicon—MediaPipe pose) | mean ± SD (lVicon—MediaPipe posel) |
|---|---|---|---|---|---|
| Stance time (second) | L | 0.80 ± 0.10 | 0.73 ± 0.12 | 0.07 ± 0.09 | 0.08 ± 0.07 |
| | R | 0.80 ± 0.11 | 0.64 ± 0.16 | 0.17 ± 0.18 | 0.18 ± 0.17 |
| Swing time (second) | L | 0.40 ± 0.04 | 0.47 ± 0.10 | – 0.08 ± 0.08 | 0.08 ± 0.07 |
| | R | 0.39 ± 0.04 | 0.56 ± 0.20 | – 0.17 ± 0.18 | 0.17 ± 0.17 |
| Double support time (second) | L2R | 0.20 ± 0.04 | 0.19 ± 0.07 | 0.01 ± 0.07 | 0.06 ± 0.04 |
| | R2L | 0.20 ± 0.03 | 0.19 ± 0.08 | 0.01 ± 0.07 | 0.05 ± 0.04 |
| Step time (second) | L | 0.60 ± 0.07 | 0.65 ± 0.07 | – 0.05 ± 0.05 | 0.05 ± 0.04 |
| | R | 0.59 ± 0.07 | 0.56 ± 0.08 | 0.03 ± 0.05 | 0.04 ± 0.04 |

## 34.4 Conclusion

The markerless MediaPipe Pose model has emerged as a valuable tool for automating frontal gait analysis, eliminating the need for human involvement. It offers accurate calculation of gait parameters, positioning itself as a reliable alternative to the widely acclaimed Vicon motion capture system. The model's effectiveness in assessing frontal plane gait is supported by consistently low mean absolute error values, ranging from 0.04 to 0.18 s. There is potential to expand the analysis by incorporating additional spatiotemporal gait parameters and evaluating lower joint kinematics. This approach, when combined with satellite technology, greatly enhances e-health capabilities. It enables remote patient monitoring, supports rehabilitation, facilitates teleconsultations and remote diagnostics, and extends healthcare services to disaster-stricken or remote regions.

**Acknowledgements** This research received financial support from the International Collaboration Fund under the Grant Number of IF1020Q1322 provided by the Ministry of Science, Technology, and Innovation (MOSTI) in Malaysia.

## References

1. J. Richards, D. Levine, M.W. Whittle, *Whittle's Gait Analysis*, 6th ed. (Elsevier, Amsterdam, 2022)
2. A.M.L. Ong, S.J. Hillman, J.E. Robb, Reliability and validity of the Edinburgh Visual Gait Score for cerebral palsy when used by inexperienced observers. Gait Posture **28**, 323–326 (2008)
3. H. Hobara, Y. Kobayashi, T. Nakamura, N. Yamasaki, K. Nakazawa, M. Akai, T. Ogata, Lower extremity joint kinematics of stair ascent in transfemoral amputees. Prosthetics Orthot. Int. **35**, 467–472 (2011)
4. P. Fernández-González, A. Koutsou, A. Cuesta-Gómez, M. Carratalá-Tejada, J.C. Miangolarra-Page, F. Molina-Rueda, Reliability of Kinovea® Software and agreement with a three-dimensional motion system for gait analysis in healthy subjects. Sensors **20**, 3154 (2020)
5. M.M. Rahman, K.B. Gan, N.A.A. Aziz, A. Huong, H.W. You, Upper Limb joint angle estimation using wearable IMUs and personalized calibration algorithm. Mathematics **11**, 970 (2023)
6. M.M. Rahman, K.B.Gan, Range of motion measurement using single inertial measurement unit sensor: a validation and comparative study of sensor fusion techniques, in *2022 IEEE 20th Student Conference on Research and Development (SCOReD)*. IEEE (2022), pp. 114–118
7. M.M. Rahman, K.B. Gan, N.A. Aziz, A Review on challenges in telerehabilitation and human activity recognition approaches during covid-19 pandemic. J. Kejuruter **35**, 577–586 (2023)
8. P.-H. Chen, R.-L. Wang, D.-J. Liou, J.-S. Shaw, Gait disorders in Parkinson's disease: assessment and management. Int. J. Gerontol. **7**, 189–193 (2013)
9. Z. Cao, G. Hidalgo, T. Simon, S.-E. Wei, Y. Sheikh, OpenPose: realtime multi-person 2D pose estimation using part affinity fields. IEEE Trans Pattern Anal. Mach. Intell. **43**,172–186 (2019)
10. D. Maji, S. Nagori, M. Mathew, D. Poddar, YOLO-Pose: enhancing YOLO for multi person pose estimation using object keypoint similarity Loss, in *2022 IEEE/CVF Conference on Computer Vision and Pattern Recognition Workshops (CVPRW)*. IEEE (2022), pp. 2636–2645

11. V. Bazarevsky, I. Grishchenko, K. Raveendran, T. Zhu, F. Zhang, M. Grundmann, BlazePose: on-device real-time body pose tracking (2020)
12. C.S.T. Hii, K.B. Gan, N. Zainal, N. Mohamed Ibrahim, S. Azmin, N. Abd Shattar, Marker free gait analysis using pose estimation model, in *2022 IEEE 20th Student Conference on Research and Development (SCOReD)*. IEEE (2022), pp. 109–113
13. A. Viswakumar, V. Rajagopalan, T. Ray, P. Gottipati, C. Parimi, Development of a robust, simple, and affordable human gait analysis system using bottom-up pose estimation with a smartphone camera. Front. Physiol. **12** (2022)
14. J. Stenum, C. Rossi, R.T. Roemmich, Two-dimensional video-based analysis of human gait using pose estimation. PLOS Comput. Biol. **17**, e1008935 (2021)
15. N. Abd Shattar, K.B. Gan, N.S. Abd Aziz, Experimental setup for markerless motion capture and landmarks detection using OpenPose during dynamic gait index measurement, in *2021 7th International Conference on Space Science and Communication (IconSpace)*. IEEE (2021), pp. 286–289
16. C.S.T. Hii, K.B. Gan, N. Zainal, N. Mohamed Ibrahim, S. Azmin, S.H. Mat Desa, B. van de Warrenburg, H.W. You, Automated gait analysis based on a marker-free pose estimation model. Sensors **23**, 6489 (2023)
17. B. Kwolek, A. Michalczuk, T. Krzeszowski, A. Switonski, H. Josinski, K. Wojciechowski, Calibrated and synchronized multi-view video and motion capture dataset for evaluation of gait recognition. Multimed Tools Appl. **78**, 32437–32465 (2019)

# Chapter 35
# A Perfect Metamaterial Absorber for Sensing Application of Edible Oil

**Muhammad Amir khalil, Wong Hin Yong, Mohammad Tariqul Islam, Ahasanul Hoque, Md. Rashedul Islam, and Md. Shabiul Islam**

**Abstract** The reliable characterization of cooking oils poses challenges due to the presence of minor variations in their dielectric behaviors. This research paper introduces a newly developed sensor that utilizes metamaterial technology for the purpose of analyzing cooking oils. The sensor design integrates a square split-ring resonator (S.S.R.R.) and a microstrip transmission line. The design exhibits properties of negative permittivity and permeability within the resonance frequency range. This study presents an assessment of the performance of the sensor's permittivity and absorption coefficient. The sensitivity values corresponding to the frequencies of 3.04 and 3.4 GHz are 0.64% and 0.21%, respectively. The absorption values for the frequencies are 99.9 and 99.7%, correspondingly. The q-factor values corresponding to the frequencies of 3.04 and 3.44 GHz are 1410.29 and 1148.16, respectively. A metamaterial absorber with double negative (D.N.G.) characteristics exhibits the capability to be utilized in various applications, including but not limited to microwave shielding, satellite remote sensing, and the evaluation of dielectric properties in materials.

M. A. khalil · W. H. Yong · Md. S. Islam
Faculty of Engineering (FOE), Multimedia University (MMU), 63100 Cyberjaya, Selangor, Malaysia

M. T. Islam · Md. R. Islam
Department of Electrical, Electronic and Systems Engineering, Universiti Kebangsaan Malaysia, 43600 Bangi, Malaysia

Md. R. Islam
Department of Electrical, Electronic and Communication Engineering, Pabna University of Science and Technology, Pabna, Bangladesh

A. Hoque (✉)
Pusat Sains Angkasa (ANGKASA), Institut Perubahan Iklim, Universiti Kebangsaan Malaysia, 43600 Bangi, Malaysia
e-mail: ahasanul@ukm.edu.my



M. T. Islam et al. (eds.), *Proceedings of the 8th International Conference on Space Science and Communication*, Springer Proceedings in Physics 303,
https://doi.org/10.1007/978-981-97-0142-1_35

## 35.1 Introduction

In recent years, metamaterials have gained widespread interest because of the breadth of their potential electromagnetic applications. They are defined as man-made structures whose properties are derived from the unit structure rather than the materials themselves. A property of this material that has been recently discovered is its ability to absorb electromagnetic waves perfectly. The simplicity of design, ease of fabrication, and perfect broadside absorption capabilities of metamaterial absorbers make them ideal for remote sensing applications that range from microwaves to optics [1]. Although Landy was the first to introduce MM absorbers, numerous designs and applications have since been introduced by other researchers. Metamaterials are widely used for various applications, including cloaking, super lenses, energy harvesting, and antennas [2–4]. This has led to the development of many different sizes, shapes, materials, and geometries. Sensing technology is one of the most prevalent applications of MM absorbers, which provides passive countermeasures for electronic warfare.

We have also made some progress fabricating such structures, primarily electrical metamaterials with sensing applications. For example, adulteration of edible oils is a worldwide problem, especially severe in Asian countries. Edible oil adulteration can be detected using a variety of techniques, such as density measurement, fiber grating sensor technology, emission testing, filter paper method, gas chromatography, and ash contamination determination. MM configurations with split-ring resonators (S.R.R.) are the most common due to their intriguing electromagnetic characteristics. In response to E.M. excitation, the S.R.R. produces an intense magnetic resonance as the inductor stores magnetic energy and the capacitor stores electrical energy. Numerous recent studies have demonstrated that metamaterials can be used efficiently as electrochemical sensors.

This paper aims to present a novel metamaterial with a high potential for electromagnetic (E.M.) absorption under its perfect D.N.G. property, which guarantees the trapping of propagated waves. Furthermore, the proposed absorber also has a wide bandwidth within a specified frequency band for sensing purposes.

### *35.1.1 Metamaterial Absorber Design and Theory*

The design is based on an FR-4 substrate with a thickness of 1.575 mm, dielectric constant of 4.3, and tangent loss of 0.025 mm [5]. Figure 35.1 depicts an absorber structure in the form of a square split-ring resonator (S.S.R.R.) unit cell developed on FR-4 substrate material. The FR-4 composite material consists of epoxy resin and woven fiberglass filaments and exhibits both flame resistance and self-extinguishing properties [6]. In order to analyze the structural integrity of the absorber, however, a finite-integration technique (F.I.T.)-based simulation method

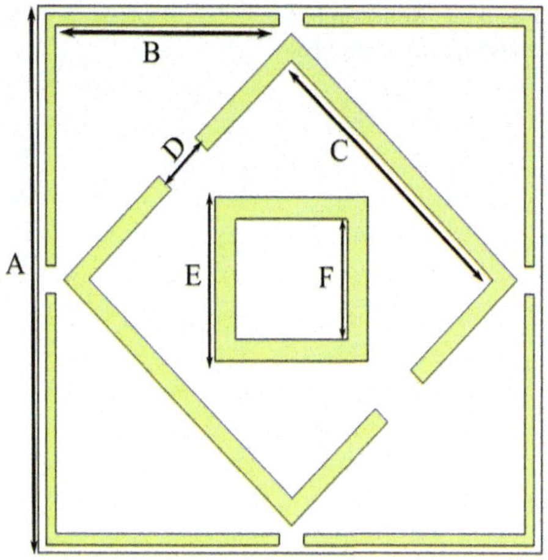


**Fig. 35.1** Layout of the proposed metamaterial unit cell

**Table 35.1** Dimension of parameters MTM Sensor

| Parameter | Dimension (mm) | Parameter | Dimension (mm) |
|---|---|---|---|
| A | 20 | B | 12.37 |
| C | 9.05 | D | 2 |
| E | 6 | F | 4 |

implemented in the recently released 2022 version of Computer Simulation Technology (C.S.T.) microwave studio software was used. But first, a quick primer on what to expect from such a design. An electromagnetic signal is trapped in a microstrip patch when it is used to produce a perfect resonance circuit [7]. The first step was to determine whether a resonance was present in the desired frequency band by using the transmission line model depicted in Fig. 35.1. Based on the lumped elements presented in [8], we have arrived at the following design, which is shown in Table 35.1 with geometrical legends. Q-factor and resonance frequency are important properties because they were formed using lumped gearbox line components. Based on these design considerations, the MM absorber was calculated using the principle described in [9]. A number of metamaterial properties, such as permittivity, permeability, refractive index, and others, can be determined by simulation; however, impedance and admittance parameters must also be considered to achieve accurate results. The value of the reflection and transmission coefficient is depicted in Fig. 35.2.

### *35.1.2 Experimental Measurement and Transmission*

In order to provide the reader with a clear understanding of why the proposed design operates on a specific band with a high absorption rating, the E-field and H-field characteristics at the resonance frequency of 2.88 GHz with the current distribution

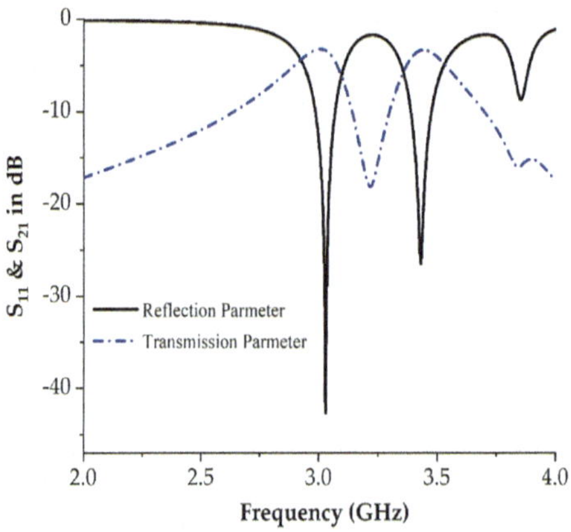


**Fig. 35.2** Reflection and transmission coefficient

are presented one after the other. A simulation was conducted for the values of permeability and permittivity represented in Fig. 35.3 in order to verify the left-handed characteristic of the proposed sensor. S-parameters were determined by using the Nicolson-Ross-Weir (NRW) technique. Above 98% absorption from 3.04 and 3.44 GHz is achieved in the simulation environment (Fig. 35.4). In the proposed architecture, the waveguided ports of the coaxial cable are sandwiched between the vector network analyzer and its waveguided ports. In order to determine the scattering parameters, the VNA and ports must be set up as depicted in Figs. 35.5 and 35.6.

In the proposed architecture, the waveguided ports of the coaxial cable are sandwiched between the vector network analyzer and its waveguided ports. In order to determine the scattering parameters, the VNA and ports must be set up as depicted in Figs. 35.5 and 35.6. Ultimately, the measured data contained in the PRN file is subjected to processing in order to facilitate the creation of graphs. The diagram presented below depicts a comparative analysis between the reflection coefficients

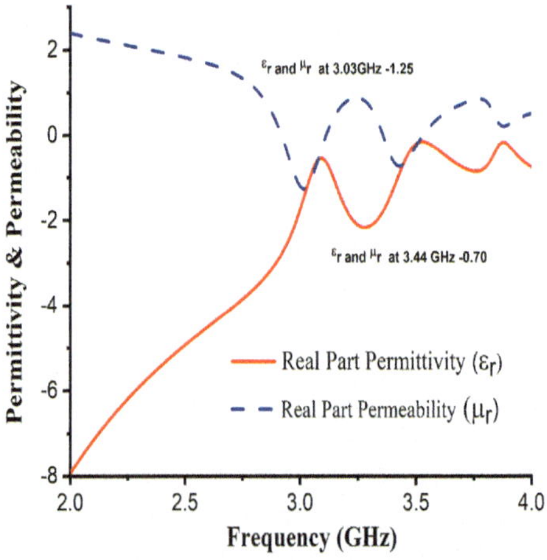


**Fig. 35.3** D.N.G. permittivity and permeability values at resonance frequencies 3.03 and 3.4 GHz

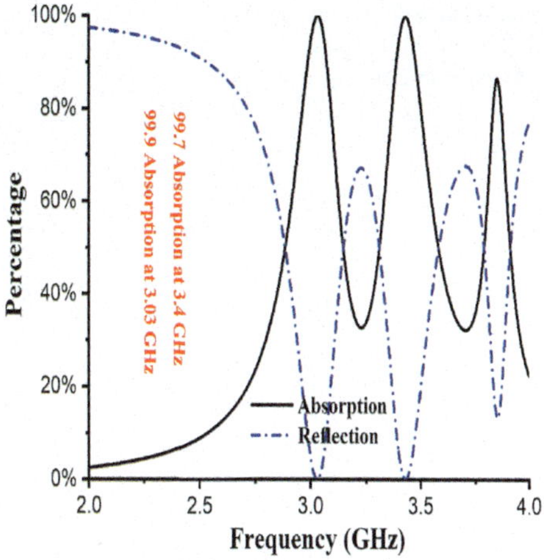


**Fig. 35.4** Absorption versus reflection values at resonance frequencies 3.03 and 3.40 GHz

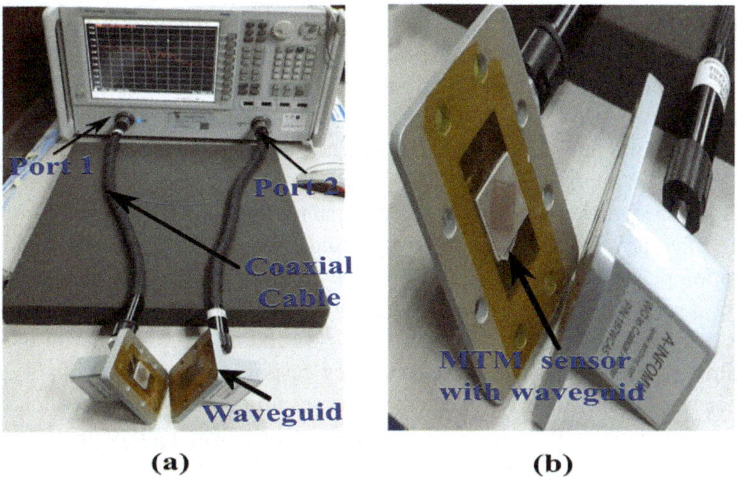


**Fig. 35.5** **a** Procedure for measurement of reflection coefficient. **b** The proposed sensor is placed between waveguide ports

obtained through experimental measurements and those simulated through computational models. The provided figure clearly demonstrates a discernible disparity between the reflection coefficients obtained from simulation and those obtained from measurement. The simulation results indicate that the reflection coefficient ($S_{11}$) values at frequencies of 3.04 GHz and 3.44 GHz are − 42.04 dB and − 26.8 dB, respectively. The reflection coefficient values at the resonance frequencies of 2.84 GHz and 3.54 GHz are measured to be 37.07 GHz and 15.09 GHz, respectively, as illustrated in Fig. 35.6. A minor disparity was observed between the

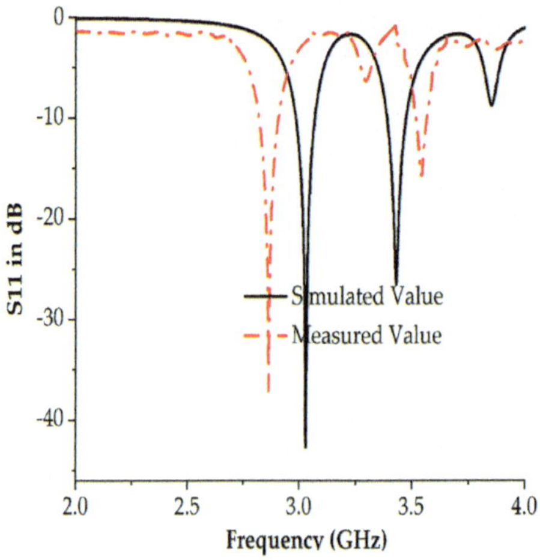


**Fig. 35.6** Simulation versus measured value of the reflection coefficient

outcomes of the simulation and experimental analyses, which can be attributed to fabrication errors and the influence of coupling effects.

As a result of its exceptional ability to absorb electromagnetic fields, the M.M.A. sensor is highly versatile. It is also possible to use metamaterial absorbers in a variety of sensors. A complete simulation setup with a scientifically installed sensor is shown in Fig. 35.7a. An interstitial liquid material fills the space between the substrate materials. Analyses were conducted on different pabulum oils in order to determine their sensing efficiency. In Fig. 35.7b, different liquids' absorption responses are depicted. Absorption peaks move lower as permittivity increases, as shown in Fig. 35.5. We are considering the lower frequency band for liquid sensing in this case. A comparison is provided in Table 35.2 of the studies we intend to conduct with those that have already been published. The frequency response, sensitivity, quality factor, absorbency, and usefulness of the sensors are examined in order to evaluate them. As a result of analyzing a variety of structures and performing parametric analyses, it became evident that our proposed design was the most effective.

## 35.2 Conclusions

The paper presented a D.N.G. metamaterial absorber that exhibits promising absorption for the S-band frequency and could be applied to sensing applications. Its sensitivity makes it possible for the sensor to detect even trace amounts of liquid in a small sample. In addition, the sensors offer micro-quantity, rapid sensing, and repeatability, all of which bode well for the potential widespread use of the product.

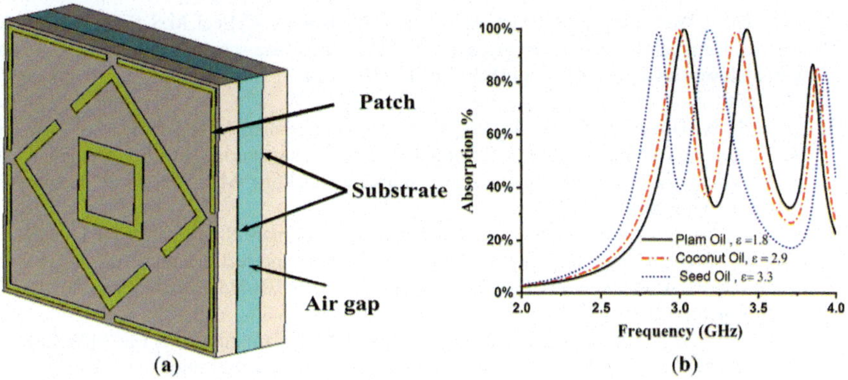


**Fig. 35.7** **a** Liquid material sensing, **b** absorption characteristics of different liquid

**Table 35.2** Performance comparison between the proposed sensor and previous works

| Refs. | Frequency (GHz) | | Sensitivity % | | Absorption % | | Material Sensing | Published in |
|---|---|---|---|---|---|---|---|---|
| | CH1 | CH2 | CH1 | CH2 | CH1 | CH2 | | |
| [10] | 2.4 | – | 0.044 | – | – | – | Measurement of the magnetodielectric materials | IEEE microwave and wireless components letters, 2020 |
| [8] | 1.564 | – | 0.037 | – | – | – | Characterization of dielectric materials | Aces Journal, 2021 |
| [11] | 5.76 | 7.85 | 0.28 | 0.3 | – | – | Liquid sensing | Science Direct, Measurement, 2022 |
| [12] | 1.96 | – | 0.012 | – | – | – | Biomedical applications | Micro and nanostructures, 2022 |
| [13] | 9.06 | 11.16 | 0.012 | 0.01 | – | – | Material sensing | Science Direct, Measurement, 2022 |
| Work | 3.04 | 3.44 | 0.64 | 0.21 | 99.9 | 99.7 | Material characterization and sensing | – |

## References

1. A. Alex-Amor, Á. Palomares-Caballero, C. Molero, 3-D metamaterials: trends on applied designs, computational methods and fabrication. Techniques (2022). https://doi.org/10.3390/electronics11030410

2. M.A. Khalil, W.H. Yong, M.T. Islam, A. Hoque, Md.S. Islam, C.C. Leei, M.S. Soliman, Double-negative metamaterial square enclosed Q.S.S.R For microwave sensing application in S-band with high sensitivity and Q-factor. Sci Rep. **13**, 7373 (2023). https://doi.org/10.1038/s41598-023-34514-z
3. S. Sakib, A. Hoque, S.K.B.A. Rahim, M. Singh, N.M. Sahar, M.S. Islam, M.S. Soliman, M.T. Islam, A Central spiral split rectangular-shaped metamaterial absorber surrounded by polarization-insensitive ring resonator for S-Band applications. Materials **16** (2023). https://doi.org/10.3390/ma16031172
4. N. Ullah, Md.S. Islam, A. Hoque, W.H. Yong, M.S. Soliman, S. Albadran, M.T. Islam, A compact complementary split ring resonator (CSRR) based perfect metamaterial absorber for energy harvesting applications. Eng. Sci. Technol. Int. J. **45**, 101473 (2023). https://doi.org/10.1016/j.jestch.2023.101473
5. S.A. Qureshi, Z.Z. Abidin, A.Y.I. Ashyap, H.A. Majid, M.R. Kamarudin, M. Yue, M.S. Zulkipli, J. Nebhen, Millimetre-wave metamaterial-based sensor for characterisation of cooking oils. Int. J. Antennas Propag. (2021). https://doi.org/10.1155/2021/5520268
6. R. Ravindaran, S.K. Kandasamy, et al., Design and characterization of polymer based microstrip patch antenna, in *2023 IEEE 8th International Conference for Convergence in Technology (I2CT)*. IEEE (2023), pp. 1–5. https://doi.org/10.1109/I2CT57861.2023.10126488
7. N.I. Landy, S. Sajuyigbe, J.J. Mock, D.R. Smith, W.J. Padilla, A perfect metamaterial absorber (2008). https://doi.org/10.1103/PhysRevLett.100.207402
8. H.Y. Gan, W.S. Zhao, D.W. Wang, J. Wang, Q. Liu, G. Wang, High-q active microwave sensor based on microstrip complementary split- ring resonator (mcsrr) structure for dielectric characterization. Appl. Comput. Electromagn. Soc. J. **36**, 922–927 (2021). https://doi.org/10.47037/2021.ACES.J.360715
9. A. Hoque, M.T. Islam, A.F. Almutairi, T. Alam, M.J. Singh, N. Amin, A polarization independent quasi-TEM metamaterial absorber for X and ku band sensing applications. Sensors (Switzerland) **18** (2018). https://doi.org/10.3390/s18124209
10. H.Y. Gan, W.S. Zhao, L. He, Y. Yu, K. Xu, F. Wen, L. Dong, G. Wang, A CSRR-loaded planar sensor for simultaneously measuring permittivity and permeability. IEEE Microwave Wirel. Compon. Lett. **30**, 219–221 (2020). https://doi.org/10.1109/LMWC.2019.2957657
11. S. Kiani, P. Rezaei, M. Navaei, Dual-sensing and dual-frequency microwave SRR sensor for liquid samples permittivity detection. Measurement (Lond) **160** (2020). https://doi.org/10.1016/j.measurement.2020.107805
12. M.-R. Nickpay, M. Danaie, A. Shahzadi, Graphene-based tunable quad-band fan-shaped split-ring metamaterial absorber and refractive index sensor for THz spectrum. Micro Nanostruct. **173**, 207473 (2023). https://doi.org/10.1016/j.micrna.2022.207473
13. M. Rashedul Islam, M. Tariqul Islam, M.S. Soliman, B. Bais, M. Jit Singh, H. Alsaif, M. Shabiul Islam, Metamaterial sensor based on reflected mirror rectangular split ring resonator for the application of microwave sensing. Measurement (Lond) **198** (2022). https://doi.org/10.1016/j.measurement.2022.111416

# Chapter 36
# Concentric Square Rings with Phase Delay Line Design for BST Reconfigurable Reflectarray Technology

**Siti Fatimah Harun, Mohd Fais Mansor, and Norbahiah Misran**

**Abstract** This paper presents a reconfigurable reflectarray unit cell design that incorporates Barium Strontium Titanate (BST) technology for satellite applications. The design proposes three concentric square rings, two of which are divided by a gap and equipped with phased delay lines. This configuration aims to achieve a broader phase range and lower reflection loss. The BST layer, a thin film, is capacitively loaded with a narrow gap within the slot of the innermost square ring, while a few phase delay lines are connected to the outermost square ring. By tuning a DC-bias voltage across the gap, the permittivity of the BST layer can be modified as an integrated capacitor. The paper presents the step-by-step design process of this unit cell, including the selection of substrate material and thickness, analysis of design to increase the phase range, and analysis of the effective biasing line design. The achieved overall phase range is 290.54°, with a maximum reflection loss of 13.653 dB at 13.4 GHz (Ku band), making it suitable for satellite applications.

## 36.1 Introduction

A reconfigurable reflectarray antenna has emerged as a potential replacement for parabolic and phased array antennas due to its ability to combine the advantages of both, including a simple structure, lightweight, high gain, and low losses. It offers beam scanning capabilities, allowing for the reconfiguration of the beam's steering when the satellite or radar changes position [1]. Moreover, its lightweight and uncomplicated structure facilitates the satellite launch process, indirectly reducing costs [1].

S. F. Harun · M. F. Mansor (✉) · N. Misran
Fakulti Kejuruteraan and Alam Bina, Universiti Kebangsaan Malaysia, Bangi, Malaysia
e-mail: m.mansor@ukm.edu.my

S. F. Harun
e-mail: p115826@siswa.ukm.edu.my

N. Misran
e-mail: bahiah@ukm.edu.my

 
M. T. Islam et al. (eds.), *Proceedings of the 8th International Conference on Space Science and Communication*, Springer Proceedings in Physics 303,
https://doi.org/10.1007/978-981-97-0142-1_36

Considering its relevance to space applications, the Ku-band frequency range was chosen for this research.

Various technologies have been developed to harness the potential of reflectarray antennas, including mechanical, lumped element, tunable material, and optical technologies. However, mechanical technology involving moving elements or antennas is impractical due to cost, size, and overall structural mass [2]. Optical technology is still considered immature in this context. Although adding electronic phase shifters to each element is an option for low-frequency applications, it negatively impacts the quality factor (Q-factor) in high-frequency scenarios, particularly in mm-wave applications [3, 4] The binary concept used in diode and MEMS electronic phase shifters limits beam scanning to a certain angle and introduces phase quantization errors. Consequently, tunable materials have dominated high-frequency applications. Materials, such as liquid crystal, ferroelectric, graphene, and electrochromic materials are known for their phase-shifting capabilities [5–7].

Liquid crystal tunable material presents drawbacks, including complicated integration with reflectarray antennas, slow switching speed, and moderate dielectric change [8, 9]. The integration of electrochromic materials with reflectarray antennas is still in its early stages, and slow switching speeds have been observed [10]. Graphene materials integrated with reflectarray antennas are more suitable for THz applications but are not suitable for satellite applications. Hence, ferroelectric materials were selected for this research due to their good switching speed performance and simple fabrication structure.

Ferroelectric materials possess a property known as ferroelectricity, enabling them to exhibit spontaneous electrical polarization that can be reversed by applying voltage [11]. By varying the supplied voltage, the permittivity of the ferroelectric material can be modified, acting as a phase shifter for the reflectarray element and enabling beam shifting [12]. Among various types of ferroelectric materials, Barium Strontium Titanate (BST) was chosen due to its high permittivity values, fast tuning response, and good power handling capability [13].

The investigation of integrating this tunable material with a reconfigurable reflectarray antenna will involve two stages: the element and the antenna. At the element level, the main parameters to be studied include the phase range and return loss, while at the antenna level, the focus will be on beam shifting and gain. Additionally, designing a biasing line for the precondition voltage at the element level is crucial to ensure minimal impact on the performance of the unit cell due to the introduction of another metallic line [14].

This paper will present the design process and performance analysis of concentric square rings with phase delay line design for the reconfigurable reflectarray element integrated with BST technology. The research will commence with material and thickness selection, followed by the design of a unit cell with a large phase range and low reflection loss, and finally, the design of effective biasing lines that have minimal impact on the unit cell's performance.

## 36.2 Unit Cell Design

### 36.2.1 Material and Thickness Selection

There are several materials known for their ferroelectric properties, including $KTaO_3$, $SrTiO_3$, and (Ba, Sr)$TiO_3$, which are commonly used in tunable microwave applications [15]. However, for this research, the material $Ba_xSr_{1-x}TiO_3$ (BST) was chosen due to its high permittivity, tunability, good power handling, and low dielectric loss [15, 16]. The material and dielectric properties of BST depend on factors such as material composition, microstructure, crystalline structure, and thickness [15]. In this study, $Ba_{0.5}Sr_{0.5}TiO_3$ composition was selected, which provides a variation in permittivity ranging from 250 to 500 and a tangent loss ranging from 0.004 to 0.16, with a direct proportionality [3].

The choice of substrate material is important to ensure proper lattice matching with the ferroelectric material, as it affects the dielectric performance of BST [17]. Substrates such as sapphire, MgO, $LaAlO_3$, and $Al_2O_3$ are commonly used in microwave applications due to their good lattice matching with ferroelectric tunable materials [18]. Among these options, alumina ($Al_20_3$) stands out for its excellent mechanical and dielectric properties and lower cost compared with MgO and sapphire [19]. Therefore, the TMM10i material from Rogers was chosen as its properties (permittivity $= 9.8 \pm 0.245$ and loss tangent $= 0.002$) are similar to alumina, while also facilitating the fabrication process.

The thickness of the ferroelectric thin film has a significant influence on its tunable properties. In order to achieve the desired variation in permittivity as mentioned in [3], a thickness of 400 nm was selected for the BST layer in this research. The standard thickness of the TMM10i material was provided by the manufacturer (Rogers), and the selection process involved evaluating the performance of different thicknesses in terms of phase range and reflection loss. Figure 36.1a, b depict the performance comparison for various provided thicknesses, including 0.635, 0.762, 1.27 mm, and 1.524 mm. The results demonstrate that thicker substrates yield lower loss but struggle to achieve a large phase range. On the other hand, thinner substrates offer a higher phase range but contribute to higher losses. Taking these factors into consideration, a substrate thickness of 0.762 mm was ultimately chosen for its favorable performance in terms of phase range and reflection loss compared with the other options.

### 36.2.2 Design in Increasing Phase Range

The design of the unit cell plays a crucial role in achieving a wide phase range and low return loss in reflectarray antennas. Previous researchers have analyzed various designs with the aim of enlarging the phase range of the reflectarray unit cell. For instance, it has been noted that a square ring design can yield a larger phase range compared with a circular ring [20]. The same authors also mentioned

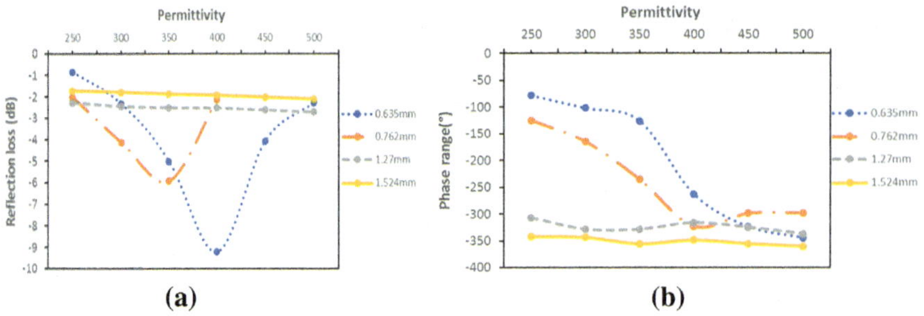


**Fig. 36.1** Performance for every thickness of the substrate in terms of **a** reflection loss, **b** phase range

that incorporating multiple rings can contribute to an increased phase range. Other researchers have proposed techniques such as reducing the width of the ring [21] and utilizing a concentric split ring design [22]. Additionally, the addition of a phase delay line has been suggested as a technique to achieve a wider phase swing [23, 24].

In this research, a combination of these techniques was proposed, as depicted in Fig. 36.2. The design started with a basic square ring loaded with BST in the center of the patch. The performance analysis showed a phase shifting of 48.091% with a maximum reflection loss of − 7.34 dB, as given in Table 36.1. By adding two concentric split rings to the design, a further 19.40% phase shift was achieved, widening the bandwidth. However, this addition also contributed to increased loss (see Fig. 36.3a). The inclusion of a few phase delay lines resulted in an additional 12.845% increment in the phase range. The loss remained unaffected, but the bandwidth was reduced. With this unit cell configuration, a phase range of 289.22° was achieved, with a maximum reflection loss of − 10.145 dB at 13.4 GHz, which is within the Ku-band application.

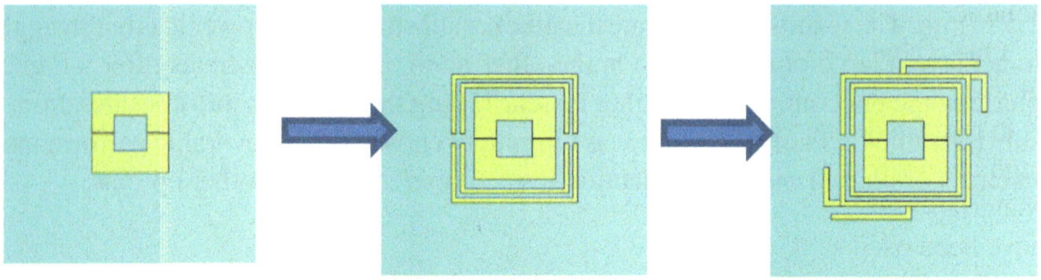

**Fig. 36.2** Formation of concentric square rings with phase delay line design

**Table 36.1** Performance formation of unit cell design

| Name of design | Maximum reflection loss (dB) | Phase range (°) | Percentage phase range over 360° (%) | Percentage of phase range increased (%) |
|---|---|---|---|---|
| Square ring | − 7.341 | 173.128 | 48.091 | 48.091 |
| Concentric square rings | − 10.494 | 242.980 | 67.494 | 19.403 |
| Concentric square rings with phase delay line | − 10.145 | 289.223 | 80.339 | 12.845 |

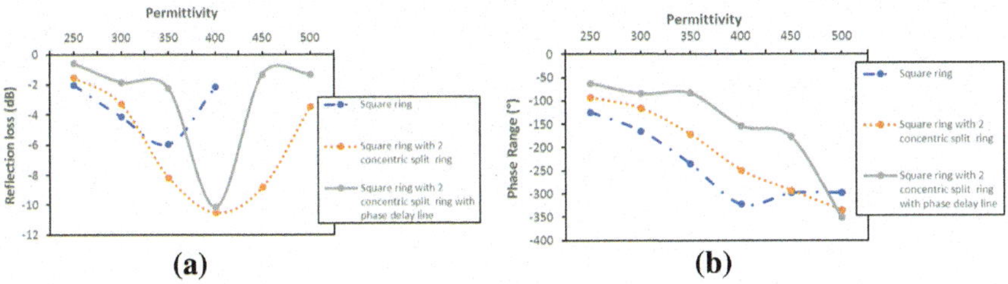


**Fig. 36.3** Performance for every stage of design in terms of **a** reflection loss, **b** phase range

### *36.2.3 Design of Biasing Line*

The biasing line serves as the medium to supply DC voltage, enabling the tunability of the BST material. It is a critical component in the design of reconfigurable reflectarray antennas as it introduces a new connection to the unit cell, which can impact the overall antenna performance. The presence of a metallic biasing line typically has a significant effect on the reconfigurable antenna [14]. Several studies have been conducted to minimize the impact of the biasing circuit on the radiator.

One technique proposed is the use of cross-layer via that connects from the back of the PCB onto the surface of the radiating patch [23, 25, 26]. Another approach is to replace the metallic bias line with a resistive conductive material, allowing for high attenuation of electromagnetic waves along the line with minimal impact [14]. Materials such as chromium, silicon chromium, and resistive conductive adhesive have been used as resistive lines [14].

A simpler technique involves connecting the metallic biasing line at the minimum electric field (E-field) point of the radiator [3]. This technique requires conducting an analysis to identify the minimum E-field point of the patch, typically done using computer simulation technology (CST) software. The biasing line is then connected to this point, as shown in Fig. 36.4. For comparison purposes, different positions were analyzed, including connecting the bias lines to the edge, middle, and minimum E-field point on each patch. Figure 36.5a, b demonstrates that connecting the bias lines to the minimum E-field point has the least impact on the overall performance of the

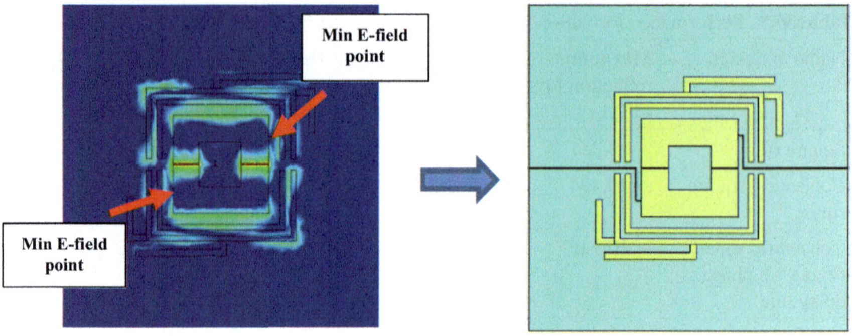


**Fig. 36.4** Connecting biasing line on the minimum E-field point of the patch

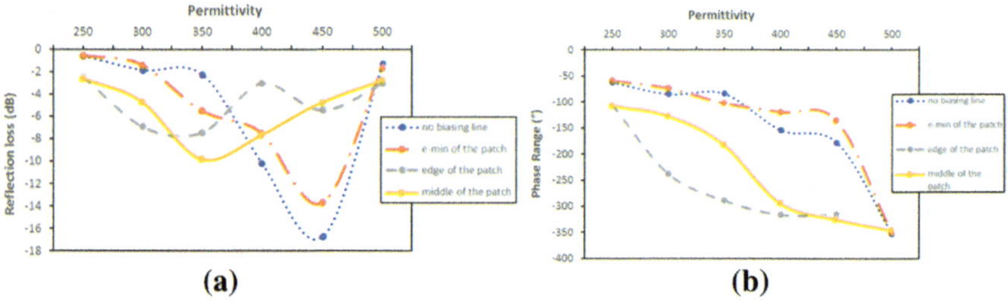


**Fig. 36.5** Performance of bias line connected position at each patch in terms of **a** Reflection loss, **b** phase range

unit cell, resulting in a phase range of 290.54° and a maximum reflection loss of -13.653 dB at 13.4 GHz.

## 36.3 Conclusion

In conclusion, this paper presented an analysis of the design process and performance evaluation of a reconfigurable reflectarray unit cell integrated with BST technology. The design involved the selection of appropriate materials, such as the substrate, considering the chemical properties of the tunable material BST. The substrate thickness was determined, and the formation of the patch was optimized to increase the phase range and reduce reflection loss. Furthermore, the optimal location for the biasing line was determined to minimize its impact on the unit cell performance. As a result, the designed reflectarray unit cell achieved a phase range of 290.54° with a maximum reflection loss of – 13.653 dB at 13.4 GHz, which falls within the Ku band and is suitable for satellite applications.

## References


1. M. Ramli, Antena Tatasusun Pantulan Boleh Konfigurasi Semula Terintegrasi Diod Varaktor Dengan Imbasan Alur Pada Jalur-Ku, Thesis Ph.D. Universiti Kebangsaan Malaysia (2018)
2. M. Mirmozafari et al., Mechanically reconfigurable, beam-scanning reflectarray and transmitarray antennas: a review. Appl. Sci. **11**(15), 6890 (2021)
3. K.K. Karnati et al., A BST-integrated capacitively loaded patch for Ka- and X-band beam-steerable reflectarray antennas in satellite communications. IEEE Trans. Antennas Propag. **63**, 1324–1333 (2015)
4. M. Zhang, et al., K-band tunable phase shifter with microstrip line structure using BST technology, i*n Asia-Pacific Microwave Conference Proceedings*, APMC 2 (2016), pp. 11–13
5. S.V. Hum, J. Perruisseau-Carrier, Reconfigurable reflectarrays and array lenses for dynamic antenna beam control: a review. IEEE Trans. Antennas Propag. **62**, 183–198 (2014)
6. P. Nayeri, F. Yang, A.Z. Elsherbeni, Beam-scanning reflectarray antennas: a technical overview and state of the art. IEEE Antennas Propag. Mag. **57** (2015)
7. Electrochormic Reflectarray Antenna. US Patent Application, pp. 1–14 (2018)
8. Perez-Palomino, G., et al., Liquid crystal based beam scanning reflectarrays and their potential in SATCOM antennas, *in 2017 11th European Conference on Antennas Propagation*, EUCAP 2017 (2017), pp. 3428–3431. https://doi.org/10.23919/EuCAP.2017.7928474
9. G. Perez-Palomino et al., Design and Demonstration of an electronically scanned reflectarray antenna at 100 GHz Using multiresonant cells based on liquid crystals. IEEE Trans. Antennas Propag. **63**, 3722–3727 (2015)
10. M. Noroozarab et al., Variable Temperature broadband microwave and millimeter-wave characterization of electrochromic (WO3/LiNbO3/NiO) thin films. IEEE Trans. Microw. Theory Tech. **66**, 1070–1080 (2018)
11. M.Y. Ismail, M.H. Dahri, W.N. Zaihasra, Characterization of material properties for tunable reflectarray antenna design. AIP Conf. Proc. **1528**(1), 237–242 (2013)
12. G. Vélu, et al., A 360 BST phase shifter with moderate bias voltage at 30 GHz. **55**, 438–444 (2007)
13. B. Noren, Thin film barium strontium titanate (BST) for a new class of tunable RF components. Microw. J. **47**, 210–220 (2004)
14. J.M. Kovitz, H. Rajagopalan, Y. Rahmat-Samii, Practical and cost-effective bias line implementations for reconfigurable antennas. IEEE Antennas Wirel. Propag. Lett. **11**, 1552–1555 (2012)
15. Ketkar, S.A. Synthesis, characterization and applications of barium strontium titanate thin film structures. Theses, University of South Florida (2013)
16. Electrical 4u homepage www.electrical4u.com/ferroelectric-materials. Ferroelectric Materials : What Are They ? Last accessed 31 May 2023
17. A.K. Tagantsev, V.O. Sherman, K.F. Astafiev, J. Venkatesh, N. Setter, Ferroelectric materials for microwave tunable applications. J. Electroceramics **5**, 66 (2004)
18. A.K. Tagantsev, V.O. Sherman, K.F. Astafiev, J. Venkatesh, N. Setter, Ferroelectric materials for microwave tunable applications. J. Electroceram. **11**, 5–66 (2003)
19. A. Tumarkin, E. Sapego, A. Gagarin, S. Senkevich, Enhanced tunability of batixsn1−xo3 films on dielectric substrate. Appl. Sci. **11** (2021)
20. M.E. Bialkowski, K.H. Sayidmarie, Phasing characteristics of a single layer microstrip reflectarray employing various basic element shapes, in C*onference Proceedings 2008 IEEE International Workshop on Antenna Technology: Small Antennas and Novel Metamaterials, IWAT 2008* (2008), pp. 79–82. https://doi.org/10.1109/IWAT.2008.4511295
21. S.H. Yusop, N. Misran, M.T. Islam, M.Y. Ismail, Design of high performance dual frequency concentric split ring square element for broadband reflectarray antenna. Appl. Comput. Electromagn. Soc. J. **27**, 334–339 (2012)
22. M. Ramli, N. Misran, M.F. Mansor, M.T. Islam, Scattering phase performance of varactor diode across ring's gap reconfigurable reflectarray element, in *International Conference on*

*Space Science and Communication (Iconsp)* (2013), pp. 272–275. https://doi.org/10.1109/IconSpace.2013.6599479.
23. Y. Li, A. Abbosh, Reconfigurable reflectarray antenna using single-layer radiator controlled by PIN diodes. IET Microwaves Antennas Propag. **9**, 664–671 (2015)
24. C. Han, Y. Zhang, Q. Yang, A novel single-layer unit structure for broadband reflectarray antenna. IEEE Antennas Wirel. Propag. Lett. **16**, 681–684 (2017)
25. M.E. Trampler, R.E. Lovato, Dual-resonance continuously beam-scanning X-band reflectarray antenna. IEEE Trans. Antennas Propag. **68**, 6080–6087 (2020)
26. H. Yang, F. Yang, S. Xu, M. Li, X. Cao, Experimental study of a 1-bit 10×10 reconfigurable reflectarray antenna. IEEE Antennas and Propagation Society (AP-S) International Symposium (2015), pp. 2153–2154

# Chapter 37
# AMC-Based Slotted Ground Circular Polarized Antenna for C-Band Satellite Communications

**Mohammad Lutful Hakim, Touhidul Alam, Rezaul Azim, Mohammad Tariqul Islam, Mohd Hafiz Baharuddin, and Norsuzlin Mohd Sahar**

**Abstract** This article proposed a planar compact circular polarized (CP) slotted ground antenna for C-band satellite applications. CP antennas are well suited for satellite communication, where robust signal reception, resistance to interference, and efficient frequency utilization are essential. They contribute to reliable and high-quality communication links between ground-based terminals and satellites, supporting various applications, including broadcasting, telecommunication, remote sensing, and satellite-based navigation systems. The proposed antenna achieves CP features by circular stubs in the ground plane. The FR-4 substrate material with a dielectric constant of 4.3 is utilized to design the antenna, and the antenna size is 20 mm$^2$ $\times$ 20 mm$^2$. A square ring-based artificial magnetic conductor (AMC) is designed, which has a reflection phase of 90° to 90° from 2.44 to 6.88 GHz. The utilization of AMC enhances the antenna gain by around 1.5 dBi. The proposed AMC-based antenna shows S11 $< -$ 10 dB from 3.68 to 6.6 GHz and 3 dB > axial ratio from 3.90 to 4.15 GHz. The efficiency of the antenna also shows above 90% at the operating frequency. Finally, the compact size, high gain, and CP features make the proposed antenna for C-band satellite applications.

---

M. L. Hakim · T. Alam (✉) · N. M. Sahar (✉)
Pusat Sains Angkasa (ANGKASA), Institut Perubahan Iklim, Universiti Kebangsaan Malaysia, 43600 UKM Bangi, Selangor, Malaysia
e-mail: touhidul@ukm.edu.my

N. M. Sahar
e-mail: norsuzlin@ukm.edu.my

R. Azim
Department of Physics, University of Chittagong, Chattogram 4331, Bangladesh

M. T. Islam · M. H. Baharuddin
Department of Electrical, Electronic and Systems Engineering, Faculty of Engineering and Built Environment, Universiti Kebangsaan Malaysia, 43600 UKM Bangi, Selangor, Malaysia

 
M. T. Islam et al. (eds.), *Proceedings of the 8th International Conference on Space Science and Communication*, Springer Proceedings in Physics 303,
https://doi.org/10.1007/978-981-97-0142-1_37

## 37.1 Introductions

Circularly polarized (CP) antennas are commonly used in various applications, including wireless communication systems, satellite communication, radio astronomy, and RFID (Radio Frequency Identification) systems [1]. They offer several advantages over linearly polarized antennas, such as improved signal reception in the presence of multipath interference, better resistance to signal fading, and compatibility with circularly polarized signals [2]. CP antennas can be designed in different configurations, including patch antennas, helical antennas, and crossed-dipole antennas [3]. These antennas generate a circularly polarized electromagnetic wave by creating a phase shift of 90° between the two orthogonal components of the electric field. The patch antennas are commonly used in wireless communication systems, including Wi-Fi networks, Bluetooth devices, RFID systems, satellite communication, and many other applications where compact size, directional radiation, and cost-effectiveness are important considerations. Circularly polarized patch antennas find extensive use in satellite communication systems, where they are employed for both transmission and reception [4]. They are also commonly used in radio-controlled aircraft and drones, where they help maintain a more stable and reliable signal link between the transmitter and receiver. The metamaterial becomes attractive for different applications like absorber [5–10], antenna [11–14], sensor [15–18], EMI shielding [19], etc. Usually, the metamaterial is positioned on the antenna radiating element or behind the defected ground antenna [20].

Different configurations of the antenna are designed for CP behavior. In [21], a meandered ring antenna is designed, whose size is 50 $mm^2 \times 50$ $mm^2$. The antenna operated at 1.6 GHz frequency and achieved 3 dB > AR from 1.56 to 1.59 GHz with a gain value of 3.13 dBi. A dual-band CP antenna is presented in [22]. The size of the antenna is 93.6 $mm^2 \times 96$ $mm^2$. The 3 dB > AR is very narrow, achieved at 2.84, 5.24 GHz. In [23], a slot monopole antenna is presented with a dimension of 42 $mm^2 \times 42$ $mm^2$. The antenna resonates at 4.5 GHz frequency and shows a wide range of 3 dB > AR from 2.1 to 5.6 GHz, but the gain is poor. Another slotted antenna is designed in [24], which has a 45 $mm^2 \times 43.6$ $mm^2$ dimension. The 3 dB > AR is achieved from 2.09 to 7.07 GHz, where the gain is above 2.3 dB. Therefore, a high gain compact size CP antenna is still a great demand for C-band applications.

This paper proposed a CP AMC-based slotted ground antenna for C-band satellite applications. The antenna shows a 3 dB > axial ratio from 3.90 to 4.15 GHz. The antenna gain of 1.5 dBi is improved by AMC.

## 37.2 Circular Polarized Antenna Design

### *37.2.1 Antenna Design*

Figure 37.1 shows the proposed CP slotted antenna. The antenna consists of a microstrip line on the top layer and a slotted ground on the back layer of the FR-4 substrate material (thickness 1.6), whose dielectric constant and loss tangent are 4.3 and 0.025. The CP of the proposed antenna achieves by circular stubs in the ground plane. The antenna design parameters are $L = 20$ mm, $L_1 = 8.40$ mm, $R_1 = 8.75$ mm, $R_2 = 3.5$ mm, $W_1 = 3.5$ mm. The CST microwave studio is utilized to design and simulate the antenna feed by a 50 Ω port. The reflection coefficient (S11), AR, and gain of the normal antenna are presented in Fig. 37.2. The normal antenna shows − 10 dB bandwidth from 3.70 to 6.4 GHz. The 3 dB > axial ratio is archived from 3.98 to 4.37 GHz. The gain of the antenna is 2.5–3.4 dBi. The efficiency of the normal antenna is above 80% at the operating frequency presented in Fig. 37.3. The 2D and 3D radiation pattern is presented in Fig. 37.4. The surface current distribution of the slotted ground antenna is presented in Fig. 37.5 at 4 GHz frequency. The current allocation in the circular stub is at different time instants ($\omega t = 0°$, 90°, 180°, and 270°), which is rotating in the anticlockwise direction. These correspond to right-hand circularly polarized (RHCP) behavior.

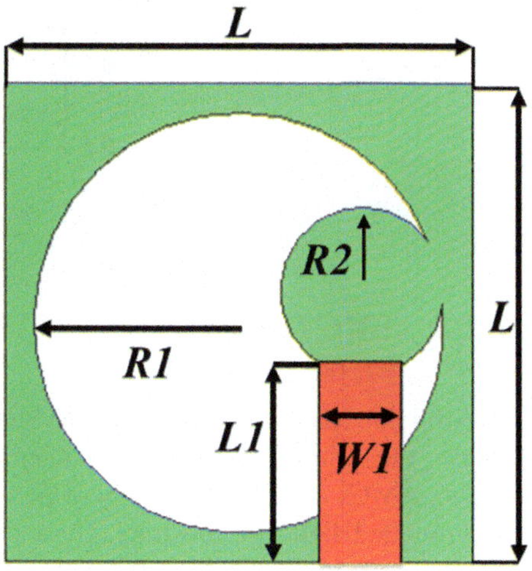


**Fig. 37.1** Layout of the proposed slotted patch antenna

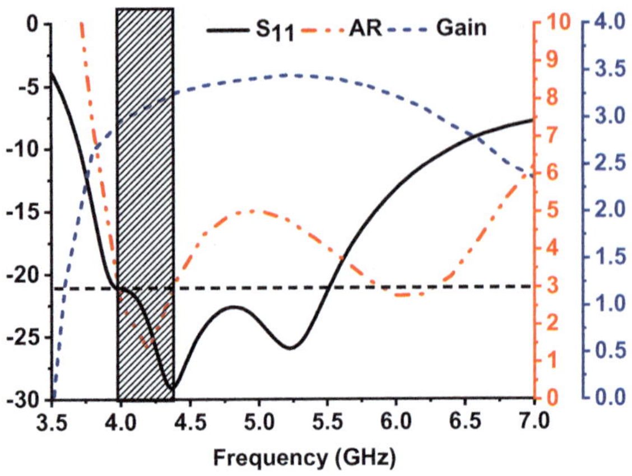


**Fig. 37.2** S11, AR, and the proposed slotted ground antenna gain

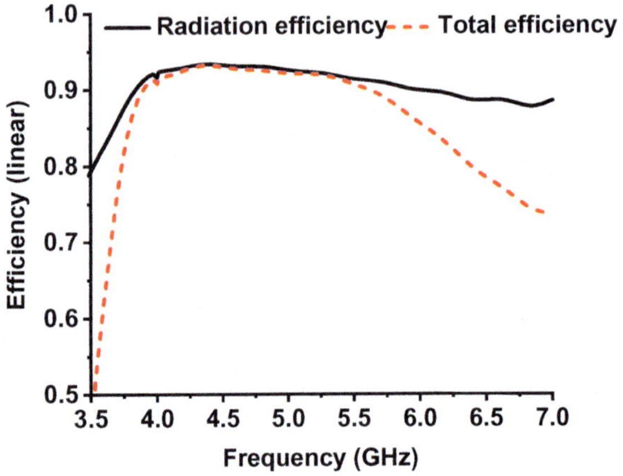


**Fig. 37.3** Efficiency of the normal slotted ground antenna

### *37.2.2 AMC Design*

An Artificial Magnetic Conductor (AMC) is designed, shown in Fig. 37.6a, to enhance the gain of an antenna by manipulating the electromagnetic wave propagation. AMCs consist of an array of periodic metallic patches or elements placed on a grounded dielectric substrate. These patches are usually designed to behave as a perfect magnetic conductor, reflecting and redirecting the incident electromagnetic waves. By strategically placing the AMC structure around the antenna, the reflected waves interfere constructively, resulting in increased radiation in the desired direction shown in Fig. 37.6b [12]. The proposed AMC is simulated by CST microwave studio using unit cell boundary conditions along the *x* and *y* where EM wave is applied along the *z*-axis. Figure 37.6c presents the reflection phase value, where 90° to 90° bandwidth is achieved at the antenna's operating frequency.

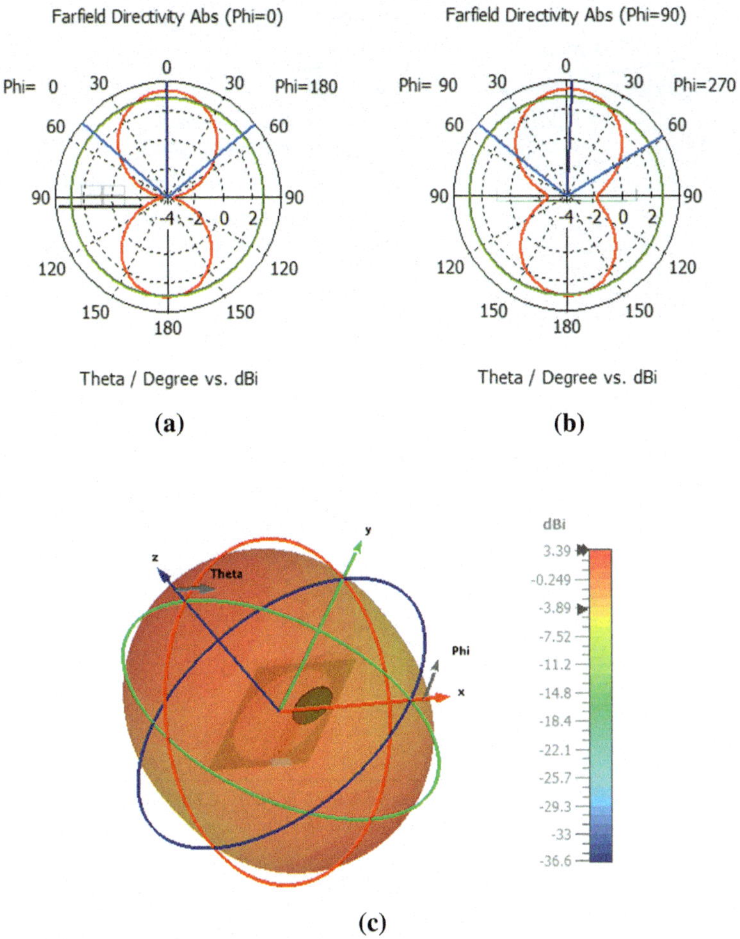


**Fig. 37.4** 2D radiation pattern at 4 GHz frequency **a** phi = 0°, **b** phi = 90°, **c** fairfield radiation pattern

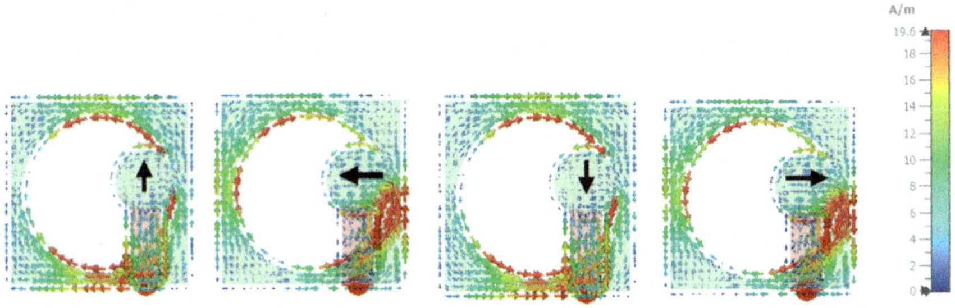


**Fig. 37.5** Surface current distributions of the normal antenna at 4 GHz **a** ωt = 0°, **b** ωt = 90°, **c** ωt = 180°, and **d** ωt = 270°

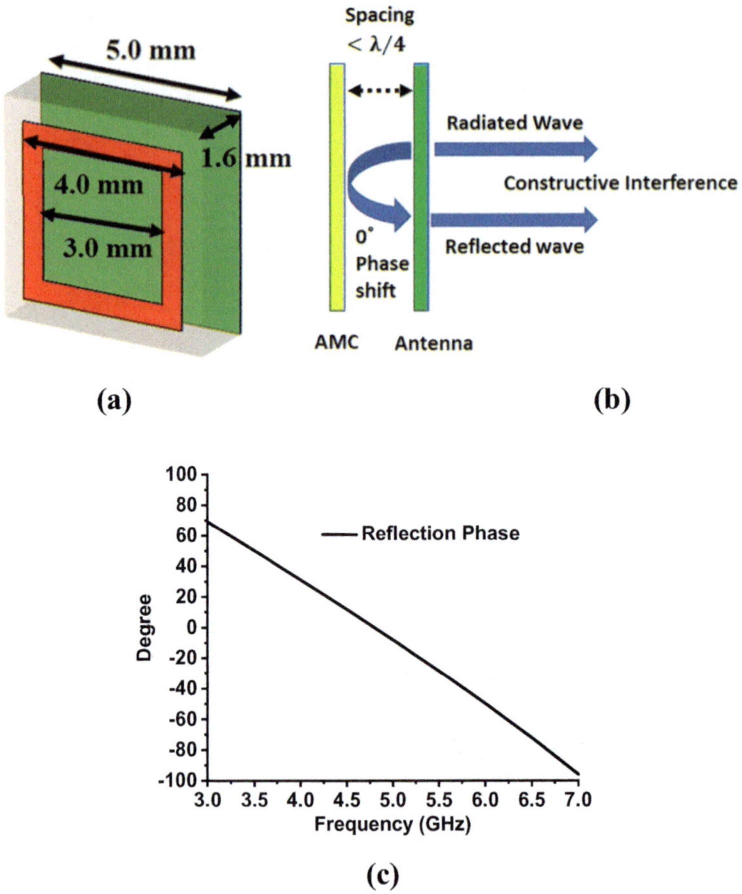


**Fig. 37.6** **a** AMC unit cell, **b** constructive wave interference, **c** reflection phase of AMC in degree

## 37.3 AMC-Based Slotted Antenna and Results Analysis

Figure 37.7 shows the AMC-based slotted antenna where a 4 × 4 AMC array is placed behind the antenna with an air gap (ha = 19.97 mm). The AMC array reflects the back radiation of the antenna and makes a constructive EM wave interference with the forward-radiated EM wave. The reflection coefficient, AR, and gain of the proposed antenna are presented in Fig. 37.8. The − 10 dB operating frequency of the antenna achieved from 3.68 to 6.6 GHz and a 3 dB axial ratio from 3.90 to 4.15 GHz. The gain value increases significantly, and peak gain is achieved at 5.93 dBi. The efficiency of the antenna is presented in Fig. 37.9, which shows above 90% at the operating frequency. The surface current distribution of the proposed antenna at 4 GHz frequency is illustrated in Fig. 37.10 for different time instants (ωt = 0°, 90°, 180°, and 270°) which are rotating in the anticlockwise direction.

**Fig. 37.7** AMC-based slotted ground antenna

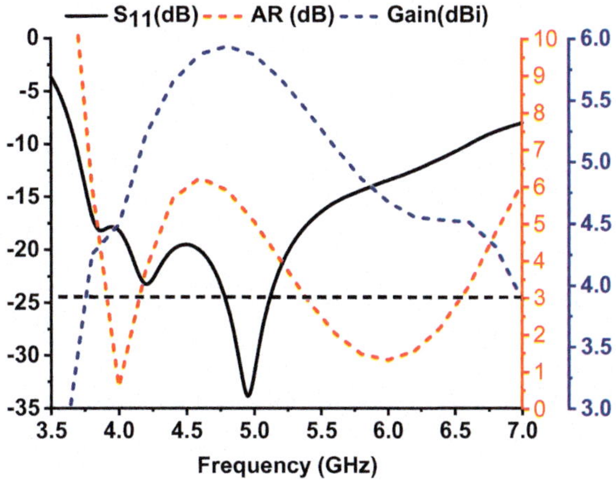


**Fig. 37.8** S11, AR, and gain of the proposed AMC-based slotted ground antenna

These correspond to right-hand circularly polarized (RHCP) behavior. The radiation pattern of the proposed antenna is presented in Fig. 37.10, which shows the AMC mitigate the backward radiation of the antenna and high radiation intensity in the forward direction of the antenna.

## 37.4 Compression

A detailed comparison of the proposed antenna with the existing CP antenna is presented in Table 37.1. Different types of antennas, like meandered ring antenna in [21], slotted patch or ground antenna in [22–24], are designed to achieve the CP

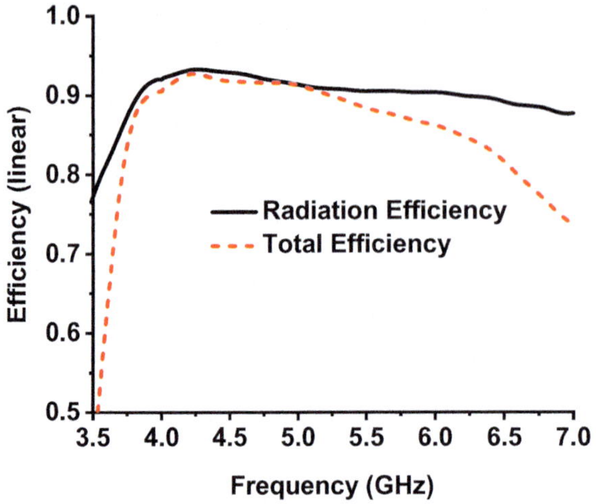


**Fig. 37.9** Efficiency of the AMC-based slotted ground antenna

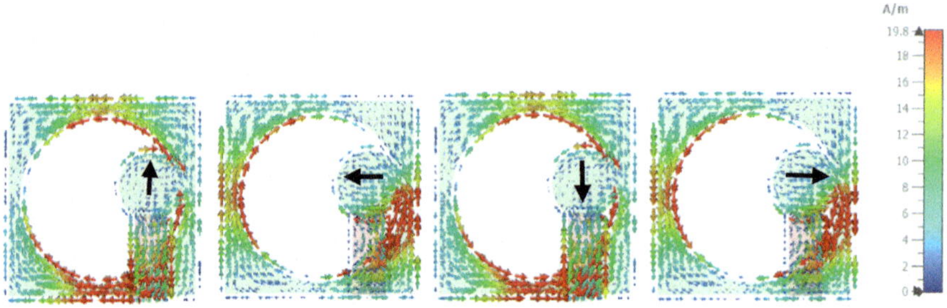


**Fig. 37.10** Surface current distributions of AMC-based antenna at 4 GHz **a** ωt = 0°, **b** ωt = 90°, **c** ωt = 180°, and **d** ωt = 270°

attributes, but all of them have a larger size than the proposed antenna. The gain value of the proposed AMC-based antenna is also higher than the listed antenna in Table 37.1. Finally, the proposed antenna has a 250 MHz 3 dB AR bandwidth, making it a potential candidate for C-band satellite applications.

**Table 37.1** Comparison with the existing CP antenna

| Ref. | Antenna configurations | Size (mm$^2$) | Frequency | 3 dB AR | Gain (dBi) |
|---|---|---|---|---|---|
| [21] | Meandered ring antenna | 50 × 50 | 1.6 | 1.56–1.59 | 3.13 |
| [22] | Slotted antenna | 93.6 × 96 | 2.84, 5.24 | 2.84, 5.24 | N/A |
| [23] | Slotted antenna | 42 × 42 | 4.5 | 2.1–5.6 | 1–3.2 |
| [24] | Slotted antenna | 45 × 43.6 | 4.3 | 2.09–7.07 | 2.3–5.5 dB |
| Proposed | AMC-based slotted antenna | 20 × 20 | 4.0 | 3.90–4.15 | 4.5 |

## 37.5 Conclusion

This paper presents a low profile, small-size AMC-based slotted CP antenna for C-band applications. The dimension of the antenna is 20 $mm^2 \times 20$ $mm^2$. The antenna shows a 3 dB axial ratio from 3.90 to $-$ 4.15 GHz and a gain of 4.5 dBi at 4 GHz frequency. The antenna's efficiency also shows above 90% at the operating frequency. The CP featured is achieved by integrating circular tubs on the ground plane. The antenna is suitable for use in satellites due to its CP, lighter weight, and smaller size.

**Funding** The authors acknowledge the Fundamental Research Grant Scheme (FRGS), Grant Number FRGS/1/2021/TK0/UKM/02/36 funded by the Ministry of Higher Education (MOHE), Malaysia. The authors are also grateful to Universiti Kebangsaan Malaysia.

## References

1. U. Banerjee, A. Karmakar, A. Saha, A review on circularly polarized antennas, trends and advances. Int. J. Microwave Wirel. Technol. **12**(9), 922–943 (2020)
2. R. Nalanagula, N.K. Darimireddy, R. Kumari, C.-W. Park, R.R. Reddy, Circularly polarized hybrid dielectric resonator antennas: a brief review and perspective analysis. Sensors **21**(12), 4100 (2021)
3. N.A. Abd Rahman et al., A review of circularly polarized dielectric resonator antennas: recent developments and applications. Micromachines **13**(12), 2178 (2022)
4. V. Akan, E. Yazgan, Antennas for space applications: a review, in *Advanced Radio Frequency Antennas for Modern Communication and Medical Systems* (2020), pp. 139–171
5. A. Hanif, M.L. Hakim, T. Alam, M.T. Islam, H. Alsaif, M.S. Soliman, Polarization insensitive oblique incident angle stable ultra-thin nano ring resonator-based metamaterial absorber for visible and near-infrared region applications. Opt. Laser Technol. **164**, 109494 (2023)
6. A. Musa et al., Broadband plasmonic metamaterial optical absorber for the visible to near-infrared region. Nanomaterials **13**(4), 626 (2023)
7. M.L. Hakim et al., Triple-band square split-ring resonator metamaterial absorber design with high effective medium ratio for 5G sub-6 GHz applications. Nanomaterials **13**(2), 222 (2023)
8. M.L. Hakim et al., Ultrawideband polarization-independent nanoarchitectonics: a perfect metamaterial absorber for visible and infrared optical window applications. Nanomaterials **12**(16), 2849 (2022)
9. A. Musa et al., Polarization independent metamaterial absorber with anti-reflection coating nanoarchitectonics for visible and infrared window applications. Materials **15**(10), 3733 (2022)
10. M.L. Hakim et al., Wide-oblique-incident-angle stable polarization-insensitive ultra-wideband metamaterial perfect absorber for visible optical wavelength applications. Materials **15**(6), 2201 (2022)
11. M.L. Hakim et al., Metamaterial physical property utilized antenna radiation pattern deflection for angular coverage and isolation enhancement of mm-wave 5G MIMO antenna system. Radiat. Phys. Chem. **209**, 110998 (2023)
12. T. Alam et al., Metamaterial based Ku-band antenna for low earth orbit nanosatellite payload system. Nanomaterials **13**(2), 228 (2023)
13. M. Hakim, T. Alam, M. Baharuddin, M. Islam, Frequency integration of dual-band hexagonal metamaterial resonator antenna for Wi-Fi and 5G wireless communication. J. Phys. Conf. Ser. **2250**(1), 012001(2022). IOP Publishing
14. M.L. Hakim, T. Alam, N.M. Sahar, N. Misran, M.F. Mansor, Elliptical slot metasurface high gain microstrip line antenna for sub-6 GHz 5G wireless communication, in *2021 7th*

*International Conference on Space Science and Communication (IconSpace)* (IEEE, 2021), pp. 156–160
15. M.L. Hakim, T. Alam, M.T. Islam, H. Alsaif, M.S. Soliman, Polarization-independent fractal square splits ring resonator (FSSRR) multiband metamaterial absorber/artificial magnetic conductor/sensor for Ku/K/Ka/5G (mm-Wave) band applications. Measurement **210**, 112545 (2023)
16. M.L. Hakim, T. Alam, M.T. Islam, M.H. Baharuddin, A. Alzamil, M.S. Islam, Quad-band polarization-insensitive square split-ring resonator (SSRR) with an inner Jerusalem cross metamaterial absorber for Ku-and K-Band sensing applications. Sensors **22**(12), 4489 (2022)
17. M.L. Hakim et al., Polarization insensitive symmetrical structured double negative (DNG) metamaterial absorber for Ku-band sensing applications. Sci. Rep. **12**(1), 479 (2022)
18. M.L. Hakim, T. Alam, A.F. Almutairi, M.F. Mansor, M.T. Islam, Polarization insensitivity characterization of dual-band perfect metamaterial absorber for K band sensing applications. Sci. Rep. **11**(1), 17829 (2021)
19. M.L. Hakim, T. Alam, M.T. Islam, Polarization insensitive and oblique incident angle stable miniaturized conformal FSS for 28/38 GHz mm-wave band 5G EMI shielding applications. IEEE Antennas Wirel. Propag. Lett. (2023)
20. R. Sonak, M. Ameen, R.K. Chaudhary, High gain dual-band open-ended metamaterial antenna utilizing CRR with broadside radiation characteristics based on left-handed AMC and PEC. Mater. Res. Exp. **6**(5), 055811 (2019)
21. J. Yuan, J. Zheng, Z.D. Chen, A compact meandered ring antenna loaded with parasitic patches and a slotted ground for global navigation satellite systems. IEEE Trans. Antennas Propag. **66**(12), 6835–6843 (2018)
22. Y. Xu, L. Zhu, N.-W. Liu, Design approach for a dual-band circularly polarized slot antenna with flexible frequency ratio and similar in-band gain. IEEE Antennas Wirel. Propag. Lett. **21**(5), 1037–1041 (2022)
23. H.H. Tran, N. Nguyen-Trong, A.M. Abbosh, Simple design procedure of a broadband circularly polarized slot monopole antenna assisted by characteristic mode analysis. IEEE Access **6**, 78386–78393 (2018)
24. H.F. Huang, Z. Zhang, A small single fed broadband circularly polarized slot antenna. Int. J. RF Microwave Comput.-Aided Eng. **30**(4), e22122 (2020)

# Chapter 38
# Exploration of Wi-Fi-Based Indoor Positioning System Using Linear Regression and K-Nearest Neighbour

**Mohtady Ehab Hasan Aly Barakat, Gwo Chin Chung, Wai Leong Pang, Murman Dwi Prasetio, Mardeni bin Roslee, and Kah Yoong Chan**

**Abstract** In recent years, the use of Wi-Fi signals and machine learning techniques for indoor positioning has shown promising results. However, challenges such as privacy issues and resource constraints such as the availability of Wi-Fi access points (APs) and cost, need to be addressed to further improve the accuracy and applicability of these systems. In this paper, we aim to explore the performance of the indoor positioning system (IPS) using Wi-Fi signals. Three common machine learning models, linear regression, decision trees, and K-nearest neighbour (KNN) have been proposed for indoor positioning using the received signal strength indicator (RSSI). This study also includes the finding of the $K$ value for the optimum performance of the KNN model in IPS. By comparing the experimental results of these two models, the optimised KNN model with $K = 19$ outperformed the linear regression and decision trees models with an accuracy of 96% in just 0.9 min, compared with an accuracy of 70 and 82% for the linear regression and decision trees models respectfully in three minutes. This suggests that the optimised KNN model is more effective and efficient in predicting the position of a device based on RSSI values in an IPS.

## 38.1 Introduction

Indoor positioning system (IPS) has become an integral part of modern technology, with applications spanning navigation, security, emergency services, and even retail for tracking customer behaviour. Unlike outdoor positioning, which has been largely

M. E. H. A. Barakat · G. C. Chung (✉) · W. L. Pang · M. Roslee · K. Y. Chan
Faculty of Engineering, Multimedia University, Jalan Multimedia, 63100 Cyberjaya, Selangor, Malaysia
e-mail: gcchung@mmu.edu.my

M. D. Prasetio
Fakultas Rekayasa Industri, Telkom University, Landmark Tower (TULT), Jalan Telekomunikasi No 1, Terusan Buah Batu, Bandung, Jawa Barat 40257, Indonesia

 
M. T. Islam et al. (eds.), *Proceedings of the 8th International Conference on Space Science and Communication*, Springer Proceedings in Physics 303,
https://doi.org/10.1007/978-981-97-0142-1_38

addressed by the global positioning system (GPS), indoor positioning presents unique challenges. These challenges stem from signal attenuation and multipath propagation caused by walls and other obstructions [1]. Among the various technologies explored for IPS, such as infrared, ultrasonic, and radio frequency, Wi-Fi has gained significant attention. The ubiquity of Wi-Fi in indoor environments and the ability to leverage existing infrastructure make it a cost-effective solution for indoor positioning [2]. The core principle behind Wi-Fi-based IPS involves measuring signal strength from multiple Wi-Fi access points (APs) to determine a device's position.

Fingerprinting, a technique often used in Wi-Fi-based IPS, has shown considerable promise due to its relative immunity to multipath propagation and signal attenuation. It involves creating a radio map of the environment, where each point is associated with a unique fingerprint of signal strengths from various APs. The position of a device is then determined by matching the observed signal strengths with the fingerprints in the radio map. Machine learning techniques have been extensively used to improve the accuracy of fingerprinting-based IPS [3]. The K-nearest neighbour (KNN) algorithm, an instance-based learning technique, has been successfully applied to IPS. In this context, the $K$ nearest fingerprints in the radio map are used to estimate a device's position. Linear regression is another machine learning technique applied to IPS to predict a device's position based on observed signal strengths.

As stated earlier, one of the key advancements in this area is the use of machine learning algorithms for indoor positioning. Yin and Lin [4] proposed a machine learning-based IPS using ultra-wideband (UWB) signals. They developed a mathematical strategy to create data for fingerprint-based indoor localisation systems, reducing the need for extensive measurements. The study reported that the average error of most test locations could be reduced to less than 150 mm using the best model. Another UWB positioning proposal by Kim et al. [5] enhanced the positioning performance by classifying channel conditions with channel impulse response (CIR) using long short-term memory (LSTM) training of the observed channel status.

Meanwhile, Bellavista-Parent et al. [6] conducted a systematic review of works that use Wi-Fi signals for positioning and machine learning models to estimate indoor position. They found that indoor positioning based on Wi-Fi trends uses neural network-based models, as evaluated in empirical experiments. However, many works still conduct an assessment in small areas, which can influence the quality of the results presented. Menon et al. [7] presented DIY-IPS, an open-source real-time indoor positioning mobile application that detects users' indoor position by employing dual-band RSSI fingerprinting of available Wi-Fi access points. The app can be used without additional infrastructural costs to detect users' indoor positions in real time.

Nevertheless, Etiabi et al. [8] proposed a federated learning (FL) framework for hierarchical 3D indoor localisation using a deep neural network. They found that by adopting a hierarchical learning scheme, they could improve the localisation accuracy by up to 24.06% compared with the non-hierarchical approach. Alhomayani and Mahoor [9] provided a comprehensive review of deep learning methods in indoor positioning. They discussed the advantages and disadvantages of various fingerprint types for indoor positioning and analysed the performance of different deep learning methods. They noted that while incorporating deep learning into fingerprinting has resulted in significant improvements, it has also introduced new challenges such as complexity, additional cost, computational time, reliability, lack of resources, etc.

Despite the promising performance of indoor positioning, new challenges imposed by machine learning need to be addressed and explored to further improve the accuracy and practicability of the IPS. Hence, in this paper, we propose to explore the use of three well-known machine learning techniques, linear regression, decision trees and KNN in a Wi-Fi-based IPS. Received signal strength indicator (RSSI) values are used as input to train these two models for positioning purposes. A layout is designed to ensure a robust dataset that accurately represents the variations in signal strength across different locations within the indoor space. Upon testing the trained models, performance accuracy is compared among the linear regression, decision trees, and KNN algorithms, and the results will be discussed technically. The time it takes to reach their maximum performance is also measured. The novelty of this research is to develop robust, accurate, and cost-effective IPS that can meet the diverse needs of various applications using low complexity machine learning algorithms.

This paper is organised as follows: Sect. 38.2 presents the research methodology by explaining the system flow of the research in detail. It also discusses the concepts of linear regression, decision trees, and KNN algorithms. Section 38.3 compares the performance accuracy between the above-mentioned models. It includes finding the best parameter value for the KNN model to achieve the highest accuracy. Lastly, Sect. 38.4 comprises a summary of the exploration of Wi-Fi-based IPS using the above-mentioned machine learning techniques as well as future recommendations.

## 38.2 Research Methodology

The methodology for this study involves comparing three different machine learning models, linear regression, decision trees, and KNN, in the context of an IPS. The process begins with data collection, followed by data preprocessing, model training, and finally, model comparison, as shown in Fig. 38.1.

Linear regression is a statistical method that has been widely used for predictive modelling, including in the context of indoor positioning. Linear regression models the relationship between two or more variables by fitting a linear equation to the observed data. In IPS, linear regression can be used to predict the position of a device based on the observed signal strengths from various Wi-Fi access points. The coefficients of the linear equation represent the weights or importance of each signal strength in determining the position [10].

A decision tree is a straightforward yet powerful machine learning model for classification and regression tasks. They use a flowchart-like structure, where each node represents a feature, each branch a decision rule, and each leaf an outcome. Selecting which feature to use as a node is crucial, often solved by optimization techniques looking for the most informative features. Criteria such as Gini impurity,

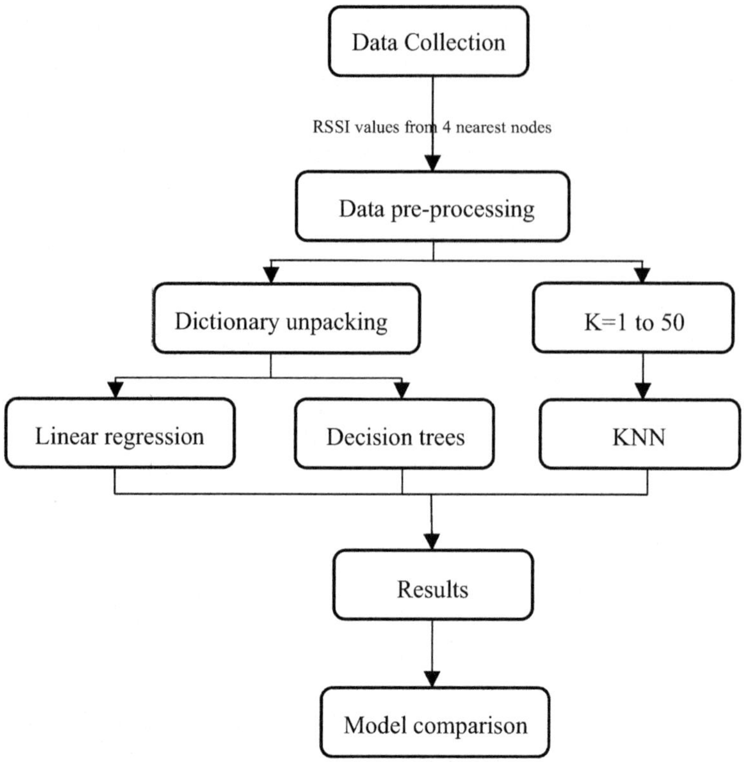


**Fig. 38.1** Process flow of research

information gain, and variance reduction are used to measure the quality of a split. One main issue with decision trees is overfitting, which can be mitigated by setting minimum samples per leaf, maximum tree depth, and tree pruning. Advanced algorithms like random forests and gradient boosting machines use decision trees as their base, improving performance and generalisation [11]. Whereas, KNN algorithm is a type of instance-based learning that is widely used in various fields, including indoor positioning. The fundamental principle behind KNN is classification by the majority vote of its neighbours. In other words, an object is assigned to the class most common among its $K$-nearest neighbours [12]. In the context of indoor positioning, KNN is used to estimate the position of a device based on the $K$ nearest fingerprints in the radio map. These fingerprints are selected based on a distance metric, such as Euclidean distance, which measures the similarity between the observed signal strengths and the fingerprints in the radio map.

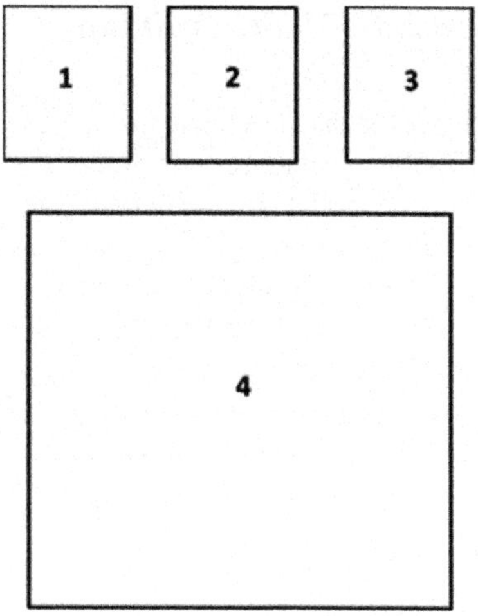


**Fig. 38.2** Floor layout

### *38.2.1 Data Collection*

The first step in the methodology is data collection. The data is collected from the four nearest nodes, represented by four different rooms in an indoor layout shown in Fig. 38.2. The RSSI values, a measure of the power level, are received [13] by the mobile device located at each of the four nodes. The data collected is in the form of a dictionary, where the keys are the names of the nodes (e.g. "X1", "X2", "X3", "X4"), and the values are the corresponding RSSI values (e.g. RSSI1, RSSI2, RSSI3, RSSI4).

### *38.2.2 Data Preprocessing*

The next step is data preprocessing. Dictionary unpacking is required for the linear regression model, and decision trees, then it is used to input the data into the model. Dictionary unpacking is a technique in Python [14] that allows you to pass items from a dictionary as keyword arguments to a function. On the other hand, searching for the best $K$ value is needed only for the KNN model. By varying $K$ in the range from 1 to 50, the value of $K$ that results in the highest accuracy is selected for the final KNN model in the next step.

### 38.2.3 Model Training

Once the data has been preprocessed, it is used to train the machine learning models. For the linear regression model, the unpacked RSSI values are used as input features, and the model is trained to predict the position of the device based on these features.

Furthermore, for decision trees the same unpacked RSSI values are fed to the model to later predict the position.

For the KNN model, the position of the device based on the $K$ nearest fingerprints in the radio map is classified using the chosen best $K$ value. Both models are trained in a real-time situation using Wi-Fi, and the time to reach their highest accuracy will be recorded, which is an indicator of the number of data points required by the models for the IPS system.

### 38.2.4 Model Comparison

The final step in the methodology is model comparison. The linear regression, decision trees, and KNN model prediction results are compared to evaluate their performance. The model that achieves the highest accuracy in predicting the position of the device is selected as the best model for the IPS.

## 38.3 Results and Discussion

The results of this study provide an in-depth analysis of three distinct machine learning models, linear regression, decision trees, and KNN, in the context of an IPS. The performance of the models has been evaluated based on their accuracy in predicting the position of a device using RSSI values from the four nearest nodes.

The data collection process is meticulously carried out in four different rooms. In each room, data is collected for three minutes, with one sample being read every second. This process ensures a robust dataset that accurately represents the variations in signal strength across different locations within the rooms. The data collection process is designed to capture a comprehensive snapshot of the Wi-Fi signal landscape in each room, considering potential signal strength variations due to physical obstructions, distance from the access point, and interference from other devices.

### 38.3.1 Selection of K Value (KNN Model)

For the KNN model, different values of $K$ in the range of 1–50 are tried since the choice of $K$ can significantly impact the performance of KNN. Based on the

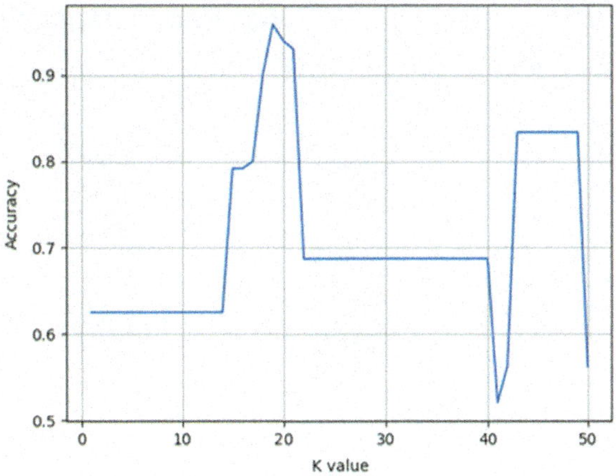


**Fig. 38.3** Performance of KNN model with different $K$ values

result shown in Fig. 38.3, the highest accuracy for KNN has been achieved at $K = 19$. Moreover, only a small range of $K$ values ($K = 18$ to 21) is able to generate accuracies higher than 90%. This indicates that the KNN model is very sensitive to the selection of $K$ in IPS.

## *38.3.2 Performance of Indoor Positioning*

Figure 38.4 presents the accuracy comparisons of three different machine learning models. First of all, the linear regression model achieves an accuracy of 70%, which indicates that it can only correctly predict the position of the device based on the RSSI values 70% of the time. The assumption made by this model, which is a linear relationship between the RSSI values and the position, may not always hold in complex indoor environments with multipath propagation and signal attenuation. Despite this limitation, the model serves as a useful baseline for indoor positioning and can be effective in environments where signal propagation is relatively stable and predictable.

The second model tested is decision trees, the uniqueness of its approach proved better results than linear regression, but it is still not sufficient. Decision trees achieve 82% accuracy compared with 70% with linear regression. Further alternations to the model may prove useful.

On the other hand, the KNN model is able to achieve a significantly higher accuracy of 96% in predicting the position of the device. This result is obtained with $K = 19$, which has been found to be the optimal value of $K$ in this study. The high accuracy of the KNN model suggests that it is highly effective in classifying the position of the device based on the $K$ nearest fingerprints in the radio map. The KNN model does not make any assumptions about the relationship between the RSSI values and

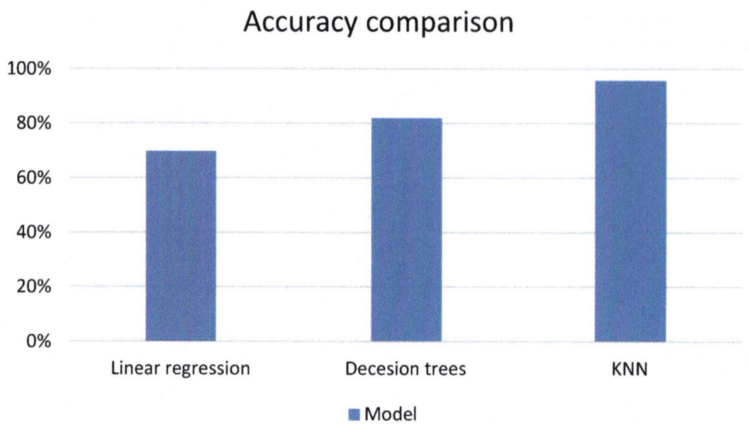


**Fig. 38.4** Accuracy comparison between the linear regression, decision trees, and KNN models

the position, which makes it more flexible and capable of capturing complex patterns in the data.

Nevertheless, another significant breakthrough in this study is the development of an optimised KNN model that only requires 0.9 min (approximately 54 s) to reach its maximum accuracy of 96%, as shown in Fig. 38.5. This means that only around 30% of the data is needed for the model to reach its saturation point, beyond which the accuracy does not increase. This finding is crucial as it can help optimise the data collection process, saving time and computational resources, and making the IPS more efficient.

**Fig. 38.5** Time taken for the KNN model to reach maximum performance

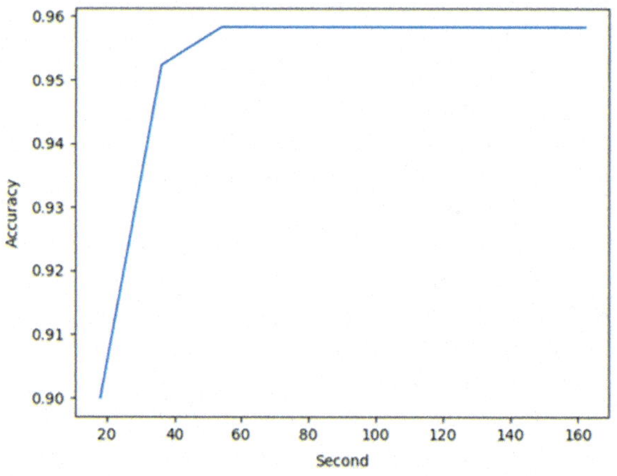

## 38.4 Conclusion

Indoor positioning has become a critical component of modern technology, finding applications in diverse fields such as navigation, security, emergency services, and even retail for tracking customer behaviour. This study focused on comparing three machine learning models, linear regression, decision trees, and KNN, in the context of an IPS using Wi-Fi signals. The experiment was carried out practically in four different rooms, and the real-time dataset was obtained by collecting the RSSI values from each node in the rooms. From the results obtained, it is clear that the KNN model outperformed the linear regression model with much higher accuracy and in a shorter period of time in predicting the position of the device. However, it is important to note that the choice of model should be based on the specific requirements and constraints of the application. For example, the KNN model may require more computational resources than the linear regression model, especially for large values of $K$. Therefore, the trade-off between accuracy and computational resources should be considered when choosing a model for an IPS.

In conclusion, this study provides valuable insights into the application of machine learning models for indoor positioning using Wi-Fi signals. The results demonstrate the potential of these models for improving the accuracy of IPS, with the KNN model showing particular promise. However, further research is needed to address the challenges associated with these techniques and explore other potential machine learning models for indoor positioning. The findings in this study can also serve as an indicator for further exploration of multi-floor indoor positioning and the selection of Aps' locations.

**Acknowledgements** This research project is fully sponsored by Internal Research Fund (MMUI/220061), Multimedia University.

## References

1. G.C. Chung, M. Kamarudin, I.E. Lee, S.F. Tan, Identification and mitigation of non-line-of-sight path effect using repeater for hybrid ultra-wideband positioning and networking system. Turk. J. Electr. Eng. Comput. Sci. **30**(3), 1084–1096 (2022)
2. F. Liu, J. Liu, Y. Yin, W. Wang, D. Hu, P. Chen, Q. Niu, Survey on WiFi-based indoor positioning techniques. IET Commun. **14**(9), 1372–1383 (2020)
3. A. Nessa, B. Adhikari, F. Hussain, X.N. Fernando, A survey of machine learning for indoor positioning. IEEE Access **8**, 1372–1383 (2020)
4. Y. Anqi, L. Zihuai, Machine learning aided precise indoor positioning (2022). http://arxiv.org/abs/2204.03990v1
5. D.H. Kim, A. Farhad, J.Y. Pyun, UWB positioning system based on LSTM classification with mitigated NLOS effects. IEEE Internet Things J. **10**(2), 1822–1835 (2022)
6. V. Bellavista-Parent, J. Torres-Sospedra, A. Perez-Navarro, New trends in indoor positioning based on WiFi and machine learning: a systematic review, in *2021 International Conference on Indoor Positioning and Indoor Navigation (IPIN)* (2021), pp. 1–8

7. R. Menon, A. Lakhdari, A. Abusafia, Q. He, A. Bouguettaya, DIY-IPS: towards an off-the-shelf accurate indoor positioning system, in *Proceedings of the 28th Annual International Conference on Mobile Computing and Networking* (2022), pp. 778–780
8. Y. Etiabi, W. Njima, E.M. Amhoud, DIY-IPS: federated learning based hierarchical 3D indoor localization, in *2023 IEEE Wireless Communications and Networking Conference (WCNC)* (2023), pp. 1–6
9. F. Alhomayani, M.H.Y. Mahoor, Deep learning methods for fingerprint-based indoor positioning: a review. J. Location Based Serv. **14**(3), 129–200 (2020)
10. D.C. Montgomery, E.A. Peck, G.G. Vining, *Introduction to Linear Regression Analysis* (Wiley, 2021)
11. T. Hastie, R. Tibshirani, J. Friedman, *The Elements of Statistical Learning* (2009)
12. Z. Zhang, Introduction to machine learning: K-Nearest Neighbors. Ann. Transl. Med. **4**(11), 218 (2016)
13. H.P. Mistry, N.H. Mistry, RSSI based localization scheme in wireless sensor networks: a survey, in *2015 Fifth International Conference on Advanced Computing and Communication Technologies* (2015), pp. 647–652
14. S. Raschka, V. Mirjalili, *Python Machine Learning: Machine Learning and Deep Learning with Python, scikit-learn, and TensorFlow 2* (Packt Publishing Ltd, 2019)

# Chapter 39
# Investigations of User Positioning Errors by Using Local DFMC SBAS Correction with Higher-Order Ionospheric Delay Mitigation in Thailand

**Somkit Sophan, Pornchai Supnithi, and Lin M. M. Myint**

**Abstract** Dual-frequency multi-constellation (DFMC) satellite-based augmentation system (SBAS) is essential to support airplane navigation. Typically, the global navigation satellite system (GNSS) with the L1 (1575.42 MHz) and L5 (1176.45 MHz) frequencies are utilized to remove the ionospheric delays based on the ionosphere-free (IF) combination (L1-L5), nevertheless, the errors due to higher-order ionospheric (HOI) delays still need to be corrected. With the local DFMC SBAS correction with a HOI mitigation in Thailand had not been considered and investigated yet. Therefore, we investigate local DFMC SBAS corrections with the HOI mitigation which are generated from the base stations in Thailand. The corrections are approximated from the local total electron content (TEC) values based on the Klobuchar model. Firstly, the local HOI days are estimated by using the observed pseudorange, and consequently, the local DFMC SBAS corrections are generated based on the minimum sum-variance technique with the IF carrier smoothing code. The GPS, Galileo, and QZSS data from the reference stations network in Thailand are utilized. In the preliminary study, the user positioning errors on the quiet days are evaluated based on the single point positioning (SPP) algorithm. The results show that both horizontal and vertical position errors are reduced by the local DFMC SBAS corrections with the HOI mitigation. The position error improvements on the quiet days can be experienced by 11%.

---

S. Sophan · P. Supnithi (✉) · L. M. M. Myint
School of Engineering, King Mongkut's Institute of Technology Ladkrabang, Bangkok 10520, Thailand
e-mail: pornchai.su@kmitl.ac.th

S. Sophan
e-mail: 61601026@kmitl.ac.th

L. M. M. Myint
e-mail: linminmin.my@kmitl.ac.th

 
M. T. Islam et al. (eds.), *Proceedings of the 8th International Conference on Space Science and Communication*, Springer Proceedings in Physics 303,
https://doi.org/10.1007/978-981-97-0142-1_39

## 39.1 Introduction

Dual-frequency multi-constellation (DFMC) satellite-based augmentation system (SBAS) has been developed after the global navigation satellite system (GNSS) can broadcast the GNSS navigation via the (1575.42 MHz) and L5 (1176.45 MHz) frequencies. Typically, airplane navigation from enroute to approach phases can be supported by DFMC SBAS similar to L1 SBAS (first generation), whereby many satellite systems can be supported, not only the global positioning system (GPS) [1–3]. Particularly, in DFMC SBAS, the errors due to first-order ionospheric (FOI) delays are removed by the ionosphere-free (IF) combination using L1 and L5 frequencies. However, the residual errors of the pseudoranges between a satellite and a receiver still include higher-order ionospheric (HOI) delays [4, 5]. Hence, the residual errors include the satellite clock-offset errors and the HOI delays. Consequently, the DFMC SBAS corrections generated from the residual errors can be affected by those HOI delays.

In [6], the researchers estimate the HOI values by the code pseudorange with the International Geomagnetic Reference Field 13th generation (IGRF-13) model and the Global Ionospheric Maps (GIM) model. Then, they analyze HOI effects on the Beidou Navigation Satellite (BDS) through precise orbit determination (POD). In addition, the HOI effects on POD of the low-earth orbit (LEO) satellite and the GPS data are also investigated in [7]. The International Reference Ionosphere-2016 (IRI-2016) and the IGRF-13 models are utilized to observe the HOI delays. Typically, HOI values can be approximated by the total electron content (TEC) as explained in [8] and the TEC values are estimated through the geometry-free combination with the raw GNSS data [9]. The maximum HOI delays are about 10 cm, [5]. On the other hand, for the GPS L1 frequency, the TEC value of 1 TEC unit (TECU equals $10^{16}$ electrons/$m^2$) is approximately equal to a delay of 0.16 m. The local HOI delays can be approximated by using the local TEC values based on the Klobuchar model, [10]. This technique is a sample method for preprocessing. However, the DFMC SBAS correction with the HOI effects, computed from the local data, has not been explored in many regions including Thailand, and the preliminary study of those effects should be investigated before realization of DFMC SBAS.

Therefore, we investigate the DFMC SBAS corrections with the HOI mitigation, locally computed in Thailand. Firstly, the HOI delays are estimated from the local TEC estimations considering the equivalent of the TEC value (1 TECU) to the ionospheric delay (0.16 m) for the L1 frequency. Particularly, to compute the local HOI delays, the local ionospheric delays are calculated using the Klobuchar model and then, they are subtracted from the residual errors. After that, the locally computed DFMC SBAS correction is generated from the residual errors of the local reference stations. The minimum sum-variance and IF carrier smoothing code techniques are also employed. In the preliminary study, we employ the single point positioning (SPP) algorithm to evaluate the user positioning errors during the days characterized by locally quiet conditions. These quiet days are determined using a local parameter, the rate of TEC index (ROTI) values. Hence, the related theories are explained in

Sect. 39.2, followed by the methodology in Sect. 39.3, and the dataset in Sect. 39.4. Afterward, the results and discussions are described in Sect. 39.5, and the conclusion is given in Sect. 39.6.

## 39.2 Related Theories

This section explains how the DFMC SBAS corrections are locally computed from the residual errors of the local reference stations using the minimum sum-variance and IF carrier smoothing code techniques. Then we will explain the preliminary investigation of the user positioning accuracies with the locally computed DFMC SBAS corrections. All related theories are described as follows.

### *39.2.1 Pseudorange Measurements*

Typically, the IF carrier smoothing code technique is used to reduce the high noises of the code pseudorange for DFMC SBAS. The technique originates from the Hatch filter [4]. Thus, the smoothed pseudoranges between the $i$th satellite and the $j$th receiver ($R^i_j$) at time $t$ can be expressed as

$$R^i_j(t) = \gamma P^i_{j,\mathrm{IF}} + (1-\gamma) \times \left(R^i_j(t-1) + \Phi^i_{j,\mathrm{IF}}(t) - \Phi^i_{j,\mathrm{IF}}(t-1)\right), \tag{39.1}$$

where $\gamma$ is the weight of the filter function, i.e., $\left(t_{\text{previous}} - t_{\text{current}}\right)/\alpha$, $\alpha$ is the constant weight (in sec), $P^i_{j,\mathrm{IF}}$ is the code pseudorange computed from the IF combination between the $i$th satellite and the $j$th receiver (in meters), and $\Phi^i_{j,\mathrm{IF}}$ is the carrier-phase pseudorange computed from the IF combination between the $i$th satellite and the $j$th receiver (in meters). Based on the DFMC SBAS standard [1], with $\alpha = 100$ s, the error in a delay of 0.25 m can be achieved by the carrier smoothing code technique within 200 s after initialization.

Hence, the pseudorange from the $i$th satellite to the $j$th receiver at each time can be calculated from

$$R^i_j = \rho^i_j + c\left(\delta t_j - \delta t^i\right) + T^i_j + \delta I^i_{j,\mathrm{IF}} + e^i_j, \tag{39.2}$$

where $\rho^i_j$ is the geometric range (in meters), $\delta t_j$ is the receiver clock-offset (in sec), $\delta t^i$ is the satellite clock-offset (in sec), $c$ is the speed of light in the vacuum (in meters/sec), $T^i_j$ is the tropospheric delay (in meters), $e^i_j$ is the combination of the code bias and smoothed error, and $\delta I^i_{j,\mathrm{IF}}$ is the HOI delay of the code (in meters). The Klobuchar model is used to estimate the local ionospheric delays as explained in [10]. The HOI delays of code can be estimated by using the local TEC values as described in [8]. As the TEC value (1 TECU) is equal to the ionospheric delay (0.16

m) for the L1 frequency, the HOI delays can be directly converted from the TEC values.

### 39.2.2 Local DFMC SBAS Correction Estimation

Here, the DFMC SBAS correction with the HOI mitigation is locally generated through three steps. In the first step, the residual errors of the pseudoranges between each pair of satellite $i$th and receiver $j$th ($\delta R_j^i$) at each time can be computed by

$$\delta R_j^i = R_j^i - \|\mathbf{r}_j^i - \mathbf{r}_j\| + c\widehat{\delta t}_j^i - \hat{T}_j^i - \widehat{\delta I}_{j,IF}^i = c\left(\delta t_j - \Delta\delta t_j^i\right) + e_j^i, \quad (39.3)$$

where $\mathbf{r}_j^i$ is the coordinate vectors $\left(\left[X_j^i, Y_j^i, Z_j^i\right]\right)$ of the $i$th satellite with the $j$th receiver, $\mathbf{r}_j$ is the precise coordinate vectors $\left(\left[X_j, Y_j, Z_j\right]\right)$ of the $j$th receiver, $\|\cdot\|$ is the Euclidean distance function, $\widehat{\delta t}_j^i$ is the $i$th satellite clock-offset interpolated by the pseudorange of the $j$th receiver, $\hat{T}_j^i$ is the tropospheric delay of the $i$th satellite based on the minimum operational performance specification (MOPS [3]) with the $j$th receivers, $\widehat{\delta I}_{j,\mathrm{IF}}^i$ is the local HOI delays after the IF combination of the $i$th satellite with the $j$th receiver approximated by the local TEC values based on the Klobuchar model [8, 10], and $\Delta\delta t_j^i$ is the residual error of the $i$th satellite clock-offset at the $j$th receiver (in sec).

In the second step, the satellite clock-offset errors are approximated using the minimum sum-variance technique with the triple difference (TD) observations [11]. This technique approximates the $\delta t_j$ term and consequently, the $\Delta\delta t_j^i$ term. Given that the reference receiver is fixed, the $j$th receiver clock-offset approximation ($\widehat{\delta t}_j$) is considered as a constant for a short interval time (5 min). Thus, the clock-offset errors of each satellite $i$th ($\hat{b}^i$) at time $t$ are estimated by

$$\hat{b}^i(t) \approx \min_{K_{min} \le \mathscr{b}^i \le K_{max}} \left( \sum_{j=1}^{M_r} \sum_{\tau=1}^{N_\tau} \frac{\left(\hat{b}_j^i(t-\tau) - \mathscr{b}^i\right)^2}{M_r + N_\tau} \right), \quad (39.4)$$

where $\hat{b}_j^i$ is the remaining term of $\delta R_j^i - c\widehat{\delta t}_j$ at time $t$, $K_{\min}$ and $K_{\max}$ are the minimum and maximum values of the $\hat{b}_j^i(t)$, where $j \in \{1, 2, \ldots, M_r\}$, $\mathscr{b}^i$is the arbitrary mean values, $M_r$ is the number of receivers, and $N_\tau$ is the approximated time interval.

In the final step, according to the DFMC SBAS's draft standard version [1], the satellite ephemeris prediction error in the $X$-, $Y$-, and $Z$-directions can be caused by the residual satellite clock-offset errors. These errors are included in the pseudorange utilized for interpolating the satellite positions based on the Earth-Center Earth-Fixed (ECEF). Thus, the satellite position error correction vectors ($\delta^i = \left[\delta X^i, \delta Y^i, \delta Z^i\right]$) can be generated from the $\hat{b}^i$ values projected by unit vectors of the virtual satellite coordinate. Therefore, the local DFMC SBAS correction included the HOI delay mitigation at each time is calculated by

$$\delta^i = \frac{\mathbf{r}^i}{\left\|\mathbf{r}^i\right\|} \times \hat{b}^i, \tag{39.5}$$

where $\mathbf{r}^i$ is the virtual coordinate vectors ($\left[X^i, Y^i, Z^i\right]$) of the $i$th satellite approximated from the pseudoranges at the master station. Additionally, the ECEF coordinate system is the World Geodetic System 1984 (WGS-84) [12].

In addition, the satellite position error correction vectors from the message type format (DFMC SBAS) are utilized for correcting the satellite position errors estimated by the pseudorange of the user. For many satellites, the position error corrections of the satellite $i$th at the user ($\mathbf{r}_u^i$) for each time are expressed as

$$\mathbf{r}_u^i = \hat{\mathbf{r}}_u^i + \delta^i, \tag{39.6}$$

where $\hat{\mathbf{r}}_u^i$ is the estimated coordinate vectors ($\left[\hat{X}_u^i, \hat{Y}_u^i, \hat{Z}_u^i\right]$) of the $i$th satellite interpolated by the pseudorange of the user. Note that the $\hat{\mathbf{r}}_u^i$ vectors are dependent on the user's location. If the users are away from the selected reference station in Eq. (39.4), the satellite position error corrections might be degraded.

### 39.2.3 Accuracy Issue

The GNSS positioning errors are calculated through differencing between the actual position and the estimated position. The probability of the estimated position is determined by at least 95th percent position error, which is the accuracy requirement of the International Civil Aviation Organization (ICAO, [2]). If the precision approach (PA) applications of the APV-I, APV-II, and CAT-I categories are considered, the accuracy requirement can be summarized as given in Table 39.1. Additionally, in this work, the user positioning errors with the local DFMC SBAS correction are evaluated, the availability is not included.

**Table 39.1** ICAO SARP's accuracy requirement on the signal-in-space [2]

| Typical operation | Accuracy of horizontal 95% (m) | Accuracy of vertical 95% (m) |
|---|---|---|
| APV-I | 16.0 | 20 |
| APV-II | 16.0 | 8.0 |
| CAT-I | 16.0 | 6.0–4.0 |

## 39.3 Methodology

The local DFMC SBAS correction with the HOI mitigation computed in Thailand is investigated by six steps as shown in Fig. 39.1.

The FOI delays are subtracted based on the IF combination method with the GPS (L1-L5), Galileo (E1-E5a), and QZSS (L1-L5) data. Then, the method cancels the multipath noises through the elevation mask parameter. It is used to cut off any satellites with an elevation angle below the set value. After that, it reduces the high noises of the IF code pseudorange with the IF carrier smoothing code technique, Eq. (39.1). In the fourth step, after subtracting the HOI delays using Eq. (39.3), the residual error of the pseudoranges between the satellite and the receiver are obtained. Then, the local DFMC SBAS correction is generated by Eqs. (39.4) to (39.5), this correction consists of the satellite position error correction vectors. Finally, the horizontal positioning error (HPE) and vertical positioning error (VPE) of the user position are computed based on the SPP algorithm [12] with the Haversine formula. Additionally, the position error correction of the satellite is described in Eq. (39.6).

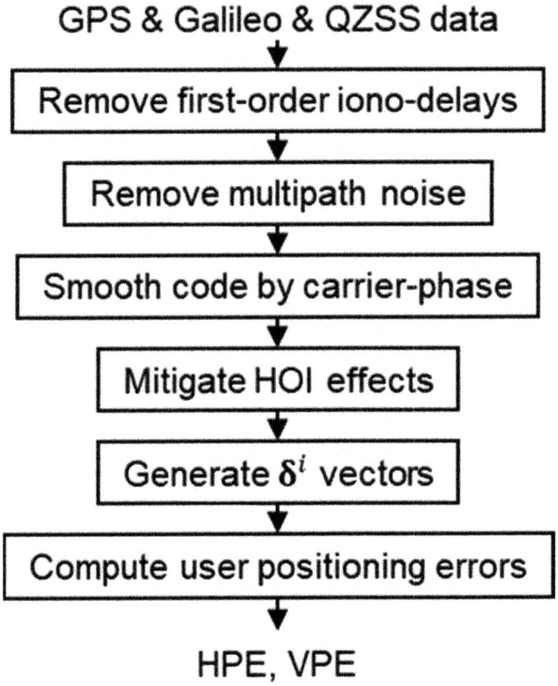


**Fig. 39.1** Investigated diagram of the local DFMC SBAS corrections

## 39.4 Dataset

The important parameters for the experimental setup are explained in Table 39.2. In this work, the six stations are considered, the local GNSS reference stations, and the two GNSS stations are represented as the users. The geodetic position: latitude ($\varphi$), longitude ($\lambda$), and height ($h$) are used for each station. Since the virtual coordinate vectors in Eq. (39.5) can be approximated by only one reference station, in this study, DPT9 is considered the master station. The distances between the master station and user stations (CNBR and UTTD) are approximately 60 and 432 km, respectively. For investigating the positioning performance of the HOI delay mitigation from the local DFMC SBAS correction, the quiet days are determined based on the low ROTI values. Here, the GNSS observations of four days (non-consecutive) are considered.

Normally, the data sampling period is 1 s for supporting a preprocessing of the SBAS service because the SBAS correction data is updated at least every 6 s. The elevation mask of 15° is defined to minimize the multipath noise. Moreover, the $N_\tau$ term of 300 samples (for 5 min) is considered to approximate the $\hat{b}_j$ values in the preprocessing. Similarly, the ROTI calculation uses the $N_\tau$ of 5 min (the time window of the 5-min interval).

The days, the local ROTI values below 0.5 TECU/min, are identified as the quiet days [13]. Figure 39.2 shows the ROTI values on quiet days computed from the UTTD station. The ROTI values are less than 0.5 TECU/min with a small fluctuation. The different colors represent each PRN (GPS, Galileo, and QZSS). The ROTI is computed by defining the elevation masks of 30° with the 5-min interval. These parameters are typically employed for the ROTI computations.

**Table 39.2** Simulation parameters

| Items | Values | | |
|---|---|---|---|
| References | $\varphi$ (°) | $\lambda$ (°) | $h$ (m) |
| CHAN | 12.6103 | 102.1024 | 7.1469 |
| CHMA | 18.8353 | 98.9700 | 294.9409 |
| DPT9 | 13.7568 | 100.5732 | 37.2459 |
| NKSW | 15.6906 | 100.1141 | 20.3220 |
| PJRK | 11.8116 | 99.7963 | -12.5090 |
| UDON | 17.4127 | 102.7807 | 149.3229 |
| *Users* | | | |
| CNBR | 13.4060 | 100.9977 | -7.5523 |
| UTTD | 17.6301 | 100.0963 | 50.2470 |
| Observed year | 2022 | | |
| Quiet days | 5th, 7th, 14th, and 19th (February) | | |
| Sampling period | 1 s | | |
| Elevation mask | 15° | | |
| $N_\tau$ | 300 samples (5 min) | | |

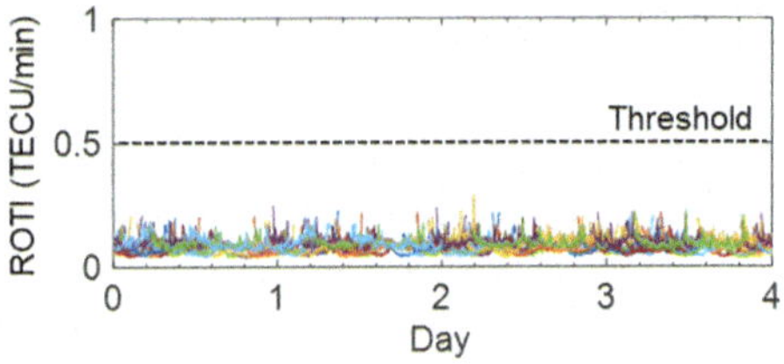


**Fig. 39.2** Local ROTI observed at the UTTD station for four quiet days

## 39.5 Results and Discussions

In this work, the user positioning errors are investigated on quiet days. For the preliminary study, the local reference stations in Table 39.2 are considered to generate local DFMC SBAS corrections and then evaluate the user position accuracy based on the ICAO SARPs requirement described in Table 39.1. The simulation results are demonstrated and discussed below.

### 39.5.1 *Local DFMC SBAS Correction*

Here, the local DFMC SBAS corrections are generated using the residual errors computed from the local reference stations in Thailand. Then, the correction for each satellite is evaluated without encoding to the DFMC SBAS message type format [1]. Figure 39.3 displays the satellite position correction vectors using the HOI mitigation in time series, the universal time clock (UTC) for four days. The separate color lines represent each PNR of satellites such as G04 and G10 for GPS, E02 and E03 for Galileo, and J02 and J03 for QZSS. Each line is computed by Eq. (39.4) with the virtual coordinate vectors approximated by the pesudorange at the DPT9 station.

In Fig. 39.3, the $\delta X^i$, $\delta Y^i$, and $\delta Z^i$ values of J02 and J03 are mostly small fluctuations ($\pm 2$ m) due to the quasi-zenith satellite orbit. The moved trajectory of QZSS is slower than GPS and Galileo, they are the medium Earth's orbit. Some spike values below $-10$ m can be caused by the outlier of each reference station. Since the receivers are installed on the ground, it is possibly affected by the multipath noise, especially the satellite with the elevation angle below 30°.

In addition, the local DFMC SBAS correction with and without the HOI delays are compared to investigate the local HOI effects. For the correction with the HOI delays, the $\widehat{\delta I}^{i}_{j,\mathrm{IF}}$ term in Eq. (39.2) is deleted. Figure 39.4 shows the 95th percent difference on four quiet days. The distinct color graphs represent the absolute difference of the satellite position error correction vectors with and without the HOI delays for four days, with the $Y$-axis that is shown in the logarithm scale. The different results above 10 cm (maximum HOI) are the J03 satellites with $\delta X^i$, and G04 and E03 satellites with $\delta Y^i$. Note that some different results are more than the maximum HOI delays.

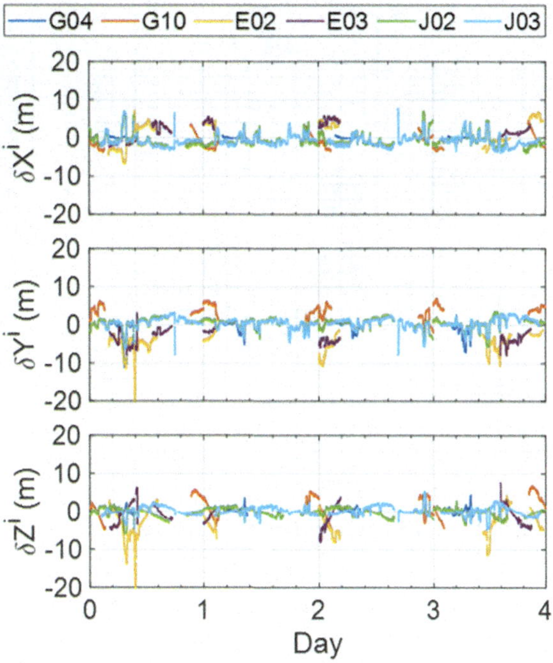


**Fig. 39.3** Satellite position correction vectors with the HOI mitigation for four quiet days

Possibly, some outliers in the correction vectors with the HOI mitigation can be filtered by applying the minimum sum-variance technique on the TD observations.

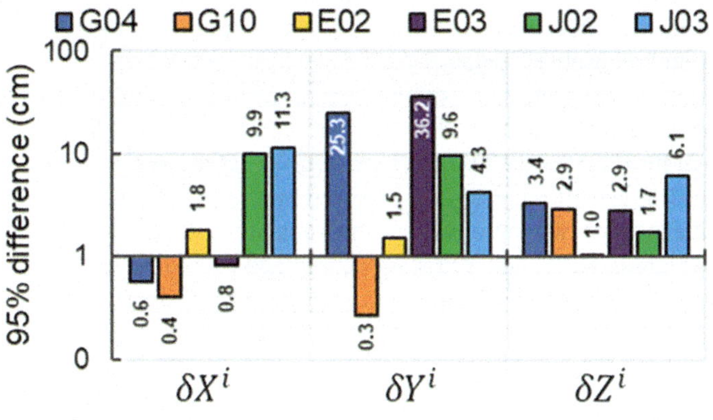


**Fig. 39.4** Differences of the satellite position correction vectors between without and with the HOI mitigation for four quiet days

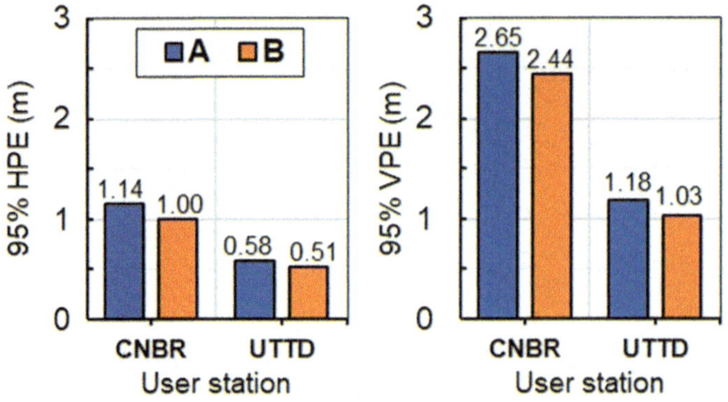


**Fig. 39.5** Comparisons between the 95th percentile positioning errors without and with the HOI mitigation in terms of HPE (left panel) and VPE (right panel) during the quiet days at CNBR and UTTD

### *39.5.2 User Positioning Errors*

Finally, the user positioning errors are evaluated to investigate the local DFMC SBAS correction without and with the local HOI mitigation. Figure 39.5 shows the 95th percent HPEs and VPEs at the user stations (CNBR and UTTD). The legends of '*A*' and '*B*' represent the user position estimated by the DFMC SBAS using locally generated corrections without and with the HOI mitigation, respectively. In the HPE, the position errors without HOI are less than with HOI about 0.14 and 0.07 m at CNBR and UTTD, respectively. Likewise, in the VPE, the position errors without HOI are also less than with HOI which are about 0.21 and 0.15 m at CNBR and UTTD, respectively. As expected, the results of HPE are less than VPE due to the geography and ionospheric delays.

In addition, we compute the position error improvements of the local DFMC SBAS corrections without and with the HOI mitigation. The HPE and VPE at CNBR can be improved by 12.28% and 7.92%, $(1-(B/A)) \times 100\%$, respectively. Similarly, at UTTD, the HPE and VPE can be also achieved by 12.07% and 12.71%, respectively. Note that the average of position error improvements is about 11%. Moreover, when the 95th percentile HPEs and VPEs are determined according to the ICAO SARPs requirement, the typical operation of all categories can be done in Thailand.

## 39.6 Conclusion

In this work, we attempt to investigate the local DFMC SBAS correction without and with the HOI mitigation in Thailand. The user positioning errors are evaluated on four quiet days. From the simulations, the local DFMC SBAS correction with the HOI mitigation can be used to reduce the user positioning errors. The position error improvements can be experienced by 11%. In future work, the availability of the local DFMC SBAS correction will be evaluated.

**Acknowledgements** The GNSS data is supported by Department of Public Works and Town and Country Planning (DPT) of Thailand. This work is partially funded by the NSRF via the Program Management Unit for the Human Resources and Institutional Development, Research and Innovation (grant no. B05F640197), King Mongkut's Institute of Technology Ladkrabang (grant no. KDS2018/003), and the ASEAN IVO project (http://www.nict.go.jp/en/asean_ivo/index.html).

## References

1. DFMC SBAS SARPS—Baseline draft for validation. Int. Civil Aviat. Org. (2018). [Online]. Available: https://www.icao.int/airnavigation/Pages/DFMC-SBAS.aspx. Last accessed 2022/06/05
2. ICAO SARPS Annex 10: International Standards and Recommended Practices, Aeronautical Telecommunications. Int. Civil Aviat. Org. Montréal, QC, Canada (2006)
3. RTCA DO-229D, *Minimum Operational Performance Standards for Global Positioning System/Wide Area Augmentation System Airborne Equipment* (RTCA Inc., Washington, DC 20036 USA, 2006)
4. J. Subirana, J. Zornoza, M. Hernández-Pajares, *GNSS Data Processing-Volume I: Fundamentals and Algorithms* (European Space Agency, 2013). [Online]. Available: https://gssc.esa.int/navipedia/GNSS_Book/ESA_GNSS-Book_TM-23_Vol_I.pdf. Last accessed 2022/06/05
5. Z. Liu, Y. Li, J. Guo, F. Li, Influence of higher-order ionospheric delay correction on GPS precise orbit determination and precise positioning. Geodesy Geodyn. **7**(5), 369–376 (2016)
6. K. Xi, X. Wang, Higher order ionospheric error correction in BDS precise orbit determination. Adv. Space Res. **67**(12), 4054–4065 (2021)
7. J. Guo, L. Qi, X. Liu, X. Chang, B. Ji, F. Zhang, High-order ionospheric delay correction of GNSS data for precise reduced-dynamic determination of LEO satellite orbits: cases of GOCE, GRACE, and SWARM. GPS Solutions **27**(13) (2023)
8. H.A. Marques, J.F. Monico, M. Aquino, RINEX_HO: second-and third-order ionospheric corrections for RINEX observation files. GPS Solutions **15**(3), 305–314 (2011)
9. A. Krypiak-Gregorczyk, P. Wielgosz, Carrier phase bias estimation of geometry-free linear combination of GNSS signals for ionospheric TEC modeling. GPS Solutions **22**(45) (2018)
10. P.D.T. Setti, D.B.M. Alves, C.M.D. Silva, Klobuchar and Nequick G ionospheric models comparison for multi-GNSS single-frequency code point positioning in the Brazilian region. Boletim de Ciências Geodésicas **25**(0) (2019)
11. J.V. Sickle, J.A. Dutton, GEOG 862: GPS and GNSS for geospatial professionals. Penn State's College of Earth and Mineral Sciences (2020). [Online]. Available: https://www.e-education.psu.edu/geog862/print/book/export/html/1727. Last accessed 27 Apr. 2022
12. T. Takasu, RTKLIB ver. 2.4. 2 Manual, RTKLIB: an open source program package for GNSS positioning (2013)
13. X. Pi, A.J. Mannucci, U.J. Lindqwister, C.M. Ho, Monitoring of global ionospheric irregularities using the worldwide GPS network. Geophys. Res. Lett. **24**(18), 2283–2286 (1997)

# Chapter 40
# Low-Profile Wideband MIMO Antenna with Low ECC and High Efficiency for Sub-6 GHz 5G NR/Wi-Fi 5 Applications

**Md. Mhedi Hasan, Mohammad Tariqul Islam, Touhidul Alam, Badariah Bais, and Norbahiah Misran**

**Abstract** A low-profile wideband two-element multiple input multiple output (MIMO) antenna with high efficiency and high isolation covering in sub-6 GHz 5G new radio (NR) n77/78/79/Wi-Fi 5 is proposed for 5G communications. Antenna geometry is constructed by wide hexagonal slot patch and ground structure with a simple inset-fed microstrip line. The antenna wideband feature is greatly influenced by the hexagonal slot size and position. Two single antennas have been used to design MIMO system with a small interelement distance of 18 mm, having a size of 80 $mm^3$ $\times$ 40 $mm^3$ $\times$ 1.575 $mm^3$. The proposed MIMO antenna achieves wideband coverage of 53% (3.02–5.2 GHz) and high-stable total efficiency of > 87% within the operating range. Moreover, the developed MIMO configuration shows a high isolation of > 14.6 dB and a high gain of 5.4 dBi. In addition, the designed MIMO system exhibited a very low-envelope correlation coefficient (ECC) of < 0.009 and near 10 dB diversity gain (DG) (greater than 9.96 dB), which are remarkably better than acceptable limit. Thus, the outstanding performances of proposed MIMO configuration support its potentiality for 5G NR/Wi-Fi 5 communications.

Md. Mhedi Hasan · M. T. Islam (✉) · B. Bais · N. Misran
Department of Electrical, Electronic and Systems Engineering, Faculty of Engineering and Built Environment, Universiti Kebangsaan Malaysia (UKM), 43600 UKM Bangi, Selangor, Malaysia
e-mail: tariqul@ukm.edu.my

Md. Mhedi Hasan
e-mail: p106258@siswa.ukm.edu.my

B. Bais
e-mail: badariah@ukm.edu.my

Md. Mhedi Hasan
Department of Information and Communication Technology (ICT), Faculty of Engineering, Comilla University, Cumilla 3506, Bangladesh

T. Alam
Pusat Sains Angkasa (ANGKASA), Institut Perubahan IklimUniversiti Kebangsaan Malaysia (UKM), 43600 UKM Bangi, Selangor, Malaysia

 
M. T. Islam et al. (eds.), *Proceedings of the 8th International Conference on Space Science and Communication*, Springer Proceedings in Physics 303,
https://doi.org/10.1007/978-981-97-0142-1_40

## 40.1 Introduction

The evolving wireless technology demands improved signal quality and system capacity. The globally accelerating fifth generation (5G) technology unlocked new research opportunities that transform our world smarter, faster, and safer. Multiple input multiple output (MIMO) technique with multi-antennas is a core technology of modern high-speed wireless communications, supporting high data rate and low latency and improving channel capacity [1–4]. In a compact MIMO system, multiple antennas are integrated within a limited space, therefore the MIMO components are coupled closely, leading to high coupling energy between closely integrated antenna components. This mutual coupling energy reduces antenna individual performance by affecting input impedance, radiation patterns, and port isolation, thus the low-profile highly isolated wideband antenna array with high efficiency and high gain is a major concern of the scientific community [2, 5, 6]. The metasurface, frequency selective surface (FSS)-based antennas, and dielectric resonator (DR) antennas provide wideband performance, but they are high profile, which hinders their applicability for 5G compact smart devices [7]. The patch antenna is a promising candidate for high-speed compact smart devices owing to its low-profile, lightweight, and low-cost characteristics [8, 9]. A broadband compact highly isolated MIMO antenna with 3.6 dBi gain and 69.7% radiation efficiency has been designed in [5] for 5G sub-6 GHz band applications. Another 5G operating wideband MIMO antenna is developed in [10] with low isolation (> 13 dB). Recently, a shared radiator-based MIMO antenna is developed by Tian et al. with > 12.5 dB isolation and 40.3–48.5% total efficiency that can be potentially useful in 5G compatible smartphones [1]. Similarly, a 5G terminal antenna is developed in [11], which operates partially in n78 and n79 5G bands with a maximum efficiency of 42.6% and ECC < 0.42. A narrow band metamaterial-inspired 5G MIMO antenna is designed with a minimum isolation of 15 dB [12]. Also, another triband MIMO antenna with better than 14.2 dB isolation is proposed by Maurya et al. for IoT applications [13]. This antenna shows 2.9 and 4.7% impedance bandwidth (BW) at the lower two operating bands. From the above-published recent related studies, it is noticed that the low-profile, highly efficient MIMO patch antenna with a broad working range is crucial for 5G wireless applications.

Accordingly, this article presents a low-profile MIMO antenna with wideband and high-efficiency features for sub-6 GHz 5G/Wi-Fi 5 applications. The key specialties of the proposed MIMO configuration are its wide coverage area (3.02–5.2 GHz) 53%, short spacing between two radiators (18 mm), compact size (80 $mm^3$ × 40 $mm^3$ × 1.575 $mm^3$), better isolation, and high stable total efficiency (maximum 95%). Moreover, the designed MIMO achieves high isolation (> 14.6 dB) with excellent realized gain (5.4 dBi). The diversity indices of the developed MIMO configuration were also investigated by ECC and DG and obtained excellent performance compared with the standard limit.

## 40.2 Single Antenna Geometry

Figure 40.1a exhibits the single antenna configuration with an indication of its details design specifications. The antenna patch and feed line dimensions are predicted using the well-established mathematical formula [2]. Initially, a microstrip feed integrated rectangular patch is designed on top of a 40 mm$^3$ × 40 mm$^3$ × 1.575 mm$^3$ low loss Rogers 5880 substrate (dielectric constant = 2.2). The 0.035 mm thick copper (annealed) is coated on the back side of the substrate. The EM simulator CST is utilized to design and analyze antenna structures and the simulated outcomes are plotted in Fig. 40.1b. No resonances are observed in the initial stage, as depicted in Fig. 40.1b. Two insets are introduced into the feed line in the next step by shifting the feed line to the left of the patch. In this stage, the antenna resonates at 4.15 GHz with a narrow − 10 dB BW as demonstrated in Fig. 40.1b. Finally, the hexagonal-shaped structures are etched in the center and diagonal corners of the radiator. Also, the antenna ground is etched by the wide hexagonal slot, which is demonstrated in Fig. 40.1a. The hexagonal slot on both sides of the antenna offers large coverage in the sub-6 GHz 5G n77, n78, n79, and 5G Wi-Fi bands (Fig. 40.1b). These slots size and location have a significant influence on the wideband characteristics.

Figure 40.2a presents the gain (realized gain) and efficiency (total efficiency) for initial antenna and shows outstanding radiation characteristics. It is observed from the total efficiency plot (Fig. 40.2a), more than 90% total efficiency is seen within the entire operating band (2.8–5.2 GHz) with the highest total efficiency of 99%. The maximum gain is recorded 5.2 dBi with > 4 dBi gain within the operating band, as depicted in Fig. 40.2a, supporting its suitability for n77/n78/n79/Wi-Fi 5 applications.

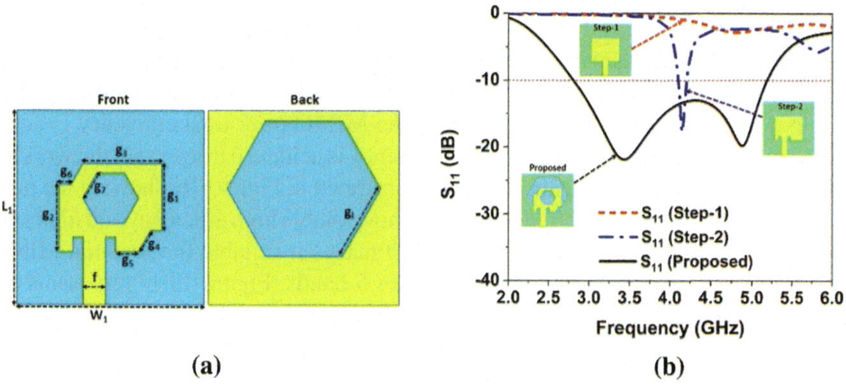


**Fig. 40.1** **a** Antenna geometry with details specifications ($W_1, L_1 = 40, f = 4.7, g_1, g_2 = 13.67, g_3 = 17, g_4 = 5, g_5 = 4.9, g_6 = 2.5, g_7 = 6, g_l = 16.1$, unit mm) and **b** $S_{11}$ for various design stages

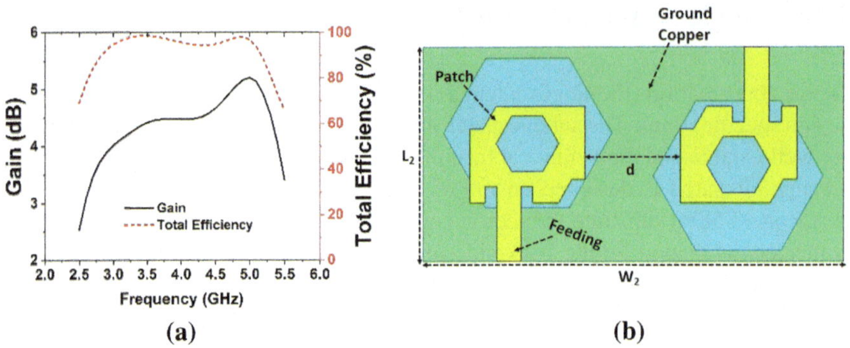


**Fig. 40.2** **a** Total efficiency and gain curve for single antenna, **b** proposed MIMO configuration ($L_2 = 40$ mm, $W_2 = 80$ mm, and $d = 18$ mm)

## 40.3 MIMO Configuration

The designed MIMO geometrical configuration is shown in Fig. 40.2b, where two similar single antennas are utilized to develop the proposed MIMO system. Moreover, the ground hexagonal slots are positioned like a single antenna, which confirms its design simplicity. The two MIMO excitation ports are placed oppositely as displayed in Fig. 40.2b. The developed MIMO system size is 40 mm$^3$ × 80 mm$^3$ × 1.575 mm$^3$ with a low interelement spacing of $d = 18$ mm, outperforming the related recent MIMO antennas. Figure 40.3a reveals the scattering performance of the developed MIMO configuration, showing nearly identical impedance bandwidth to the initial antenna and high isolation between two ports. From the reflection curve ($S_{11}$), it is obvious that the antenna bandwidth extends 3.02–5.2 GHz, which covers sub-6 GHz 5G/Wi-Fi 5 frequencies. Figure 40.3a also plotted the isolation ($S_{21}$) curve; notably, greater than 14.6 dB isolation is achieved within the working spectrum.

Figure 40.3b illustrates the developed MIMO total efficiency and gain for port 1, while port 2 results are similar to port 1. The MIMO peak total efficiency is noted 95% and almost stable, more than 87% efficiency is achieved throughout the working band. The MIMO realized gain is also demonstrated in Fig. 40.3b, showing the peak gain of 5.4 dBi. Thus, the designed antenna shows stable high efficiency and high gain in the wide working frequency range, which makes it suitable for potential MIMO applications in the sub-6 GHz 5G and Wi-Fi 5 bands. Figure 40.4a represents the antenna patterns for single and MIMO configurations (proposed) at 4.5 GHz. The single and MIMO radiation patterns are quite identical for both planes. Noticeably, the E-plane pattern resembles omnidirectional, whereas the H-plane pattern looks dumbbell shape.

MIMO Diversity Performance. In MIMO arrangement, the correlation effects among MIMO antennas can be anticipated using envelope correlation coefficient (ECC), which is a key performance indicator of any MIMO antenna. The ECC of a two-port antenna can be computed using Eqs. (40.1) and (40.2) [2, 14].

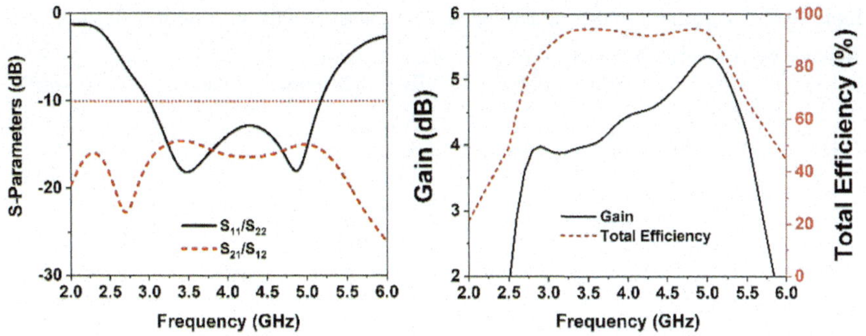


(a) (b)

**Fig. 40.3** **a** $S_{11}$ and $S_{21}$ responses for proposed MIMO configuration and **b** MIMO gain and total efficiency

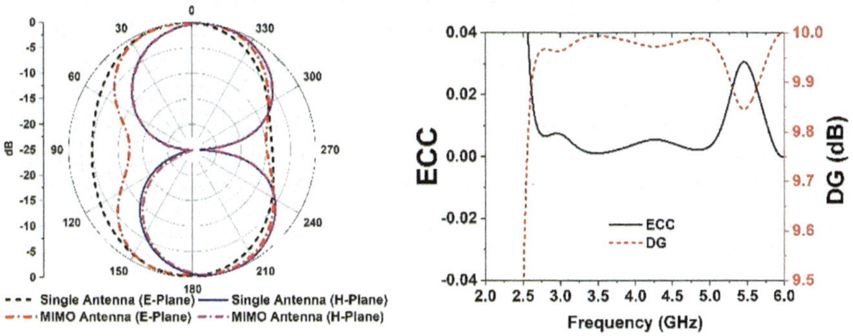


(a) (b)

**Fig. 40.4** **a** Radiation pattern for single and MIMO antenna, **b** MIMO diversity (ECC and DG) curves

$$\mathrm{ECC} = \frac{\left|S_{11}^{*}S_{11} + S_{21}^{*}S_{22}\right|^{2}}{\left(1 - |S_{11}|^{2} - |S_{21}|^{2}\right)\left(1 - |S_{22}|^{2} - |S_{12}|^{2}\right)} \tag{40.1}$$

$$\mathrm{ECC} = \frac{\left|\iint_{4\pi}\left[\overrightarrow{E}_{1}(\theta,\varphi).\overrightarrow{E}_{2}(\theta,\varphi)\right]\mathrm{d}\Omega\right|^{2}}{\iint_{4\pi}\left|\overrightarrow{E}_{1}(\theta,\varphi)\right|^{2}\mathrm{d}\Omega\iint_{4\pi}\left|\overrightarrow{E}_{2}(\theta,\varphi)\right|^{2}\mathrm{d}\Omega} \tag{40.2}$$

Figure 40.4b plotted the ECC curve for the designed two-element MIMO configuration and shows a very low value of $< 0.009$, which is much smaller from acceptable limit (0.5). Another notable MIMO performance indicator is diversity gain (DG), which can be determined from expression (40.3) [14]:

**Table 40.1** Comparative studies of developed MIMO antenna with other related antennas

| Reference | Antenna size ($mm^2$) | Antenna profile | BW (GHz) | Gain (dBi) | Total efficiency | Least isolation (dB) | ECC | DG (dB) |
|---|---|---|---|---|---|---|---|---|
| [14] | 80 × 40 | Low | 3.48–3.8 | 4.3 | 73% | > 12 | < 0.03 | 9.81 (ave) |
| [15] | 100 × 50 | Low | 3.21–3.28 | 3.3 | 95% (radiation) | > 22 | < 0.001 | > 9.9 |
| [10] | 100 × 50 | High | 2.7–3.6 | 4.5 | 92% | > 13 | < 0.009 | Not given |
| This work | 80 × 40 | Low | 3.02–5.2 | 5.4 | 95% | > 14.6 | < 0.009 | > 9.96 |

$$DG = 10\sqrt{1 - |ECC|^2} = 10\sqrt{1 - |\rho|^2} \tag{40.3}$$

The simulated DG values are plotted in Fig. 40.4b, showing a value of > 9.96 dB between MIMO antennas throughout the operating spectrum. Table 40.1 tabulated the comparative studies of developed MIMO antenna with other related antennas, showing better performance in respect of bandwidth, size, efficiency, ECC, DG, and isolation. Although the antenna [15] demonstrated high isolation and low ECC compared with proposed antenna, it has a very large dimension, narrow bandwidth, and low gain. Therefore, the proposed MIMO configuration is a promising choice for sub-6 GHz NR 5G/Wi-Fi 5 applications.

## 40.4 Conclusion

This work designed a low-profile, highly efficient MIMO configuration with wideband properties for n77/n78/n79 5G bands and 5G Wi-Fi applications. Two identical single patch antennas are deployed in the proposed MIMO configuration using 18 mm spacing between two radiators, confirming its compactness. The working bandwidth of each antenna component is within 3.02–5.2 GHz (BW = 2.18 GHz, 53%), covering n77/n78/n79 and 5G Wi-Fi frequency spectrums. Moreover, the designed MIMO structure demonstrated high efficiency, high isolation between two ports as well as excellent gain. The MIMO diversity characteristics have also been studied and found outstanding ECC value of less than 0.009 and DG of better than 9.96 dB, which are much better than the acceptable limit. Therefore, the above promising performance proves that the designed MIMO configuration can be suitable for 5G sub-6 GHz and Wi-Fi 5 applications.

**Acknowledgements** The authors acknowledge the support provided by Universiti Kebangsaan, Malaysia (grant code: DIP-2021-011). The authors also acknowledge the Bangabandhu Science and Technology Fellowship Trust, Ministry of Science and Technology, Bangladesh.

## References


1. X. Tian, Z. Du, Wideband shared-radiator four-element MIMO antenna module for 5G mobile terminals. IEEE Trans. Antennas Propag. **71**(6), 4799–4811 (2023)
2. M. Hasan, M.T. Islam, M. Samsuzzaman, M.H. Baharuddin, M.S. Soliman, A. Alzamil, I.I. Abu Sulayman, M. Islam, Gain and isolation enhancement of a wideband MIMO antenna using metasurface for 5G sub-6 GHz communication systems. Sci. Rep. **12**, 1–17 (2022)
3. M. Rahman, M.S. Islam, M.T. Islam, S.S. Al-Bawri, W.H. Yong, Metamaterial-based compact antenna with defected ground structure for 5G and beyond. Comput. Mater. Contin. **71**, 2383–2399 (2022)
4. M.S.J. Singha, W.S.W. Saleha, A.T. Abedc, M.A. Fauzid, A review on massive MIMO antennas for 5G communication systems on challenges and limitations. Jurnal Kejuruteraan **35**, 95–103 (2023)
5. K.V. Babu, S. Das, S.S. Ali, M. El Ghzaoui, B.T.P. Madhav, S.K. Patel, Broadband sub-6 GHz flower-shaped MIMO antenna with high isolation using theory of characteristic mode analysis (TCMA) for 5G NR bands and WLAN applications. Int. J. Commun. Syst. **36**, e5442 (2023)
6. M.L. Hakim, T. Alam, M.T. Islam, N.B.M. Sahar, M.S.J. Singh, H. Alsaif, M.S. Soliman, Metamaterial physical property utilized antenna radiation pattern deflection for angular coverage and isolation enhancement of mm-wave 5G MIMO antenna system. Radiat. Phys. Chem. **209**, 110998 (2023)
7. J. Fang, J. Li, P. Xiao, J. Dong, G. Li, S. Du, W.T. Joines, Coupling mode transformation based dielectric surface and metasurface for antenna decoupler. IEEE Trans. Antennas Propag. (2022)
8. J. Tiang, M. Islam, N. Misran, J. Mandeep, Slot loaded circular microstrip antenna with meandered slits. J. Electromagn. Waves Appl. **25**, 1851–1862 (2011)
9. R. Azim, M.T. Islam, N. Misran, Dual polarized microstrip patch antenna for Ku-band application. Informacije MIDEM-J. Microelectron. Electron. Compon. Mater. **41**, 114–117 (2011)
10. H.T. Chattha, 4-port 2-element MIMO antenna for 5G portable applications. IEEE Access **7**, 96516–96520 (2019)
11. L. Chang, H. Wang, Dual-band four-antenna module covering N78/N79 based on PIFA for 5G terminals. IEEE Antennas Wirel. Propag. Lett. **21**, 168–172 (2021)
12. F. Ali, M. Salih, M. Ilyas, MIMO patch antenna with metamaterial 3.5 GHz for 5G applications, in *Proceedings of the 2022 Second International Conference on Advances in Electrical, Computing, Communication and Sustainable Technologies (ICAECT)* (2022), pp. 1–4
13. N.K. Maurya, R. Bhattacharya, Design of compact dual-polarized multiband MIMO antenna using near-field for IoT. AEU-Int. J. Electron. Commun. **117**, 153091 (2020)
14. S.S. Jehangir, M.S. Sharawi, A single layer semi-ring slot Yagi-like MIMO antenna system with high front-to-back ratio. IEEE Trans. Antennas Propag. **65**, 937–942 (2016)
15. K.N. Patil, G. Akshay, T. Chaitanya, A.K. Dwivedi, N.K. Narayanaswamy, V.A. Singh, Two-element MIMO antenna with a small footprint for 5G connectivity, in *Proceedings of the 2022 Workshop on Microwave Theory and Techniques in Wireless Communications (MTTW)* (2022), pp. 110–113

# Chapter 41
# Rotational Symmetric Spider Net-Shaped Multi-Band ENG Metamaterial for C- and X-band Applications

**Md. Salah Uddin Afsar, Mohammad Rashed Iqbal Faruque, and Sabirin Abdullah**

**Abstract** A rotational symmetric spider net-shaped multiband-band epsilon-negative (ENG) metamaterial is proposed in this book chapter. This is an assembly of three modified interlinked octagonal split rings with a central octagonal close ring (OCRR) resonator which are surrounded by a square ring (SSR). 0.1608λ mm$^2$ × 0.1608λ mm$^2$ is the electrical dimension of the suggested metamaterial unit cell; the smallest resonance frequency 4.02 GHz is used to estimate the concerned wavelength (λ). A dielectric material RT 6002 of customary thickness 1.524 mm is castoff as substrate. In numerical simulation, it yields five resonance peaks at 4.02, 5.68, 6.72, 8.48 and 11.66 GHz with magnitudes − 50.50 dB − 21.27 dB, − 39.86 dB, − 36.17 dB and − 32.31 dB, respectively. Besides, the negative value of permittivity ($\epsilon$) is observed at the frequency ranges 4.02–4.61 GHz, 6.72–7.03 GHz and 8.48–8.72 GHz. In addition, negative refractive index (NRI) shows at the frequency ranges 4.03–5.21 GHz, 5.69–5.76 GHz, 6.73–7.53 GHz and 8.49–9.62 GHz negative. The presented multi-band metamaterial is designed for C- and X-bands applications for the wireless electromagnetic communication system.

## 41.1 Introduction

Metamaterial is an engineered structure that can attain some advantageous exotic electromagnetic properties [1, 2]. The two fundamental effective characteristics of metamaterials are defined by the permittivity ($\varepsilon$) and permeability ($\mu$) [3, 4]. The electrical permittivity is the ratio of electric field strength ($E$) and displacement current ($D$). It indicates the ability of a material to accumulate electrical potential energy under the bias of an electric field to the vector point in the opposite directions

Md. S. U. Afsar · M. R. I. Faruque (✉) · S. Abdullah
Space Science Centre (ANGKASA), Institute of Climate Change (IPI), Universiti Kebangsaan Malaysia, UKM, 43600 Bangi, Selangor D. E., Malaysia
e-mail: rashed@ukm.edu.my

 
M. T. Islam et al. (eds.), *Proceedings of the 8th International Conference on Space Science and Communication*, Springer Proceedings in Physics 303,
https://doi.org/10.1007/978-981-97-0142-1_41

for negative permittivity. On the other hand, the ratio of speed of EM wave in the vacuum with speed of EM wave in any medium is known as refractive index ($\eta$) of any material, and NRI material means that the direction of electric lines of force density is opposite to the direction of electric field intensity.

A theoretical metamaterial with ENG and MNG features which is also termed as epsilon-negative and mu-negative metamaterial was introduced by Veselago [5]. Scientists all over world are showing great interest in the research of MM arena due its novel characteristics for electromagnetic communication system [6, 7]. An array of metal-dielectric-metal or reverse arrangement functioned as a metamaterial as single unit cell cannot act as an MM [8].

A numerous application is observed in the microwave application-based devices like stealth coating or cloaking [9], filtering [10], MM absorber design [11], bandwidth boosting [12], SAR reduction [13], etc. Faruque et al. [14] offered an LC patch-shaped MM for satellite band applications. This NRI metamaterial yielded dual resonance peaks which was located at 5.4 to 6.3 and 10.5 to 13.5 GHz.

Recently, a dual parallel T-shaped metamaterial was proposed for C-, X- and Ku-band applications which were founded on silicon substrate. The compact size of the proposed MM unit cell was $0.385\lambda0 \times 0.385\lambda \times 0.032\lambda$ [15]. Islam et al. designed an SRR inspired MM of pie-shaped metallic patches. It was a compact size triple-band MM in microwave band spectra that covered S-, C- and X-bands [16]. A complementary split ring resonator (CSRR)-based MM of size 5 $mm^2 \times 5$ $mm^2$ was proposed in 2017 [17] for WLAN/WiMAX applications. It was an MM of EMR 8 which resonance peak was located at 7.5 GHz. In the same year, a single negative MM on FR4 substrate was designed for improving the performance of microstrip transmission line [18]. It yielded dual resonance peaks 13.9 and 27.5 GHz. Recently, a quad-band MM of S-, C- and Ku-bands application was offered in [19]. It was an arrangement of altered three split ring resonator (SRR) which produced quad-resonance peaks at 2.02, 6.77 and 13.22. In the same year, a quad-band ENG metamaterial with enriched EMR was reported by Afsar et al. for satellite band applications [20]. This compact size MM exhibited quadruple resonance peaks at 2.35, 7.72, 9.23 and 10.68 GHz.

This new research writes up presented an innovative MM which is consist of three interlaced adjusted octagonal split rings and an central octagonal close ring resonator (OCRR) which are encircled by a square ring resonator. This spider net-shaped MM possesses multi-band resonance peaks at 4.02, 5.68, 6.72, 8.48 and 11.66 GHz which covers C- and X-bands. Furthermore, it also reveals negative permittivity ($\epsilon$) with negative refractive index ($n$) simultaneously.

## 41.2 Design Parameters and Simulation Setup

The unit cell of the proposed MM is labeled with parameters symbol which is illustrated in Fig. 41.1a. This is a new arrangement of three modified octagonal split rings with a central octagonal close ring which are surrounded by a square ring resonator

(SSR). Commercially available material RT 6002 of regular width 1.524 mm with the dimension 12 $mm^2 \times 12$ $mm^2$ is utilized as a substrate. The thermal conductivity and loss tangent of dielectric RT 6002 are 0.6 W/K/m and 0.0012, respectively. The copper (annealed) of thickness ($t_1$) 0.035 mm with thermal conductivity 5.96 × 107 $Sm^{-1}$ is used for the all patches of the upper layer and ground plate. The outer and inner radiuses of the first octagon are $R_1 = 4.6$ mm and $R_2 = 4.0$ mm, whereas the radius of the second octagon is $r_1 = 3.7$ mm and $r_2 = 3.1$ mm, respectively. The outer and inner radius of the third octagon is $r_3 = 2.8$ mm and $r_4 = 2.2$ mm, respectively. Outer and inner radius of the central smallest octagon is $r_5 = 1.8$ mm and $r_6 = 1.2$ mm. These three interlinked octagons (OSRRs) are placed at the center of SSR of dimension 11.40 $mm^2 \times 11.40$ $mm^2$. This is noteworthy that the width ($w$) of all octagonal resonating patches is 0.60 mm, whereas the width of the SSR is 0.50 mm. The split width ($g$) of three octagonal ring is 0.50 mm.

The perspective outlook and the simulation arrangement of suggested metamaterial is shown in Fig. 41.1b, c, respectively. Table 41.1 illustrates the values of design structure.

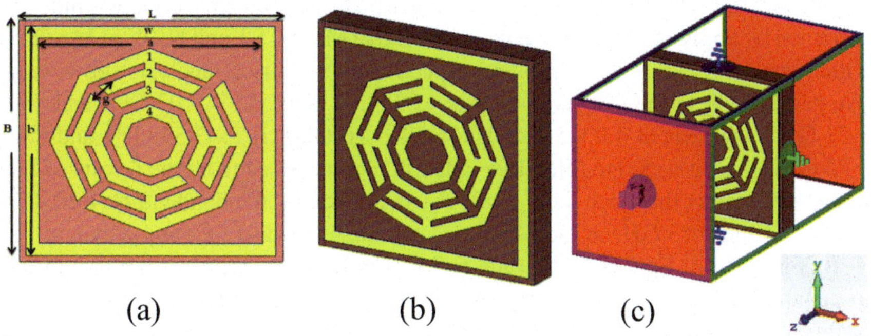


**Fig. 41.1** **a** Front view with parameters, **b** perspective view and **c** simulation setup of proposed MM unit cell

**Table 41.1** Configured design structure parameters

| Parameter symbol | Value (mm) | Parameter symbol | Value (mm) |
|---|---|---|---|
| $L$ | 12 | $R_1$ | 4.60 |
| $B$ | 12 | $R_2$ | 4.0 |
| $a$ | 11.40 | $r_1$ | 3.7 |
| $b$ | 11.40 | $r_2$ | 3.1 |
| $g$ | 0.50 | $r_3$ | 2.80 |
| $w$ | 0.50 | $r_4$ | 2.20 |
| $t$ | 1.60 | $r_5$ | 1.80 |
| $t_1$ | 0.035 | $r_6$ | 1.20 |

Appropriate boundary conditions are inserted to harvest the anticipated outcome from the offered MM design structure. The incident EM wave travel through the *z*-co-ordinate while the perfect electric condition and the perfect magnetic condition applied parallel to the *x*- and *y*-directions, respectively.

## 41.3 Results and Discussion

FIT-based EM simulator CST microwave studio is used to characterize the suggested metamaterial unit cell. The numerical analysis was performed within the frequency range 3–12 GHz to investigate the performance of the suggested MM unit cell. The results of reflection coefficient ($S_{11}$) and transmission coefficient ($S_{21}$) are depicted in Fig. 41.2. Numerical simulation yields five resonance peaks at 4.02, 5.68, 6.72, 8.48 and 11.66 GHz with magnitudes − 50.50 dB − 21.27 dB, − 39.86 dB, − 36.17 dB and − 32.31 dB, respectively which covered C- and X-bands of satellite band applications.

The electric permittivity ($\epsilon$), magnetic permeability ($\mu$) and refractive index ($n$) extracted by applying robust retrieval method (RRM) in CST and NRW in MATLAB which are depicted in Fig. 41.3a–c, respectively. Numerical analysis shows negative permittivity at the frequency ranges 4.02–4.61 GHz, 6.72–7.03 GHz and 8.48–8.72 GHz, whereas at 4.03–5.21 GHz, 5.69–5.76 GHz, 6.73–7.53 GHz and 8.48–8.72 GHz exhibit negative refractive index. It can be stated from the results of scattering parameters that each peak of the $S_{21}$ is tracked by the bandwidths 1.26, 0.07, 0.68, 0.57 and 0.43 GHz at the concerned resonance peaks of C- and X-bands.

It is seen from Fig. 41.3a that the value of permittivity varies from the positive to negative. Moreover, the magnitude of transmission coefficient started with the change of the permittivity from maximum to minimum value. This is also noticed

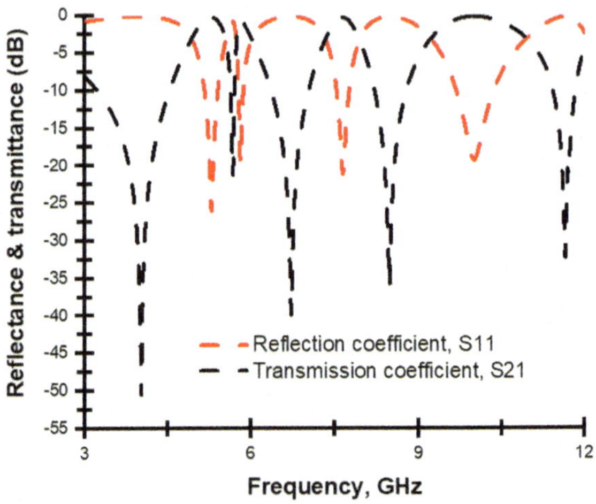


**Fig. 41.2** Scattering parameters ($S_{11}$ and $S_{21}$) of the proposed metamaterial

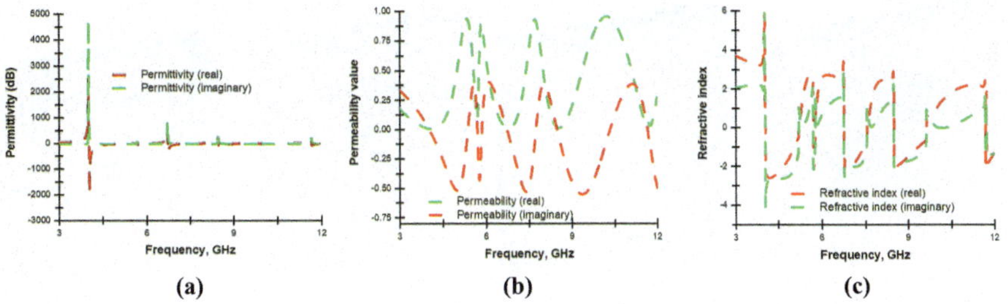


**Fig. 41.3** Materials characteristics, **a** permittivity ($\varepsilon$), **b** permeability ($\mu$), **c** refractive index

from Fig. 41.3b that the minimum positive value of permeability ($\mu$) begins at the minimum resonance frequency, and it is continued over the full frequency range.

Figures 41.4, 41.5 and 41.6 exhibit the H-field, E-field and surface current distribution for five resonance frequencies 4.02, 5.68, 6.72, 8.48 and 11.66 GHz, respectively. The amount of current flow as well as flow direction is concerned with the polarity and intensity of magnetic field distribution which is depicted in Fig. 41.4. Magnetic field becomes more intensified at the position of resonating patches where the flow of current is also high.

The intensity of E-field is observed at the reverse position of H-field which is depicted in Fig. 41.5. These changing patterns of two fields ($E$ and $H$) and surface current density can be described with Eqs. (41.1) and (41.2) [21].

$$\nabla \times \boldsymbol{H} = \boldsymbol{J} + \frac{\partial \boldsymbol{D}}{\partial t} \tag{41.1}$$

$$\nabla \times \boldsymbol{E} = \frac{\partial \boldsymbol{H}}{\partial t} \tag{41.2}$$

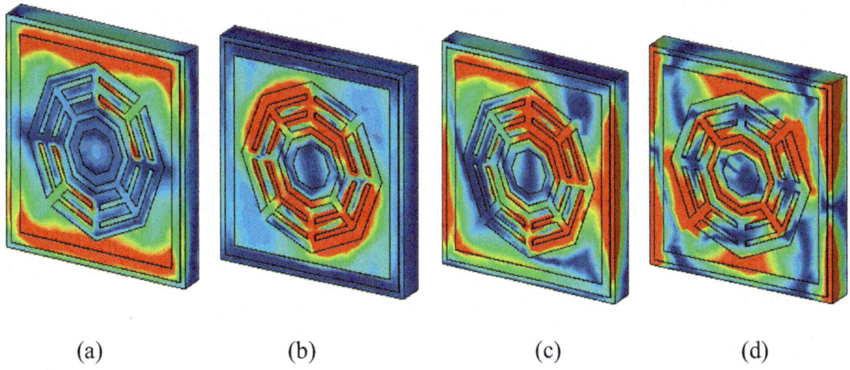


**Fig. 41.4** Magnetic field distribution at **a** 3.63 GHz, **b** 6.26 GHz, **c** 7.65 GHz and **d** 10.34 GHz

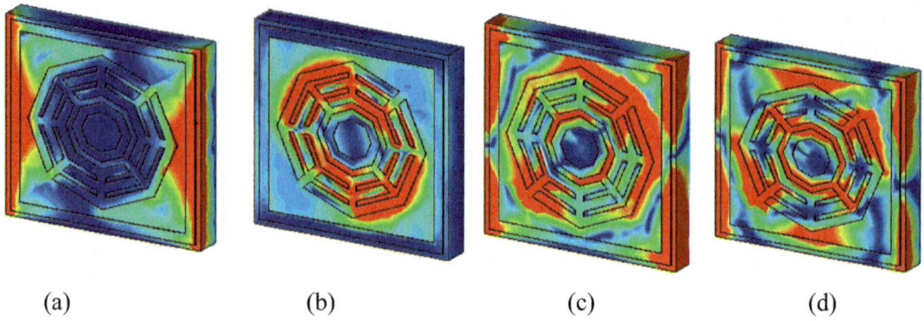


**Fig. 41.5** Electric field distribution at **a** 3.63 GHz, **b** 6.26 GHz, **c** 7.65 GHz and **d** 10.34 GHz

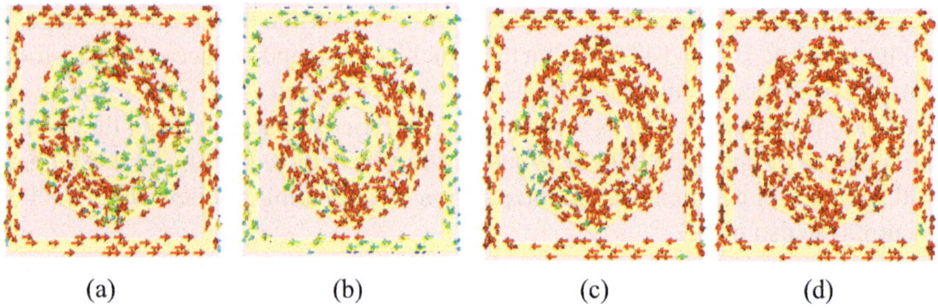


**Fig. 41.6** Surface current distribution at **a** 3.63 GHz, **b** 6.26 GHz, **c** 7.65 GHz and **d** 10.34 GHz

where vector operator $\nabla = \hat{\boldsymbol{i}}\frac{\partial x}{\partial t} + \hat{\boldsymbol{j}}\frac{\partial y}{\partial t} + \hat{\boldsymbol{k}}\frac{\partial z}{\partial t}$.

Additionally, each split gap in the resonators of the suggested MM structure played as a capacitor that contributes for the increment of the E-field strength in the split gaps.

Rotational symmetry is an important characteristic for being polarization insensitive of a metamaterial. The resonators of the MM unit cell are rotated 90° from its normal position (0°). Figure 41.7 depicts that at 0° and 90° rotation, the MM unit cell exhibits the same result. Therefore, the proposed metamaterial is rotational symmetric.

To validate the consistency of the outcome of proposed MM two types of parametric studies are carried out. Figure 41.8 exhibits the results of split gap analysis that shows the best response for the split gap 0.50 mm as yielding at 4.02, 5.68, 6.72, 8.48 and 11.66 GHz with magnitudes – 50.50 dB – 21.27 dB, – 39.86 dB, – 36.17 dB and – 32.31 dB, respectively. An appropriate material selection is another important task to make the offered MM efficient and cost-effective. The dielectric RT 6002 is selected for substrate as it exhibits expected outcome which is demonstrated in Fig. 41.9.

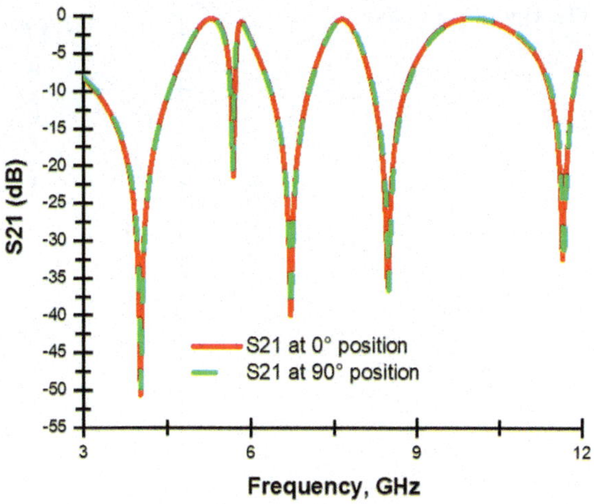


**Fig. 41.7** Transmission coefficient at different rotation angles

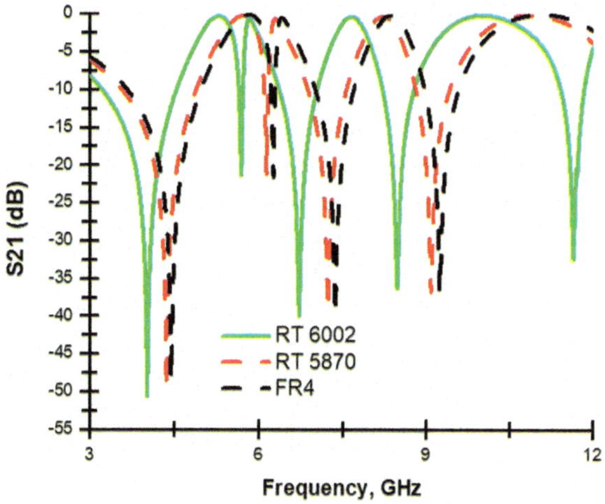


**Fig. 41.8** $S_{21}$ for various materials

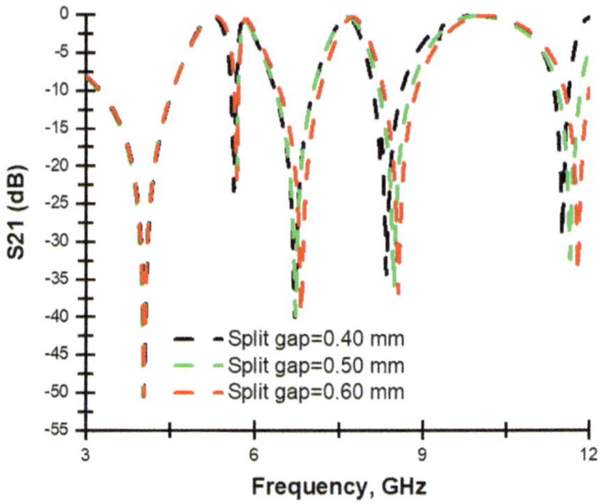


**Fig. 41.9** $S_{21}$ at various split gaps

## 41.4 Conclusion

In this paper, a multi-band spider net designed epsilon-negative metamaterial for satellite bands (C- and X-band) application is offered. For the numerical analysis of the performance of proposed MM, a commercial EM simulator CST microwave studio is deployed that exhibits five resonance peaks at 4.02, 5.68, 6.72, 8.48 and 11.66 GHz with magnitudes − 50.50 dB − 21.27 dB, − 39.86 dB, − 36.17 dB and − 32.31 dB, respectively. The simulated results are achieved through some renowned parametric studies such as split gaps, effect of dielectric materials. The E-field, H-field and surface current distribution are analyzed to comprehend of absorption mechanism of incident EM wave. The material characteristics permittivity ($\epsilon$), permeability ($\mu$) and refractive index ($n$) are extracted by using MATLAB. This new MM can be commissioned to enhance the efficiency of various EM communication devices for its exotic NRI and ENG properties. Meanwhile, the C- and X-bands are frequently used for satellite and radar applications [22].

**Acknowledgements** The authors acknowledge to the research Universiti Grant, Universiti Kebangsaan Malaysia, Geran Translasi, UKM-TR2022-05 conduct the research work.

## References

1. S.U. Afsar, M.R.I. Faruque, M.J. Hossain, M.U. Khandaker, H. Osman, S. Alamri, Modified hexagonal split ring resonator based on an epsilon-negative metamaterial for triple-band satellite communication. Micromachines **12**(8), 878 (2021)
2. P. Alitalo, A.E. Culhaoglu, A.V. Osipov, S. Thurner, E. Kemptner, S.A. Tretyakov, Bistatic scattering characterization of a three-dimensional broadband cloaking structure. J. Appl. Phys. **111**(3), 034901 (2012)

3. L.T. Govindaraman, A. Arjunan, A. Baroutaji, J. Robinson, A.-G. Olabi, Metamaterials for energy harvesting, in *Encyclopedia of Smart Materials* (Elsevier, 2022), pp. 522–534
4. H. Nakano, Categorization of natural materials and metamaterials, in *Low-Profile Natural and Metamaterial Antennas* (Wiley, Hoboken, NJ, USA, 2016), pp. 1–10
5. G. Veselago, The electrodynamics of substances with simultaneously negative values of $\epsilon$ and $\mu$. Soviet Phys. Uspekhi **10**(4), 509–514 (1968)
6. M. Martín, P. Vélez, A. Karami-Horestani, F. Medina, C. Fumeaux, F. Martín, L. Zhu, J.F. Hong, F. Medina, *Balanced Microwave Filters* (Wiley, Hoboken, NJ, USA, 2018), pp. 353–371
7. S.R. Thummaluru, R.K. Chaudhary, Mu-negative metamaterial filter-based isolation technique for MIMO antennas. Electron. Lett. **53**(10), 644–646 (2017)
8. R.W. Ziolkowski, Metamaterial-based antennas: research and developments. IEICE Trans. Electron **E89-C**(9), 1267–1275 (2006)
9. S. Islam, M. Faruque, M. Islam, A near zero refractive index metamaterial for electromagnetic invisibility cloaking operation. Materials **8**(8), 4790–4804 (2015)
10. W. Fu, Y. Han, J. Li, H. Wang, H. Li, K. Han, X. Shen, T. Cui, Polarization insensitive wide-angle triple-band metamaterial bandpass filter. J. Phys. D Appl. Phys. **49**(28), 285110 (2016)
11. K.V. Sreekanth, M. ElKabbash, Y. Alapan, A.R. Rashed, U.A. Gurkan, G. Strangi, A multiband perfect absorber based on hyperbolic metamaterials. Sci. Rep. **6**(1), 26272 (2016)
12. N. Nasimuddin, Z.N. Chen, X. Qing, Bandwidth enhancement of a single-feed circularly polarized antenna using a metasurface: metamaterial-based wideband CP rectangular microstrip antenna. IEEE Antennas Propag. Mag. **58**(2), 39–46 (2016)
13. Y. Wang, J. Bao, Y. Tian, Z. Wang, N. Li, Design of high gain wearable antenna based on wireless body area network communications. J. Phys. Conf. Ser. **1971**(1), 012018 (2021)
14. M.R.I. Faruque, A.M. Siddiky, E. Ahamed, M.T. Islam, S. Abdullah, Parallel LC shaped metamaterial resonator for C and X band satellite applications with wider bandwidth. Sci. Rep. **11**(1), 16247 (2021)
15. M. Bennaoum, M. Berka, A. Bendaoudi, A.Y. Rouabhi, Z. Mahdjoub, Investigation of a near-perfect quad-band polarization-insensitive metamaterial absorber based on dual-T circular shaped resonator array designed on a silicon substrate for C-, X-and Ku-bands applications. Silicon. **15**(2), 699–712 (2023)
16. M.R. Islam, M. Samsuzzaman, N. Misran, G.K. Beng, M.T. Islam, A tri-band left-handed meta-atom enabled designed with high effective medium ratio for microwave based applications. Results Phys. **17**, 103032 (2020)
17. W. Ali, E. Hamad, M. Bassiuny, M. Hamdallah, Complementary split ring resonator based triple band microstrip antenna for WLAN/WiMAX applications. Radioengineering **26**(1), 78–84 (2017)
18. A.R. Azeez, T.A. Elwi, Z.A. Abed AL-Hussain, Design and analysis of a novel concentric rings based crossed lines single negative metamaterial structure. Eng. Sci. Technol. Int. J. **20**(3), 1140–1146 (2017)
19. M.B. Hossain, M.R.I. Faruque, S. Abdullah, Modified combination of triple split ring resonator based triple band single negative metamaterial for satellite applications. IOP Conf. Ser. Earth Environ. Sci. **1167**(1), 012041 (2023). IOP Publishing
20. M.S.U. Afsar, M.R.I. Faruque, S. Abdullah, M.T. Islam, M.U. Khandaker, K.S. Al-Mugren, An innovative compact split-ring-resonator-based power tiller wheel-shaped metamaterial for quad-band wireless communication. Materials **16**(3), 1137 (2023)
21. S. Hannan, M.T. Islam, S.H. Almalki, M.R.I. Faruque, M.S. Islam, Rotational symmetry engineered, polarization and incident angle-insensitive, perfect metamaterial absorber for X and Ku band wireless applications. Sci. Rep. **12**(1), 3740 (2022)
22. P. Ranjan, C. Barde, A. Choubey, R. Sinha, A. Jain, K. Roy, A wideband metamaterial cross polarizer conversion for C and X band applications. Frequenz **76**(1–2), 63–74 (2022)

# Chapter 42
# Ionospheric Critical Frequency Observation from Ionosonde Data for GPS Positioning Improvement Over Malaysia

**Karthigesu Nagarajoo**

**Abstract** This research work was done to determine the critical frequency, *fc*, variation over Advanced Telecommunication Research Center (ATRC) UTHM by identifying the variability of the ionosphere from Ionosonde. The ionospheric variability was analyzed on diurnal variation basis at Low Solar Activity (LSA) and High Solar Activity (HSA). Ionosonde that was placed in ATRC was used to detect the *fc* of the ionosphere. Observation data in the form of Ionogram than was used to identify the *fc* and then the plotted using MATLAB. The comparison was done for two diurnal solar activity variation data for two consecutive months from the year 2011. The results then were compared with the International Reference Ionosphere's (IRI) *fc*. Overall, the solar activity has direct effect to the *fc* during daytime than after sunset over the equatorial region. By knowing the ionospheric condition, this will be helpful to improve the position of train when applying the European Rail Traffic Management System (ERTMS) in Malaysia later.

## 42.1 Introduction

The Earth's atmosphere is divided in few layers, and ionospheric layer is ionized by radiation from the Sun [1]. Ionosphere layer is situated from about 60–2000 km from the surface of the Earth. The ionosphere consists of free electrons and ions that being produced by the extreme ultraviolet radiation from the Sun. This layer is divided further into *D* layer, E layer and F-layer (which can be split further into F1 and F2 layers).

K. Nagarajoo (✉)
Department of Transportation Engineering Technology, Faculty of Engineering Technology, Universiti Tun Hussein Onn, 86400 Batu Pahat, Johor, Malaysia
e-mail: karthi@uthm.edu.my

 
M. T. Islam et al. (eds.), *Proceedings of the 8th International Conference on Space Science and Communication*, Springer Proceedings in Physics 303,
https://doi.org/10.1007/978-981-97-0142-1_42

The composition of the ionosphere is strongly influenced by the solar activity, and the amount of electron density depends on latitude, longitude and the altitude [2]. Since the solar activity is very active on the equatorial region, the amount of electron density is always greatest over this region. The five main variations are diurnal variations, seasonal variations, latitudinal variations, altitudinal variations and sunspot number. In order to study the ionospheric variations, all these five elements have to be considered.

In this work, the critical frequency, $f_c$, of the ionosphere has been determined using Ionosonde (Canadian Advanced Digital Ionosonde (CADI)) in ATRC UTHM. Ionosonde is pulse radar that operates in between 1 and 20 MHz of frequency range. The $f_c$ is equal to the maximum reflection frequency at vertical incidence from the ionospheric layer [4]. In other word, the $f_c$ correlates very well with the solar activity, and it is the measure of maximum electron density in the ionospheric layer [5].

### 42.1.1 Critical Frequency

The ionospheric $f_c$ varies with time of a day, altitude, season and location, and it correlates with the maximum amount of electron density. The $f_c$ is equal to the vertical incidence reflection from the ionospheric layer due to the maximum frequency [7]. Normally, the highest frequency signal can be obtained from the reflected point at the transmission location, which is located at the F layer which sometimes termed as $f_oF_2$ (critical frequency reflected from the F2 layer).

The $f_c$ depends on the ultraviolet (UV) radiation intensity from the Sun. It varies continuously with the time of the day, which correlates well with the local activity of the Sun. Since the $f_c$ correlates with the solar activity, it can be used to calculate the electron density composition in the ionospheric. This is where the role of Ionosonde comes in. The Ionosonde is used to transmit sequence of varying frequency pulses, and the reflection signal frequency will be used to measure the composition of the ionosphere in the unit of $f_c$. For this, we need a proper structured antenna farm which we have at UTHM.

From here, by using the Ionogram, the maximum electron density and the $f_c$ (in the function of height) can be obtained. It depends very much on the reflection frequency value, which represents the $f_c$ value. However, if the transmission frequency value is very high, the refraction index will be finite. Due to this, there is no any vertical reflection. Therefore, rather than using higher frequency value, a minimum value is sufficient to obtain the $f_c$ at which the vertical reflection will take place to determines the maximum electron density from the ionospheric layer. The reflection height is determined by the frequency of the Ionosonde signal that being transmitted up in the sky [8]. The vertical asymptote corresponds to the critical frequency of the F2 layer, $f_oF_2$.

Every ionospheric layers has its own $f_c$, and it correlates with the maximum electron density from the ionospheric layer. The relationship between electron density and the $f_c$ can be obtained using Eq. (42.1) below:

$$f_c = 9 \times 10^{-6}\sqrt{N_{\max}} \tag{42.1}$$

where $f_c$ = in MHz, and $N_{\max} = N_m F_2 = electron$ per $m^3$.

Other than that, the variation of $f_c$ can be obtained by comparing it with the International Reference Ionosphere's (IRI) database [9]. IRI is a database that being developed to determine the ionospheric data at any part of latitude, longitude and altitude. Most of the time, it correlates very well with the observed values.

## 42.2 Methodology

To conduct this research work, the understanding and the knowledge on how to 'read' (interpret) the Ionogram (in this case, CADI diagram) are very important. CADI is a low cost, state of art, flexible, full featured Ionosonde ideal for both routine ionospheric monitoring and research. This system provides sounding capability using high-power RF pulses at vertical incident. The reflection frequency range of CADI is from 1 to 20 MHz.

### *42.2.1 Interpretation of the Ionogram*

The transmitter provides power amplification of the RF signals. The required transmitter power is only 600 watts used to send the frequency range 1–20 MHz. Barker pulse coding provides an 11 dB S/N improvement which is equivalent to having 13 times more transmitter power [10]. The amplifier unit is a solid state and includes monitors for forward and reverse power.

The receiver cards amplify and demodulate the return signals. The system can incorporate from one to four receivers for spaced antenna measurements. The receiving signal will be displayed in the form of Ionogram and save data storage in the hard disk.

To operate the Ionosonde, two main programs are needed;

***CADIRUN***—to check the function of the CADI system to make sure it works properly. When this program executes, it reads the CADI configuration data from the file LOCATION.INI.

***CADITEST***—the program used for collecting data. When this program executes, the instructions will be given for the system to identify which data to collect and when to collect.

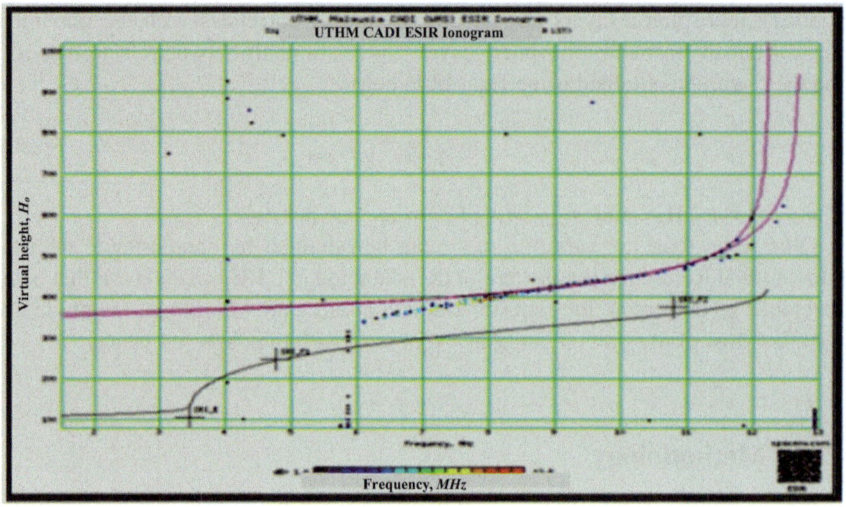


**Fig. 42.1** Sample of ESIR Ionogram from ATRC UTHM

The Expert System for Ionogram Reduction (ESIR) CADI Analysis software allows CADI Ionograms to be scaled. ESIR is an approach to automate Ionogram scaling. It uses a combination of knowledge-based pattern recognition and physics-based modeling techniques to extract traditional ionospheric parameters such as $f_oF_2$ and hmF2. ESIR also generates electron density profiles. If the Ionogram has adequate quality, the virtual height trace will be utilized to obtain a true height for given electron density profile.

The analysis of the Ionograms provides an extended database of EDP that contains the ionospheric variability as a function of latitude, local time (UT) and season. Figure 42.1 shows the sample of Ionogram, and from here, the $f_c$ and the virtual reflection height can be determined.

## 42.3 Results and Analysis

In this work, the $f_c$ of the ionosphere over equatorial region was observed, estimated and analyzed. The analysis was done for two levels of solar activity: local time LSA and HSA. The approach that was made in this work is that LSA is when the solar activity is not active at local time and HSA is when the solar activity is active at local time. So, to fulfill this, two different time periods have been chosen, which is 10:00–14:50 LT (HAS) and 01:00–04:50 LT (LSA). The Ionogram data that was obtained are for the months of October and November 2011 respectively. Only 2 months of data have been taken for this work to show the relationship between consecutive dataset. Data from 2011 was chosen because 2011 was the year where the solar activity was in up-trend mode.

Although the ESIR software can record the data at 15-min interval, but for this project, a 30-min interval is selected for analysis purposes. The difference between the $f_oF_2$ observations and the $f_oF_2$ values from IRI model were computed using Eq. (42.1). Analysis has been made by showing the similarity in $f_oF_2$ between observed (Ionosonde) and measured (IRI).

### *42.3.1 Variation Due to Solar Activity on October 2011*

The comparison between the $f_c$ in four days in October 2011 is shown in Fig. 42.2 during HSA. It can been seen that the $f_c$ develops gradually during the day between 10:00 and 14:50 h LT.

The observed and model $f_oF_2$ rise and reach the maximum value on 17 October 2011 at 13.05 PM (LT) compared to 14, 15 and 16 October 2011. This is due to the photoionization process that can occur during the day when the Sun is above the horizon. The $f_c$ of ionospheric layer will be greater at higher level solar activity.

On 14 October 2011, $f_oF_2$ is much lower than other observation days. It happens because when the Sun is disturbed (due to solar flares and high solar wind streams), it can affect the ionosphere. Most of the solar flares occur during high solar activity, and it will cause only a small effect to the ionosphere.

Figure 42.3 shows the comparison between the $f_oF_2$ observed and the IRI value. It has been proved that the IRI's $f_oF_2$ are very close to the observed $f_oF_2$ values.

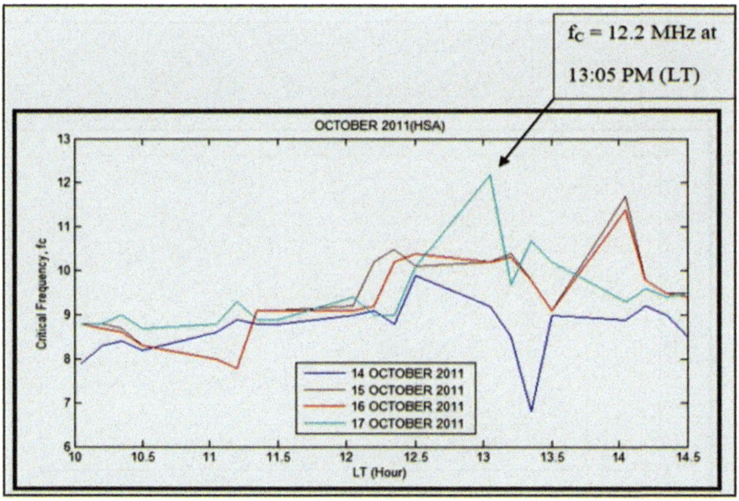


**Fig. 42.2** $f_oF_2$ values on October 2011 from the Ionogram

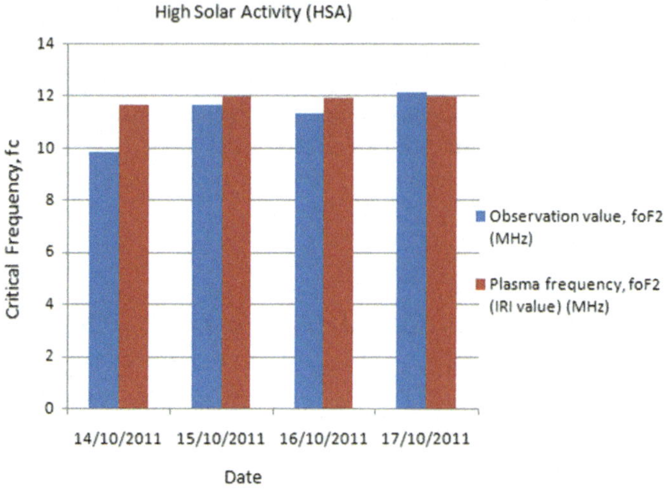


**Fig. 42.3** Comparisons of the $f_oF_2$ between the observed and the IRI during HSA in October 2011

Figure 42.4 shows the comparison between the $f_c$ for four days in October 2011 at LSA. The frequency of the F2 layer varies with the time of the day at 1:00 AM until 4:50 AM (19:00 UTC to 22:50 UTC). The maximum value of $f_c$ on 14 October 2011 is 5.9 MHz.

From the observation, $f_c$ is 5.9 MHz. However, from the IRI model, the $f_c$ is 5.04 MHz. The variation between these two data is very small than the calculated $f_oF_2$. The $f_c$ also very low (from the plot) after the sunset. This phenomena could be

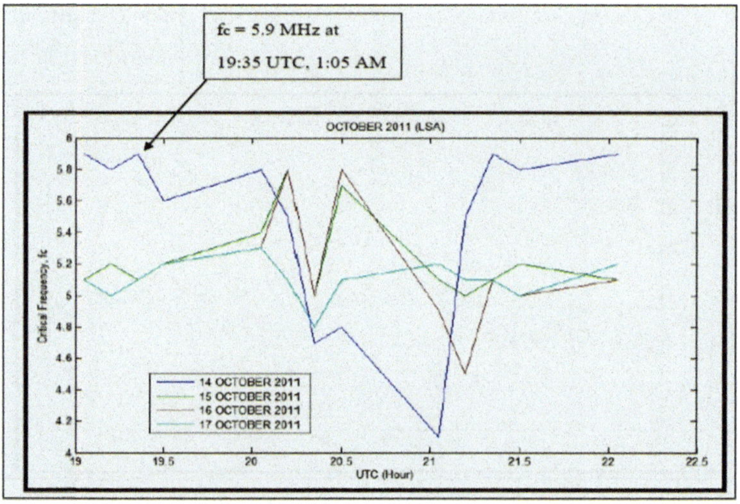


**Fig. 42.4** $f_oF_2$ values on October 2011 from the Ionogram

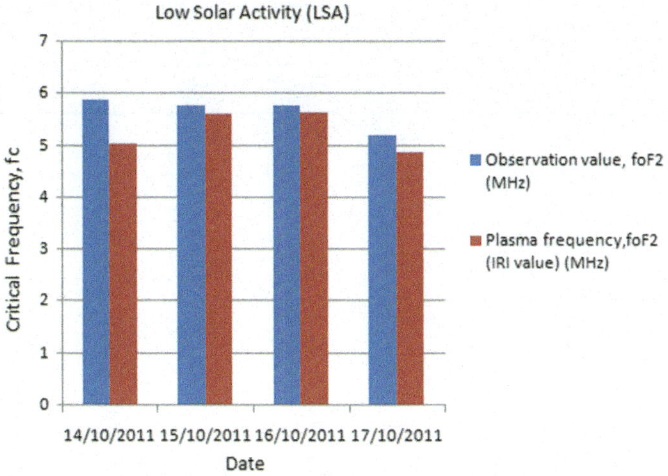


**Fig. 42.5** Comparisons of the $f_oF_2$ between the observed and the IRI during LSA in October 2011

due to the recombination process that takes place in the ionospheric layer. Figure 42.5 shows that the IRI's $f_oF_2$ value which is quite closer to the observed $f_oF_2$ values.

### *42.3.2 Variation Due to Solar Activity on November 2011*

Figure 42.6 shows the $f_oF_2$ for four consecutive days. It is shown that the value of the $f_c$ develops gradually in the day where the peak value falls in between 10:00 AM and 14:50 PM. The maximum $f_c$ is on 16 November 2011 at 10:35 AM LT. For this condition, the $f_c$ varies with time and correlates closely with the solar activity. Other than that, increased in ultraviolet (UV) radiation also would increase the $f_c$ of the F layer.

Figure 42.7 shows the correlation between IRI's $f_oF_2$ and the observed $f_oF_2$. Both of them show very good correlation.

Figure 42.8 shows the comparison between $f_c$ from 14 until 17 November 2011. The $f_c$ is lower and weak at this time could be because it is the period of time without sunshine.

During night, the density of the ionosphere is low. It will continue to decrease, reaching its lowest point just before dawn. So, the $f_c$ is usually well below 6 MHz.

Figure 42.9 also shows good correlation between the IRI's $f_oF_2$ and the observed $f_oF_2$ values.

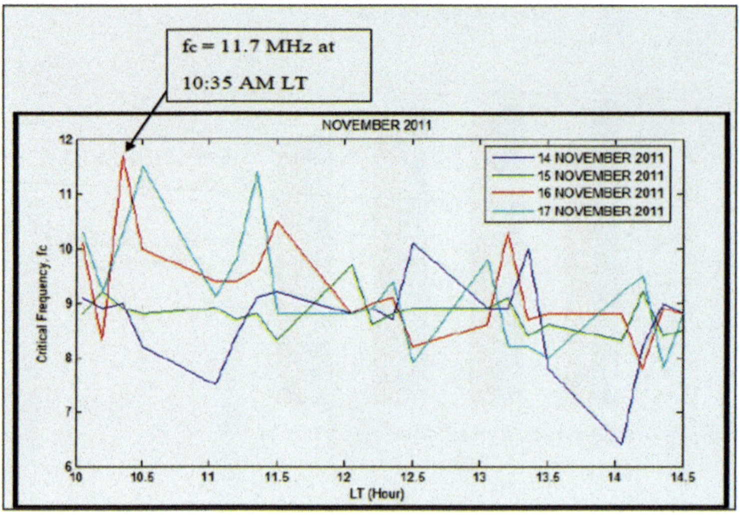


**Fig. 42.6** $f_oF_2$ values on November 2011 from the Ionogram

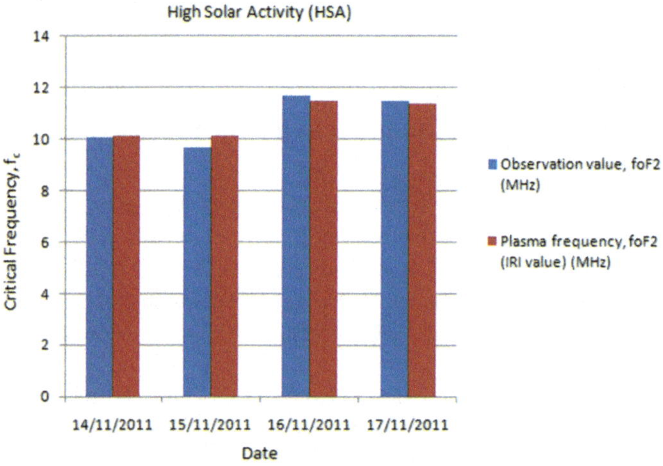


**Fig. 42.7** Comparisons of the $f_oF_2$ between the observed and the IRI during HSA in November 2011

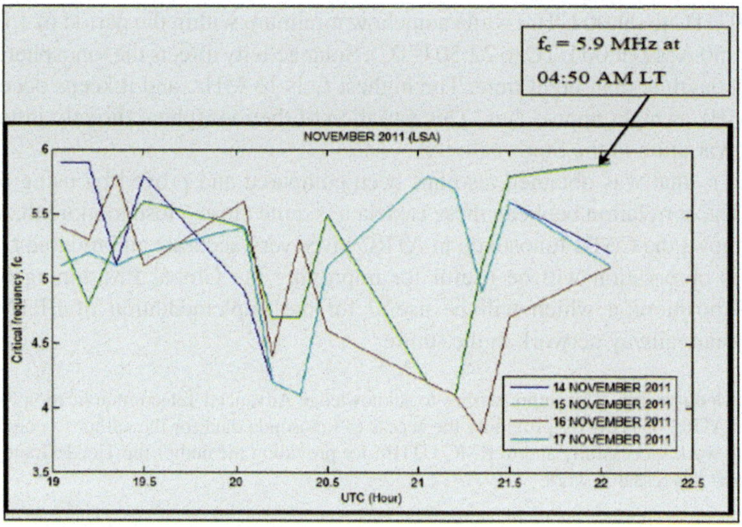


**Fig. 42.8** $f_oF_2$ values on November 2011 from the Ionogram

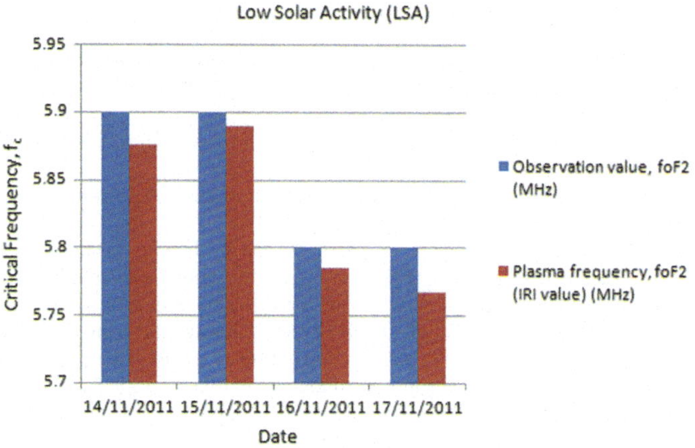


**Fig. 42.9** Comparisons of the $f_oF_2$ between the observed and the IRI during LSA in October 2011

## 42.4 Conclusion

In this work, the Ionogram variation at different solar activity has been shown. The Ionosonde data was analyzed based on short-term (diurnal) variability on hourly, daily and monthly basis to determine the $f_c$ at local time (UTC) for the month of October and November 2011.

The results showed that on October and November 2011, the value of the $f_c$ of the ionosphere is somehow maximum within the period of 10:00 AM to 3:00 PM

(04:00 UTC to 09:00 UTC) while somehow minimum within the period of 1:00 AM until 4:50 AM (19:00 UTC to 22:50 UTC). Solar activity affects the ionosphere more during daytime than night time. The highest $f_c$ is 13 MHz, and it keeps decreasing to 6 MHz as night approaches. This variation of the ionosphere directly influenced by the variation in the Sun's activity.

The $f_c$ that was obtained also has been compared and proved by using the IRI data. The correlation between these two data is quite close, close to more than 90%. That shows the CADI Ionosonde in ATRC gives very accurate and reliable $f_c$ data.

This observation will be useful for improving the Global Positioning System (GPS) positioning which will be useful for the implementation of ERTMS for Malaysian railway network in the future.

**Acknowledgements** The author wishes to acknowledge Advanced Telecommunication Research Center (ATRC) UTHM for providing the access to Ionosonde data for the author to conduct this research work successfully and to RMC UTHM for providing the author the Tier 1 Grant (H936) to support this research work.

## References

1. K. Davies, *Ionospheric Radio* (Peter Peregrinus Ltd., London, United Kingdom, 1990)
2. R.W. Schunk, A.F. Nagy, *Ionospheres Physics, Plasma Physics and Chemistry*, 2nd ed. (Cambridge University Press, 2000), pp. 75–135
3. L.F. McNamara, *Radio Atmospheric Guide to the Ionosphere* (Krieger Publishing Company, Malabar, Florida, 1994)
4. G.A. Zhbankov, P.F. Denisenko, V.V. Sotsky, Correction of ionosphere models based on ground ionosonde data. Geomagn. Aeron. **59**, 704–712 (2019)
5. K.M. Bi, Z.J. Louis, Z. Christian, O. Frédéric, L. De Recherche, D. Espace, U.N. Zongo, Variability of the critical frequency foF2 for equatorial regions during solar cycle's minima and maxima at Ouagadougou and manila stations (2019), pp. 833–844
6. A. Kumar, S. Kumar, Solar flare effects on D-region ionosphere using VLF measurements during low- and high-solar activity phases of solar cycle 24. Earth Planets Space **70**, 29 (2018)
7. M. Pezzopane, V.G. Pillat, P.R. Fagundes, Automatic scaling of critical frequency foF2 from ionograms recorded at São José dos Campos, Brazil—a comparison between Autoscala and UDIDA tools. Acta Geophys. **65**, 173–187 (2017)
8. S.A. Jilani, K. Khursheed, K. Mukhtar, Comparative study of F2—layer critical frequency. Solar cycle 22. **55**(1), 35–46 (2018)
9. D. Bilitza, IRI: the international standard for the ionosphere. Adv. Radio Sci. **16**, 1–11 (2018)
10. Scientific Instrumentation Limited, *Canadian Advanced Digital Ionosonde (CADI) System* (User's Manual Book, Canada, 2009)
11. N.M.R. Al-ubaidi, H.H.A. Al-dahlaki, Z. Fadel, Studying the variation of ionospheic parameters with earth geographic latitudes and sunspot number. Int. J. Sci. Nat. **9**(1), 35–46 (2018)
12. I.E. Zakharenkova, A. Krankowski, D. Bilitza, I.V. Cherniak, I.I. Shagimuratov, R. Sieradzki, Comparative study of foF2 measurements with IRI-2007 model predictions during extended solar minimum. Adv. Space Res. **51**(4), 620–629 (2013)

**SPRINGER NATURE**

# GPSR Compliance

*The European Union's (EU) General Product Safety Regulation (GPSR) is a set of rules that requires consumer products to be safe and our obligations to ensure this.*

*If you have any concerns about our products, you can contact us on ProductSafety@springernature.com*

In case Publisher is established outside the EU, the EU authorized representative is:

Springer Nature Customer Service Center GmbH
Europaplatz 3
69115 Heidelberg, Germany

The manufacturer's authorised representative in the EU is Springer Nature Customer Service Centre GmbH, Europaplatz 3, 69115 Heidelberg, Germany. If you have any concerns regarding our products, please contact ProductSafety@springernature.com

Printed and bound by CPI Group (UK) Ltd, Croydon, CR0 4YY
01/10/2026
02225870-0016